The Emergence of Whales

Evolutionary Patterns in the
Origin of Cetacea

ADVANCES IN VERTEBRATE PALEOBIOLOGY

Series Editors

Ross D. E. MacPhee
American Museum of Natural History, New York, New York

Hans-Dieter Sues
Royal Ontario Museum, Toronto, Ontario, Canada

THE EMERGENCE OF WHALES
Evolutionary Patterns in the Origin of Cetacea
Edited by J. G. M. Thewissen

The Emergence of Whales

Evolutionary Patterns in the
Origin of Cetacea

Edited by

J. G. M. Thewissen

Northeastern Ohio Universities
College of Medicine
Rootstown, Ohio

Plenum Press • New York and London

Library of Congress Cataloging-in-Publication Data

The emergence of whales : evolutionary patterns in the origin of
Cetacea / edited by J.G.M. Thewissen.
        p.   cm. -- (Advances in vertebrate paleobiology)
    Includes bibliographical references and index.
    ISBN 0-306-45853-5
    1. Cetacea--Evolution. 2. Cetacea, Fossil.   I. Thewissen, J. G.
M. II. Series.
    QL737.C4E335 1998
    599.5'138--dc21   98-36333
                                                                     CIP

ISBN 0-306-45853-5

© 1998 Plenum Press, New York
A Division of Plenum Publishing Corporation
233 Spring Street, New York, N.Y. 10013

http://www.plenum.com

Printed in the United States of America

Contributors

Sunil Bajpai   Department of Earth Sciences, University of Roorkee, Roorkee 247 667, Uttar Pradesh, India

Martine Bérubé   Department of Population Biology, Copenhagen University, Copenhagen Ø, DK-2100, Denmark. *Present address:* Evolutionary Genetics, Free University of Brussels (ULB), 1050 Brussels, Belgium

Emily A. Buchholtz   Department of Biological Sciences, Wellesley College, Wellesley, Massachusetts 02181

Frank E. Fish   Department of Biology, West Chester University, West Chester, Pennsylvania 19383

John Gatesy   Laboratory of Molecular Systematics and Evolution, Department of Ecology and Evolutionary Biology, University of Arizona, Tucson, Arizona 85721

Jonathan H. Geisler   Department of Vertebrate Paleontology, American Museum of Natural History, New York, New York 10024-5192

Philip D. Gingerich   Museum of Paleontology, University of Michigan, Ann Arbor, Michigan 48109

Richard C. Hulbert, Jr.   Department of Geology and Geography, Georgia Southern University, Statesboro, Georgia 30460-8149

S. Taseer Hussain   Department of Anatomy, Howard University College of Medicine, Washington, D.C. 20059

Zhexi Luo   Section of Vertebrate Paleontology, Carnegie Museum of Natural History, Pittsburgh, Pennsylvania 15213-4080

Sandra I. Madar   Department of Biology, Hiram College, Hiram, Ohio 44234

William A. McLellan   Biological Sciences and Center for Marine Science Research, University of North Carolina at Wilmington, Wilmington, North Carolina 28403

Michel C. Milinkovitch   Evolutionary Genetics, Free University of Brussels (ULB), 1050 Brussels, Belgium

Maureen A. O'Leary   Department of Cell Biology, New York University School of Medicine, New York, New York 10016. *Present address:* Department of Anatomical Sciences, Health Sciences Center, State University of New York at Stony Brook, Stony Brook, New York 11794-8081

James R. O'Neil   Department of Geological Sciences, University of Michigan, Ann Arbor, Michigan 48109

D. Ann Pabst   Biological Sciences and Center for Marine Science Research, University of North Carolina at Wilmington, Wilmington, North Carolina 28403

Per J. Palsbøll   Department of Ecology and Evolutionary Biology, University of California, Irvine, Irvine, California 92697-2525. *Present address:* Evolutionary Genetics, Free University of Brussels (ULB), 1050 Brussels, Belgium

Jay Quade   Desert Laboratory and Department of Geosciences, University of Arizona, Tucson, Arizona 85721

Lois J. Roe   Division of Ecosystem Sciences, University of California, Berkeley, Berkeley, California 94720

Sentiel A. Rommel   Marine Mammal Pathobiology Laboratory, Florida Marine Research Institute, Florida Department of Environmental Protection, St. Petersburg, Florida 33711

Ashok Sahni   Centre for Advanced Studies in Geology, Panjab University, Chandigarh 160-014, India

J. G. M. Thewissen   Department of Anatomy, Northeastern Ohio Universities College of Medicine, Rootstown, Ohio 44272

Mark D. Uhen   Cranbrook Institute of Science, Bloomfield Hills, Michigan 48303-0801

Ellen M. Williams   Department of Anatomy, Northeastern Ohio Universities College of Medicine, Rootstown, Ohio 44272

Preface

Zoologists have long known that whales, dolphins, and porpoises are mammals: In the fourth century B.C., Aristotle pointed out that whales breathe air, have hair, and nurse their young. Nevertheless, nonscientists throughout the centuries have been hard to convince that cetaceans are not fish. The attitude of whale hunter Ishmael in Herman Melville's *Moby Dick* was common when the novel was written (1851) and is still widely held among the public. Ishmael lists Linnaeus's reasons for considering whales as mammals: "their warm bilocular heart, their lungs, their movable eyelids, their hollow ears, penem intrantem feminam, mammis lactantem." He rejects these reasons on the basis of habitat: "sharks and shad, alewives and herring, against Linnaeus's express edict, were still found dividing the possession of the same seas with the Leviathan." Ishmael defines a whale as "a spouting fish with a horizontal tail."

This discrepancy about cetaceans between the scientist's views and those of the public is easy to understand, because the morphology and ecology of modern whales seem to belie their genealogy. This alone indicates that the transformation of a four-footed terrestrial mammal into a fully aquatic cetacean must have been one of the most remarkable events in evolutionary history.

Although ancient whales have been known for more than a century, it was not until the early 1980s that fragmentary remains of the earliest of whales were recovered. In the 1990s the earliest whales became well-known skeletally and it became possible to track the acquisition of aquatic adaptations and the changing morphologies in cetacean evolution.

Research in whale origins is now in an explosive phase, with a cascade of discoveries adding to our understanding of the evolutionary pattern and a suite of new techniques being applied to address new questions. The objective of this volume is to provide a snapshot of this explosion. The volume paints the scene with a broad brush, as I have not forced the individual chapters to be consistent in details of interpretation, simply because individual authors have different views, and all of these are reasonable based on available evidence. Taken together, though, these chapters clearly indicate that cetacean origins is a field that is dynamic, multidisciplinary, and that the end of the explosive phase is not in sight. I hope that this volume provides an accessible summary of ongoing discoveries and that it identifies areas where future research will prove most fruitful.

J. G. M. Thewissen
Rootstown, Ohio

Contents

Chapter 9 • Homology and Transformation of Cetacean Ectotympanic Structures

Zhexi Luo

Chapter 10 • Biomechanical Perspective on the Origin of Cetacean Flukes

Frank E. Fish

Chapter 11 • Implications of Vertebral Morphology for Locomotor Evolution in Early Cetacea

Emily A. Buchholtz

Chapter 12 • Structural Adaptations of Early Archaeocete Long Bones

Sandra I. Madar

Chapter 13 • Evolution of Thermoregulatory Function in Cetacean
Reproductive Systems

D. Ann Pabst, Sentiel A. Rommel, and William A. McLellan

Chapter 14 • Isotopic Approaches to Understanding the Terrestrial-to-Marine
Transition of the Earliest Cetaceans

Lois J. Roe, J. G. M. Thewissen, Jay Quade, James R. O'Neil, Sunil Bajpai,
Ashok Sahni, and S. Taseer Hussain

Chapter 15 • Paleobiological Perspectives on Mesonychia, Archaeoceti,
and the Origin of Whales

Philip D. Gingerich

Chapter 16 • Cetacean Origins: Evolutionary Turmoil during the Invasion
of the Oceans

J. G. M. Thewissen

The Emergence of Whales

Evolutionary Patterns in the
Origin of Cetacea

CHAPTER 1

Synopsis of the Earliest Cetaceans

Pakicetidae, Ambulocetidae, Remingtonocetidae, and Protocetidae

ELLEN M. WILLIAMS

1. Introduction

The story of cetacean origin, early evolution, diversification, and dispersal has dramatically changed in the last 20 years, related in large part to discoveries made in Pakistan, India, and the southeastern Coastal Plain of the United States. These discoveries have helped to document an extraordinary progression in whales from a predatory terrestrial ancestor to highly specialized open marine dwellers. Although this story is far from complete, it is useful to review the known diversity of early cetaceans in order to provide a framework for future discoveries and paleobiological studies. This chapter will give a synopsis of the systematic paleontology of the four earliest cetacean families and their occurrence both geologically and geographically.

Historically, the most primitive whales were included in a single family, the Protocetidae, and cetacean origins were placed in Africa with the Egyptian protocetid *Protocetus atavus* as the oldest and most primitive cetacean known (Kellogg, 1936; Barnes and Mitchell, 1978). Cetacean specimens collected in Asia during the 1970s (Sahni and Mishra, 1972, 1975) gave the first indication that cetaceans may have arisen on the Indian subcontinent. However, the belief that whales originated in and dispersed from Africa was so firmly entrenched in the literature that the possibility of an Asian descent was not given full consideration until the 1980s and 1990s when an abundance of material collected in Pakistan and India forced a reevaluation of the evolution of early whales. During this period three families were erected: the Remingtonocetidae (Kumar and Sahni, 1986), Pakicetidae (Gingerich and Russell, 1990), and Ambulocetidae (Thewissen *et al.*, 1996). The most primitive of these, the terrestrial pakicetids, clearly established Asia as the center of earliest cetacean evolution.

Our understanding of early cetaceans has itself evolved since the beginning of this cen-

ELLEN M. WILLIAMS • Department of Anatomy, Northeastern Ohio Universities College of Medicine, Rootstown, Ohio 44272.

The Emergence of Whales, edited by Thewissen. Plenum Press, New York, 1998.

tury. The fossil record has unfolded before us a story of one family that has grown to four, with the number of known taxa from 1 to 26. Twenty-one of those taxa have been described since the 1970s with 10 coming to light in just the last 6 years. This growth can only be indicative of the depth of material yet to be discovered and described.

Institutional Abbreviations

ChM	Charleston Museum, Charleston, South Carolina
GSI-K	Geological Survey of India–Kachchh, Calcutta, India
GSM	Georgia Southern Museum, Statesboro, Georgia
GSP-UM	Geological Survey of Pakistan–University of Michigan, Islamabad, Pakistan
H-GSP	Howard University–Geological Survey of Pakistan, Islamabad, Pakistan
IPHG	Institut für Paläontologie und Historische Geologie, Munich, Germany
LUVP	Lucknow University Vertebrate Paleontology collection, Lucknow, India
NHML	Natural History Museum, London, England
RUSB	Department of Earth Sciences, University of Roorkee, Roorkee, India
SMNS	Staatliches Museum für Naturkunde, Stuttgart, Germany
VPL	Vertebrate Paleontology Laboratory, Panjab University, Chandigarh, India

2. Systematic Paleontology

This review of the four earliest cetacean families is predominantly based on the following authors: Thewissen and Hussain (1998) for pakicetids, Thewissen *et al.* (1996) for ambulocetids, Kumar and Sahni (1986) for remingtonocetids, and Hulbert *et al.* (1998) for protocetids. Both pakicetids and protocetids are diagnosed primarily on the basis of primitive characters, and at present all families except remingtonocetids may represent paraphyletic grades.

Following Bajpai and Thewissen (this volume), *Gaviacetus* (Gingerich *et al.*, 1995a) is not included in the Protocetidae. The elongated vertebrae of *Gaviacetus* and the lack of an M^3 may place it in the Basilosauridae. The biostratigraphy of Kachchh follows Biswas (1992) who gives the cetacean-bearing Harudi Formation a Lutetian age. Gingerich *et al.* (1997) reinterpreted Kachchh stratigraphy to be Bartonian in age. A complete list of referred specimens is given in the Appendix.

Class MAMMALIA Linnaeus, 1758
Order CETACEA Brisson, 1762
Family PAKICETIDAE Gingerich and Russell, 1990

Pakicetinae Gingerich and Russell, 1990, p. 17
Pakicetidae Thewissen *et al.*, 1996, p. 70; Thewissen and Hussain, 1998

Type Genus.—Pakicetus Gingerich and Russell, 1981.
Referred Genera.—Ichthyolestes Dehm and Oettingen-Spielberg, 1958; *Nalacetus* Thewissen and Hussain, 1998.

Age and Distribution.—Ypresian (early Eocene), Kuldana Formation, northern Pakistan and Subathu Formation, northwestern India.

Diagnosis.—Cetaceans characterized by the presence of palatine fissures, external nares dorsal to the incisors, a hypoglossal foramen well separated from the jugular foramen, and a small mandibular foramen. The paraconid and metaconid of the molars are present but smaller than the protoconid, and the only cusp on the talonid is the hypoconid. The P^4 has three roots and a single cusp which is higher than the paracone of the molars (Gingerich and Russell, 1981; Thewissen and Hussain, 1998).

Pakicetus Gingerich and Russell, 1981

Protocetus (in part) West, 1980, p. 515
Pakicetus Gingerich and Russell, 1981, p. 238; 1990, p. 3; Thewissen and Hussain, 1998
Pakicetus sp. Luo, this volume, Fig. 4A–C

Type Species.—*Pakicetus inachus* Gingerich and Russell, 1981.
Referred Species.—*Pakicetus attocki* West, 1980.
Age and Distribution.—Ypresian (early Eocene), redbeds of the Kuldana Formation, northern Pakistan, and Subathu Formation, northwestern India.
Diagnosis.—Unlike *Nalacetus*, *Pakicetus* has a smaller molar paraconid, a larger basal outline of P^4 than M^1, a higher trigonid, and a higher P^4 paracone than that on the molar. *Pakicetus* differs from both *Ichthyolestes* and *Nalacetus* in having a larger paracone than metacone in the upper molars and a wider trigonid basin in the lower molars (Gingerich and Russell, 1981; Thewissen and Hussain, 1998).

Pakicetus inachus Gingerich and Russell, 1981

Ichthyolestes pinfoldi West 1980, p. 516, Pl. 2, Fig. 1
Pakicetus inachus Gingerich and Russell, 1981, p. 238, Figs. 2–4; Gingerich *et al.*, 1983, p. 404; Gingerich and Russell, 1990 (in part), p. 3, Figs. 1–5, 6A–H, M–O; Thewissen and Hussain, 1998, Fig. 6

Holotype.—GSP-UM 84, a braincase with complete right auditory bulla. This specimen has suffered considerable damage since its original description.
Type Locality.—Four kilometers north-northwest of Chorlakki Village, Kohat District, Northwest Frontier Province, Pakistan.
Referred Specimens.—Several upper and lower molars and a mandible with premolars.
Age and Distribution.—Ypresian (early Eocene), redbeds of the Kuldana Formation, northern Pakistan.
Diagnosis.—*P. inachus* has low and slender protocones, weak cristae, and lacks metacristae in the upper molars (Thewissen and Hussain, 1998).

Pakicetus attocki (West, 1980)

Protocetus attocki West, 1980, p. 515, Pl. 1, Fig. 5
Protocetidae indet. (in part) West, 1980, p. 516, Pl. 1, Fig. 6
Pakicetus attocki Gingerich and Russell (in part), 1981, p. 242, Fig. 5; 1990, p. 14; Thewissen and Hussain, 1998, Figs. 2A,B, 3B, 4A–E, I–J; O'Leary, this volume, Figs. 4A,B, 5A
Ichthyolestes pinfoldi Kumar and Sahni, 1985, p. 166, Fig. 7S–X
Pakicetus sp. Thewissen and Hussain, 1993, p. 361, Figs. 1, 2; Thewissen, 1993, p. 125, Fig. 9

Holotype.—H-GSP 1694, mandible with P_{3-4} (not P_4–M_1 as stated in the type description) and alveoli or roots for several other teeth (Thewissen and Hussain, 1998).

Type Locality.—H-GSP Locality 62, Ganda Kas Area, Kala Chitta Hills, Punjab, Pakistan.

Referred Specimens.—Several maxillary and mandibular fragments with teeth, several isolated premolars and molars, a single incisor and canine, two left tympanics, and an incus.

Age and Distribution.—Ypresian (early Eocene), redbeds of the Kuldana Formation, northern Pakistan, and Subathu Formation, northwestern India.

Diagnosis.—*P. attocki* is distinguished by large and bulbous protocones, strong cristae, and a wide trigonid basin produced by paracristae which are stronger than the metacristae in upper molars (West, 1980; Thewissen and Hussain, 1998).

Ichthyolestes Dehm and Oettingen-Spielberg, 1958

Ichthyolestes Dehm and Oettingen-Spielberg, 1958, p. 15; Szalay and Gould, 1966, p. 151; Thewissen and Hussain, 1998

Type and Only Species.—*Ichthyolestes pinfoldi* Dehm and Oettingen-Spielberg, 1958.

Age and Distribution.—Ypresian (early Eocene), redbeds of the Kuldana Formation, northern Pakistan.

Diagnosis.—*Ichthyolestes* differs from *Pakicetus* in having upper molar protocones close to labial cusps, a narrow premolar protoconid, a larger M^{1-2} metacone, and a sharply angled P_2 crown. *Ichthyolestes* is distinguished from *Nalacetus* by a higher P^4 paracone and a smaller paraconid. It differs from *Pakicetus attocki* by a slender protocone. A smaller lingual bulge on the P^4 and a longer (anteroposteriorly) trigonid differentiate it from both *Pakicetus* and *Nalacetus* (Thewissen and Hussain, 1998).

Ichthyolestes pinfoldi Dehm and Oettingen-Spielberg, 1958

Ichthyolestes pinfoldi Dehm and Oettingen-Spielberg, 1958, p. 15, Pl. 1.5, Fig. 2; Szalay and Gould, 1966, p. 151; Gingerich and Russell, 1990, p. 16, Fig. 9; Thewissen and Hussain, 1998, Figs. 4F,H, 5; Luo, this volume, Figs. 4D,E, 5, 6

Creodontium gen. et spec. indet. (*Ichthyolestes?*) Dehm and Oettingen-Spielberg, 1958, p. 17, Pl. 1.1

Basilosauridae indet. West, 1980, p. 517, Pl. 2, Fig. 2

Protocetidae indet. West, 1980, p. 516, Pl. 2, Fig. 3

Pakicetus attocki (in part) Gingerich and Russell, 1981, p. 243; 1990, p. 14

Pakicetus inachus (in part) Gingerich and Russell, 1990, p. 3, Figs. 6I–L, 7

Ichthyolestes or *Pakicetus* Gingerich and Russell, 1990, p. 3, Fig. 1E

Pakicetus sp. Thewissen, 1993, p. 125, Fig. 9

Holotype.—IPHG 1956 II 7, maxilla with M^{2-3}.

Type Locality.—Locality 21 of Dehm and Oettingen-Spielberg (1958), Ganda Kas Area, Kala

Chitta Hills, Punjab, Pakistan. West and Lukacs (1979) identified this as H-GSP Locality 220.

Referred Specimens.—Mandibles with teeth, several isolated upper and lower premolars, lower deciduous premolars, and a single lower molar.

Age and Distribution.—Ypresian (early Eocene), redbeds of the Kuldana Formation, northern Pakistan.

Diagnosis.—As for genus.

Nalacetus Thewissen and Hussain, 1998

Nalacetus Thewissen and Hussain, 1998

Type and Only Species.—Nalacetus ratimitus, Thewissen and Hussain, 1998.
*Age and Distribution.—*Ypresian (early Eocene), redbeds of the Kuldana Formation, northern Pakistan.
Diagnosis.—Nalacetus is characterized by a robust molar protocone, similarly sized metacone and paracone, and a narrow width (mediolaterally) between labial and lingual cusps in the upper molars. *Nalacetus* differs from *Ichthyolestes* in a strong molar paraconid and a P^4 paracone half the height of the M^1 paracone (Thewissen and Hussain, 1998).

Nalacetus ratimitus Thewissen and Hussain, 1998

Pakicetus sp. Thewissen, 1993, p. 125, Fig. 9; Maas and Thewissen, 1996, p. 1155, Fig. 1
Nalacetus ratimitus Thewissen and Hussain, 1998, Figs. 2C,D, 3C, 4G,K,L; O'Leary, Chapter 5, this volume, Figs. 4C,D, 5B

*Holotype.—*H-GSP 18521 (field number 92132) maxillary fragment with P^4–M^1.
*Type Locality.—*H-GSP Locality 62, Ganda Kas Area, Kala Chitta Hills, Punjab, Pakistan.
*Referred Specimens.—*A mandible with fragmentary teeth, a single lower deciduous premolar, premolar, and molar.
*Age and Distribution.—*Ypresian (early Eocene), redbeds of the Kuldana Formation, northern Pakistan.
*Diagnosis.—*As for genus.

Family AMBULOCETIDAE Thewissen, Madar, and Hussain, 1996

Ambulocetidae Thewissen *et al.*, 1996, p. 9

Type Genus.—Ambulocetus Thewissen *et al.*, 1994.
Referred Genus.—Gandakasia Dehm and Oettingen-Spielberg, 1958.
*Age and Distribution.—*Lutetian (middle Eocene), Kuldana Formation, northern Pakistan.
*Diagnosis.—*Cetaceans characterized by a loss of the molar metaconid and the presence of a large parastyle and a large mandibular foramen that does not extend the full depth of the jaw. The pterygoid processes are fused ventrally producing a nasopharyngeal duct that is ossified caudal to the M^3, and the height of the pterygoid process is equal to that of the braincase (Thewissen *et al.*, 1996).

Ambulocetus Thewissen, Hussain, and Arif, 1994

Ambulocetus Thewissen *et al.*, 1994, p. 212, Fig. 1; 1996, p. 10

Type and Only Species.—Ambulocetus natans Thewissen *et al.*, 1994.
*Age and Distribution.—*Lutetian (middle Eocene), uppermost Kuldana Formation of the Kala Chitta Hills, northern Pakistan.
Diagnosis.—Ambulocetus is larger than *Gandakasia* and possesses a higher trigonid (Thewissen *et al.*, 1996).

Ambulocetus natans Thewissen, Hussain, and Arif, 1994

Basilosauridae indet. West, 1980, p. 517, Pl. 2, Fig. 4
Ambulocetus natans Thewissen *et al.*, 1994, p. 212, Fig. 1; 1996, p. 10, Figs. 2, 4–10, 12–23

Holotype.—H-GSP 18507, skull and partial skeleton, several vertebrae and ribs, radius, ulna and manus, femur, proximal tibia and pes.
Type Locality.—H-GSP Locality 9209, Ganda Kas Area, Kala Chitta Hills, Punjab, Pakistan.
Referred Specimens.—Several caudal vertebrae, a proximal manual phalanx, two lower premolars, and a left mandibular fragment with a fragmentary premolar.
Age and Distribution.—Lutetian (middle Eocene), uppermost Kuldana Formation of the Kala Chitta Hills, northern Pakistan.
Diagnosis.—As for genus.

Gandakasia Dehm and Oettingen-Spielberg, 1958

Gandakasia Dehm and Oettingen-Spielberg, 1958, p. 11

Type and Only Species.—*Gandakasia potens* Dehm and Oettingen-Spielberg, 1958.
Age and Distribution.—Lutetian (early middle Eocene), uppermost Kuldana Formation of the Kala Chitta Hills, northern Pakistan.
Diagnosis.—*Gandakasia* differs from *Ambulocetus* in its smaller size and much lower trigonid (Thewissen *et al.*, 1996).

Gandakasia potens Dehm and Oettingen-Spielberg, 1958

Gandakasia potens Dehm and Oettingen-Spielberg, 1958, p. 11, Pl. 1, Figs. 3a–e, 4; West, 1980, p. 512, Pl. 1, Figs. 2–4; Gingerich and Russell, 1990, p. 17, Fig. 10

Holotype.—IPHG 1956 II 4 and 5, fragmentary mandible with M_{1-2} and possibly associated P_4 and M_3.
Type Locality.—Locality 13 of Dehm and Oettingen-Spielberg (1958), Ganda Kas Area, Kala Chitta Hills, Punjab, Pakistan. West and Lukacs (1979) identified this as H-GSP Locality 58.
Referred Specimens.—Two lower molars, a partial second premolar, and a third premolar.
Age and Distribution.—Lutetian (middle Eocene), uppermost Kuldana Formation of the Kala Chitta Hills, northern Pakistan.
Diagnosis.—As for genus.

Family REMINGTONOCETIDAE Kumar and Sahni, 1986

Remingtonocetidae Kumar and Sahni, 1986, p. 330

Type Genus.—*Remingtonocetus* Kumar and Sahni, 1986.
Referred Genera.—*Andrewsiphius* Sahni and Mishra, 1975; *Dalanistes* Gingerich *et al.*, 1995a; *Attockicetus* Thewissen and Hussain, in press.
Age and Distribution.—Lutetian (middle Eocene), Harudi Formation, northwestern India, and Domanda Formation, central Pakistan.
Diagnosis.—Cetaceans characterized by the presence of a supraorbital shield, small orbits,

a mediolaterally convex plate between the molars, oval auditory bullae, and an ear region positioned laterally to the basioccipital (Kumar and Sahni, 1986; Gingerich *et al.*, 1995a; Bajpai and Thewissen, this volume).

Remingtonocetus Kumar and Sahni, 1986

Protocetus Sahni and Mishra, 1975, p. 20
Remingtonocetus Kumar and Sahni, 1986, p. 330; Gingerich *et al.*, 1995a, p. 310, Figs. 12–15

Type Species.—*Protocetus harudiensis* Sahni and Mishra, 1975.
Referred Species.—*Remingtonocetus sloani* Sahni and Mishra, 1972.
Age and Distribution.—Lutetian (middle Eocene), Harudi Formation of Kachchh District, western India, and Domanda Formation of the Sulaiman Range, central Pakistan.
Diagnosis.—*Remingtonocetus* differs from *Andrewsiphius* in having a wider snout, more posterior position of the external nares, a mandibular symphysis extending to P_4 (to M_3 in *Andrewsiphius*, and P_3 in *Dalanistes*), and the presence of a three-rooted M^3, a falcate process, and a molar protocone. It differs from *Dalanistes* in having external nares above P^1 rather than C^1; and from *Attockicetus* in the smaller size of the molar protocone and orbits positioned posterior to M^3 (Sahni and Mishra, 1975; Kumar and Sahni, 1986; Gingerich *et al.*, 1995a; Thewissen and Hussain, in press; Bajpai and Thewissen, this volume).

Remingtonocetus harudiensis Sahni and Mishra, 1975

Protocetus harudiensis Sahni and Mishra, 1975, p. 21, Pl. 4, Figs. 4–7
Protosiren fraasi (in part) Sahni and Mishra, 1975, p. 27, Fig. 4, Pl. 6, Fig. 1
Cf. Moeritheriid Sahni and Mishra, 1975, p. 29, Fig. 5, Pl. 6, Fig. 2
Remingtonocetus harudiensis Kumar and Sahni, 1986, p. 330, Figs. 3–10; Bajpai and Thewissen, this volume, Figs. 2A–D, 3A
Indocetus ramani (in part) Gingerich *et al.*, 1993, p. 399, Figs. 4–8, 9A,B, 10A, 12B, 13A–C

Holotype.—LUVP 11037, fragmentary skull with roots for P^4–M^3, isolated cusps of upper teeth, left mandibular ramus with roots for P_4–M_3, crowns of left M_{1-2}, and right mandibular fragment with roots for P_4–M_2.
Type Locality.—Rato Nala, 2 km north of Harudi, Hachchh District, Gujarat, India.
Paratypes.—VPL 15001, skull with partial dentition and left periotic; 15002, fragment of skull; LUVP 11132, mandibular fragment with alveoli for P_{1-4}.
Referred Specimens.—Partial skulls, associated skull fragments with postcrania, two braincases, and rostrum fragments with teeth.
Age and Distribution.—Lutetian (middle Eocene), Harudi Formation of Kachchh District, western India, and Domanda Formation of the Sulaiman Range, central Pakistan.
Diagnosis.—Unlike *R. sloani*, *R. harudiensis* has flat nasals over the M^2 and a narrow rostrum over the M^{1-2}. *R. harudiensis* has a larger body size than *R. sloani* (Sahni and Mishra, 1975; Kumar and Sahni, 1986; Bajpai and Thewissen, this volume).

Remingtonocetus sloani Sahni and Mishra, 1972

Protocetus sloani Sahni and Mishra, 1972, p. 491, Pl. 97, Figs. 1–7, 1A,B,C; Sahni and Mishra, 1975, p. 20, Pl. 5, Figs. 1,2

Remingtonocetus sloani Kumar and Sahni, 1986, p. 341, Fig. 8K, Bajpai and Thewissen, this volume, Figs. 2E–G, 3B, 4A–C, 5A

Holotype.—LUVP 11002, an anterior mandibular fragment with symphysis and alveoli for P_2 and P_4.

Type Locality.—Rato Nala, 2 km north of Harudi, Kachchh District, Gujarat, India.

Referred Specimens.—Three partial skulls, several rostral fragments with teeth, and a mandibular fragment with alveoli.

Age and Distribution.—Lutetian (middle Eocene), Harudi Formation of Kachchh District, western India.

Diagnosis.—*R. sloani* is distinguished from *R. harudiensis* by its smaller body size, domed nasals over the M^2, and a wider rostrum over the M^{1-2} (Sahni and Mishra, 1975; Kumar and Sahni, 1986; Bajpai and Thewissen, this volume).

Andrewsiphius Sahni and Mishra, 1975

Andrewsiphius Sahni and Mishra, 1975, p. 23; Kumar and Sahni, 1986, p. 329

Type Species.—*Andrewsiphius kutchensis* Sahni and Mishra, 1975.

Referred Species.—*Andrewsiphius minor* Sahni and Mishra, 1975.

Age and Distribution.—Lutetian (middle Eocene), Harudi Formation of Kachchh District, western India.

Diagnosis.—*Andrewsiphius* differs from *Remingtonocetus* and *Dalanistes* in its narrower snout, more anterior position of the external nares, a mandibular symphysis extending to the M_3, the presence of a double-rooted M^3, and the absence of falcate process and molar protocone. Orbits positioned posterior to the M^3 and a narrower palate distinguish *Andrewsiphius* from *Attockicetus* (Sahni and Mishra, 1975; Kumar and Sahni, 1986; Thewissen and Hussain, in press; Bajpai and Thewissen, this volume).

Andrewsiphius kutchensis Sahni and Mishra, 1975

Andrewsiphius kutchensis Sahni and Mishra, 1975, p. 23, Pl. 5, Fig. 6; Bajpai and Thewissen, this volume, Figs. 3C, 4D, 6, 7A,B

Holotype.—LUVP 11060, mandible with alveoli for I_2–M_1.

Type Locality.—Southwest of Nareda, Kachchh District, Gujarat, India.

Referred Specimens.—A partial skull and two fragmentary rostra with teeth.

Age and Distribution.—Lutetian (middle Eocene), Harudi Formation of Kachchh District, western India.

Diagnosis.—*A. kutchensis* is larger than *A. minor*, and its mandibles diverge more gradually posterior to P_3 (Sahni and Mishra, 1975).

Andrewsiphius minor Sahni and Mishra, 1975

Andrewsiphius minor Sahni and Mishra, 1975, p. 25, Pl. 5, Fig. 7

Holotype.—LUVP 11165, incomplete mandibular fragment with alveoli of P_1–M_1.

Type Locality.—Rato Nala, 2 km north of Harudi, Kachchh District, Gujarat, India.

Age and Distribution.—Lutetian (middle Eocene), Harudi Formation of Kachchh District, western India.

Diagnosis.—*A. minor* is smaller than *A. kutchensis*, and its mandibles diverge more abruptly at P_3 (Sahni and Mishra, 1975).

Dalanistes Gingerich, Arif, and Clyde, 1995a

Type and Only Species.—*Dalanistes ahmedi*, Gingerich *et al.*, 1995a.
Age and Distribution.—Lutetian (middle Eocene), Domanda Formation of the Sulaiman Range, central Pakistan.
Diagnosis.—*Dalanistes* is larger than all other remingtonocetid genera. It differs from *Remingtonocetus* in having external nares above C^1 rather than P^1, higher sagittal and nuchal crests, and a mandibular symphysis extending to P_3 rather than P_4 (to M_3 in *Andrewsiphius*). It does not retain the primitive large molar protocone or orbits positioned above the M^3 as in *Attockicetus* (Gingerich *et al.*, 1995a; Kumar and Sahni, 1986; Thewissen and Hussain, in press).

Dalanistes ahmedi Gingerich, Arif, and Clyde, 1995a

Dalanistes ahmedi Gingerich *et al.*, 1995a, p. 317, Figs. 17–20

Holotype.—GSP-UM 3106, partial skull and skeleton.
Type Locality.—Basti Ahmed in Dalana Nala drainage, south of Takra Valley, Sulaiman Range, Punjab, Pakistan.
Referred Specimens.—Several vertebrae and fragmentary skulls, a sacrum, an innominate, and a distal femur.
Age and Distribution.—Lutetian (middle Eocene), Domanda Formation of the Sulaiman Range, central Pakistan.
Diagnosis.—As for genus.

Attockicetus, Thewissen and Hussain, in press

Attockicetus Thewissen and Hussain, in press

Type and Only Species.—*Attockicetus praecursor* Thewissen and Hussain (in press).
Age and Distribution.—Lutetian (middle Eocene), uppermost Kuldana–lower Kohat Formations of the Kala Chitta Hills, northern Pakistan.
Diagnosis.—*Attockicetus* is distinguished by its large molar protocone and orbits positioned dorsal to the M^3 (Thewissen and Hussain, in press).

Attockicetus praecursor, Thewissen and Hussain, in press

Attockicetus praecursor Thewissen and Hussain, in press, Figs. 1,2

Holotype.—H-GSP 96232, fragmentary skull including partial rostrum and endocast of the braincase.
Type Locality.—H-GSP Locality 9604, Ganda Kas Area, Kala Chitta Hills, Punjab, Pakistan.
Age and Distribution.—Lutetian (middle Eocene), uppermost Kuldana or lowermost Kohat Formation, Ganda Kas Area, Kala Chitta Hills, northern Pakistan.
Diagnosis.—As for genus.

Family PROTOCETIDAE Stromer, 1908

Protocetidae Stromer, 1908, p. 148; Kellogg, 1936, p. 231, Hulbert *et al.*, 1998

Type Genus.—Protocetus Fraas, 1904.

Referred Genera.—Eocetus Fraas, 1904; *Pappocetus* Andrews, 1920; *Indocetus* Sahni and Mishra, 1975; *Babiacetus* Trivedy and Satsangi, 1984; *Rodhocetus* Gingerich *et al.*, 1994; *Takracetus* Gingerich *et al.*, 1995a; *Georgiacetus* Hulbert *et al.*, 1998.

Age and Distribution.—Lutetian and Bartonian (middle Eocene), central Pakistan, western India, northern Egypt, southern Nigeria, southern Togo, and southeastern United States.

Diagnosis.—Cetaceans possessing less than four fused sacral vertebrae, reduced but functional hind limbs, shortened cervical vertebrae, and laterally facing orbits covered by a supraorbital shield (Kellogg, 1936; Sahni and Mishra, 1975; Barnes and Mitchell, 1978; Hulbert *et al.*, 1998; Thewissen, this volume).

Protocetus Fraas, 1904

Protocetus Fraas, 1904, p. 201; Kellogg, 1936, p. 235; Barnes and Mitchell, 1978, p. 585

Type and Only Species.—Protocetus atavus Fraas, 1904.

Age and Distribution.—Lutetian (middle Eocene), Mokattam Formation, northern Egypt. Kellogg (1936) also described *Protocetus* sp. from the Cook Mountain Formation of Texas.

Diagnosis.—Protocetus is distinguished from contemporary protocetids (*Indocetus, Takracetus,* and *Rodhocetus*) by external nares dorsal to the P^1; and from *Indocetus* and *Rodhocetus* by a double-rooted P^1. It differs from *Takracetus* in the presence of a three-rooted M^3 and from *Babiacetus* by the absence of a M^3 metacone. The absence of accessory cusps on molars separates *Protocetus* from the later protocetids *Eocetus, Pappocetus,* and *Georgiacetus* (Kellogg, 1936; Sahni and Mishra, 1975; Gingerich *et al.*, 1993, 1995a,b; Hulbert *et al.*, 1998).

Protocetus atavus Fraas, 1904

Protocetus atavus Fraas, 1904, p. 20, Pl. 10–12; Stromer, 1908, p. 108, Pl. 5, Figs. 20, 21; Kellogg, 1928, p. 39, Fig. 1; 1936, p. 235, Pl. 34, Figs. 1, 2, Pl. 35, Figs. 1, 2; Slijper, 1936, p. 385, Fig. 199; Barnes and Mitchell, 1978, p. 585

Holotype.—SMNS 11084, a partial skull lacking the anterior portion of rostrum, parts of the zygomatic arches, and supraoccipital region.

Type Locality.—Gebel Mokattam, Cairo, Egypt.

Referred Specimens.—Several vertebrae and rib fragments.

Age and Distribution.—Lutetian (middle Eocene), Lower Building Stone Member of the Mokattam Formation, Gebel Mokattam, northern Egypt.

Diagnosis.—As for genus.

Eocetus Fraas, 1904

Mesocetus Fraas, 1904, p. 217; preoccupied by *Mesocetus* Van Beneden, 1883
Eocetus Fraas, 1904, p. 374; Kellogg, 1936, p. 231; Barnes and Mitchell, 1978, p. 587

Type and Only Species.—Eocetus schweinfurthi (Fraas, 1904).

*Age and Distribution.—*Bartonian (middle Eocene), Guishi Formation, northern Egypt.

Diagnosis.—Eocetus is distinguished from *Indocetus* and *Rodhocetus* by having a double-rooted P^1. It differs from *Takracetus* in having a three-rooted M^3; and from *Babiacetus* by the presence of accessory cuspules on cheek teeth (also missing in *Protocetus*) and the lack of an M^3 metacone. Large embrasure pits, a longer diastema between the canine and incisor, a shorter diastema between the anterior premolars, more posterior termination of the premaxilla–maxilla facial process, a posterolateral supraorbital foramen opening laterally, and the lack of parietal ridge distinguish it from *Georgiacetus*. It is the largest protocetid discovered thus far (Kellogg, 1936; Sahni and Mishra, 1975; Gingerich *et al.*, 1993, 1995a,b; Hulbert *et al.*, 1998).

Eocetus schweinfurthi Fraas, 1904

Zeuglodon macrospondylus Stromer, 1903, p. 83, Fig. 1
Mesocetus schweinfurthi Fraas, 1904, p. 201, Pl. 10, Fig. 3, Pl. 11, Figs. 10,11
Eocetus schweinfurthi Fraas, 1904, p. 374; Andrews, 1906, p. 240; Stromer, 1908, p. 106; Kellogg, 1936, p. 232, Pl. 33, Fig. 1; Barnes and Mitchell, 1978, p. 587

*Holotype.—*SMNS 10986, a poorly preserved and crushed skull lacking most of the occiput and the left cheek region.

*Type Locality.—*Gebel Mokattam, Cairo, Egypt.

*Referred Specimens.—*Two lumbar vertebrae. Barnes and Mitchell (1978) and Hulbert *et al.* (in press) note that these vertebrae are basilosaurine-like and question the association.

*Age and Distribution.—*Bartonian (middle Eocene), Guishi Formation, Gebel Mokattam, northern Egypt.

*Diagnosis.—*As for genus.

Pappocetus Andrews, 1920

Pappocetus Andrews, 1920, p. 309; Kellogg, 1936, p. 243; Barnes and Mitchell, 1978, p. 585

Type and Only Species.—Pappocetus lugardi Andrews, 1920.

*Age and Distribution.—*Lutetian (middle Eocene), Ameki Formation of the Ombialla District, southern Nigeria.

*Diagnosis.—*The presence of a steplike notch on the ventral margin of the mandible below M$_2$ and M$_3$ distinguishes *Pappocetus* from all other known protocetid genera. *Pappocetus* differs from *Indocetus* and *Rodhocetus* in having a double-rooted (d)P$_1$; and from *Protocetus* and *Babiacetus* by the presence of accessory cuspules. An unfused mandibular symphysis terminating at the anterior margin of (d)P$_3$ differentiates it from *Babiacetus*. It is similar in body size to *Eocetus*, and the overall morphology of the molars is similar to Georgiacetus (Andrews, 1920; Kellogg, 1936; Sahni and Mishra, 1975; Gingerich *et al.*, 1993, 1995b; Hulbert *et al.*, 1998).

Pappocetus lugardi Andrews, 1920

Pappocetus lugardi Andrews, 1920, p. 309, Fig. 1, Pl. 1; Kellogg, 1928, p. 38; 1936, p. 243; Barnes and Mitchell, 1978, p. 585, Fig. 29.1; Halstead and Middleton, 1974, p. 82, Figs. 1–5

Holotype.—NHML M-11414, an incomplete mandible with symphysis an deciduous premolars and unerupted molars.

Type Locality.—Port Harcourt railroad cut, Ameki, Ombialla District, Nigeria.

Referred Specimens.—A left mandible with incomplete dentition, an axis vertebra, several associated vertebrae, and a proximal rib fragment.

Age and Distribution.—Lutetian (middle Eocene), Ameki Formation of the Ombialla District, southern Nigeria.

Diagnosis.—As for genus.

Indocetus Sahni and Mishra, 1975

Indocetus Sahni and Mishra, 1975, p. 18

Type and Only Species.—*Indocetus ramani* Sahni and Mishra, 1975.

Age and Distribution.—Lutetian (middle Eocene), Harudi Formation of Kachchh District, western India.

Diagnosis.—A prominent molar protocone distinguishes *Indocetus* from *Protocetus, Eocetus, Babiacetus*, and *Georgiacetus*. Like *Rodhocetus, Indocetus* has a single-rooted P^1. *Indocetus* also has a narrower tympanic bulla unlike *Protocetus* and *Georgiacetus* (Kellogg, 1936; Sahni and Mishra, 1975; Gingerich *et al.*, 1993, 1995b; Hulbert *et al.*, 1998).

Indocetus ramani Sahni and Mishra, 1975

Indocetus ramani Sahni and Mishra, 1975, p. 18, Pl. 4, Figs. 1–3; Bajpai *et al.*, 1996, p. 582, Fig. 1; Bajpai and Thewissen, this volume, Figs. 3D, 5B, 8A,B

Holotype.—LUVP 11034, a partial skull including the frontal shield and the right occipital region with right auditory bulla.

Type Locality.—Rato Nala, 2 km north of Harudi, Kachchh District, Gujarat, India.

Referred Specimens.—A partial skull, right tympanic, right maxillary fragment with teeth, and two cranial endocasts, one with nasal cavity endocast and alveoli for molars.

Age and Distribution.—Lutetian (middle Eocene), Harudi Formation of Kachchh District, western India.

Diagnosis.—As for genus.

Babiacetus Trivedy and Satsangi, 1984

Type and Only Species.—*Babiacetus indicus* Trivedy and Satsangi, 1984.

Age and Distribution.—Lutetian (middle Eocene), Harudi Formation of Kachchh District, western India, and Drazinda Formation of the Sulaiman Range, central Pakistan.

Diagnosis.—*Babiacetus* is distinguished from *Protocetus, Eocetus*, and *Georgiacetus* in the presence of an M^3 metacone. It differs from *Indocetus* and *Rodhocetus* in having a reduced molar protocone; and from *Rodhocetus* and *Takracetus* by external nares dorsal to P^1. *Babiacetus* has a fused mandibular symphysis terminating at P$_2$, unlike *Papocetus* and *Georgiacetus*; and narrow tympanic bulla, unlike *Protocetus, Rodhocetus*, and *Georgiacetus* (Kellogg, 1936; Sahni and Mishra, 1975; Trivedy and Satsangi, 1984; Gingerich *et al.*, 1993, 1995a,b; Hulbert *et al.*, 1998).

Babiacetus indicus Trivedy and Satsangi, 1984

Babiacetus indicus Trivedy and Satsangi, 1984, p. 322; Gingerich *et al.*, 1995b, p. 348, Figs. 12, 14, 15

Holotype.—GSI 19647, left and right mandibles with cheek teeth.
Type Locality.—Babia Hill, Kachchh District, Gujarat, India.
Referred Specimens.—A partial skull with lower jaws.
Age and Distribution.—Lutetian (middle Eocene), Harudi Formation of Kachchh District, western India, and Drazinda Formation of the Sulaiman Range, central Pakistan.
Diagnosis.—*B. indicus* is distinguished from *B. mishrai* by a longer mandible, double-rooted P_1, diastemata between P_1 and P_4, wider cheek teeth, and a larger M_3.

Babiacetus mishrai Bajpai and Thewissen, Chapter 7, this volume

Babiacetus mishrai, Chapter 7, this volume, Fig. 10.

Holotype.—RUSB 2512, left and right mandibles with left I_2, root of C, base of P_2, roots of P_3, P_4, partial M_1, M_{2-3}, and right I_2, C, P_{2-3}, base of P_4–M_1, and M_{2-3}.
Type Locality.—Babia Hill, Kachchh District, Gujarat, India.
Age and Distribution.—Lutetian (middle Eocene) Harudi Formation of Kachchh District, western India.
Diagnosis.—*B. mishrai* is distinguished from *B. indicus* by a shorter mandible, lack of a diastemata between P_1 and P_4, a single-rooted P_1, narrow cheek teeth, and a small M_3.

Rodhocetus Gingerich, Raza, Arif, and Anwar, 1994

Type and Only Species.—*Rodhocetus kasrani* Gingerich *et al.*, 1994.
Age and Distribution.—Lutetian (middle Eocene), Domanda Formation of the Sulaiman Range, central Pakistan.
Diagnosis.—*Rodhocetus* is distinguished from later protocetids by the position of the external nares dorsal to C, a mandibular symphysis terminating at the posterior margin of P_2, and a single-rooted P^1 (except *B. mishrai*). *Rodhocetus* differs from *Georgiacetus* in having six lumbar vertebrae rather than eight. A wider tympanic bulla differentiates it from *Babiacetus*. Gingerich *et al.* (1995a) noted that at present there are no known differences between *Rodhocetus* and *Indocetus* (Kellogg, 1936; Sahni and Mishra, 1975; Gingerich *et al.*, 1993, 1994, 1995a,b; Hulbert *et al.*, 1998).

Rodhocetus kasrani Gingerich, Raza, Arif, and Anwar, 1994

Indocetus ramani (in part) Gingerich *et al.*, 1993, p. 396, Figs. 2,3
Rodhocetus kasrani Gingerich *et al.*, 1994, p. 844, Figs. 1a, 3; 1995a, p. 299, Fig. 6A

Holotype.—GSP-UM 3012, a partial skull with mandibles, much of the vertebral column, ribs, left and right innominates, and a femur.
Type Locality.—Middle of Bozmar Nadi valley, Sulaiman Range, Punjab, Pakistan.
Referred Specimens.—A skull with mandibles.
Age and Distribution.—Lutetian (middle Eocene), Domanda Formation of the Sulaiman Range, central Pakistan.
Diagnosis.—As for genus.

Takracetus Gingerich, Arif, and Clyde, 1995a

Type and Only Species.—*Takracetus simus* Gingerich *et al.*, 1995a.

Age and Distribution.—Lutetian (middle Eocene), Domanda Formation of the Sulaiman Range, central Pakistan.

Diagnosis.—*Takracetus* is characterized by large orbits and a broad rostrum. The external nares located above C (like *Rodhocetus*) and a double-rooted M^3 distinguish *Takracetus* from later protocetids *Protocetus, Eocetus, Babiacetus*, and *Georgiacetus* (Gingerich *et al.*, 1995a).

Takracetus simus Gingerich, Arif, and Clyde, 1995a

Takracetus simus Gingerich *et al.*, 1995a, p. 300, Figs. 6B, 7

Holotype.—GSP-UM 3041, a partial skull preserving the anterior portion of the rostrum, middle portion of skull including the frontal shield and orbits, and much of the occipital region.

Type Locality.—West of Takra Pond, west side of Takra Valley, Sulaiman Range, Punjab, Pakistan.

Referred Specimens.—A skull and mandibles, vertebrae, and ribs.

Age and Distribution.—Lutetian (middle Eocene), Domanda Formation of the Sulaiman Range, central Pakistan.

Diagnosis.—As for genus.

Georgiacetus Hulbert, Petkewich, Bishop, Bukry, and Aleshire, 1998

Georgiacetus Hulbert *et al.*, 1998

Type and Only Species.—*Georgiacetus vogtlensis* Hulbert *et al.*, 1998.

Age and Distribution.—Bartonian (middle Eocene), "Blue Bluff Unit" (formerly McBean Formation), Georgia and the Chapel Branch member of the Santee Limestone, South Carolina.

Diagnosis.—*Georgiacetus* is distinguished from other known protocetid genera in the presence of a true pterygoid sinus fossa and a pelvis that does not articulate directly with the sacral vertebrae. An unfused mandibular symphysis terminating at the anterior margin of the P^3 differentiates it from *Babiacetus*. *Georgiacetus* has accessory cuspules on cheek teeth, unlike *Babiacetus* and *Protocetus* (Kellogg, 1936; Sahni and Mishra, 1975; Gingerich *et al.*, 1993, 1995a,b; Hulbert *et al.*, 1998).

Georgiacetus vogtlensis Hulbert, Petkewich, Bishop, Bukry, and Aleshire, 1998

Georgiacetus vogtlensis Hulbert *et al.*, 1998, Figs. 4–9, 11; Hulbert, Chapter 8, this volume, Figs. 3–12

Holotype.—GSM 350, partial skull, left mandible with teeth, edentulous right mandible, 23 vertebrae, 12 ribs, xiphisternum, a hemal arch, and left and right innominates.

Type Locality.—South of Water Intake Station, Vogtle Electrical Generating Plant, Burke County, Georgia.

Referred Specimens.—Associated thoracic vertebrae with partial ribs, a tooth crown, and lumbar vertebrae. Material collected from the Santee Limestone in South Carolina (Albright, 1996) including an incisor, premolar, and molar have been tentatively referred to *Georgiacetus* cf. *G. vogtlensis* (Hulbert *et al.*, 1998).

Age and Distribution.—Bartonian (middle Eocene), "Blue Bluff Unit" (formerly McBean Formation) of Burke County, Georgia, and the Chapel Branch member of the Santee Limestone of Berkeley County, South Carolina.

Diagnosis.—As for genus.

2.1 Unnamed Protocetids

2.1.1. "Cross Whale"

A partial protocetid skeleton was uncovered from the middle Eocene (Bartonian–Priabonian) Cross Formation in the Martin Marietta Cross Quarry of Berkeley County, South Carolina, in 1994 (Geisler *et al.*, 1996; see Geisler and Luo, Chapter 6, this volume). Material includes a partial skull, portions of both mandibles, seven complete teeth, one partial atlas vertebra, one cervical vertebra, five thoracic vertebrae, and portions of 13 ribs. Based on Geisler *et al.* (1996), preliminary phylogenetic analyses suggest that the South Carolina specimen occupies an intermediate position between the much older Asian *Rodhocetus kasrani* (Lutetian) and the African *Protocetus atavus* (Lutetian).

2.1.2. "Protocetid Whale of Uhen"

An early Bartonian (middle Eocene) protocetid from the Milam Member of the Cook Mountain Formation, Louisiana, was reported by Uhen (1996). The specimen includes 13 vertebrae: 4 thoracic, 5 lumbar, 1 sacral, and 3 of questionable assignment. This protocetid is much larger than *Protocetus* and *Rodhocetus*, and the vertebrae are unlike those assigned to *Eocetus*. It is similar in size to *Pappocetus* and *Babiacetus*, but the lack of vertebral material for both genera excludes comparison (Uhen, 1996).

2.1.3. "Protocetid Whale of McLeod and Barnes"

McLeod and Barnes (1996) discussed a *Pappocetus*-like taxon found in middle to late Eocene sediments of the Coastal Plain of North and South Carolina. Collected material includes two associated mandibles with teeth and other isolated teeth.

2.2. Unnamed Archaeocetes

2.2.1. "Archaeocete Whale of Satsangi and Mukhopadhyay"

Satsangi and Mukhopadhyay (1975) figured a partial skull (GSI-K56/157) collected from the Harudi Formation west of Babia Hill in Kachchh District, western India. The specimen preserves three rooted molars and a P^4 in the maxilla. The supraoccipital process nearly covers the orbit. Satsangi and Mukhopadhyay (1975, p. 85) state that the "size and morphological detail undoubtedly suggest that the skull belongs to a primitive whale."

2.2.2 "Habib Rahi Limestone Whale"

Gingerich (1991) reported an archaeocete from the earliest Lutetian (middle Eocene) Habib Rahi platy limestone of the Sulaiman Range, central Pakistan. The specimen (GSP-UM 1858) consists of a skull and partial skeleton including numerous thoracic vertebrae, and an articulated right scapula, humerus, and ulna. It has *Pakicetus*-like teeth, but the tympanic bulla is more massive.

2.2.3. "Togo Whale"

Gingerich *et al.* (1992) identified archaeocete remains collected from the Lutetian (middle Eocene) phosphates of the Kpogamé-Hahotoé basin in Togo. The material includes archaeocete skull fragments, two mandibles, 60 isolated teeth, several isolated bullae, vertebral centra, and isolated limb elements. These elements were not found in association, but their similar size point to a *Protocetus atavus*-sized cetacean. In particular, Gingerich *et al.* (1992) noted that the limb elements suggest the archaeocete had forelimbs modified for swimming but more conservatively adapted (terrestrial) hind limbs.

2.2.4. "Cetacea indet. of Bajpai and Thewissen"

Bajpai and Thewissen (Chapter 7, this volume) discuss and isolated premolar (possible dP) from the Subathu Formation in the Dharampur area of Himachel Pradesh, northern India. Although Bajpai and Thewissen do not find the tooth to be diagnostic of any family, it is notable as the first cetacean tooth from marine sediments in the Subathu Formation.

3. Occurrence of Early Cetaceans

Early cetaceans follow a predictable progression through the sedimentary deposits in which they are found. The most primitive whales are preserved in truly terrestrial deposits whereas later forms are recovered from strata representing marginal marine and open marine environments. The following discussion will examine the occurrence of each family in terms of geographic distribution and depositional environments (see Figs. 1 and 2).

3.1. Pakicetids

The most primitive cetaceans, pakicetids, are known exclusively from the redbeds of the lower portion of the Kuldana Formation in northwestern Pakistan and from the redbeds found in the upper part of the Subathu Formation in northwestern India (Thewissen and Hussain, 1998). The lower Kuldana section consists of red mudstones, and to a less extent limestones, sandstones, and conglomerates (West and Lukacs, 1979; Aslan and Thewissen, 1996). All pakicetid specimens have been recovered from the conglomeratic unit. This lithofacies has been interpreted as representing floodplain soil nodules reworked into river channels during episodic channel incision (Wells, 1983, 1984). In a preliminary analysis of the paleosols of the lower Kuldana Formation, Aslan and Thewissen (1996) support this con-

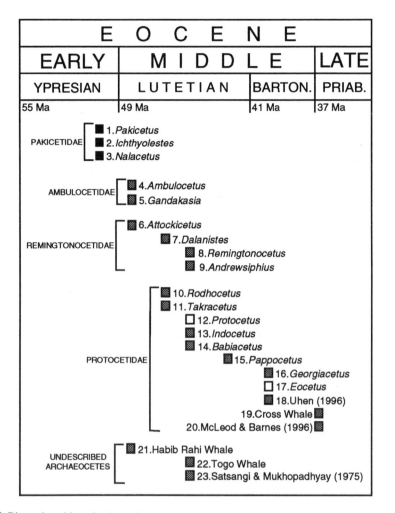

FIGURE 1. Biostratigraphic and paleoenvironmental distribution of early cetaceans. Black squares indicate fluvial environments; gray squares indicate marginal to nearshore marine environments; and white squares indicate open shallow marine environments. Geologic time scale from Berggren *et al.* (1996).

clusion and note that there are similarities between the calcareous nodules of the channel conglomerates and nodules found *in situ* in the paleosols. This sedimentological evidence suggests that pakicetids may have inhabited fluvial environments.

There has been some debate over the timing of lower Kuldana deposition in relation to deposition of the Subathu Formation of Indian and Pakistani Kashmir. Undescribed cetacean material has been recovered from presumably marine beds of the lower Subathu Formation (Kumar and Loyal, 1987; Thewissen *et al.*, 1996). These beds lie stratigraphically below the redbeds of the Subathu, which have produced a diverse assemblage of mammals including *Pakicetus attocki* (Kumar and Sahni, 1985; Kumar, 1991, 1992; Thewissen

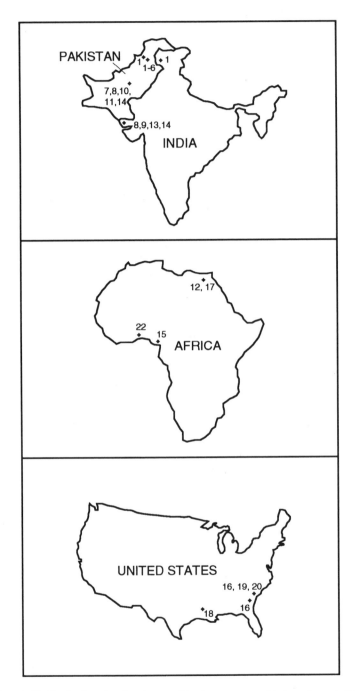

FIGURE 2. Geographic distribution of early cetaceans. Numbers correspond to taxa shown in Fig. 1. India-Pakistan, Africa, and the United States are unscaled.

et al., 1996). In adjacent Pakistan, however, there is no record of cetaceans below the redbeds of the Kuldana Formation.

Wells and Gingerich (1987) proposed that the redbeds of the Kuldana and Subathu Formations are correlative suggesting that the Subathu specimens would be the oldest cetaceans known. However, studies of fossil mammalian assemblages from the redbeds indicate that Kuldana deposits predate those of the Subathu (Bruijn *et al.*, 1982; Thewissen *et al.*, 1987, 1996; Thewissen and McKenna, 1992). Further investigation is required before this important debate can be resolved.

3.2. Ambulocetids

The ambulocetid cetaceans, *Ambulocetus natans* and *Gandakasia potens*, are found higher in the Kuldana section, near the Kuldana–Kohat formational boundary in northwestern Pakistan. The type specimen of *A. natans* was collected just below the Kuldana–Kohat contact, and *G. potens* material has been attributed to units in the upper middle portion of the section (West and Lukacs, 1979; West, 1980; Thewissen *et al.*, 1996). The upper part of the Kuldana Formation is comprised of interbedded shales, marls, and limestones with minor evaporites, and the transition from the continental redbeds of the lower Kuldana to the marginal marine deposits of the upper Kuldana is marked by variegated shales that have been interpreted as freshwater deposits (Wells, 1984). The formation is capped by an oyster ridge that is generally regarded as the highest unit of the Formation.

If the succession from the terrestrial lower Kuldana to the coastal marine deposits of the Kohat Formation is a Waltherian sequence (without paraconformities), then the transition zone from the lower to upper portions of the Kuldana must represent an extensive tidal sequence deposited in an arid environment (Allen, 1970; Middleton, 1973; Harms *et al.*, 1975; Boggs, 1986). This is supported by the interbedded nature of the pure limestones, marls, terrigenous mudstones, and evaporites, and the combined occurrence of freshwater, brackish, and marine invertebrates. This stratigraphic interpretation combined with recent isotopic studies by Roe *et al.* (this volume) that indicate freshwater signatures in the enamel apatite of *Ambulocetus* strongly suggests that ambulocetid whales occupied tidal areas with a strong freshwater influence.

3.3. Remingtonocetids and Eastern Tethyan Protocetids

Deposits of the Sulaiman Range of central Pakistan and Kachchh District of western India have yielded an abundance of remingtonocetid and protocetid material. The oldest record of a remingtonocetid, however, comes from the transitional Kuldana–Kohat deposits of northern Pakistan. *Attockicetus praecursor* was recovered as float on a lower Kohat Formation slope although its exact provenance is uncertain (Thewissen and Hussain, in press). As discussed above, the upper portion of the Kuldana Formation represents tidal deposition and the lower Kohat beds are interpreted here as representing tidal to coastal marine deposits based on the interbedded nature of the thin argillaceous limestones and calcareous shales and the presence of onshore marine invertebrates (nummulites, bivalves, gastropods; see Wells, 1984). Therefore, it is likely that the earliest remingtonocetids occupied tidal to coastal habitats similar to those of ambulocetids.

Later remingtonocetids are found in association with protocetid cetaceans in the Domanda Formation of central Pakistan and the Harudi Formation of northwestern India. The Domanda contains the remingtonocetids *Remingtonocetus harudiensis* and *Dalanistes ahmedi* and the protocetids *Takracetus simus* and *Rodhocetus kasrani*. The Harudi contains the richest abundance of remingtonocetids with *R. harudiensis, R. sloani, Andrewsiphius kutchensis,* and *A. minor* and includes the protocetids *Indocetus ramani, Babiacetus indicus,* and *B. mishrai. B. indicus* is also found in the Drazinda Formation, which lies stratigraphically above the Domanda Formation.

The Domanda is comprised of interbedded red-brown and green shales, silty marls, and thin oyster-rich limestones. The lower, middle, and upper portions of the section are differentiated by the predominance of green shales and limestones (lower), red shales (middle), and red-brown shales (upper) although there is considerable interbedding within units (Gingerich *et al.,* 1995a). Both remingtonocetid taxa come from the middle red shale unit, while the protocetids *T. simus* and *R. kasrani* are found in the lower green shales.

Gingerich *et al.* (1995a) interpret the Domanda deposits as representing offshore shallow marine (red-brown shale) to more distal carbonate-rich (green shales and limestones) environments and further point out that the remingtonocetids and protocetids may be associated with different sediments and water depths: remingtonocetids in the shallower water (brown shales) and the protocetids in the deeper water (green shales). However, the Domanda Formation contains few certain indicators of its depositional origin. The lack of diverse normal marine invertebrates (Eames, 1951, 1952; Iqbal, 1969), the abundance of oysters and benthic forams, the presence of silty marls and color variegated shales, and its overall similarity to the better-known Harudi Formation (see below) all suggest a very shallow nearshore to tidal origin for the deposit. Although the evidence for changing water depth and differential occurrence of whales within these units is equivocal, it appears that the Domanda remingtonocetids and protocetids were occupying coastal marine waters with at least some freshwater influence.

The Harudi Formation consists of interbedded green, gray, and brown shales, nummulitic, oolitic, and oyster-rich limestones, coquinas, marls, carbonaceous shales and lignites, and thin siltstones. Gypsum and glauconite are present as well as fossil plants, crocodilians, and turtles. Previously, fossil cetaceans collected in the Harudi Formation were attributed to individual beds of the Berwali Series, Babian Stage (e.g., the Chocolate or Shell Limestone of Sahni and Mishra, 1975; Tandon, 1976; Gingerich *et al.,* 1995b). Three lithologies within the Harudi (Chocolate Limestone, Gypseous Shales, and Grey Marl) have been associated with cetacean fossils (Kumar and Sahni, 1986), although Biswas (1992) does not assign names to these units. The chocolate limestone layer has yielded the remingtonocetids *R. harudiensis* and *R. sloani* and the protocetid *I. ramani.* The overlying shales (referred to as the Gypseous Shales) contain the most taxa with the remingtonocetids *R. sloani, A. kutchensis,* and *A. minor* and the protocetids *B. indicus* and *B. mishrai.* The type specimen of *A. kutchensis* is attributed to a gray marl unit overlying the gypseous shales. The sequence of strata present in the Harudi Formation is representative of lagoonal to inner shelf deposition (Biswas, 1992). It appears that Harudi cetaceans frequented environments similar to those in the Domanda Formation.

Babiacetus indicus is also known from the Drazinda Formation of Pakistan. The Drazinda overlies the Domanda and Pir Koh Limestone in the Sulaiman Range and is comprised of green to brown shales interbedded with thin fossiliferous limestones. *B. indicus* was collected in reddish-brown shales between two limestone bands. The lithological similarities between the Drazinda and Domanda Formations point to continued preservation of cetaceans in tidal to shallow nearshore marine environments.

3.4. Western Tethyan and North American Protocetids

Western Tethyan protocetids have been found in Egypt (*Protocetus* and *Eocetus*) and southern Nigeria (*Pappocetus*). The Egyptian whales were collected at Gebel Mokattam, and although the exact locations of the type localities are unknown (see Gingerich, 1992), it is likely that both sites are now obliterated by growth of the city of Cairo. The Eocene section at Gebel Mokattam includes the Mokattam, Guishi, and Maadi Formations. Unlike eastern Tethyan cetacean-bearing deposits, these formations are predominantly comprised of fossiliferous limestones, with only minor shales and marls. *Protocetus atavus* came from the Lower Building Stone Member of the Mokattam, which is described as a nummulitic limestone with bony fish remains and shark teeth (Schweinfurth, 1883; Fraas, 1904; Gingerich, 1992). *Eocetus schweinfurthi* was recovered higher in the section from the Guishi Formation, a gray-white limestone with abundant mollusks and small nummulites. Kellogg (1936) and Barnes and Mitchell (1978) both placed *Eocetus* in the Mokattam Formation.

The Gebel Mokattam Eocene section clearly represents a shallow carbonate platform environment. The Mokattam Formation is entirely comprised of limestone, nummulitic in horizons. The Guishi is a gray-white limestone with beds containing nummulites, mollusks, bryozoans, and echinoids (Gingerich, 1992), and the Maadi represents a regressional sequence with its marls and capping Ain Musa sandstone bed. In particular, the presence of echinoids is indicative of a carbonate marine environment without freshwater influence (Sprinkle and Kier, 1987). The Egyptian protocetids are the only members of the family that are preserved in beds clearly attributable to normal-marine shallow offshore shelf deposition.

The type specimen of *Pappocetus* was collected from the Port Harcourt railroad cut in the Ombialla District, southern Nigeria (Kellogg, 1936). A later description of the material (Halstead and Middleton, 1974) places *Pappocetus* in the Ameki Formation. The middle Eocene sediments of southern Nigeria consist of lagoonal and estuarine sandstones and shales that grade eastward into more open marine deposits. *Pappocetus* was recovered from the estuarine shales and was found in association with turtles, crocodilians, and mollusks (Kellogg, 1936).

North American protocetid whales are known from the Coastal Plain of South Carolina (Santee Limestone and Cross Formation), Georgia (Blue Bluff Unit, formerly the McBean Formation), and Louisiana (Milam Member of the Cook Mountain Formation). These formations represent diverse marginal and shallow marine deposits ranging from the sandy mudstones and oyster-rich marls of the Blue Bluff Unit through the calcareous and glauconitic mudstones of the Cook Mountain Formation to the marls and purer carbonates of the Santee Limestone and Cross Formation of South Carolina. Each of these units represents shallow nearshore marine, and perhaps in the case of the Blue Bluff Unit, tidal environments consistent with the occurrence of most other protocetids.

Depositional interpretations of the strata in which early cetaceans are recovered from can provide another window into cetacean evolution and migration. During the early and middle Eocene the trailing margin of the Indian subcontinent was characterized by an alluvial plain contiguous with a broad tidal to shallow marine shelf. This shelf appears to have been starved of coarser clastic debris and shows little influence of tidal currents, wave action, or sea-level fluctuation, which strongly suggests that it was rimmed and protected by reefs (Jones and Desrochers, 1992; Steckler *et al.*, 1993). This stable shallow environment provided the ideal birthplace for the Cetacea as they evolved from terrestrial predators (pakicetids) to tidal and shallow coastal marine organisms (ambulocetids, remingtonocetids, and protocetids) and finally to open marine animals (protocetids) capable of dispersal to new regions.

4. Taphonomy

In the foregoing discussion, little attention has been paid to the taphonomic factors influencing the occurrence of early cetacean fossils. Physical (depositional) biases such as higher net sediment accumulation rates in nonmarine and marginal marine environments may have enhanced the likelihood of whale fossil preservation in these deposits.

Perhaps more importantly, however, the occurrence of whale carcasses may not accurately reflect their usual habitat preferences. The fact that these whales, like modern whales, may have sought out shallow environments when distressed (Slijper, 1962) may have contributed to the subsequent preservation of their remains in onshore deposits. For example, the widespread distribution of protocetids certainly indicates oceangoing tendencies despite their predominant occurrence in shallow marine deposits.

Finally, these physical and biological biases may be magnified by paleontological (collecting) biases, as the natural tendency is for collections to be made repeatedly in horizons and localities proven to contain early cetacean fossils. In sum, the habitat preferences of living archaeocete whales may have been somewhat different from what is indicated by the fossil record, and a more complete understanding of those preferences awaits further taphonomic investigation.

Acknowledgments

I thank Dr. J. G. M. Thewissen for giving me the opportunity to contribute to this volume and for providing unpublished data and reviewing the manuscript. I also thank Dr. W. S. Bartels for reviewing several drafts of this work and Dr. R. C. Hulbert for providing unpublished data and for his generous help in reviewing protocetid paleontology. This work was funded in part by a National Science Foundation grant to J.G.M.T.

Appendix: Catalogue of Referred Specimens

Pakicetidae

Pakicetus inachus
GSP-UM specimens: 81 (mandible with P_{2-4}) 82 (M_x), 83 (M^1), 85 (M^1), 113 (trigonid), 134 (M^2, labial fragment), 147 (P^4 fragment), 751 (P^4 fragment), 1672 (M^1). Proba-

ble *P. inachus* (Gingerich and Russell, 1990): 108, 136, 749, 750, 1401, 1409, 1450, 1509, 1653, 1722, 1937, 1938.
H-GSP specimen: 1981b (M^2; listed as 1981 in West, 1980, p.516).

Pakicetus attocki
NHML specimen: M-15806 (mandible with dP_3).
H-GSP specimens: 535 (mandible with M_x fragment), 18399 (P^x fragment, possible P^2), 18410 (M_x, possible M_2), 18431 (P^x fragment, possible P^1), 18467 (rostrum fragment with partial alveolus for I^1, alveoli for I^2–C^1, and crown of I^3), 18470 (maxilla with P^4–M^3 and alveoli for P^3), 18483 (P^x fragment), 18489 (left tympanic), 18495 (fragmentary mandible with condyle and alveoli for I_1–P_3), 18519 (C^1), 91014 (M^1), 91030 (possible I^x), 91031 (M^x fragment, possible M^2), 91034 (mandible with alveoli for $I_{1–3}$, C, P_1, crowns for $dP_{2–4}$, M_1, and unerupted M_2), 91035 (left tympanic with incus).
VPL specimen: 706 ($M^{1–2}$).

Ichthyolestes pinfoldi
GSP-UM specimens: 79 (possible P_2), 110 (possible P_2), 1936 (P^x). Tentatively referred specimens described by Gingerich and Russell (1990): 1411, 1534, 1546, 1553.
H-GSP specimens: 536 (M_x), 18395 (dP_4), 18403 (partial P^4), 18485 (mandible with erupting C_1, unerupted $P_{2–4}$, and alveoli for P_1, M_1), 91015 (mandible with alveoli for $dI_{1–2}$, erupting I_3 and C_1, alveolus for P_1, and unerupted $P_{2–3}$), 91047 (mandible with ascending ramus, dP_4, unerupted M_3, P_2 fragment, and alveoli for I_3, C_1, dP_3, $M_{1–2}$), 92159 (dP_4 fragment), 1974a.
IPHG specimen: 1956 II 9 (dP_x, possible dP_3).

Nalacetus ratimitus
H-GSP specimens: 18408 (P_3, possible P_3), 91036 (mandible with fragment of C_1, $P_{3–4}$, trigonid of M_1, and alveoli for $P_{1–2}$), 91044 (dP_4), 91045 (M_x).

Ambulocetidae

Ambulocetus natans
H-GSP specimens: 498 [caudal fragment identified by West (1980) as Basilosauridae indet.], 18472 (midcaudal vertebra), 18473 (left mandibular fragment with the base of P_2), 18474 (incomplete P_3), 18497 (P_3), 92148 (caudal vertebra), 92151 (proximal manual phalanx).

Gandakasia potens
H-GSP specimens: 497 (M_x), 500 (M_x) described by West (1980).
IPHG specimens: 1956 II 6 (broken P_2 and P_3) described by Dehm and Oettingen-Spielberg (1958) and Gingerich and Russell (1990).

Remingtonocetidae

Remingtonocetus harudiensis
GSP-UM specimens: 1856 (partial skull), 3009 (skull fragments, vertebrae, sacrum, and acetabulum of pelvis), 3015 (skull fragments, vertebrae, proximal femur, and tibia),

3054 (left femur and proximal epiphyses of tibia), 3057 (skull fragments, vertebrae, and ribs), 3101 (fragmentary middle portion of a skull).

RUSB specimens: 2016 (maxilla fragment with M^{2-3} and anterior orbits), 2017 (central portion of braincase with left occipital condyle preserved), 2018 (rostral fragment with right P^2 and roots for left and right P^{1-2}).

VPL specimen: 1004 (braincase with left and right ear regions).

Remingtonocetus sloani

LUVP specimens: 11001 (partial skull), 11003 (left mandibular fragment with alveoli for P_4–M_1), 11043 (anterior portion of rostrum), 11146 (posterior portion of skull).

RUSB specimens: 2019 (premaxilla with roots for left and right I^{1-3}), 2020 (maxilla fragment with bases for left M^{2-3} and endocasts of paranasal sinuses), 2022 (rostrum with anterior orbit and bases for left and right C^1–M^1), 2025 (maxilla fragment with bases-for lft or right C^1–M^1), 2025 (maxillia fragrent with bases of M^{1-2} and endocasts of infraorbital canal and paranasal sinuses), 2026 (rostrum fragment with bases of left and right P^2).

VPL specimens: 1001 (skull with alveoli or roots for all teeth and a partial braincase), 15003 (rostrum with roots for all teeth and a left P^4 crown).

Andrewsiphius kutchensis

RUSB specimen: 2021 (rostrum with bases of all left and right molars).

VPL specimens: 1007 (fragmentary skull in three parts with complete braincase, orbital region with posterior palate and crown of M^3, and rostrum fragment with alveoli for left and right C^1–P^2 and fragmentary P^2 crown), 1019 (rostrum fragment with alveoli and partial teeth of P^3–M^3 and possible I^3–P^1).

Dalanistes ahmedi

NHML specimen: M-50719 (partial skull and cervical vertebra).

GSP-UM specimens: 11 (partial sacrum), 3045 (lumbar vertebra), 3052 (partial skull), 3089 (innominate), 3096 (cervical vertebra), 3097 (lumbar vertebra), 3099 (partial skull and associated vertebrae), 3102 (sacrum), 3109 (lumbar vertebra), 3115 (distal femur).

Protocetidae

Protocetus atavus

SMNS specimens: 11085 (axis and five cervical vertebrae), 11086 (dorsal, lumbar, and one sacral vertebra, rib fragments), 11088 (five dorsal vertebrae), 11089 (axis).

Eocetus schweinfurthi

SMNS specimen: 110934 (two lumbar vertebrae). These have been questioned by Barnes and Mitchell (1978) and Hulbert *et al.* (in press) for being basilosaurine-like.

Pappocetus lugardi

MHML specimens: M-11086 (mandible), M-11087 (tooth), M-11089 (axis).

Halstead and Middleton (1974) refer nine vertebrae and a proximal rib fragment collected in the Ameki Formation to *Pappocetus*.

Indocetus ramani
VPL specimens: 1014 (right tympanic), 1017 (endocast of nasal cavity near orbits including roots of M^{2-3} and endocast of cranial cavity), 1018 (endocast of cranial cavity), 1023 (right maxilla with fragmentary P^3–M^3).

Babiacetus indicus
GSP-UM specimen: 3005 (partial skull with lower jaws).

Rodhocetus kasrani
GSP-UM specimens: 1852 (left and right mandibles with much of the lower dentition), 1853 (skull).

Takracetus simus
GSP-UM specimen: 3070 (skull with lower jaws, cervical and thoracic vertebrae, and ribs).

Georgiacetus vogtlensis
ChM specimens: PV5037 (I_3 or C), PV5038 (P_2), PV5039 (M_1) described by Albright (1996) are referred to *Georgiacetus* cf. *vogtlensis* by Hulbert *et al.* (in press).
GSM specimens: 351 (thoracic vertebrae, portions of four ribs, and partial tooth), 352 (lumbar vertebra).

References

Albright, L. B. 1996. A protocetid cetacean from the Eocene of South Carolina. *J. Paleontol.* **70**(3):519–523.

Allen, J. R. L. 1970. *Physical Processes of Sedimentation.* Allen & Unwin, London.

Andrews, C. W. 1906. *A Descriptive Catalogue of the Tertiary Vertebrata of the Fayum, Egypt.* British Museum of Natural History, London.

Andrews, C. W. 1920. A description of new species of zeuglodont and of leathery turtle from the Eocene of southern Nigeria. *Proc. Zool. Soc. London* **22**:309–319.

Aslan, A., and Thewissen, J. G. M. 1996. Preliminary evaluation of paleosols and implications for interpreting vertebrate fossil assemblages, Kuldana Formation, northern Pakistan. *Palaeovertebrata* **25**(2–4):261–277.

Bajpai, S., Thewissen, J. G. M., and Sahni, A. 1996. *Indocetus* (Cetacea, Mammalia) endocasts from Kachchh (India). *J. Vertebr. Paleontol.* **16**(3):582–584.

Barnes, L. G., and Mitchell, E. 1978. Cetacea, in: V. J. Maglio and H. B. S. Cooke (eds.), *Evolution of African Mammals,* pp. 582–587. Harvard University Press, Cambridge, MA.

Berggren, W. A., Kent, D. V., Swisher, C. C., and Aubry, M. P. 1996. A revised Cenozoic geochronology and chronostratigraphy, in: *Geochronology, Time Scales, and Global Stratigraphic Correlation,* No. 54, pp. 129–212. SEPM Spec. Publ., Tulsa, OK.

Biswas, S. K. 1992. Tertiary stratigraphy of Kutch. *J. Palaeontol. Soc. India* **37**:1–29.

Boggs, S. 1986. *Principles of Sedimentology and Stratigraphy.* Merrill, Columbus, OH.

Bruijn, H. de, Hussain, S. T., and Leinders, J. J. M. 1982. On some early Eocene rodent remains from Barbara Banda, Kohat, Pakistan, and early history of the order Rodentia. *Proc. K. Ned. Akad. Wet. Ser. B* **85**:249–258.

Dehm, R., and Oettingen-Spielberg, T. zu. 1958. Paläontologische und geologische Untersuchungen im Tertiär von Pakistan. 2. Die mitteleocänen Saügetierre von Ganda Kas bei Basal in Nordwest Pakistan. *Abh. Bayer. Akad. Wiss. Math. Naturwiss. Kl. N.F.* **91**:1–54.

Eames, F. E. 1951. A contribution to the study of the Eocene in western Pakistan and western India: B. The description of Lamellibranchia from standard sections in the Rakhi Nala and Zinda Pir areas of the western Punjab and in the Kohat District. *Philos. Trans. R. Soc. London Ser. B* **235**:311–482.

Eames, F. E. 1952. A contribution to the study of the Eocene in western Pakistan and western India: B. A description of the Scaphopoda and Gastropoda from standard sections in the Rakhi Nala and Zinda Pir areas of western Punjab and the Kohat District. *Philos. Trans. R. Soc. London Ser. B* **236**:1–168.

Fraas, E. 1904. Neue Zeuglodonten aus dem unteren Mitteleocän vom Mokattam bei Cairo. *Geol. Paläontol. Abh. N. F.* **6**:199–220.

Geisler, J., Sanders, A. E., and Luo, Z. 1996. A new protocetid cetacean from the Eocene of South Carolina, U.S.A.; phylogenetic and biogeographic implications. *Paleontol. Soc. Spec. Bull.* **8**:139.

Gingerich, P. D. 1991. Partial skeleton of a new archaecocete from the earliest middle Eocene Habib Rahi Limestone, Pakistan. *J. Vertebr. Paleontol.* **11**(3):31A.

Gingerich, P. D. 1992. Marine mammals (Cetacea and Sirenia) from the Eocene of Gebel Mokattam and Fayum, Egypt: stratigraphy, age, and paleoenvironments. *Univ. Michigan Pap. Paleontol.* **30**:1–84.

Gingerich, P. D., and Russell, D. E. 1981. *Pakicetus inachus*, a new archaeocete (Mammalia, Cetacea) from the early-middle Eocene Kuldana Formation of Kohat (Pakistan). *Contrib. Mus. Paleontol. Univ. Michigan* **25**:235–246.

Gingerich, P. D., and Russell, D. E. 1990. Dentition of early Eocene *Pakicetus* (Mammalia, Cetacea). *Contrib. Mus. Paleontol. Univ. Michigan* **28**(1):1–20.

Gingerich, P. D., Wells, N. A., Russell, D. E., and Shah, S. M. I. 1983. Origin of whales in epicontinental remnant seas: new evidence from the early Eocene of Pakistan. *Science* **220**:403–406.

Gingerich, P. D., Cappetta, H., and Traverse, M. 1992. Marine mammals (Cetacea and Sirenia) from the middle Eocene of Kpogamé-Hahotoé in Togo. *J. Vertebr. Paleontol.* **12**(3):29–30A.

Gingerich, P. D., Raza, S. M., Arif, M., Anwar, M., and Zhou, X. 1993. Partial skeletons of *Indocetus ramani* (Mammalia, Cetacea) from the lower middle Eocene Domanda Shale in the Sulaiman Range of Punjab (Pakistan). *Contrib. Mus. Paleontol. Univ. Michigan* **38**(16):393–416.

Gingerich, P. D., Raza, S. M., Arif, M., Anwar, M., and Zhou, X. 1994. New whale from the Eocene of Pakistan and the origin of cetacean swimming. *Nature* **368**:844–847.

Gingerich, P. D., Arif, M., and Clyde, W. C. 1995a. New archaeocetes (Mammalia, Cetacea) from the middle Eocene Domanda Formation of the Sulaiman Range, Punjab (Pakistan). *Contrib. Mus. Paleontol. Univ. Michigan* **29**(11):291–330.

Gingerich, P. D., Arif, M., Bhatti, M. A., Raza, H. A., and Raza, S. M. 1995b. *Protosiren* and *Babiacetus* (Mammalia, Sirenia and Cetacea) from the middle Eocene Drazinda Formation, Sulaiman Range, Punjab (Pakistan). *Contrib. Mus. Paleontol. Univ. Michigan* **29**(12):331–357.

Gingerich, P. D., Arif, M., Bhatti, M. A., Anwar, M., and Sanders, W. J. 1997. *Basilosaurus drazindai* and *Basiloterus hussaini*, new Archaeoceti (Mammalia, Cetacea) from the middle Eocene Drazinda Formation, with revised interpretation of ages of whale-bearing strata in the Kirthar group of the Sulaiman Range, Punjab (Pakistan). *Contrib. Mus. Paleontol. Univ. Michigan* **30**(2):55–81.

Halstead, L. B., and Middleton, J. A. 1974. New material of the archaeocete whale, *Papocetus lugardi* Andrews, from the middle Eocene of Nigeria. *J. Min. Geol.* **8**:81–85.

Harms, J. C., Southard, J. B., Spearing, D. R., and Walker, R. G. 1975. Depositional environments as interpreted from primary sedimentary structures and stratiphication sequences. *Soc. Econ. Paleontol. Mineral. Short Course* **2**:1–161.

Hulbert, R. C., Petkewich, R. M., Bishop, G. A., Bukry, D., and Aleshire, D. P. 1998. A new middle Eocene protocetid whale (Mammalia: Cetacea: Archaeoceti) and associated biota from Georgia. *J. Paleontol.* **72**:905–925.

Iqbal, M. W. A. 1969. The Tertiary pelecypod and gastropod fauna from Drug, Zindapir, Vidor (District D. G. Khan), Jhalar, and Charat (District). *Mem. Geol. Surv. Pakistan* **6**:1–94.

Jones, B., and Desrochers, A. 1992. Shallow platform carbonates, in: R. G. Walker and N. P. James (eds.), *Facies Modeling: Response to Sealevel Change*, pp. 277–301. Geol. Assoc. Canada.

Kellogg, A. R. 1928. The history of whales—their adaptation to life in water. *Q. Rev. Biol.* **3**(2):174–208.

Kellogg, A. R. 1936. A review of the Archaeoceti. *Carnegie Inst. Washington Publ.* **482**:1–366.

Kumar, K. 1991. *Anthracobune ajiensis* nov. sp. (Mammalia: Proboscidea) from the Subathu Formation, Eocene from NW Himalaya, India. *Geobios* **24**:221–239.

Kumar, K. 1992. *Paratritemnodon indicus* (Creodonta: Mammalia) from the early Middle Eocene Subathu Formation, NW Himalaya, India and the kalakot mammalian community structure. *Paläontol. Zool.* **66**:387–403.

Kumar, K., and Loyal, R. S. 1987. Eocene ichthyofauna from the Subathu Formation, northwestern Himalaya, India. *J. Palaeontol. Soc. India* **32**:60–84.

Kumar, K., and Sahni, A. 1985. Eocene mammals from the upper Subathu Group, Kashmir Himalaya, India. *J. Vertebr. Paleontol.* **5**:153–168.

Kumar, K., and Sahni, A. 1986. *Remingtonocetus harudiensis*, new combination, a middle Eocene archaeocete (Mammalia, Cetacea) from western Kutch, India. *J. Vertebr. Paleontol.* **6**:326–349.

Maas, M. C., and Thewissen, J. G. M. 1995. Enamel microstructure of *Pakicetus* (Mammalia: Archaeoceti). *J. Paleontol.* **69**(6):1154–1163.

McLeod, S. A., and Barnes, L. G. 1996. The systematic position of *Pappocetus lugardi* and a new taxon from North America (Archaeoceti: Protocetidae). *Paleontol. Soc. Spec. Bull.* **8**:270.

Middleton, G. V. 1973. Johannes Walther's law of the correlation of facies. *Geol. Soc. Am. Bull.* **84**:979–988.

Sahni, A., and Mishra, V. P. 1972. A new species of *Protocetus* (Cetacea) from the middle Eocene of Kutch, western India. *Palaeontology* **15**:490–495.

Sahni, A., and Mishra, V. P. 1975. Lower Tertiary vertebrates from western India. *Monogr. Palaeontol. Soc. India* **3**:1–48.

Satsangi, P. P., and Mukhopadhyay, P. K. 1975. New marine Eocene vertebrates from Kutch. *J. Geol. Soc. India* **16**(1):84–86.

Schweinfurth, G. A. 1883. Ueber die geologische schichtentgliederung des Mokattam bei Cairo. *Z. Dtsch. Geol. Ges.* **35**:709–737.

Slijper, E. J. 1936. Die Cetaceen. Vergleichend-Anatomisch und Systematisch. *Capita Zool.* **6–7**:1–599.

Slijper, E. J. 1962. *Whales.* Basic Books, New York.

Sprinkle, J., and Kier, P. M. 1987. Phylum Echinodermata, in: R. S. Boardman (ed.), *Fossil Invertebrates*, pp. 550–611. Blackwell, Oxford.

Steckler, M. S., Reynolds, D. J., Coakley, B. J., Swift, B. A., and Jarrand, R. 1993. Modeling passive margin sequence stratigraphy. *Spec. Publ. Int. Assoc. Sedimentol.* **18**:19–41.

Stromer, E. 1903. Zeuglodon-Reste aus dem oberen Mittleocän des Fajum. *Beitr. Paläontol. Geol. Österreich-Ungarns Orients* **15**:65–100.

Stromer, E. 1908. Die Archaeoceti des Ägyptischen Eozäns. *Beitr. Paläontol. Geol. Österreich-Ungarns Orients* **21**:1–14.

Szalay, F. S., and Gould, S. J. 1966. Asiatic Mesonychidae (Mammalia, Condylarthra). *Bull. Am. Mus. Nat. Hist.* **132**:131–173.

Tandon, K. K. 1976. Biostratigraphic classification of the Middle Eocene rocks of a part of south western Kutch, India. *J. Palaeontol. Soc. India* **2**:136–148.

Thewissen, J. G. M. 1993. Eocene marine mammals from the Himalayan foothills. *Res. Explor. Natl. Geog. Soc.* **9**:125–127; erratum **9**:487.

Thewissen, J. G. M., and Hussain, S. T. 1993. Origin of underwater hearing in whales. *Nature* **361**:444–445.

Thewissen, J. G. M., and Hussain, S. T. 1998. Systematic review of the Pakicetidae, early and middle Eocene Cetacea (Mammalia) from Pakistan and India, *Bull. Carnegie Mus. Nat. Hist.* **34**:220–238.

Thewissen, J. G. M., and Hussain, S. T. In press. *Attockicetus praecursor*, a new remingtonocetid cetacean from marine Eocene sediments of Pakistan. *Nat. Hist. Mus. Los Angeles Cty. Sci. Ser.*

Thewissen, J. G. M., and McKenna, M. C. 1992. Paleobiogeography of Indo-Pakistan: a response to Briggs, Patterson, and Owen. *Syst. Biol.* **41**:248–251.

Thewissen, J. G. M., Gingerich, P. D., and Russell, D. E. 1987. Artiodactyla and Perissodactyla (Mammalia) from the early-middle Eocene Kuldana Formation of Kohat (Pakistan). *Contrib. Mus. Paleontol. Univ. Michigan* **27**:247–274.

Thewissen, J. G. M., Hussain, S. T., and Arif, M. 1994. Fossil evidence for the origin of aquatic locomotion in archaeocete whales. *Science* **263**:210–212.

Thewissen, J. G. M., Madar, S. I., and Hussain, S. T. 1996. *Ambulocetus natans*, an Eocene cetacean (Mammalia) from Pakistan. *Cour. Forsch.-Inst. Senckenberg* **191**:1–86.

Trivedy, A. N., and Satsangi, P. P. 1984. A new archaeocete (whale) from the Eocene of India. *27th Int. Geol. Congr. Abstr.* **1**:322–323.

Uhen, M. D. 1996. New protocetid archaeocete (Mammalia, Cetacea) from the late middle Eocene Cook Mountain Formation of Louisiana. *J. Vertebr. Paleontol.* **16**(3):70A.

Van Beneden, P. J. 1883. Sur quelques ossements des cétacés fossiles, recuellis dans les couches phosphatées entre l'Elbe et le Weser. *Bull Acad. R. Belg.* Ser. 3 **6**:27–33.

Wells, N. A. 1983. Transient streams in sand-poor redbeds: early-middle Eocene Kuldana Formation of northern Pakistan. *Spec. Publ. Int. Assoc. Sedimentol.* **6**:393–403.

Wells, N. A. 1984. Marine and continental sedimentation in the early Cenozoic Kohat basin and adjacent northwest Pakistan. Ph.D. dissertation, University of Michigan, Ann Arbor.

Wells, N. A., and Gingerich, P. D. 1987. Paleoenvironmental interpretation of Paleogene strata near Kotli, Azad Kashmir, northeastern Pakistan. *Kashmir J. Geol.* **5**:23–41.

West, R. M. 1980. Middle Eocene large mammal assemblage with Tethyan affinities, Ganda Kas region, Pakistan. *J. Paleontol.* **54**:508–533.

West, R. M., and Lukacs, J. R. 1979. Geology and vertebrate-fossil localities, Tertiary continental rocks, Kala Chitta Hills, Attock District, Pakistan. *Milwaukee Public Mus. Contrib. Biol. Geol.* **26**:1–20.

CHAPTER 2

Middle to Late Eocene Basilosaurines and Dorudontines

MARK D. UHEN

1. Introduction

1.1. History of Basilosaurid Discovery

The study of Eocene whales in North American began when Dr. Richard Harlan described some bones from Louisiana near the Ouachita River (Harlan, 1834). Harlan compared the bones, which included a very large and strange vertebra, with other large extinct creatures known at the time and concluded that it was a very large reptile of sorts and named it *Basilosaurus* or "king lizard."

Soon after, Owen (1842a) compared the teeth of *Basilosaurus* with those of modern cetaceans, other mammals, and reptiles and he concluded that *Basilosaurus* was instead a mammal (Owen, 1842a). Owen thought the cheek teeth looked like two simpler teeth yoked together, so he introduced the new name *Zeuglodon* (yoke tooth) for *Basilosaurus*. Unfortunately for Owen, the generic name *Zeuglodon* was a junior synonym for *Basilosaurus*. More discoveries of archaeocete cetaceans continued to be made in southeastern North America. Gibbes (1845) described *Dorudon serratus* from South Carolina. Reichenbach (1847) described *Basilosaurus kochii* [later moved to its own genus *Zygorhiza* by True (1908)] from Alabama.

Schweinfurth (1886) shifted focus from North America to Egypt when he collected some cetacean vertebrae from Fayum, Egypt, in 1879, which were later described by Dames (1883a,b, 1894). These discoveries attracted a number of other German, British, and American explorers to collect fossil whales in Fayum during the first decade of the twentieth century, where they named a number of new genera and species.

In 1936 Remington Kellogg published his monograph, *A Review of the Archaeoceti*, in which he discussed the taxonomic status of all archaeocetes known at that time. Kellogg synonymized some of the previously named species, and moved others from one genus to another. Many of the name changes were necessary as Kellogg pointed out that many of the

MARK D. UHEN • Cranbrook Institute of Science, Bloomfield Hills, Michigan 48303-0801.
The Emergence of Whales, edited by Thewissen. Plenum Press, New York, 1998.

newly described whales had been placed in the genus *Zeuglodon*, which was a junior synonym of *Basilosaurus*.

Recently a number of new archaeocetes have been described from India and Pakistan (e.g., Sahni and Mishra, 1972, 1975; West, 1980; Gingerich *et al.*, 1983). Further discoveries by Gingerich *et al.* (1994) and Thewissen *et al.* (1994) in Pakistan, and Gingerich *et al.* (1990) in Egypt have fueled interest in the study of archaeocetes once more.

1.2. Institutional Abbreviations

The following institutional abbreviations are used throughout this chapter:

ANSP   Academy of Natural Sciences of Philadelphia, Philadelphia
CGM   Cairo Geological Museum, Cairo
GSP-UM   Geological Survey of Pakistan, University of Michigan
MCZ   Museum of Comparative Zoology, Harvard University, Cambridge, Massachusetts
MNB   Museum für Naturkunde der Humboldt-Universität, Berlin
NHML   Natural History Museum, London
SMNS   Staatliches Museum für Naturkunde, Stuttgart
UF   University of Florida, Florida Museum of Natural History, Gainesville
UM   University of Michigan Museum of Paleontology, Ann Arbor, Michigan

1.3. Taxonomy of Basilosauridae

Family Basilosauridae Cope, 1868

Zeuglodontidae Bonaparte, 1849, p. 618.
Hydrarchidae Bonaparte, 1850, p. 1.
Basilosauridae Cope, 1868, p. 144.
Stegorhinidae Brandt, 1873, p. 334.
Prozeuglodontidae Moustafa, 1954, p. 87.
Basilosauridae Barnes and Mitchell, 1978, p. 582.

Type Genus.—Basilosaurus Harlan, 1834.
*Included Subfamilies.—*Basilosaurinae (Cope, 1868), Dorudontinae (Miller, 1923).
*Age and Distribution.—*The earliest known basilosaurids are *Basilosaurus drazindai* and *Basiloterus hussaini*, which are known from the middle to late Bartonian of Pakistan (Gingerich *et al.*, 1997). Vertebrae previously referred to *Eocetus schweinfurthi* (Fraas, 1904) from the Bartonian Giushi Formation of the Mokattam Hills, Egypt (Gingerich, 1992) may also be referable to *Basilosaurus drazindai* (this chapter). One specimen of a basilosaurid archaeocete is known from New Zealand, but its age is poorly constrained to the upper Eocene (Fordyce, 1985). ?*Zygorhiza wanklyni* from "the Barton Clay of the Hampshire coast" may be the earliest known basilosaurid (Seeley, 1876). The Barton Clay is latest middle Eocene and is the type section for the Bartonian. Basilosaurids from Egypt are from the latest Bartonian (late middle Eocene) Gehan-

nam and Birket Qarun Formations and earliest Priabonian (early late Eocene) Qasr el-Sagha Formation (Gingerich, 1992). Basilosaurids from North America have been recovered a number of formations that are Bartonian to Priabonian in age. Randazzo *et al.* (1990) report *Basilosaurus* from the middle Eocene Avon Park Formation of Florida. This specimen (UF 115000) is a vertebra of *Basilosaurus cetoides* that is actually from the Inglis Formation, which is either Lutetian or Bartonian in age (R. Portell, personal communication). If this formation is Lutetian, then the specimen is the earliest reported basilosaurid on record; if Bartonian, it is as old as other records of basilosaurids.

Diagnosis.—Basilosaurids are distinguished from all other archaeocetes by a number of derived features. These include the loss of M^3, the presence of multiple accessory denticles on the cheek teeth, loss of the sacrum, reduction of the hind limb, and rotation of the pelvis.

Discussion.—Moustafa's (1954) use of the name Prozeuglodontidae for the Basilosauridae is both taxonomically invalid and unnecessary as noted by Barnes and Mitchell (1978). It should also be noted that Mitchell (1989) created a new superfamily Basilosauroidea (and parallel superfamilies Protocetoidea and Remingtonocetoidea) that are not used here.

Subfamily Basilosaurinae Cope, 1868

Basilosauridae Cope, 1868, p. 144.
Zeuglodontinae Slijper, 1936, p. 540.
Basilosaurinae Barnes and Mitchell, 1978, p. 590.

Type Genus.—*Basilosaurus* Harlan, 1834.
Included Genera.—*Basilosaurus* Harlan, 1834; *Basiloterus* Gingerich *et al.*, 1997.
Age and Distribution.—The distribution in time (Bartonian to Priabonian) and space of the subfamily Basilosaurinae is the same as that of the family Basilosauridae, except that there are no basilosaurines known from New Zealand.
Diagnosis.—Basilosaurinae have all of the characters that distinguish Basilosauridae from other archaeocetes and in addition have greatly elongated posterior thoracic, lumbar, and anterior caudal vertebrae.
Discussion.—Previously Kellogg (1936) included the genus *Prozeuglodon* in the family Basilosauridae (here equivalent to Basilosaurinae). This genus has been synonymized with the genus *Dorudon*, as discussed below, and it is thus in the subfamily Dorudontinae.

Genus *Basilosaurus* Harlan, 1834

Basilosaurus Harlan, 1834, p. 397.
Zygodon Owen, 1839, p. 35.
Zeuglodon Owen, 1842a, p. 24.
Hydrargos Koch, 1845a, p. 1.
Hydrarchos Koch, 1845b, p. 1.
Zyglodon Hammerschmidt, 1848, p. 323, typographical error (Kellogg, 1936).
Hydrarchus Müller, 1849, p. 3.
Zugodon Scudder, 1882, p. 357, typographical error (Kellogg, 1936).
Alabamornis Abel, 1906, p. 450.

Type Species.—Basilosaurus cetoides Harlan, 1834.

Included Species.—Basilosaurus cetoides Harlan, 1834; *Basilosaurus isis* Andrews, 1904; and *Basilosaurus hussaini* Gingerich *et al.,* 1997.

Age and Distribution.—The range and distribution for *Basilosaurus* are the same as for the subfamily Basilosaurinae.

Diagnosis.—Same as for the subfamily.

Discussion.—Despite its inappropriateness, the generic name *Basilosaurus* has priority over *Zeuglodon.* This has been recognized for some time (Kellogg, 1936) and was recently reiterated by Gingerich *et al.* (1990).

Species *Basilosaurus cetoides* Harlan, 1834

Zeuglodon cetoides Owen, 1842b, p. 69.
Zeuglodon harlani DeKay, 1842, p. 123.
Hydrargos sillimanii Koch, 1845a, p. 1.
Hydrarchos sillimanii Koch, 1845b, p. 1.
Zeuglodon ceti Wyman, 1845, p. 65.
Hydrarchos harlani Koch, 1845a, p. 1.
Basilosaurus cetoides Geinitz and Reichenbach in Carus, 1847, p. 1.
Basilosaurus cetoides Gibbes, 1847, p. 1.
Basilosaurus harlani Hammerschmidt, 1848, p. 121.
Zeuglodon macrospondylus Müller, 1849, p. 3.
Alabamornis gigantea Abel, 1906, p. 450.
Basilosaurus cetoides Kellogg, 1936, p. 15.

Type Specimen—Vertebral centrum of a lumbar vertebra, ANSP 12944A. The specimen is figured in Harlan (1834), plate XX, Figs. 1 and 2.

Type Locality.—Bry in Harlan (1834) states that the type specimen of *Basilosaurus cetoides* was found on a hill approximately 200 yards from the Ouachita River. He also states that the hill is located about 50 miles south of Monroe by land and 110 miles by the river, in the parish of Ouachita. Kellogg (1936) interprets the locality as being in the southeastern part of Caldwell Parish, about 50 miles south of Monroe. The actual position of this locality is discussed at length by Huner (1939). He notes that Caldwell Parish was formed from part of Catahoula and Ouachita Parishes, and that the position of Bry's locality is in doubt, as there are no points 50 miles south or southeast of Monroe that lie within Caldwell Parish.

Age and Distribution.—Bartonian to early Priabonian.

Diagnosis.—Basilosaurus cetoides has the characters that distinguish *Basilosaurus* from other genera of basilosaurids, and it is slightly larger than Egyptian *Basilosaurus isis* (Gingerich *et al.,* 1990).

Discussion.—Owen (1842b) proposed the specific epithet *cetoides* to suggest that Harlan's *Basilosaurus* was a cetacean. *Basilosaurus cetoides* has not been studied in detail since Kellogg (1936). Many more specimens of *B. cetoides* have been collected since that time and need to be compared with the large number of *B. isis* specimens that have also been collected since then as well to fully characterize the differences between these two species. The numerous specimens of *B. cetoides* cover virtually the entire skeleton. Only some of the distal hind limb elements have yet to be discovered. No subadult individuals of *B. cetoides* have been described that include deciduous teeth.

Species *Basilosaurus isis* Andrews, 1906

Zeuglodon isis Andrews, 1904, p. 214.
Prozeuglodon isis Kellogg, 1936, p. 75.
Basilosaurus isis, Gingerich, Smith, and Simons, 1990, p. 154.

Type Specimen.—CGM 10208. Right dentary missing the articular condyle and angle. It includes alveoli for I_1–C_1 and broken crowns of P_1–M_3.

Type Locality.—Birket el-Qarun beds west of Birket el-Qarun (Andrews, 1904).

Age and Distribution.—Bartonian to early Priabonian. *Basilosaurus isis* is found in the Gehannam, Birket Qarun, and Qasr el-Sagha Formations, Fayum, Egypt.

Diagnosis.—*Basilosaurus isis* has the characters that distinguish *Basilosaurus* from other genera of basilosaurids, and it is slightly smaller than North American *Basilosaurus cetoides* (Gingerich *et al.*, 1990).

Discussion.—Andrews (1904) briefly described a new species of large archaeocete from the "Qasr-es-Sagha series" that he called *Zeuglodon isis* and attributed to H. J. L. Beadnell. Despite this, the species should be attributed to Andrews (1904) rather than Beadnell. In 1906, Andrews gave a more complete description of *Zeuglodon isis*. Virtually the entire skeleton of *Basilosaurus isis* is known, but many elements have yet to be described in detail. No subadult individuals of *B. isis* have been described that include deciduous teeth.

Species *Basilosaurus drazindai* Gingerich *et al.*, 1997

Basilosaurus drazindai Gingerich *et al.*, 1997, p. 57.

Type Specimen.—GSP-UM 3193, lumbar vertebra.

Type Locality.—Bari Nadi west of Satta Post of the Border Military Police, Punjab, Pakistan (Gingerich *et al.*, 1997).

Age and Distribution.—Middle to late Bartonian of Pakistan and Egypt.

Diagnosis.—Lumbar vertebrae are similar in size and centrum proportions to other members of the genus *Basilosaurus*, but the neural arch, neural spine, and transverse processes are more elongated anteroposteriorly than those of *Basilosaurus cetoides* or *Basilosaurus isis*.

Discussion.—Gingerich *et al.* (1997) described the single lumbar vertebra of the type specimen of *Basilosaurus drazindai* but did not refer any other specimens to the new species. Examination of specimens from the Mokattam Hills of Egypt collected around the turn of the century and previously referred to *Eocetus schweinfurthi* indicates that they too have the diagnostic features listed for *B. drazindai*. The specimen (SMNS 10934) consists of two lumbar vertebrae, one in poor condition and the other in good condition. The vertebra in good condition was figured by Stromer (1903, Fig. 1, p. 84).

Genus *Basiloterus* Gingerich *et al.*, 1997

Basiloterus Gingerich *et al.*, 1997, p. 62.

Type Species.—*Basiloterus hussaini*, Gingerich *et al.*, 1997.
Included Species.—*Basiloterus hussaini*, Gingerich *et al.*, 1997.
Age and Distribution.—Middle to late Bartonian of Pakistan (Gingerich *et al.*, 1997).

Diagnosis.—Gingerich *et al.* (1997) diagnose *Basiloterus* as differing from *Basilosaurus* in being smaller, having lumbar vertebrae slightly less elongate, and having more vertically oriented metapophyses.

Discussion.—*Basiloterus* (along with *Basilosaurus drazindai*) may be the earliest occurring basilosaurid archaeocetes.

Species *Basiloterus hussaini* Gingerich *et al.*, 1997

Basiloterus hussaini Gingerich *et al.*, 1997, p. 62.

Type Specimen.—GSP-UM 3190, two lumbar vertebrae.
Type Locality.—Bari Nadi west of Satta Post of the Border Military Police, Punjab, Pakistan (Gingerich *et al.*, 1997).
Age and Distribution.—Same as for the genus.
Diagnosis.—Same as for the genus.
Discussion.—Gingerich *et al.* (1997) refer a caudal centrum (NHML-M 26553) to *Basiloterus hussaini* based on its age and size. There are no processes present on the centrum that could help determine where in the caudal series the vertebra came from, and because the caudal vertebrae decrease in size from anterior to posterior, it is difficult to know whether its size is indicative of placement in *B. hussaini* or position in the vertebral column of another species of basilosaurine.

Subfamily Dorudotinae Miller, 1923

Dorudontidae Miller, 1923, p. 13.
Dorudontinae Slijper, 1936, p. 540.
Dorudontidae Kellogg, 1936, p. 100.
Dorudontinae Barnes and Mitchell, 1978, p. 588.

Type Genus.—*Dorudon* Gibbes, 1845.
Included Genera.—*Dorudon* Gibbes, 1845; *Pontogeneus* Leidy, 1852; *Zygorhiza* True, 1908; *Saghacetus* Gingerich, 1992; *Ancalecetus* Gingerich and Uhen, 1996.
Age and Distribution.—The age and distribution for the subfamily Dorudontinae are the same as for the family Basilosauridae.
Diagnosis.—Dorudontines have all of the derived characters of the Basilosauridae, but lack the elongate posterior thoracic, lumbar, and anterior caudal vertebrae of the Basilosaurinae. Dorudontine genera are very similar to one another, but each has a few autapomorphies that differentiate them.
Discussion.—Miller (1923) originally proposed a division of the Archaeoceti into three groups of equal (familial) rank based on his own work and the usage of these terms by previous authors. Slijper (1936) and Barnes and Mitchell (1978) placed Miller's Basilosauridae and Dorudontidae at the subfamilial level to emphasize the phylogenetic relationships among the subgroups of archaeocetes. This arrangement is followed here because it emphasizes the evolutionary history of these groups. If the Basilosaurinae and Dorudontinae were elevated to the familial level, there would be no taxonomic group to encompass all of the taxa that share the large number of synapomorphies currently represented by the Basilosauridae.

Genus *Dorudon* Gibbes, 1845

Dorudon Gibbs, 1845, p. 255.
Basilosaurus Gibbes, 1847, p. 5 (in part).
Doryodon Cope, 1868, p. 156.
Durodon Gill, 1872, p. 93, typographical error (Kellogg, 1936).
Prozeuglodon Andrews, 1906, p. 243.
Zeuglodon Abel, 1914, p. 204.
Prozeuglodon Kellogg, 1928, p. 40.

Type Species.—*Dorudon serratus* Gibbes, 1845.

Included Species.—*Dorudon serratus* Gibbes, 1845; *Dorudon atrox* Andrews, 1906.

Age and Distribution.—Same as for the subfamily Dorudontinae.

Diagnosis.—Members of the genus *Dorudon* can be distinguished from other archaeocetes
based on overall size and other morphological characteristics. Members of the genus
Dorudon are half again as large as members of *Saghacetus*, and about three-fourths
the size of members of the genus *Pontogeneus*. *Saghacetus* lacks the long narial
process of the frontals that is characteristic of the genus *Dorudon*. Lingual cingula on
P^{2-3} of members of the genus *Dorudon* lack the well-developed cuspules found on the
cingula of members of the genus *Zygorhiza*. Forelimbs of members of the genus
Dorudon lack the highly derived characters of *Ancalecetus*.

Discussion.—Much of Kellogg's (1936) discussion of the genus *Dorudon* is no longer accurate because much of his description was based on specimens of species now sub-
sumed within *Saghacetus osiris*. Most of what is now known of the genus *Dorudon* is
based on *Dorudon atrox*, which he had included in the genus *Prozeuglodon*, in the family Basilosauridae.

Species *Dorudon serratus* Gibbes, 1845

Dorudon serratus Gibbes, 1845, p. 255.
Basilosaurus serratus Gibbes, 1847, p. 5.
Doryodon serratus Cope, 1868, p. 144.
Zeuglodon serratum Abel, 1914, p. 204.
Dorudon serratus, Kellogg, 1936, p. 178.

Type Specimen.—MCZ 8763. Right maxilla with dP^2 to dP^4, left maxilla with dP^2?, dI^2?,
as originally described by Gibbes (1845). Left and right premaxillae, cranial fragments, and 12 caudal vertebrae collected by Gibbes at the same site and believed to be
from the same individual were added to the type specimen (Gibbes, 1847).

Type Locality.—The type locality of *Dorudon serratus* is given by Gibbes as "in a bed of
Green sand near Santee Canal, in South Carolina . . . on the plantation of R. W. Mazyck,
Esq., about three miles from the entrance of the canal from the head waters of Cooper
river" (1845). The sediments from which the type specimen was recovered are very
similar to those of the Harleyville Formation, which outcrops in this area.

Age and Distribution.—Hazel *et al.* (1977) and Zullo Harris (1987) list the age of the
Harleyville Formation as Priabonian (nannoplankton zone 19/20). A single tooth from
the Castle Hayne Formation of North Carolina has been referred to *Dorudon serratus*
by Kellogg (1936). This locality (Kellum, 1926) is Bartonian or Priabonian in age.

Diagnosis.—It is not clear what, if any, characters distinguish *Dorudon serratus* from *Zygorhiza kochii*, as most specimens of *Z. kochii* are adults and the type of *D. serratus* is a juvenile. Additional study of midsized North American archaeocetes is needed to determine if these named taxa represent one or two species.

Discussion.—Kellogg (1936) failed to recognize the type of *Dorudon serratus* as a juvenile individual. Only fragmentary cranial bones, teeth, and the caudal vertebrae of *D. serratus* are known. Some of the cranial fragments and the caudal vertebrae added to the type individual (Gibbes, 1847) have been lost. The type locality is now under Lake Moultrie (Sanders, 1974; Domning *et al.*, 1982).

Species *Dorudon atrox* Andrews, 1906

Prozeuglodon atrox Andrews, 1906, p. 243.
Dorudon intermedius Dart, 1923, p. 629.
Prozeuglodon stromeri Kellogg, 1928, p. 40.
Prozeuglodon isis Kellogg, 1936, p. 75 (in part).
Dorudon stromeri Kellogg, 1936, p. 203.
?*Protocetus isis*, Trofimov and Gromova, 1968, p. 225.
Dorudon osiris Pilleri, 1985, p. 35.
Dorudon atrox, Uhen, 1996a, p. 51.

Type Specimen.—CGM 9319. Skull with right ramus of the lower jaw.

Type Locality.—Andrews (1906) stated that the type locality is in the "Birket-el-Qurun beds (Middle Eocene): a valley about 12 kilometres W.S.W. of the hill called Gar-el-Gehannem." Gingerich (1992) noted that this places the locality in Zeuglodon Valley (Wadi Hitan).

Age and Distribution.—Bartonian to early Priabonian. Specimens of *Dorudon atrox* have been recovered from the Gehannam and Birket Qarun Formations of Fayum. The type specimen of *Dorudon stromeri*, now placed in *Dorudon atrox*, is probably from the Qasr el-Sagha Formation. The Birket Qarun beds mark the lowstand between the Tejas 4 and Tejas 5 sequence tracts (Gingerich, 1992), at the Bartonian–Priabonian boundary (Haq *et al.*, 1987).

Diagnosis.—*Dorudon atrox* can be distinguished from most other dorudontines based on its size. *D. atrox* is larger than *Saghacetus*, and smaller than *Pontogeneus*. *D. atrox* can be distinguished from *Zygorhiza* by the lack of well-developed cuspules on the lingual cingula of P$^{2-3}$. It is not clear what characters distinguish *D. atrox* from *D. serratus*, as adult specimens of *D. serratus* are poorly known.

Discussion.—Andrews (1906) differentiated a new genus and species that he called *Prozeuglodon atrox* from *Zeuglodon isis* by the presence of a "postero-internal buttress" on the two posterior premolars (his P^3 and P^4). He failed to recognize that the type specimen of *P. atrox* is a juvenile individual, and the teeth in question are actually deciduous premolars (dP^3 and dP^4). Kellogg (1928) described *Prozeuglodon stromeri*, which he later moved to the genus *Dorudon* (1936). Kellogg also (1936) reasoned that the type specimen of *Prozeuglodon atrox* was a juvenile individual, of Andrews's *Zeuglodon isis*. Kellogg combined the two into a single species that he called *Prozeuglodon isis*.

Gingerich *et al.* (1990) referred the type specimen and other smaller archaeocete specimens like the type back to *Prozeuglodon atrox*, and placed large specimens with elongated posterior thoracic and lumbar vertebrae in the genus *Basilosaurus*, as the species *Basilosaurus isis*. Gingerich (1992) also synonymized *Dorudon intermedius* with *Prozeuglodon atrox*, noting the similarities in brain size, skull length, and stratigraphic position of the type specimen of *D. intermedius* and *P. atrox*. These and other specimens have been compared with other dorudontine species and based on a comparison with *Dorudon serratus*, they are now placed in the genus *Dorudon*, making the species *Dorudon atrox*. Uhen (1996a,b) also synonymized *Dorudon stromeri* with *Dorudon atrox*. The entire dentition (both deciduous and permanent) and almost the entire skeleton of *Dorudon atrox* are known. Only the innominata and some other hind limb elements have yet to be discovered.

Genus *Pontogeneus* Leidy, 1852

Pontogeneus Leidy, 1852, p. 52.

Type Species.—Pontogeneus brachyspondylus Müller, 1849.
Included Species.—Pontogeneus brachyspondylus Müller, 1849.
Age and Distribution.—Bartonian and Priabonian deposits of southeastern United States and Fayum, Egypt.
Diagnosis.—Skull, cervical vertebrae, and anterior thoracic vertebrae are very similar in size and morphology to those of *Basilosaurus*. This large size distinguishes *Pontogeneus* from other dorudontines. The posterior thoracic vertebrae, lumbar vertebrae, and anterior caudal vertebrae are not elongated, but have proportions similar to other dorudontines, distinguishing *Pontogeneus* from *Basilosaurus*.
Discussion.—Some of the vertebrae of Koch's chimera *Hydrarchos harlani* were later identified as the vertebrae of a large archaeocete that lacked elongated vertebral centra and given the name *Zeuglodon brachyspondylus* (Müller, 1849). Kellogg (1936) stated that these vertebrae probably came from Alabama, from "the vicinity of Washington Old Court House."

Leidy (1852) named *Pontogeneus priscus* in a brief note based on a centrum from a single cervical vertebra, later designated as ANSP 13668. The only locality information given by Leidy is Ouachita (Washita) River, Louisiana. Kellogg (1936) combined Müller's *Zeuglodon brachyspondylus* and Leidy's *Pontogeneus priscus* and used the generic name *Pontogeneus* for the new combination.

Species *Pontogeneus brachyspondylus* Müller, 1849

Hydrarchos harlani Koch, 1846, p. 1 (in part).
Zeuglodon brachyspondylus Müller, 1849, p. 26.
Pontogeneus priscus Leidy, 1852, p. 52.
Zeuglodon brachyspondylum Abel, 1914, p. 203.
Pontogeneus brachyspondylus Kellogg, 1936, p. 248.

Type Specimen.—No type specimen has been designated. See discussion below.
Type Locality.—No type locality has been identified. See discussion below.
Age and Distribution.—Same as for the genus.
Diagnosis.—Same as for the genus.
Discussion.—Müller (1849) stated that many of the vertebrae in Koch's *Hydrarchos har-*

lani belong in the species *Zeuglodon brachyspondylus*, but he failed to designate any one set of vertebrae as the type specimen for the species. In 1851 Müller separated out some of these vertebrae as belonging to *Zeuglodon brachyspondylus minor* (= *Zygorhiza kochii*). Kellogg (1936) indicated that the vertebrae Müller left in *Zeuglodon brachyspondylus* belong to his new combination, *Pontogeneus brachyspondylus*, but he too did not identify a type specimen. The cervical vertebra that is the type specimen of *Pontogeneus priscus* (ANSP 13668) could be designated as the neotype of *Pontogeneus brachyspondylus*.

In addition to the specimens from the southeastern United States listed by Kellogg (1936), some specimens that can be assigned to *Pontogeneus brachyspondylus* have been collected in Egypt. A discontinuous series of cervical, thoracic, and lumbar vertebrae (SMNS 11414) and skull fragments (SMNS 11413) from the same individual are probably *P. brachyspondylus*. In addition, vertebrae collected from Fayum by the University of Michigan Museum of Paleontology (UM 97535 and UM 101222) are also probably *P. brachyspondylus*.

Genus *Zygorhiza* True, 1908

Zygorhiza True, 1908, p. 78.

Type Species.—Zygorhiza kochii Reichenbach in Carus, 1847.
Included Species.—Zygorhiza kochii Reichenbach in Carus, 1847.
Age and Distribution.—Bartonian and Priabonian of North America.
Diagnosis.—Kellogg (1936) indicated that *Zygorhiza* has strong, crenulated cingula on the anterior and posterior edges of the lingual sides of upper P^{2-4}. *Dorudon, Saghacetus*, and *Ancalecetus* lack this feature. *Zygorhiza* is smaller than *Pontogeneus*, and larger than *Saghacetus*.
Discussion.—All of the carbonate units in Alabama above the Gosport Sand, including the Moodys Branch Formation, are considered Priabonian in age, (Zullo and Harris, 1987), so the type specimen is probably Priabonian. Both *Zeuglodon wanklyni* (Seeley, 1876) and *Balaenoptera juddi* (Seeley, 1881) were transferred to the genus *Zygorhiza* by Kellogg (1936). These two species are based on vertebral material that can only confidently be assigned to Dorudontinae *incertae sedis* (Uhen, 1996a,b).

Species *Zygorhiza kochii* Reichenbach, 1847

Basilosaurus kochii Reichenbach in Carus, 1847, p. 13.
Zeuglodon hydrarchus Carus, 1849, p. 385.
Zeuglodon brachyspondylus Müller, 1849, p. 20.
Zeuglodon brachyspondylus minor Müller, 1851, p. 240.
Zygorhiza brachyspondylus minor True, 1908, p. 78.
Zeuglodon brachyspondylum Abel, 1914, p. 203.
Zygorhiza minor Kellogg, 1928, p. 40.
Zygorhiza kochii Kellogg, 1936, p. 101.

Type Specimen.—A damaged braincase, MNB 15324a-b (m. 44), figured in Müller (1849, Plates 3–5), as designated by Kellogg (1936).
Type Locality.—Ocala limestone, near Clarksville, Clark County, Alabama (from Kellogg,

1936). The carbonate unit(s) in Alabama that were formerly called the Ocala limestone [above the Gosport Sand (Frazier and Schwimmer, 1987)] are now divided in a number of ways in different localities by different authors (Zullo and Harris, 1987). The type specimen of *Zygorhiza kochii* is probably from what most authors would call the Moodys Branch Formation.

Age and Distribution.—Same as for the genus.

Diagnosis.—Same as for the genus.

Discussion.–Reichenbach (1847) gave the name *Basilosaurus kochii* to the species represented by an occipital region of a skull that was misidentified as a "Gaumenstück" (palatal fragment). Müller (1849) further prepared the specimen, recognized the true anatomical nature of the fragment, and placed it in his species *Zeuglodon brachyspondylus*, along with a number of vertebrae that were a mixture of what was later identified by Müller (1851) as *Zeuglodon brachyspondylus* (*Pontogeneus brachyspondylus*) and *Zeuglodon brachyspondylus minor* (*Zygorhiza kochii*). True (1908) named a new genus, *Zygorhiza*, for Müller's (1851) *Zeuglodon brachyspondylus minor*. Kellogg (1928) listed the specific epithet of the species of *Zygorhiza* as *minor*. Kellogg (1936) subsequently identified the type specimen of the species, as the type of Reichenbach's *Basilosaurus kochii*, and he gave the species the specific epithet *kochii*, making the species name *Zygorhiza kochii*. Most of the skeleton of *Z. kochii* has been described in detail, as well as the deciduous and permanent dentition (Kellogg, 1936). No hind limb material has been described from *Z. kochii*.

Genus *Saghacetus* Gingerich, 1992

Type Species.—*Saghacetus osiris* Dames, 1894.

Included Species.—*Saghacetus osiris* Dames, 1894.

Age and Distribution.—Priabonian of Zeuglodon Valley, Fayum, Egypt.

Diagnosis.—*Saghacetus* is about two-thirds the size of *Dorudon atrox*, which makes it much smaller than any other dorudontine. *Saghacetus* consistently shows a pinching of the nuchal crest, just dorsal to the occipital condyles. In addition, Gingerich (1992) states that it has normally proportioned thoracic vertebrae, and slightly elongated posterior lumbar and anterior caudal vertebrae.

Discussion.—*Saghacetus* is known only from the Qasr el-Sagha Formation of Zeuglodon Valley, Fayum, Egypt. This formation is considered to be entirely Priabonian in age, and has been interpreted as a lagoonal environment. No similar lagoonal deposits are known from the Bartonian of Egypt, and *Saghacetus* is not known from the Priabonian deposits of North America. This may suggest that *Saghacetus* was restricted in distribution to lagoonal environments.

Species *Saghacetus osiris* Dames, 1894

Zeuglodon osiris Dames, 1894, p. 204.
Zeuglodon zitteli Stromer, 1903, p. 82.
Zeuglodon sensitivus Dart, 1923, p. 618.
Zeuglodon elliotsmithii Dart, 1923, p. 625.
Dorudon osiris Kellogg, 1936, p. 184.
Dorudon zitelli Kellogg, 1936, p. 212.

Dorudon elliotsmithii Kellogg, 1936, p. 220.
Dorudon sensitivus Kellogg, 1936, p. 221.
Saghacetus osiris Gingerich, 1992, p. 73.

Type Specimen.—MNB 28388, portions of left and right premaxillaries, left dentary with
    I_2 and P_2 to M_3 in place, and alveoli for I_1, I_3, C_1, and P_1.
Type Locality.—♀-Berge 121/2 km im Westen vom alten Tempelbau, AAAb (Schweinfurth,
    1886): upper Qasr el-Sagha Formation, Temple Member (Gingerich, 1992).
Age and Distribution.—Same as for the genus.
Diagnosis.—Same as for the genus.
Discussion.—*Saghacetus osiris* is rather well-known skeletally. Around a half-dozen rela-
    tively complete skeletons and additional skulls are present in the collections of the Uni-
    versity of Michigan Museum of Paleontology and the Staatliches Museum für
    Naturkunde, Stuttgart. These specimens are in need of further preparation and careful
    reexamination in light of Gingerich's (1992) recent synonymy of *Dorudon zitteli,
    Dorudon sensitivus*, and *Dorudon elliotsmithii* with *Saghacetus osiris*. No hind limb
    material or deciduous teeth have been described from *S. osiris*.

Genus *Ancalecetus* Gingerich and Uhen, 1996

Type Species.—*Ancalecetus simonsi*, Gingerich and Uhen, 1996.
Included Species.—*Ancalecetus simonsi*, Gingerich and Uhen, 1996.
Age and Distribution.—Bartonian or Priabonian of Egypt. *Ancalecetus* is known only from
    the type specimen. The type specimen was recovered from the Birket Qarun Forma-
    tion, which marks the lowstand at the Bartonian/Priabonian boundary (Gingerich,
    1992).
Diagnosis.—*Ancalecetus* can be distinguished from all other dorudontines by its highly de-
    rived forelimb. The scapula is very narrow with a posteriorly projecting acromion. In
    addition, the elbow joint and intercarpal joints are fused in *Ancalecetus*, whereas they
    are free in other dorudontines.
Discussion.—See below.

Species *Ancalecetus simonsi* Gingerich and Uhen, 1996

Type Specimen.—CGM 42290.
Type Locality.—University of Michigan locality ZV-081, Zeuglodon Valley, Fayum, Egypt.
Age and Distribution.—Same as for the genus.
Diagnosis.—Same as for the genus.
Discussion.—*Ancalecetus simonsi* is known from a single specimen that includes a basi-
    cranium; both dentaries; some cervical, thoracic, lumbar, and caudal vertebrae; and
    both forelimbs. No hind limb material is known from *A. simonsi*. All skeletal elements
    except the forelimbs are very similar to those of contemporaneous *Dorudon atrox*. The
    vertebrae of *A. simonsi* are a bit smaller than the corresponding vertebrae of *D. atrox*.
    Radiographs of the forelimbs of *A. simonsi* show no evidence of injury that could ac-
    count for the differences compared with other dorudontine forelimbs (Gingerich and
    Uhen, 1996).

2. Phylogenetic Relationships of Basilosaurids

2.1. Introduction

The phylogenetic relationships among basilosaurids and the relationships of basilosaurids to earlier and later cetaceans have not been explored in detail until recently (Uhen, 1996b). Late nineteenth- and early twentieth-century authors suggested a number of possible interrelationships among basilosaurid archaeocetes, but these hypotheses are difficult to evaluate today as many of the taxa they discussed have been synonymized with other taxa and new species have been described since then. The phylogeny of Barnes and Mitchell (1978) suffers from the same problem. Barnes *et al.* (1985) showed a phylogeny of Cetacea at the family level, with a dashed line leading toward modern cetaceans that curves from Protocetidae (= Pakicetidae + Ambulocetidae + Protocetidae *sensu* Thewissen *et al.*, (1996), hinting that modern cetaceans were derived from Protocetidae, yet they stated in the text that the modern suborders are likely to be derived from the Dorudontinae (p. 19). Lastly, Hulbert (1994) performed a cladistic analysis that supported the monophyly of Basilosauridae (not considering Mysticeti + Odontoceti). Unfortunately, this work has only been published as an abstract, so a critical evaluation of the characters used is not possible.

Because all previous phylogenetic analyses of archaeocetes have some shortcoming, a new phylogenetic analysis of archaeocetes was performed. A phylogenetic analysis method that uses both morphologic and stratigraphic data called stratocladistics was developed by D. C. Fisher (1992). Stratocladistics uses morphologic data in the same way that cladistic analysis does, but it allows the relative stratigraphic occurrence of fossil taxa to be included as data in the analysis. Stratocladistic analysis evaluates alternative hypotheses using both morphologic and stratigraphic data. This method treats ad hoc hypotheses that explain away missing fossils in the same way as ad hoc hypotheses of homoplasy. Hypotheses of relationships that are minimal in overall appeals to ad hoc hypotheses are preferred by stratocladistics. Stratocladistics also allows some taxa to be evaluated as ancestors (Fisher, 1992).

The results of a stratocladistic analysis are shown in Fig. 1. The tree shows that Basilosauridae is the sister group to the Protocetidae (*sensu* Thewissen *et al.*, 1996). The conventional cladistic analysis and stratocladistic analysis indicate that Basilosauridae is derived from the paraphyletic Protocetidae and the subfamily Basilosaurinae is derived from within the Dorudontinae, as is the Odontoceti + Mysticeti. Table I lists characters that differentiate the taxonomic groups of cetaceans discussed here.

2.2. Relationships of Basilosauridae to Earlier Cetaceans

2.2.1. Mesonychid Origins

Cetacea + Mesonychia form the clade Cete (McKenna, 1975). Mesonychid mesonychians are shown here as the sister taxon to Cetacea, but it is not clear which group of mesonychians is most closely related to cetaceans; mesonychids, hapalodectids, or an-

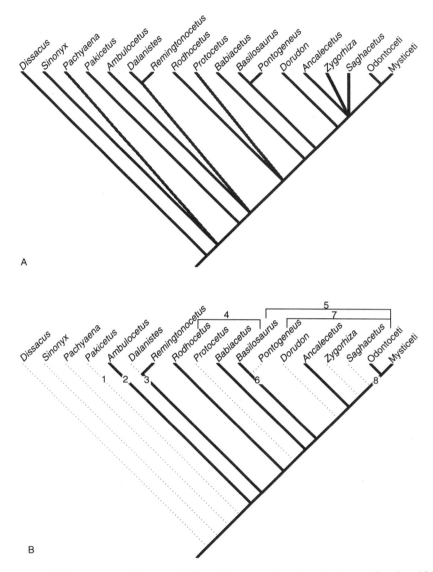

FIGURE 1. (A) Strict consensus cladogram of 18 equally most parsimonious cladograms (length = 134 steps) resulting from a branch-and-bound search in PAUP from the data matrix in the Appendix. The stratigraphic character was not included in the morphologic cladistic analysis. Higher taxa are identified below. Lack of complete resolution is present in the position of *Sinonyx* and *Pachyaena* relative to Cetacea, the position of *Ambulocetus* relative to the Remingtonocetidae, the positions of *Rodhocetus* and *Protocetus* within Protocetidae, and among dorudontines. (B) Phylogenetic tree of Basilosauridae resulting from the addition of a stratigraphic character in a stratocladistic analysis. Taxa shown in dotted lines can be parsimoniously reconstructed as ancestors at the nodes where their branches join the tree. Taxa shown in thick lines cannot be made ancestors without increasing overall tree length. Groups to note are indicated with Arabic numerals. 1: Pakicetidae, represented by *Pakicetus*. *Pakicetus* is shown as the ancestor to the rest of Cetacea, making Pakicetidae paraphyletic. 2: Ambulocetidae, represented by *Ambulocetus*. *Ambulocetus* is shown as the sister taxon to the rest of Cetacea. 3: Remingtonocetidae. Remingtonocetids share a number of highly derived features of the skull and hind limb. 4: Protocetidae. Protocetids are a paraphyletic assemblage of early cetaceans that share some advanced features with Basilosauridae. 5:

drewsarchids. A previous study (Zhou *et al.*, 1995) indicated that mesonychids were more closely related to cetaceans than hapalodectids or andrewsarchids. Another study (Thewissen, 1994) showed that *Andrewsarchus* + Mesonychidae was the sister taxon to Cetacea.

2.2.2. Early Cetaceans

Early cetaceans (Pakicetidae, Ambulocetidae, and Remingtonocetidae) all retain a number of primitive characters that they share with mesonychians. These include: large hind limbs, sacra with multiple fused vertebrae, and poorly developed aquatic auditory systems. Pakicetids are the most primitive cetaceans, with remingtonocetids and ambulocetids both being derived in different ways. Remingtonocetids have very long skulls and fused mandibular symphyses (Kumar and Sahni, 1986; Gingerich *et al.*, 1995). Ambulocetids have dorsally positioned orbits and very large hind limbs (Thewissen *et al.*, 1996).

2.2.3. Protocetidae

Thewissen *et al.* (1996) indicate that Ambulocetidae gives rise to the Protocetidae, with the position of Remingtonocetidae somewhat uncertain. This view is consistent with the phylogenetic tree shown here as long as both Remingtonocetidae and Protocetidae are derived from the common ancestor with Ambulocetidae. *Ambulocetus* cannot be supported as the direct ancestor of later cetaceans using the data in this analysis. The arrangement of these families shown here is consistent with the morphologic data, and only weakly preferred by the stratigraphic data. Because many of these taxa are poorly known, more morphologic data could easily affect this portion of the tree. Protocetidae arises from this group of early cetaceans and shares some characters with basilosaurids. These include: reduced sacra, smaller hind limbs, and more posteriorly placed narial openings. Protocetidae is a paraphyletic group that gives rise to Basilosauridae + Odontoceti + Mysticeti. It is not clear which protocetid is most closely related to basilosaurids, but likely candidates include *Babiacetus indicus*, *Eocetus schweinfurthi*, and *Georgiacetus vogtlensis* (Hulbert, 1994; Hulbert *et al.*, 1998). *Georgiacetus* has small accessory cuspules on the upper cheek teeth that are thought to be homologous with the accessory denticles of basilosaurids (Hulbert, 1996).

2.3. Relationships within Basilosauridae and Later Cetaceans

2.3.1. Relationships within Basilosauridae

Basilosaurids share a number of characters that differentiate them from earlier cetaceans, but they do not differ much from one another. The subfamily Basilosaurinae is

---

Basilosauridae. Basilosaurids share a number of derived features, many of which indicate that they were fully aquatic. 6: Basilosaurinae, represented by *Basilosaurus*. Basilosaurines have elongate posterior thoracic, lumbar, and anterior caudal vertebrae. 7: Dorudontinae. Dorudontines are a paraphyletic assemblage that give rise to Odontoceti + Mysticeti and Basilosaurinae. 8: Odontoceti + Mysticeti. Hypothetical ancestral mysticetes and odontocetes were used to show the derivation of Odontoceti + Mysticeti from Archaeoceti. Early Odontoceti + Mysticeti share a number of derived features with basilosaurids and lack the autapomorphic features of basilosaurines.

characterized by its large size and greatly elongated posterior thoracic, lumbar, and anterior caudal vertebrae. *Basilosaurus cetoides* and *Basilosaurus isis* are very similar, differing only slightly in size (Gingerich *et al.*, 1990). The other genera, which are all in the Dorudontinae, are very similar to one another, but each has one or two autapomorphies. *Zygorhiza* has cuspules on its upper premolar cingula, *Pontogeneus* is large, *Saghacetus* is small, *Ancalecetus* has a highly modified forelimb, and *Dorudon* has a particular conformation of the cranial vertex.

2.3.2. Dorudontinae with Odontoceti + Mysticeti

All dorudontines share a number of characters with Odontoceti + Mysticeti, including loss of the sacrum, rotation of the pelvis (discussed below), and reduction of the pelvic girdle and hind limb. Other character states including increase in size of the infraspinous fossa of the scaula, radius and ulna articulate with the humerus in a common trochlea, and lack of saddle-shaped articular surfaces on the carpals may also delimit Basilosauridae + Odontoceti + Mysticeti, but forelimbs are unknown for any protocetids, so the character state distributions are unknown. Almost any dorudontine could be the sister taxon to the Odontoceti + Mysticeti as no one species shares an overwhelming number of characters with Odontoceti + Mysticeti.

3. Characters Suggesting a Fully Aquatic Life of Basilosaurids

3.1. Forelimb Characters

The forelimbs of basilosaurid archaeocetes are intermediate in form between the forelimbs of mesonychids and those of modern cetaceans. Mesonychids have typically terrestrial forelimbs with narrow scapulae; gracile humeri, radii, and ulnae; mobile elbow and wrist joints; and a weight-bearing, pentadactyl manus (O'Leary and Rose, 1995). The forelimb of *Ambulocetus* is the only one known from a nonbasilosaurid archaeocete (Thewissen *et al.*, 1996). It has some of the primitive features found in mesonychids, and some of the advanced features found in basilosaurids. The shape of the scapular blade of *Ambulocetus* is unknown, as is the humerus. The radius and ulna are more robust than those of mesonychids and are generally cylindrical to ovate in cross section. The conformation of the proximal articular surfaces of the radius and ulna are such that there could be no motion of the radius relative to the ulna, and thus no pronation and supination possible at the elbow joint (Thewissen *et al.*, 1996). The carpals have large, curved articular surfaces that would allow a significant amount of flexibility in the wrist. In addition, the magnum and trapezoid are separate bones in *Ambulocetus*, as in mesonychians.

All basilosaurids have similar forelimbs, with the exception of *Ancalecetus*, which has a secondarily derived and reduced forelimb (Gingerich and Uhen, 1996). The scapula of basilosaurids is very broad and fan-shaped, like those of modern cetaceans and unlike those of mesonychids (Fig. 2). The scapula has a long acromion process which is oriented anteriorly, as is the coracoid process. This feature has been cited as a shared derived feature of cetaceans in general (Barnes, 1984), but the orientation of the acromion and coracoid processes is not known from any earlier cetaceans.

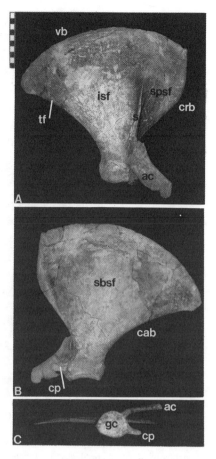

FIGURE 2. Right scapula of *Dorudon atrox* (UM 101222) in (A) lateral, (B) medial, and (C) glenoid-on (distal) view. Note the broad fan-shaped scapular blade. The lateral side is divided into the supraspinous and infraspinous portions by the nearly vertical spine. The teres fossa is along the caudal border on the lateral side. Note the orientations of the acromion and coracoid process in each of the views. Scale is in centimeters. Abbreviations: ac, acromion process; cab, caudal border; cp, coronoid process; crb, cranial border; gc, glenoid fossa; isf, infraspinous fossa; sbsf, subscapular fossa; spsf, supraspinous fossa; tf, teres fossa; vb, vertebra border.

The humerus of basilosaurids is much more robust than that of mesonychians (Fig. 3). The humerus is about the same length as the radius and ulna. The humerus of modern cetaceans is even more robust, and it is considerably shorter than the radius and ulna. The humerus of basilosaurids has a large, hemispherical head, with well-developed muscle attachment surfaces on the proximal end. It also has a very prominent deltopectoral crest, which is much more prominent distally than the deltoid crest in mesonychians (O'Leary and Rose, 1995). The distal humerus has a single trochlea that articulates with both the radius and ulna. The distal humerus of basilosaurids also lacks the supratrochlear foramen found in mesonychians.

The radius and ulna articulate with the distal trochlea of the humerus, with the radius anterior to the ulna (Fig. 4; Uhen, 1994). The radius has a head that is roughly circular in outline and slightly concave. The radius articulates with the ulna along a flat surface. This feature and the fact that the radius and ulna articulate with a common trochlea on the humerus indicate that basilosaurids, like *Ambulocetus*, were incapable of pronation and supination at the elbow. The shaft of the radius is distally flattened. The ulna is flat along its entire length and the olecranon is large and flat.

The carpals are all flattened in the plane of the flipper. The articular surfaces between

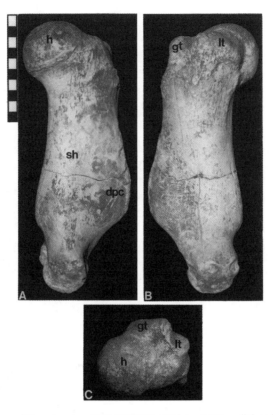

FIGURE 3. Right humerus of *Dorudon atrox* (UM 101222) in (A) lateral, (B) medial, and (C) anterior view. Note the large, hemispherical head and long robust shaft. The deltopectoral crest along the cranial margin of the humerus is greatly developed, especially at its distal end. Scale is in centimeters. Abbreviations: dpc, deltopectoral crest; gt, greater tuberosity; h, head; lt, lesser tuberosity; sh, shaft.

the carpals and between the carpals and the radius and ulna are generally flat, lacking extensions of the articular surfaces onto the dorsal and ventral surfaces of the carpals. The trapezoid and magnum are fused together in basilosaurids. These features indicate that motion in the wrist was very restricted. It was probably limited to mostly passive motion as in modern cetaceans. Basilosaurids are pentadactyl, and they appear to lack the hyperphalangy seen in modern cetaceans (Uhen, 1996a).

Many of these features described in the forelimbs of basilosaurids indicate a fully aquatic existence. The large, fan-shaped scapula is not found in terrestrial mammals. The flattening of the radius and ulna in the plane of the flipper and the restriction of the mobility of the wrist would prevent the forearm and manus from being used to support weight on land.

3.2. Vertebral Characters

Mesonychians have necks that are not exceptionally short and are typical of terrestrial mammals (Zhou *et al.*, 1992). The cervical vertebrae of *Ambulocetus* (Thewissen *et al.*,

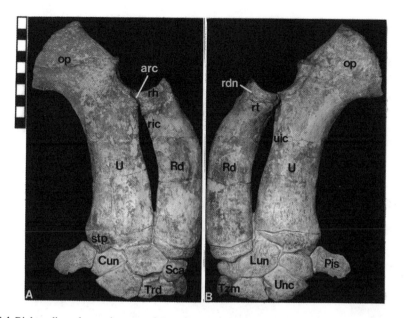

FIGURE 4. Right radius, ulna, and carpus of *Dorudon atrox* (UM 101222) in (A) lateral and (B) medial view. Note that both the radius and ulna are flattened in the plane of the flipper, especially at their distal ends. The articulated radius and ulna form a semicircular articular surface on their proximal ends for articulation with the distal humerus. Also note the very large olecranon process on the ulna. The carpals are in alternating rows, the pisiform is extremely large, and the trapezoid and magnum are fused. Scale is in centimeters. Abbreviations: arc, articular circumference; Cun, cuneiform; Lun, lunate; op, olecranon process; Pis, pisiform; Rd, radius; rdn, radial neck; rh, radial head; ric, radial interosseous crest; rt, radial tuberosity; Sca, scapoid; stp, styloid process; Trd, trapezoid + magnum; Tz, trapezium; U, ulna; uic, ulnar interosseous crest; Unc, unciform.

1996) and remingtonocetids (Gingerich *et al.*, 1995) are rather long, like those of mesonychians. Protocetids, such as *Rodhocetus* (Gingerich *et al.*, 1994) and *Protocetus* (Stromer, 1908), have somewhat shorter cervical vertebrae. Basilosaurids have distinctly shortened cervical vertebrae (Fig. 5) that are as short relative to the thoracic vertebrae as some modern cetaceans. No basilosaurids are known that have any of the cervical vertebrae fused to one another, which is a common feature of some modern cetaceans.

Another vertebral characteristic that is found in both modern cetaceans and modern sirenians is a general uniformity of vertebral length from the posterior thoracic through the anterior caudal region. Mesonychians, like other terrestrial mammals, have vertebrae that vary considerably in length down the vertebral column. The lengths of vertebrae of *Rodhocetus*, a protocetid, vary less than those of mesonychians, but more than those of a basilosaurid. A plot showing the lengths of the vertebrae in a number of basilosaurids is shown in Fig. 6. Basilosaurids also have a great number of lumbar vertebrae, usually around 20, relative to mesonychians or protocetids, which typically have between 6 and 8.

Mesonychians have sacra composed of three fused vertebrae (Zhou *et al.*, 1992). Sacra are known from a number of early cetaceans, and they vary a great deal between groups. Sacra are currently not known from pakicetids or ambulocetids, but remingtonocetids have sacra composed of four fused vertebrae (Gingerich *et al.*, 1995). *Rodhocetus* has a sacrum

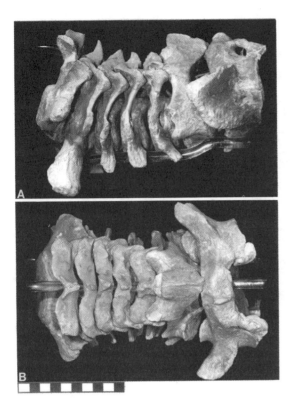

FIGURE 5. Cervical vertebrae of a juvenile *Dorudon atrox* (UM 93220) in (A) lateral and (B) dorsal view. The neural spines of C2 and C7 are both broken. Note the slight curvature of the neck when the vertebrae are articulated and the broadening of the dorsal surface of the neural arches from C2 to C7. Scale is in centimeters.

composed of four vertebrae that are not fused to one another, but rather show a large area of contact between adjacent transverse processes (Gingerich *et al.*, 1994). *Protocetus* (Stomer, 1908), *Gaviacetus* (Gingerich *et al.*, 1995), and *Natchitochia jonesi* (Uhen, 1996c) have single sacral vertebrae.

Whereas the innominates of basilosaurids do not appear to have a bony connection to the vertebral column (see below), Kellogg (1936) identified two sacral vertebrae in *Basilosaurus cetoides*. Kellogg suggested that these were sacral vertebrae, rather than posterior lumbar vertebrae, because they had dorsoventrally thickened transverse processes that were trihedral in outline, which was not the case in other lumbar vertebrae. This same condition has been observed in *Dorudon atrox* (Fig. 7). The transverse process of the last (20th) lumbar is very thick with the penultimate lumbar only slightly less. Neither these vertebrae in *Dorudon* nor those in *Basilosaurus* are considered lumbars here as only vertebrae that articulate with the innominates are considered to be the sacral vertebrae in terrestrial mammals (Flower, 1876).

The caudal regions of basilosaurids are very different from mesonychians and what little is known of earlier cetaceans. Mesonychians, like typical terrestrial mammals, have long, rod-shaped caudal vertebrae that decrease in length from cranial to caudal down the vertebral column. The midcaudal vertebra referred to *Ambulocetus* is long like those of mesonychids. Whereas the anterior caudal vertebrae of *Basilosaurus* are elongated, like the lumbar vertebrae, their posterior caudal vertebrae are short, like those of dorudontines and

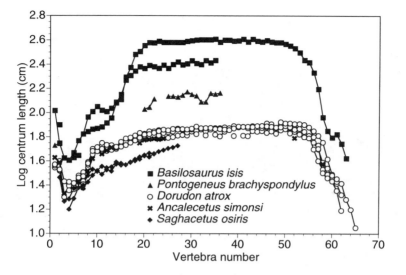

FIGURE 6. Centrum lengths of Basilosauridae. Most of the species of Egyptian basilosaurids can be distinguished based on size of individual vertebrae. *Basilosaurus isis* (composite of UM 97503, UM 97526, ZV-50; SMNS 14763) and *Pontogeneus brachyspondylus* (SMNS 11414) are much larger than *Dorudon atrox* (CGM 42183, UM 97512, UM 100146, UM 101215, UM 101222) and *Ancalecetus simonsi (CGM 42290)*, whereas *Saghacetus osiris* is much smaller. A single specimen of *Ancalecetus* has been discovered that has vertebrae that are very similar to those of *Dorudon atrox*, and are slightly smaller than the smallest specimen of *D. atrox*.

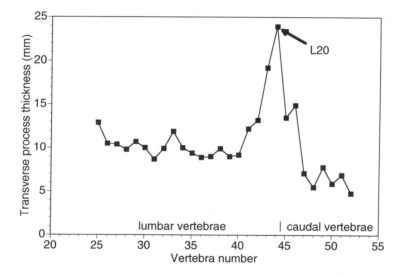

FIGURE 7. Plot of transverse process thickness of lumbar and anterior caudal vertebrae of *Dorudon atrox*, UM 101215. Note that the thickness is relatively constant in most of the lumbar region, but it increases dramatically at the end and in the first two caudal vertebrae. These vertebrae with thickened transverse processes are the ones that Kellogg (1936) identified as sacral vertebrae in *Basilosaurus cetoides*.

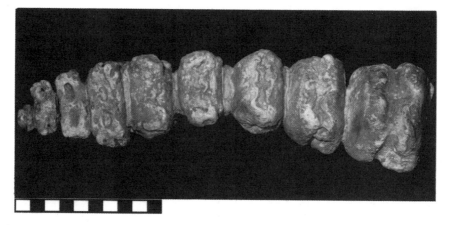

FIGURE 8. Caudal vertebrae 13 to 21 of *Dorudon atrox* (UM 101222) in dorsal view. This series was found articulated in the field and has been left as it was found. There is a slight twist in the series between caudal vertebrae 15 and 16. Note the dramatic shape change from anterior to posterior. Scale is in centimeters.

modern cetaceans. The anterior caudal vertebrae of dorudontines and modern cetaceans are similar in length to the posterior lumbar vertebrae. The caudals decrease in length from cranial to caudal, and they are all very short compared with those of mesonychids. There is a distinct shape change in the caudal region where the vertebrae become much wider than they are tall or long (Fig. 8). This is the point where the fluke inserts onto the vertebral column in modern cetaceans. This distinct shape transition can be found in the caudal regions of both basilosaurines and dorudontines (Uhen, 1991).

These vertebral characters all suggest that basilosaurids were even more aquatic than any earlier cetaceans. The shortening of the neck that was evident in protocetids was continued in the basilosaurids, and taken even further in some modern cetaceans. It has been suggested that a short neck in cetaceans is associated with stabilization of the head on the trunk of the body (Slijper, 1962). The vertebral columns of terrestrial mammals must support load placed on them at the cranial and caudal ends and at the limbs. The differing lengths of the vertebrae reflect this load-bearing function (Slijper, 1946). Cetaceans that are always supported by water and never have to venture out onto land, such as basilosaurids, do not have to have vertebrae of different sizes to distribute the load of the head and limbs. Lastly, the changes seen in the caudal region strongly suggest that basilosaurids had a tail fluke like that of modern cetaceans, and had thus adopted the cetacean mode of propulsion from a tail fluke, which would be almost useless on land.

3.3. Hindlimb Characters

The hind limbs of mesonychians have innominates that articulate with a sacrum composed of three fused vertebrae. The innominates are unremarkable in most features, and have large obturator foramina and well-developed, hemispherical acetabulae. The femora, tibiae, and fibulae of mesonychians are long and gracile. The tarsals are all free from one another, making the pes of mesonychians flexible (O'Leary and Rose, 1995).

The innominate is not known from *Ambulocetus*, but it is known from a number of remingtonocetids (Gingerich *et al.*, 1995), *Rodhocetus* (Gingerich *et al.*, 1994), and *Georgiacetus* (Hulbert and Petkewich, 1991). All of these primitive whales have large innominates that are generally similar to those of mesonychians, although *Georgiacetus* lacks an auricular surface for attachment to the sacrum. Associated femora are also known from *Ambulocetus*, remingtonocetids, and *Rodhocetus*. The femora of these archaeocetes are all somewhat different from one another, but they all have robust trochanters, well-developed trochanteric fossae, large spherical heads, and well-developed patellar grooves. No more is known of the hind limbs of early archaeocetes, except for *Ambulocetus*, which has extremely large, tetradactyl feet. Only a small portion of the astragalus is known from *Ambulocetus*, so the state of the tarsals is unknown (Thewissen *et al.*, 1996).

The hind limbs of basilosaurids are greatly reduced relative to those of earlier cetaceans and mesonychians. The innominates of *Basilosaurus* are shorter than any given lumbar vertebra. In dorudontines, where the vertebrae are not elongated as in *Basilosaurus*, the innominate is no more than twice the length of any given lumbar vertebra. A single specimen of *Basilosaurus cetoides* (USNM 12261) is known that includes hind limb elements: left and right innominates and the right femur (Lucas, 1900). Numerous specimens of *Basilosaurus isis* include innominates that are similar to that of *B. cetoides* (Gingerich *et al.*, 1990).

The innominates of *Basilosaurus* are very different from those of other archaeocetes in that the pubis is greatly elongated, such that the innominates are long and straplike. The innominates articulate with each other on the proximal end of the pubis, and show no evidence of articulation with the vertebral column. The reconstruction of the hind limb in Fig. 9 shows the orientation of the innominate reconstructed by Gingerich *et al.* (1990) for *Basilosaurus*. The orientation is unusual because the pubis is anterior to the obturator foramen, which is anterior to the acetabulum. For this to be the case, the innominate would have had to have rotated caudally in conjunction with the elongation of the pubis. This rotation scenario was previously put forth by Abel (1901) who was studying modern cetacean innominates. All known innominates of *Basilosaurus* show patent obturator foramina. The acetabula of the innominates of *Basilosaurus* range from well-formed but shallow to those that are very irregular and filled with bone.

The femora of *Basilosaurus* are small and gracile. They are usually a bit shorter than the innominate. The head of the femur is sometimes well-formed with a circular head, whereas at other times it is somewhat malformed. The distal femur has a well-developed patellar groove. The patella is ovate and thick. The tibia and fibula are fused to one another, with the tibia about twice as robust as the fibula. The tarsals are variably fused together (Gingerich *et al.*, 1990). The pes of *Basilosaurus* is tetradactyl, missing the first digit like *Ambulocetus*, and the second digit is reduced to a tiny splint of metatarsal II (Gingerich *et al.*, 1990).

Hind limb elements known from *Dorudon atrox* include a proximal femur (missing the head), patella, and much of an astragalus (Gingerich *et al.*, 1990; Uhen, 1996a). These compare favorably with *Basilosaurus isis*, and are taken to be representative of the condition of these bones in Dorudontinae in general.

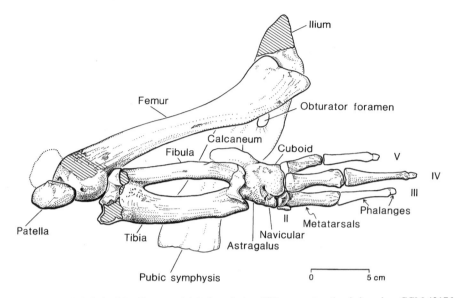

FIGURE 9. Left hind limb of *Basilosaurus isis* in lateral view. This reconstruction is based on CGM 42176 and UM 93231. Note the very small size of the hind limb, considering that it is associated with an animal that was around 16 m long. Also note the straplike form and rotated orientation of the innominate. The tarsals are almost completely fused in this reconstruction, but some are free in some individuals. Digit I is lost, and digit II is represented only by a tiny metatarsal II. Reprinted with permission from *Science*, Vol. 249, Gingerich, Smith, and Simons. Copyright © 1990, American Association for the Advancement of Science.

A specimen of an as yet undescribed new genus and species of dorudontine from South Carolina includes most of the left innominate and fragments of the right innominate. The innominate shows a greatly elongated pubis and a very reduced ilium. Much of the ischium is broken. The innominates are very similar to those of *Basilosaurus*, but are proportionally smaller, as they are from a smaller dorudontine.

These features of the hind limb of basilosaurids are the most salient evidence that basilosaurids are fully aquatic. The extremely reduced hind limbs were not large enough to support the bodies of even the smallest basilosaurids on land. Even if basilosaurids could come out on land, the tiny hind limbs could not help support the body mass as they are not connected to the vertebral column via any bony connection.

4. Summary and Conclusions

Basilosaurid archaeocetes are the most advanced group of archaeocete cetaceans currently known. Basilosaurids can be distinguished from other archaeocetes by the development of extensive peribullary air sinuses (but see Thewissen *et al.*, 1996), loss of M^3, and the presence of accessory denticles on the premolars and molars, although this condition is incipient in *Georgiacetus* (Hulbert, 1996). In addition to these cranial and dental characters, basilosaurids have a number of advanced features that are associated with the adoption of a completely aquatic lifestyle. Some of these features are shared with early mysticetes and odontocetes.

The aquatic features found in basilosaurids either are very similar to those found in mysticetes and odontocetes, such as scapular shape and vertebral size uniformity, or are taken even further in mysticetes and odontocetes, such as forelimb motion restriction and hind limb reduction. Most of the characters that differentiate mysticetes and odontocetes from archaeocetes are associated with the skull, dentition, and thorax.

The crania of modern mysticetes and odontocetes are very different from one another and from basilosaurids. Modern mysticetes have telescoped the bones of the skull by extending the occipital region forward over the braincase, whereas odontocetes have extended the bones of the rostrum and face backwards over the braincase. Basilosaurids show no evidence of telescoping.

Modern mysticetes and odontocetes are both considered to be monophyodont (Fordyce, 1982) as odontocetes only erupt a single set of teeth and mysticetes display a single set of teeth that are resorbed *in utero* (Karlsen, 1962). The teeth of modern cetaceans are also generally homodont and simple in form. Basilosaurids are both diphyodont and heterodont. Early mysticetes and odontocetes have teeth that are very basilosaurid in form, with multiple accessory denticles (Tuomey, 1847; Mitchell, 1989; Barnes and Sanders, 1996). It is not yet clear whether these early mysticetes and odontocetes were monophyodont.

The divergence of mysticete and odontocete dentitions following their origin from dorudontines is profound and is probably related to different styles of feeding. Modern mysticetes have lost erupted teeth altogether and have acquired a new structure, baleen, with which they strain food from water. Odontocetes are more similar in feeding strategy to that hypothesized for archaeocetes, which is eating fish (Swift and Barnes, 1996) and/or squid. The difference is that basilosaurids retain cheek teeth that are capable of shearing prey into smaller pieces and wear facets indicate that they chewed their food. Odontocetes on the other hand usually swallow their food whole or tear off pieces, but they do not chew as such. The shift toward loss of teeth entirely in mysticetes and toward a single set of simple teeth in odontocetes reflects these changes in feeding strategy.

All of these further changes in the cranium and dentition of mysticetes and odontocetes were possible after basilosaurids became full aquatic. The changes in odontocete skulls are almost certainly related to the development of their highly developed echolocation system, which would be ineffective on land. The development of baleen was also tied directly to achievement of a fully aquatic existence in basilosaurids as mysticetes usually travel long distances in the water to find high-density food sources that are worth their effort to feed on. Additional study of basilosaurids, particularly dorudontines, is needed to further explore the details of the transition from archaeocetes to mysticetes and odontocetes.

Appendix: Character Matrix Codes

Taxa are listed in the matrix as follows: DIS, *Dissacus*; PAC, *Pachyaena*; SIN, *Sinonyx*; PAK, *Pakicetus*; REM, *Remingtonocetus*; ROD, *Rodhoceteus*; BAS, *Basilosaurus;* PON, *Pontogeneus;* SAG, *Saghacetus;* ZYG, *Zygorhiza*; DOR, *Dorudon*; ANC, *Ancalecetus*; PRO, *Protocetus*; AMB, *Ambulocetus*; ODO, Odontoceti; Mys, Mysticeti; BAB, *Babiacetus*; DAL, *Dalanistes*. Some size characters are relative to the breadth across the occipital condyles, which is taken as a measure indicative of overall body size (Character 67).

1. frontal shield size relative to condylar breadth, 0: < 2, 1: $> 2 < 2.5$, 2: > 2.5.
2. posterior frontal border, 0: angled back, 1: straight, 2: curved.
3. rostrum breadth at $P^4 > 2\times$ breadth at P^2, 0: yes, 1: no.
4. rostrum length relative to condylar breadth, 0: < 3, 1: > 3.
5. relative skull length, 0: < 7, 1: 7–8, 2: > 8.
6. embrasure pits, 0: absent, 1: present.
7. orbit height relative to condylar breadth, 0: < 0.3, 1: > 0.3.
8. palate narrows at, 0: post to M^3, 1: M^3, 2: M^2, 3: M^1, 4: P^4.
9. palate shape, 0: flat, 1: convex, 2: concave.
10. falcate process of the basioccipital, 0: absent, 1: small, 2: large.
11. vomer covers basioccipital/bassisphenoid suture, 0: no, 1: yes.
12. pachyosteosclerotic bulla, 0: no, 1: yes.
13. medial bulla articulation, 0: broad, 1: short, 2: absent.
14. lateral wall of pterygoid sinus, 0: absent, 1: present.
15. sigmoid process of the auditory bulla, 0: absent, 1: small, 2: large.
16. hypoglossal foramen in jugular, 0: no, 1: yes.
17. anterior palatine foramen, 0: absent, 1: present.
18. postorbital process of the frontal touches the zygomatic process of squamosal, 0: no, 1: yes.
19. nuchal crest orientation, 0: angled back, 1: slightly angled back, 2: vertical, 3: absent.
20. position of posterior nares, 0: C^1, 1: P^1, 2: P^2, 3: top of skull.
21. nasal breadth, 0: narrow, 1: moderate, 2: broad, 3: reduced.
22. posterior medial maxilla contacts, 0: nasal, 1: frontal, 2: premaxilla.
23. mandibular foramen, 0: small, 1: moderate, 2: large.
24. posterior end of the mandibular symphysis position, 1: P_1, 2: P_2, 3: P_3, 4: P_4.
25. mandibular symphysis fused, 0: no, 1: yes.
26. incisors in tooth row, 0: no, 1: yes.
27. cheek teeth with many denticles, 0: no, 1: incipient, 2: yes.
28. number of upper molars, 2: two, 3: three.
29. # M^1 roots, 0: 3, 1: incipient 3, 2: 2.
30. # M^2 roots, 0: 3, 1: incipient 3, 2: 2.
31. # M^3 roots, 0: 3, 1: incipient 3, 2: 2.
32. tooth replacement, 0: diphyodont, 1: monophyodont.
33. number of P_1 roots, 1: one, 2: two.
34. reentrant groove on lower molars, 0: absent, 1: present.
35. vertebral profile arched, 0: no, 1: yes.
36. cervical vertebrae compressed, 0: none, 1: moderate, 2: a lot.
37. number of thoracics, 0: <10, 1: 10–15, 2: 16–20, 3: >20.
38. number of lumbars, 0: <10, 1: 10–15, 2: >15.
39. lumbars elongate, 0: no, 1: yes.
40. lumbar zygapophyses, 0: revolute, 1: curved, 2: flat, 3: absent.
41. # sacrals solidly ankylosed, 1: none, 2: two, 3: three, 4: four.
42. # sacrals loosely joined, 1: none, 2: two, 3: three, 4: four.

43. # sacrals articulate with pelvis, 0: none, 1: one, 2: two.
44. posterior caudals dorsoventrally compressed, 0: no, 1: yes.
45. anterior caudals elongate, 0: no, 1: yes.
46. infraspinous fossa on scapula, 0: small, 1: moderate, 2: large.
47. coracoid process oriented anteriorly, 0: no, 1: yes.
48. acromion process oriented anteriorly, 0: no, 1: yes, 2: folded back.
49. humeral shaft anteroposteriorly thick, 0: no, 1: yes.
50. distal humeral articulation divided into trochlea and capitulum, 0: no, 1: yes.
51. radius and ulna flat, 0: no, 1: yes.
52. broad olecranon process, 0: no, 1: yes.
53. distal ulna, 0: pointed, 1: broad.
54. saddle-shaped carpal articulations, 0: no, 1: yes.
55. trapezoid and magnum, 0: separate, 1: fused.
56. carpals in alternating rows, 0: no, 1: yes.
57. pisiform, 0: absent, 1: small and round, 2: big and flat.
58. os centrale, 0: absent, 1: present.
59. hyperphalangy, 0: not present, 1: present.
60. pelvis, 0: nonfunctional, 1: small, 2: moderate, 3: large.
61. pelvis rotation, 0: no, 1: yes.
62. femur, 0: absent, 1: nonfunctional, 2: small, 3: moderate, 4: large.
63. tarsals, 0: separate, 1: fused.
64. tibia/fibula, 0: separate, 1: fused.
65. sternum form, 0: rodlike, 1: big and heavy, 2: flat and light.
66. anterior ribs distally expanded, 0: no, 1: yes.
67. body size, 0: small, 1: moderate, 2: large, 3: very large.
68. stratigraphic position, 0: Thanetian, 1: Ypresian, 2 to 6: Lutetian (2: sea level cycle TA 3.1, 3: 3.2, 4: early 3.3, 5: late 3.3, 6: 3.4), 7: early Bartonian (TA 3.5), 8: late Bartonian (TA 3.6), 9: early Priabonian (TA 4.1 & 4.2), A: late Priabonian, B: post-Priabonian.

Character Matrix

	DIS	PAC	SIN	PAK	REM	ROD	BAS	PON	SAG	ZYG	DOR	ANC	PRO	AMB	ODO	MYS	BAB	DAL
1	0	0	0		0	1	2		2	2	2		2	0	1	2		0
2	0	0	0		0	1	1		1	1	1		1	0	1	2	1	0
3	1		1	1	0	1	1		1	1	1		1				1	1
4	0	0	0	0	1	1	1		0	1	1		1			1		2
5			0	0	2	1	2	1		1	1		1	1	0	0		1
6	0		0	0	1	1	1		1	1	1		1		0	0	1	0
7	1	1	1	0	0	1	1		4	3	4	0	2	1	0	0	3	1
8	1	1	1	1	1	0	0		0	0	0	2	0		0	2	0	1
9	0	0	0	0	2	2	2	2	2	2	2	2	2	2	2	2	2	2
10	0	0	0	1		1	1		1	1	1	1	1		1	0	1	
11	0	0	0	0	1		1	1	1	1	1			0	1	1	2	
12				1		2	2	2	2	2	2	2		2	2	2	2	
13			0	0	1	0	1		1	1	1	1	0	1	1	1	0	
14	0	0	0	1		2	2	2	2	2	2	2	2	1	2	2	2	
15	0	0	0	0	1	0	1	2	2	2	2	2	2	1	2	2	0	
16	0		1	0		2	2		2	2	2	2	1	0	1	2	2	
17			1	1	0	1	1		1	1	1	1	1		1	1	1	
18	0	0	0	0	0	0	0		0	0	0	0	0	0	0	0	0	0
19	0	0	0		0	0	0		0	0	0	0	1		1	1		0
20	0	0	0		1	1	2	2	2	2	2	2	0	0	3	3	1	0
21	2	2	2		0	0	1	2	2	2	2	2	0	0	3	3	2	0
22	1		1		0	0	0	0	0	0	0		0	0	0&1	2	0	0
23	0	0	0	0	0	2	2	2	2	2	2	2		0	2	2	2	3
24	3	3	3		4	2	2	2	2	2	2	2		1	2	2	2	0
25	0	0	0	0	1	0	0	0	0	0	0	0	1	1	0	1	1	1
26	0	0	0	1	1	1	1	1	1	1	1	1	1	1	1	1	1	0
27	0	0	0	0	0	0	2	2	2	2	2	2	0	0	2	2	0	0
28	3	3	3	3	3	3	2	2	2	2	2	2	3	3			3	3
29	0	0	0	0	0	0	2	2	2	2	2	2	0	0			0	0
30	0	0	0	0	1	0	2	2	2	2	2	2	0	0			0	1
31	0	0	0	0	2	0	2	2	2	2	2	2	1	0			0	2
32	0	0	0	0	0	0	0	0	0	0	0	0	0	0	1	1	0	0

	33	34	35	36	37	38	39	40	41	42	43	44	45	46	47	48	49	50	51	52	53	54	55	56	57	58	59	60	61	62	63	64	65	66	67	68
4	2	1		0				4	-	2		1															3	0	4					2	4	
6	2	1																																2	6	
B			0	2	3	2	0	3	-		0		0	2				0				0					0		0				2	0		B
B			0	2	3	2	0	3	-		0		0	2				0				0					0		0	0			2	0		B
5			0	2	2	0	0	2	-	-	-																						2			5
8&9	2	-	0	2	2	2	0	2			0		0				0		0		0				2	0	0					1	0	2		8&9
8&9	2	-	0	2	2	2	0	2	-		0		0		2						0				2	0	0		2					2		8&9
8-A	2	-	0	2	2	2	0	2	-		0		0		2						0				2	2				0				2		8-A
9			0	2	2	2		2			0																									9
8-A		-		0	2			0		2			0						0		-													3		8-A
8-A			0	2	2	2		2			0				2				0		-				2		0				2				3	8-A
4	2		0			0	0		4				0														3	0	3						1	4
4	2	1		0						4	-	2		1													3	0	4		0			2	4	
2	2	1																															0	2		
0&1													0	1				0	0		1	0		1			0	3	0	4	0	0			0	0&1

References

Abel, O. 1901. Les daupins longirostres du Boldérien (Miocène supérieur) des environs d'Anvers. *Mem. Mus. R. Hist. Nat. Belg.* **1**:1–95.

Abel, O. 1906. Ueber den als Beckengürtel von Zeuglodon beschriebenen Schultergürtel eines Vogels aus dem Eocän von Alabama. *Zentralbl. Mineral. Geol. Paläontol.* **15**:450–458.

Abel, O. 1914. Die vorfahren der bartenwale. *Denkschr. Kaiserlichen Akad. Wiss. Math. Naturwiss. Kl.* **90**:155–224.

Andrews, C. W. 1904. Further notes on the mammals of the Eocene of Egypt. *Geol. Mag. London Ser. 5.* **1**:211–215.

Andrews, C. W. 1906. *A Descriptive Catalogue of the Tertiary Vertebrata of Fayum, Egypt.* British Museum of Natural History, London.

Barnes, L. G. 1984. Whales, dolphins and porpoises: origin and evolution of the Cetacea, in: T. W. Broadhead (ed.), *Mammals, Notes for a Short Course*, pp. 139–154. University of Tennessee Department of Geological Sciences, Studies in Geology **8**(1–4).

Barnes, L. G., and Mitchell, E. 1978. Cetacea, in: V. J. Maglio and H. B. S. Cooke (eds.), *The Evolution of African Mammals*, pp. 582–602. Harvard University Press, Cambridge, MA.

Barnes, L. G., and Sanders, A. E. 1996. The transition from archaeocetes to mysticetes: late Oligocene toothed mysticetes from near Charleston, South Carolina. *Sixth North American Paleontological Convention Abstracts of Papers. Paleontolog. Soc. Spec. Pub.* **8**:24.

Barnes, L. G., Domning, D. P., and Ray, C. E. 1985. Status of studies on fossil marine mammals. *Mar. Mammal Sci.* **1**:15–53.

Bonaparte, C. L. 1849. [Classification of Havapatedyrene i Pinnipedia, Cete og Sirenia]. *Forhandlinger ved de skandinaviske Naturforskeres femte Møde Kjøbenhaven* **1847**:618.

Brandt, J. F. 1873. Untersuchungen über die fossilen und subfossilen cetaceen Europa's. *Mem. Acad. Imp. Sci. Saint-Petersbourg, Ser. 7* **20**:1–372.

Carus, C. G. 1847. *Resultate geologischer, anatomischer und zoologischer Untersuchungen über das unter dem Namen Hydrarchos von Dr. A. C. Koch, zuerst nach Europa gebrachte und in Dresden ausgestellte grofse fossile Skelett*, pp. 1–15. Arnoldische Buchhandlung, Dresden.

Carus, C. G. 1849. Das Kopfskelet des Zeuglodon hydrarchos. *Nova Acta Leopold.* **22**:371–390.

Cope, E. D. 1867(1868). An addition to the VERTEBRATE FAUNA of the Miocene period, with a synopsis of the extinct CETACEA of the United States. *Proc. Acad. Nat. Sci. Philadelphia* **19**:138–156.

Dames, W. B. 1883a. Über eine tertiäre Wirbelthierfauna von der westlichen Insel des Birket-el-Qurun im Fajum (Aegypten). *Sitzungsber. K. Preuss. Akad. Wiss.* **6**:129–153.

Dames, W. B. 1883b. Ein Epistropheus von *Zeuglodon* sp. *Sitzungsber. Ges. Naturforsch. Freunde* **1883**:3.

Dames, W. B. 1894. Über Zeuglodonten aus Aegypten und die Beziehungen der Archaeoceten zu den übrigen Cetaceen. *Paläontol. Abh. Jena N. F.* **1**:189–222.

Dart, R. A. 1923. The brain of the Zeuglodontidae (Cetacea). *Proc. Zool. Soc. London* **42**:615–654.

DeKay, J. E. 1842. *Zoology of New York, Natural History of New York.* B. Appleton & Co. and Wiley & Putnam, New York.

Domning, D. P., Morgan, G. S., and Ray, C. E. 1982. North American Eocene sea cows (Mammalia, Sirenia). *Smithson. Contrib. Paleobiol.* **52**:1–69.

Fisher, D. C. 1992. Stratigraphic parsimony, in: W. P. Maddison and D. R. Maddison, *MacClade 3.0 Manual*, pp. 124–129. Sinauer Associates, Sunderland, MA.

Flower, W. H. 1876. *An Introduction to the Osteology of the Mammalia.* Macmillan & Co., London.

Fordyce, R. E. 1982. Dental anomaly in a fossil squalodont dolphin from New Zealand, and the evolution of polydonty in whales. *N. Z. J. Zool.* **9**:419–426.

Fordyce, R. E. 1985. Late Eocene archaeocete whale (Archaeoceti: Dorudontinae) from Waihao, South Canterbury, New Zealand. *N. Z. J. Geol. Geophys.* **28**:351–357.

Frazier, W. J., and Schwimmer, D. R. 1987. *Regional Stratigraphy of North America.* Plenum Press, New York.

Gibbes, R. W. 1845. Description of the teeth of a new fossil animal found in the Green Sand of South Carolina. *Proc. Acad. Nat. Sci. Philadelphia* **2**:254–256.

Gibbes, R. W. 1847. On the fossil genus *Basilosaurus*, Harlan, (*Zeuglodon*, Owen) with a notice of specimens from the Eocene Green Sand of South Carolina. *J. Acad. Nat. Sci. Philadelphia* **1**:2–15.

Gill, T. 1872. Arrangement of the families of mammals. *Smithson. Misc. Collect.* **11**:1–97.

Gingerich, P. D. 1992. Marine mammals (Cetacea and Sirenia) from the Eocene of Gebel Mokattam and Fayum, Egypt: stratigraphy, age and paleoenvironments. *Univ. Michigan Mus. Paleontol. Pap. Paleontol.* **30**:1–84.

Gingerich, P. D., and Uhen, M. D. 1996. *Ancalecetus simonsi*, a new dorudontine archaeocete (Mammalia, Cetacea) from the early late Eocene of Wadi Hitan, Egypt. *Contrib. Mus. Paleontol. Univ. Michigan* **29**:359–401.

Gingerich, P. D., Wells, N. A., Russell, D. E., and Shah, S. M. I. 1983. Origin of whales in epicontinental remnant seas: new evidence from the early Eocene of Pakistan. *Science* **220**:403–406.

Gingerich, P. D., Smith, B. H., and Simons, E. L. 1990. Hind limbs of Eocene *Basilosaurus isis:* evidence of feet in whales. *Science* **229**:154–157.

Gingerich, P. D., Raza, S. M., Arif, M., Anwar, M., and Zhou, X. 1994. New whale from the Eocene of Pakistan and the origin of cetacean swimming. *Nature* **368**:844–847.

Gingerich, P. D., Arif, M., and Clyde, W. C. 1995. New archaeocetes (Mammalia, Cetacea) from the middle Eocene Domanda Formation of the Sulaiman Range, Punjab (Pakistan). *Contrib. Mus. Paleontol. Univ. Michigan* **29**(11):291–330.

Gingerich, P. D., Arif, M., Bhatti, M. A., Anwar, M., and Sanders, W. J. 1997. *Basilosaurus drazindai* and *Basiloterus hussaini*, new Archaeoceti (Mammalia, Cetacea) from the middle Eocene Drazinda Formation, with a revised interpretation of ages of whale-bearing strata in the Kirthar Group of the Sulaiman Range, Punjab (Pakistan). *Contrib. Mus. Paleontol. Univ. Michigan* **30**:291–330.

Hammerschmidt, C. E. 1848. Resultate geologischer, anatomischer and zoologischer Untersuchungen [über *Hydrarchos*, Koch]. *Haidinger's Ber. Mitt. Freunden Naturwiss. Wien* **3**:322–327.

Haq, B. U., Hardenbol, J., and Vail, P. R. 1987. Chronology of fluctuating sea levels since the Triassic. *Science* **235**:1156–1167.

Harlan, R. 1834. Notice of fossil bones found in the Tertiary formation of the state of Louisiana. *Trans. Am. Philos. Soc. Philadelphia* **4**:397–403.

Hazel, J. E., Bybell, L. M., Christopher, R. A., Fredericksen, N. O., May, F. E., McLean, D. M., Poore, R. Z., Smith, C. C., Sohl, N. F., Valentine, P. C., and Witmer, R. J. 1977. Biostratigraphy of the deep corehole (Clubhouse Crossroads corehole 1) near Charleston, South Carolina. *U. S. Geol. Surv. Prof. Pap.* **1028**:71–89.

Hulbert, R. C. 1994. Phylogenetic analysis of Eocene whales ("Archaeoceti") with a diagnosis of a new North American protocetid genus. *J. Vertebr. Paleontol.* **14**:30A.

Hulbert, R. C. 1996. Dental and basicranial anatomy of a late middle Eocene protocetid cetacean from the southeastern United States. *Sixth North American Paleontological Convention Abstracts of Papers. Paleontol. Soc. Spec. Pap.* **8**:186.

Hulbert, R. C., and Petkewich, R. M. 1991. Innominate of a middle Eocene (Lutetian) protocetid whale from Georgia. *J. Vertebr. Paleontol.* **11**:36A.

Hulbert, R. C., Petkewich, R. M., Bishop, G. A., Bukry, D., and Aleshire, D. P. 1998. A new middle Eocene protocetid whale (Mammalia: Cetacea: Archaeoceti) and associated biota from Georgia. *J. Paleontol.* **72**:905–925.

Huner J. J. 1939. Geology of Caldwell and Winn Parishes. *State of Louisiana Department of Conservation Geological Bulletin* **15**:1–356.

Karlsen, K. 1962. Development of tooth germs and adjacent structures in the whalebone whale (*Balaenoptera physalus* (L.)). *Hvalradets Skr. Sci. Results Mar. Biol. Res.* **45**:1–56.

Kellogg, R. 1928. The history of whales—their adaptation to life in the water. *Q. Rev. Biol.* **3**:29–76.

Kellogg, R. 1936. A review of the Archaeoceti. *Carnegie Inst. Washington Publ.* **482**:1–366.

Kellum, L. B. 1926. Paleontology and stratigraphy of the Castle Hayne and Trent Marls in North Carolina. *U. S. Geol. Surv. Prof. Pap.* **143**:1–56.

Koch, A. C. 1845a. *Description of the Hydrarchos harlani: (Koch).* B. Owen, New York.

Koch, A. C. 1845b. *Description of the Hydrargos sillimanii: (Koch) gigantic fossil reptile, or sea serpent.* A. C. Koch, New York.

Koch, A. C. 1846. *Kurze Beschreibung des Hydrarchos harlani (Koch) eines reisenmässigen Meerungeheuers und dessen Entdeckung in Alabama in Nordamerika im Frühjahr 1845.* Druck der Königl. Hofbuchdruckerei von C. C. Meinhold und Söhnen, Dresden.

Kumar, K., and Sahni, A. 1986. *Remingtonocetus harudiensis*, new combination, a middle Eocene archaeocete (Mammalia, Cetacea) from western Kutch, India. *J. Vertebr. Paleontol.* **6**:326–349.

Leidy, J. 1852. [Description of *Pontogeneus priscus*]. *Proc. Acad. Nat. Sci. Philadelphia* **6**:52.

Lucas, F. A. 1900. The pelvic girdle of zeuglodon *Basilosaurus cetoides* (Owen), with notes on other portions of the skeleton. *Proc. U. S. Nat. Mus.* **23**:327–331.

McKenna, M. C. 1975. Toward a phylogenetic classification of the Mammalia, in: W. P. Luckett and F. S. Szalay (eds.), *Phylogeny of the Primates*, pp. 21–46. Plenum Press, New York.

Miller, C. S. J. 1923. The telescoping of the cetacean skull. *Smithson. Misc. Collect.* **76**:1–71.

Mitchell, E. D. 1989. A new cetacean from the late Eocene La Meseta Formation, Seymour Island, Antarctic Peninsula. *Can. J. Fish. Aquat. Sci.* **46**:2219–2235.

Moustafa, Y. S. 1954. Additional information on the skull of *Prozeuglodon isis* and the morphological history of the Archaeoceti. *Proc. Egypt. Acad. Sci.* **9**:80–89.

Müller, J. 1849. Über die fossilen Reste der Zeuglodonten von Nordamerica. Berlag von G. Reimer, Berlin.

Müller, J. 1851. Neue Beiträge zur Kenntniss der Zeuglodonten. *Ber. Bekanntmach. Verh. Königlichen Preufs. Akad. Wiss. Berlin* **April 28**:236–246.

O'Leary, M. A., and Rose, K. D. 1995. Postcranial skeleton of the early Eocene mesonychid *Pachyaena* (Mammalia: Mesonychia). *J. Vertebr. Paleontol.* **15**:401–430.

Owen, R. 1839. Geological Society. *The Athenaeum* **585**:35–36.

Owen, R. 1842a. Observations on the teeth of the Zeuglodon, Basilosaurus of Dr. Harlan. *Proc. Geol. Soc. London* **3**:23–28.

Owen, R. 1842b. Observations on the *Basilosaurus* of Dr. Harlan (*Zeuglodon cetoides*, Owen). *Trans. Geol. Soc. London* **6**:69–79.

Pilleri, G. 1985. Record of *Dorudon osiris* (Archaeoceti) from Wadi-el-Nuturn, lower Nile Valley. *Invest. Cetacea* **17**:35–37.

Randazzo, A. F., Kosters, M., Jones, D. S., and Portell, R. W. 1990. Paleoecology of shallow-marine carbonate environments, middle Eocene of peninsular Florida. *Sediment. Geol.* **66**:1–11.

Reichenbach, L. 1847. Systematisches, in: C. G. Carus, *Resultate geologischer, anatomischer und zoologischer untersuchungen über das unter den Namen Hydrarchos von Dr. A. C. Koch zuerst nach Eurpa gebrachte und in Dresden augestelte grofse fossile Skelett.* Arnoldische Buchhandlung, Dresden.

Sahni, A., and Mishra, V. P. 1972. A new species of *Protocetus* (Cetacea) from the middle Eocene of Kutch, western India. *Palaeontology* **15**:490–495.

Sahni, A., and Mishra, V. P. 1975. Lower Tertiary vertebrates from western India. *Monogr. Palaeontol. Soc. India* **3**:1–48.

Sanders, A. E. 1974. A paleontological survey of the Cooper Marl and Santee Limestone near Harleyville, South Carolina preliminary report. *Geol. Notes* **18**:4–12.

Schweinfurth, G. A. 1886. Reise in das Depressionsgebiet im Umkreise des Fajum im Januar 1886. *Z. Ges. Erdkurde Berlin* **21**:96–149.

Scudder, S. H. 1882. Universal index to genera in zoology. *Bull. U.S. Nat. Mus.* **19**:1–340.

Seeley, H. G. 1876. Notice of the occurrence of remains of a British fossil zeuglodon (*Z. wanklyni*, Seeley) in the Barton Clay of the Hampshire coast. *Q. J. Geol. Soc. London* **32**:428–432.

Seeley, H. G. 1881. Note on the caudal vertebra of a cetacean discovered by Prof. Judd in the Brockenhurst beds, indicative of a new type allied to *Balaenoptera (Balaenoptera juddi). Q. J. Geol. Soc. London* **37**:709–712.

Slijper, E. J. 1936. Die Cetacean. *Capita Zool.* **7**:1–590.

Slijper, E. J. 1946. Comparative biologic-anatomical investigations on the vertebral column and spinal musculature of mammals. *Verh. K. Ned Akad. Wet. Afd. Natuurkd.* **42**:1–128.

Slijper, E. J. 1979. *Whales* (2nd ed.). Cornell University Press, Ithaca, NY.

Stromer, E. 1903. Zeuglodon-reste aus dem Oberen Mittelocän des Fajum. *Beitr. Paläontol. Geol. Österreich-Ungarns Orients* **15**:65–100.

Stromer, E. 1908. Die Archaeoceti des Ägyptischen Eozäns. *Beitr. Paläontol. Geol. Österreich-Ungarns Orients* **21**:106–177.

Swift, C. C., and Barnes, L. G. 1996. Stomach contents of *Basilosaurus cetoides:* implications for the evolution of cetacean feeding behavior, and evidence for vertebrate fauna of epicontinental Eocene seas. *Sixth North American Paleontological Convention Abstracts of Papers. Paleontol. Soc. Spec. Pap.* **8**:380.

Thewissen, J. G. M. 1994. Phylogenetic aspects of cetacean origins: a morphological perspective. *J. Mamm. Evol.* **2**:157–184.

Thewissen, J. G. M., Hussain, S. T., and Arif, M. 1994. Fossil evidence for the origin of aquatic locomotion in archaeocete whales. *Science* **263**:210–212.

Thewissen, J. G. M., Madar, S. I., and Hussain, S. T. 1996. *Ambulocetus natans*, an Eocene cetacean (Mammalia) from Pakistan. *Cour. Forsch.-Inst. Senckenberg* **191**:1–86.

Trofimov, B. A., and Gromova, V. I. 1968. Order Cetacea, in: V. I. Gromova (ed.), *Fundamentals of Paleontology: Mammals*, pp. 225–241. Israel Program for Scientific Translations, Jerusalem.

True, F. W. 1908. The fossil cetacean, *Dorudon serratus* GIBBES. *Bull. Mus. Comp. Zool.* **52**:5–78.

Tuomey, M. 1847. Notice of the discovery of a cranium of the *Zeuglodon (Basilosaurus)*. *J. Acad. Nat. Sci. Philadelphia* **1**:16–17.

Uhen, M. D. 1991. Vertebral proportions as indicators of locomotor style in mammals. *J. Vertebr. Paleontol.* **11**:59A.

Uhen, M. D. 1994. Forelimb of late middle Eocene *Prozeuglodon atrox* (Mammalia, Cetacea) from Fayum, Egypt. *J. Vertebr. Paleontol.* **14**:51A.

Uhen, M. D. 1996a. *Dorudon atrox* (Mammalia, Cetacea): form, function, and phylogenetic relationships of an archaeocete from the late middle Eocene of Egypt. Ph.D. dissertation, University of Michigan, Ann Arbor, 608 pp.

Uhen, M. D. 1996b. Composition and characteristics of the subfamily Dorudontinae (Archaeoceti, Cetacea). *Sixth North American Paleontological Convention Abstracts of Papers. Paleontol. Soc. Spec. Pub.* **8**:403.

Uhen, M. D. 1996c. New protocetid archaeocete (Mammalia, Cetacea) from the late middle Eocene Cook Mountain Formation of Louisiana. *J. Vertebr. Paleontol.* **16**:70A.

West, R. M. 1980. Middle Eocene large mammal assemblage with tethyan affinities, Ganda Kas region, Pakistan. *J. Paleontol.* **54**:508–533.

Wyman, J. 1845. [Communication on skeleton of *Hydrarchos sillimani*]. *Proc. Boston Soc. Nat. Hist.* **2**:65–68.

Zhou, X., Sanders, W. J., and Gingerich, P. D. 1992. Functional and behavioral implications of vertebral structure in *Pachyaena ossifraga* (Mammalia, Mesonychia). *Contrib. Mus. Paleontol. Univ. Michigan* **28**:289–319.

Zhou, X., Zhai, R., Gingerich, P. D., and Chen, L. 1995. Skull of a new mesonychid (Mammalia, Mesonychia) from the late Paleocene of China. *J. Vertebr. Paleontol.* **15**:387–400.

Zullo, V. A., and Harris, W. B. 1987. Sequence stratigraphy, biostratigraphy, and correlation of Eocene through lower Miocene strata in North Carolina. *Cushman Found. Foraminiferal Res. Spec. Pub.* **24**:197–214.

CHAPTER 3

Molecular Evidence for the Phylogenetic Affinities of Cetacea

JOHN GATESY

1. Introduction

The phylogenetic affinities of Cetacea have not been clearly resolved by either molecular or morphological characters. The rapidly growing molecular data base should, in theory, complement anatomical evidence from the spectacular fossil discoveries of the past 15 years. Unfortunately, recent phylogenetic analyses show more conflict than compromise between molecules and morphology.

There is a marked discontinuity in morphology between extant cetaceans and their closest living relatives, but recently described fossils have closed this anatomical gap. Gross anatomical data clearly demonstrate that whales are highly transformed ungulates (Prothero *et al.*, 1988; Geisler and Luo, this volume; O'Leary, this volume). "Mesonychians" from the Paleocene and Eocene are considered the closest terrestrial relatives of Cetacea (Fig. 1A; Van Valen, 1966; Prothero *et al.*, 1988; Thewissen, 1994), and various "archaeocete" genera compose the stem lineage of modern whales (Gingerich *et al.*, 1983, 1990, 1994; Thewissen, 1994; Thewissen *et al.*, 1994).

Among extant ungulates, the cladistic analyses of Prothero *et al.* (1988) and Novacek (1989) position Cetacea closer to Perissodactyla and Paenungulata (Sirenia + Hyracoidea + Proboscidea) than to Artiodactyla (Fig. 1B). Other interpretations of morphology suggest a sister-group relationship between Artiodactyla and Cetacea (Fig. 1C; Slijper, 1962; Gingerich *et al.*, 1990; Geisler and Luo, this volume), a view that is more consistent with molecular results (e.g., Fitch and Beintema, 1990; Queralt *et al.*, 1995; Stanhope *et al.*, 1996).

Oddly, many molecular analyses insert Cetacea within Artiodactyla (Fig. 1D; e.g., Goodman *et al.*, 1985; Irwin *et al.*, 1991; Graur and Higgins, 1994; Shimamura *et al.*, 1997), and some genes specifically link cetaceans with hippopotamid artiodactyls (e.g., Irwin and Arnason, 1994; Gatesy, 1997; Milinkovitch *et al.*, this volume). These results contradict morphological evidence in support of artiodactyl monophyly (Gentry and Hooker, 1988;

JOHN GATESY • Laboratory of Molecular Systematics and Evolution, Department of Ecology and Evolutionary Biology, University of Arizona, Tucson, Arizona 85721.
The Emergence of Whales, edited by Thewissen. Plenum Press, New York, 1998.

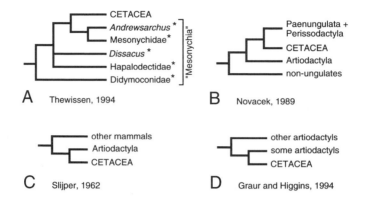

FIGURE 1. Different hypotheses of Cetacea's phylogenetic relationships. Asterisks identify extinct taxa.

Prothero, 1993) and demand extensive gaps in the fossil record of the Paleocene or remarkable evolutionary convergence between mesonychians and early cetaceans (Gatesy *et al.*, 1996). Sharp conflicts between molecules and morphology, as well as incongruence among different molecular trees (e.g., Wyss *et al.*, 1987), have led to skepticism of the molecular evidence for cetacean phylogeny.

The purpose of this chapter is twofold. First, previous molecular systematic studies that place Cetacea relative to other eutherian mammals are summarized. Sixty trees from 44 studies are presented to illustrate the diversity of molecular data, phylogenetic methods, and systematic results. In the second half of the chapter, new DNA sequences for 12 S mitochondrial (mt) ribosomal (r) DNA, 16 S mt rDNA, nuclear (nu) β-casein exon 7, nu β-casein intron 7, nu κ-casein exon 4, nu γ-fibrinogen exons 2–4/introns 2–3, and nu protamine P1 are aligned with published sequences. These alignments in combination with published sequences of mt cytochrome *b* are used to: (1) place Cetacea relative to the other extant ungulate orders, (2) assess artiodactyl paraphyly, and (3) test the putative sister-group relationship between Hippopotamidae and Cetacea.

2. Previous Molecular Hypotheses

Cetacea diverged from other extant mammalian orders minimally in the early Eocene (49.0–52.5 Ma; Gingerich *et al.*, 1994). This ancient split precludes the use of some types of molecular data for inferring the relationship of Cetacea to other mammals. For example, allozymes are generally useless at this phylogenetic distance. Five types of molecular data have been used to examine cetacean phylogeny:

1. Immunological reactions. An immune response is initiated by injecting protein from one species into a rabbit, chicken, or other lab animal. The antiserum is collected and then combined with proteins from other species. The strength of the antigen–antibody reaction is thought to be inversely related to the amount of amino acid divergence between the proteins of different species. Various methods for quanti-

fying the strength of reaction offer different levels of precision in immunological analyses (Maxson and Maxson, 1990).

2. DNA–DNA hybridization. Radioactively labeled "single-copy" DNA from one species is hybridized to an excess of "total" DNA from another species. The annealed interspecies DNA molecules, heteroduplexes, are then slowly heated. The amount of DNA liberated with each increase in temperature is measured and compared with scores for homoduplexes, double-stranded DNA from a single individual. The differences between homoduplex and heteroduplex melting curves reflect the amount of divergence between the single-copy DNA of different species. The net result is an overall measure of genetic distance between pairs of species (Werman *et al.*, 1990).

3. Amino acid sequences. The amino acid sequences of orthologous proteins from different species are determined, aligned, and analyzed phylogenetically (e.g., De Jong *et al.*, 1993).

4. DNA sequences. The DNA sequences of orthologous genes from different species are determined, aligned, and analyzed phylogenetically (e.g., Irwin and Arnason, 1994).

5. Retroposons. Retropositional insertions are used as molecular markers (e.g., Shimamura *et al.*, 1997).

Sixty trees from 44 previous molecular systematic studies of Cetacea are summarized in Figs. 2–7. Some of these trees were constructed in an effort to resolve the phylogenetic relations of Cetacea to other mammals (e.g., Boyden and Gemeroy, 1950; Milinkovitch, 1992). However, the majority of trees in Figs. 2–7 were originally used to address other issues. Many of the phylogenies based on amino acid sequences were attempts at resolving higher-level relationships among eutherian orders (e.g., Miyamoto and Goodman, 1986; Wyss *et al.*, 1987; De Jong *et al.*, 1993). Some of the recent DNA studies were undertaken to assess the reliability of different molecules and methods for phylogenetic analysis (e.g., Cao *et al.*, 1994; Allard and Carpenter, 1996; Zardoya and Meyer, 1996) or define relationships within Cetacea (e.g., Milinkovitch *et al.,* 1993; Arnason and Gullberg, 1996). Other workers assumed a close relationship between Artiodactyla and Cetacea and tested the monophyly of Artiodactyla (e.g., Graur and Higgins, 1994; Hasegawa and Adachi, 1996; Shimamura *et al.*, 1997). Finally, some of these reports stress the molecular evolution of a certain gene, gene product, or gene family but are also useful from a phylogenetic perspective (e.g., Baba *et al.*, 1981; Queralt *et al.*, 1995).

The studies summarized in Figs. 2–7 span almost 50 years and vary widely in methods of phylogenetic analysis utilized (phenetic, cladistic, maximum likelihood), taxonomic coverage (from 3 eutherian orders to 18 eutherian orders), and the types of data used (DNA sequences, amino acid sequences, DNA–DNA hybridizations, immunological reactions, retroposons). Some molecules have been analyzed multiple times. For example, over the past 6 years, DNA sequence data from mt cytochrome *b* have been reanalyzed by numerous authors with a bewildering array of phylogenetic results (Irwin *et al.*, 1991; Cao *et al.*, 1994; Graur and Higgins, 1994; Irwin and Arnason, 1994; Milinkovitch *et al.*, 1994, 1995, 1996; Philippe and Douzery, 1994; Honeycutt *et al.*, 1995; Krettek *et al.*, 1995; Adachi and Hasegawa, 1996; Allard and Carpenter, 1996; Arnason and Gullberg, 1996; Gatesy *et al.*, 1996; Hasegawa and Adachi, 1996; Randi *et al.*, 1996; Xu *et al.*, 1996; Zardoya and

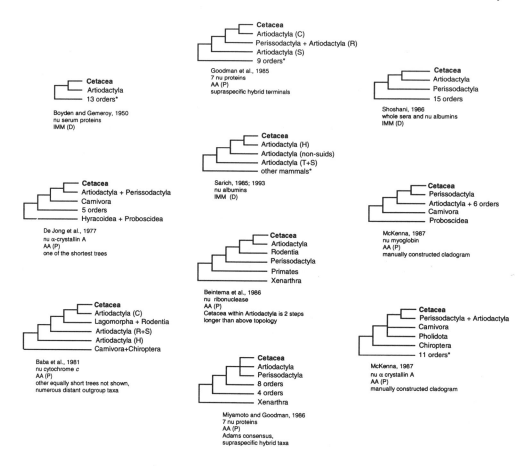

FIGURE 2. A chronological display of some previous molecular trees that relate Cetacea to other eutherian orders (1950–1987). Under each tree, the following information is given: (1) authors and date of publication, (2) genes or proteins analyzed, (3) type of data (DNA = DNA sequences, AA = amino acid sequences, DNA to AA = DNA translated to amino acids, IMM = immunological comparisons, DNA/DNA hybridization, transposons) and type of phylogenetic analysis (P = parsimony/cladistic, ML = maximum likelihood, D = distance/phenetic), (4) comments and notes. In cases where Artiodactyla is not monophyletic, the different lineages of Artiodactyla are noted (R = ruminants, C = camelids, H = hippopotamids, S = suids, T = tayassuids). "Orders" refers to eutherian orders as in Novacek (1989). Outgroups that are not monophyletic are marked with an asterisk. Taxa outside Eutheria are not shown in the summary topologies.

Meyer, 1996; Gatesy, 1997; Montgelard *et al.*, 1997). Similarly, the protein data base has been analyzed and reanalyzed over the past 20 years with the addition of new sequences, the correction of old sequences, and the development of more efficient computer algorithms (Goodman *et al.*, 1985; Miyamoto and Goodman, 1986; Czelusniak *et al.*, 1990; Stanhope *et al.*, 1993).

The 60 topologies in Figs. 2–7 do not represent 60 independent estimates of phylogeny. Regardless, some general patterns are evident from the molecular data base:

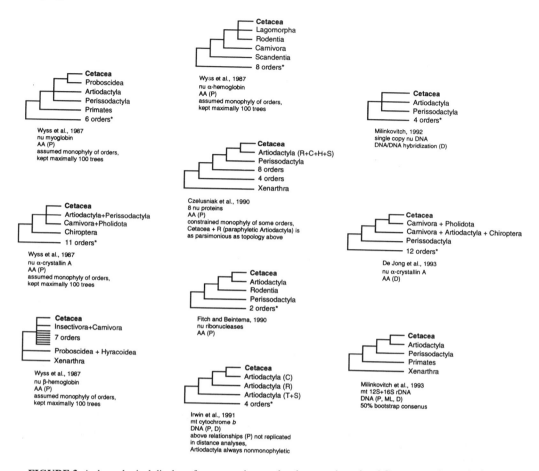

FIGURE 3. A chronological display of some previous molecular trees that related Cetacea to other eutherian orders (1987–1993). Symbols and abbreviations are as in Fig. 2.

1. Artiodactyls are the closest relatives of Cetacea in the majority of trees. No other eutherian order is consistently resolved as the extant sister group of Cetacea. Perissodactyla groups with Cetacea to the exclusion of other mammals twice (Fig. 2, McKenna, 1987; Fig. 4, Stanhope *et al.*, 1993), and Paenungulata is rarely close to Cetacea (but see Fig. 3, Wyss *et al.*, 1987).

The molecular evidence for a Cetacea–Artiodactyla association comes from diverse sources. Immunological comparisons (Boyden and Gemeroy, 1950; Sarich, 1985, 1993; Shoshani, 1986), amino acid sequences (Baba *et al.*, 1981; Goodman *et al.*, 1985; Beintema *et al.*, 1986; Czelusniak *et al.*, 1990; Fitch and Beintema, 1990; Stanhope *et al.*, 1993; Philippe and Douzery, 1994), individual mt genes (Irwin *et al.*, 1991; Springer and Kirsch, 1993; Irwin and Arnason, 1994; Philippe and Douzery, 1994; Honeycutt *et al.*, 1995; Springer *et al.*, 1995; Arnason and Gullberg, 1996; Lavergne *et al.*, 1996; Randi *et al.*, 1996), tandem alignments of mt genes (Milinkovitch *et al.*, 1993; Cao *et al.*, 1994; Krettek

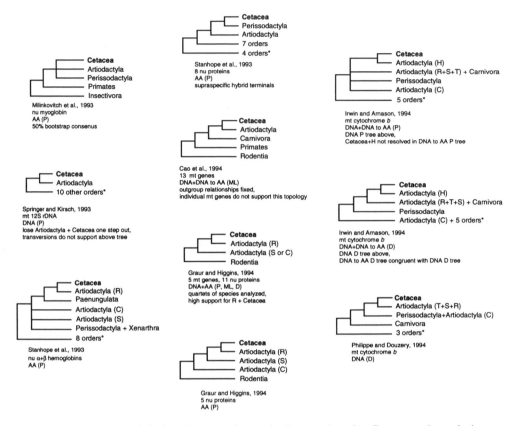

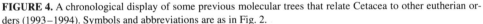

FIGURE 4. A chronological display of some previous molecular trees that relate Cetacea to other eutherian orders (1993–1994). Symbols and abbreviations are as in Fig. 2.

et al., 1995; Xu *et al.*, 1996; Allard and Carpenter, 1996; Zardoya and Meyer, 1996; Montgelard *et al.*, 1997), nu genes (Queralt *et al.*, 1995; Gatesy *et al.*, 1996; Stanhope *et al.*, 1996; Gatesy, 1997), and retroposons (Buntjer *et al.*, 1997) support close phylogenetic ties between cetaceans and artiodactyls.

It should be noted that few of the above studies sampled all six extant ungulate orders (Cetacea, Artiodactyla, Perissodactyla, Sirenia, Proboscidea, and Hyracoidea), and still fewer incorporated data for the majority of extant eutherian orders. One exception is Czelusniak *et al.* (1990; Fig. 3) in which amino acid sequences of eight nu proteins from 16 eutherian orders were strung together in a giant tandem alignment. In this study, Cetacea + Artiodactyla was resolved but not well supported. The clade collapses in cladograms one step longer than the preferred topology of Czelusniak *et al.* (1990). Furthermore, the most recent analysis of these data, with additional sequences and global branch swapping, did not support a Cetacea–Artiodactyla clade (Fig. 4, Stanhope *et al.*, 1993).

Taxonomic representation is also extensive for some immunological data sets. Boyden and Gemeroy (1950; Fig. 2) sampled 15 eutherian orders, and Shoshani (1986; Fig. 2)

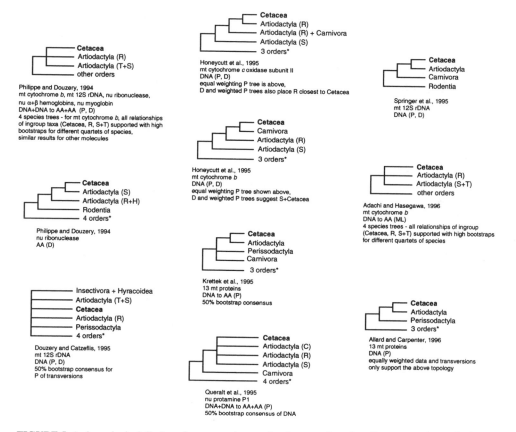

FIGURE 5. A chronological display of some previous molecular trees that relate Cetacea to other eutherian orders (1994–1996). Symbols and abbreviations are as in Fig. 2.

analyzed 18 orders. However, in these studies, all reciprocal pairwise comparisons among taxa were not completed. Even given a complete data matrix, interpretations of these vague, phenetic data are controversial. Shoshani (1986) described immunodiffusion, the mode of analysis used in his study, as "a crude method with which to study phylogeny" (see Sarich, 1993, for an alternative viewpoint).

Perhaps the most convincing evidence for a Cetacea–Artiodactyla association is the interphotoreceptor retinoid binding protein (IRBP) data of Stanhope *et al.* (1996). In analyses of IRBP DNA sequences from 17 mammalian orders, Stanhope *et al.* found overwhelming support for a Cetacea–Artiodactyla clade (Fig. 7). Both cladistic and phenetic topologies show bootstrap scores of 100% for the Cetacea–Artiodactyla node (Stanhope *et al.*, 1996). However, only two artiodactyls and one cetacean were sampled, and the stability of the IRBP trees to more complete taxonomic sampling within orders is not clear.

2. Perissodactyla often groups with Artiodactyla and Cetacea to the exclusion of other ungulates. Amino acid sequences (e.g., Miyamoto and Goodman, 1986), immunology (Shoshani, 1986), and nu DNA sequences (e.g., Stanhope *et al.*, 1996) suggest this grouping.

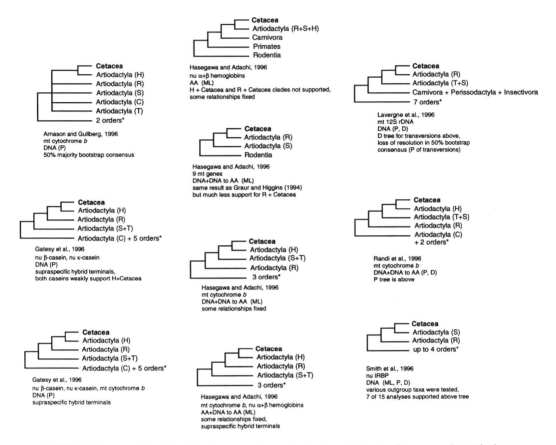

FIGURE 6. A chronological display of some previous molecular trees that relate Cetacea to other eutherian orders (1996). Symbols and abbreviations are as in Fig. 2.

3. A monophyletic Ungulata (Cetacea + Artiodactyla + Perissodactyla + Sirenia + Proboscidea + Hyracoidea) is inconsistent with most molecular evidence (e.g., Miyamoto and Goodman, 1986; Shoshani, 1986; Czelusniak *et al.*, 1990; Stanhope *et al.*, 1993, 1996). Perissodactyla, Artiodactyla, and Cetacea are often separated from Paenungulata (Proboscidea, Sirenia, and Hyracoidea).

4. Artiodactyla is often paraphyletic with Cetacea clustered as an artiodactyl subclade. Baba *et al.* (1981; Fig. 2), Goodman *et al.* (1985; Fig. 2), and Irwin *et al.* (1991; Fig. 3) presented evidence that Artiodactyla is not monophyletic. In 1994, Graur and Higgins stressed the significance of these inferences. They constructed molecular trees for *Bos taurus* (Ruminantia, Artiodactyla), *Sus scrofa* (Suidae, Artiodactyla), a cetacean, and a rodent. *Bos* and the cetacean grouped consistently to the exclusion of *Sus*. This result was solidly supported by both nu and mt data (Fig. 4).

The strength of Graur and Higgins's (1994) study is the large quantity of molecular data examined (Fig. 4), but their analysis has been criticized for its weak taxonomic sam-

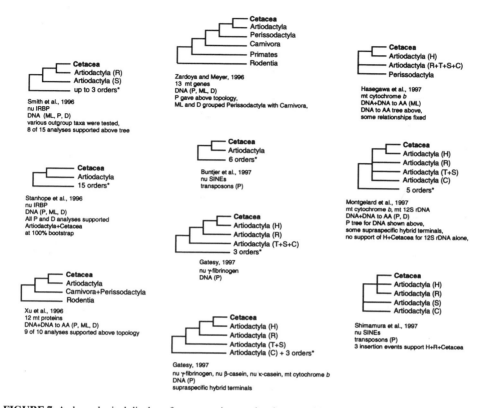

FIGURE 7. A chronological display of some previous molecular trees that relate Cetacea to other eutherian orders (1996–1997). Symbols and abbreviations are as in Fig. 2.

pling. This criticism has taken two forms. First, given the limited taxonomic scope of Graur and Higgins's study and the single distant outgroup, the accuracy of their results has been questioned. Second, if the tree advocated by Graur and Higgins is accurate, and Cetacea is indeed a sublineage of Artiodactyla, the precise position of Cetacea within Artiodactyla is not clear. These authors presented ample molecular evidence from Ruminantia (antelope, deer, and kin) and Suidae (pigs) as well as some data for Camelidae (camels and llamas; see Fig. 4). However, two other major clades of artiodactyls, Tayassuidae (peccaries) and Hippopotamidae (hippos), were excluded. Where do these lineages fit relative to the putative Ruminantia–Cetacea clade?

Subsequent papers addressed the accuracy and stability of Graur and Higgins's hypothesis. Hasegawa and Adachi (1996; Fig. 6) demonstrated that support for the Ruminantia–Cetacea clade is reduced if "more realistic" models of nucleotide substitution are used in phylogenetic reconstruction. Philippe and Douzery (1994; Fig. 5) showed that given the four taxon trees of Graur and Higgins, the choice of taxonomic exemplars has profound implications for phylogenetic results. By sampling different species quartets, Philippe and Douzery recorded high support for each resolution of the ingroup: Ruminantia–Suidae,

Suidae–Cetacea, and Ruminantia–Cetacea. Adachi and Hasegawa (1996; Fig. 5) came to similar conclusions with maximum likelihood analyses of species quartets.

Others remedied some of the deficiencies in taxonomic sampling by collecting new data (e.g., Irwin and Arnason, 1994; Douzery and Catzeflis, 1995; Arnason and Gullberg, 1996; Gatesy *et al.*, 1996; Randi *et al.*, 1996; Smith *et al.*, 1996; Gatesy, 1997; Montgelard *et al.*, 1997; Shimamura *et al.*, 1997). The inclusion of taxa not sampled by Graur and Higgins had a surprising effect. Irwin and Arnason (1994), Gatesy *et al.* (1996), Randi *et al.* (1996), Gatesy (1997), Montgelard *et al.* (1997), Stanhope and Milinkovitch (nu IRBP sequences; personal communication), and Milinkovitch *et al.* (this volume) demonstrated that nu and mt genes support a Hippopotamidae–Cetacea sister-group relationship. Immunological comparisons also suggest that cetaceans "very likely have hippos as their sister group" (Sarich, 1985, 1993), but amino acid sequences do not support this relationship (e.g., Baba *et al.*, 1981; Fitch and Beintema, 1990; Hasegawa and Adachi, 1996).

3. An Analysis of Cetacean Phylogeny Based on Seven Genes

The purpose of the second half of this chapter is to test some of the general phylogenetic patterns seen in the preceding review. New and published DNA sequences from Cetacea, Artiodactyla, and other eutherian mammals are used to:

1. Relate Cetacea to the five other extant ungulate orders
2. Assess the monophyly of Ungulata
3. Test whether Cetacea is a sublineage of Artiodactyla
4. Determine which of the five major extant artiodactyl clades is most closely related to Cetacea (if Cetacea is indeed a sublineage of Artiodactyla)

3.1. Materials and Methods

3.1.1. Data Collection

Seven genes were chosen as phylogenetic probes. These genes are characterized by a wide range of evolutionary properties (Fig. 8) and are hopefully a fair representation of the mt and nu genomes. Nu protein-coding regions, mt protein-coding regions, structural RNAs, introns, 5′ noncoding sequences, and 3′ noncoding sequences compose the battery of DNA sequences in this study. For each DNA segment, methods for PCR amplification, cloning, and sequencing are briefly summarized below.

DNA and tissue samples were acquired from zoos, field workers, and other collaborators. Taxa sequenced for each DNA fragment and published sequences reanalyzed here are listed in Figs. 9–14.

γ-fibrinogen is a blood protein that interacts with the related α and β fibrinogen chains in the blood coagulation process (Rixon *et al.*, 1985). Fragments of γ-fibrinogen that extend from the 3′ end of exon 2 to the 5′ end of exon 4 were PCR amplified, cloned, and sequenced as in Gatesy (1997).

Protamine P1 is a protein involved in the condensation of the nucleus during spermatogenesis (Queralt *et al.*, 1995). The entire protamine P1 gene was amplified as in Quer-

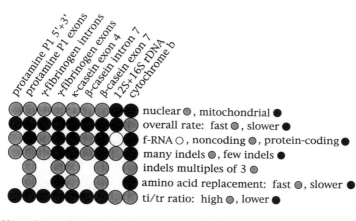

FIGURE 8. DNA regions analyzed in this chapter and a crude characterization of their evolutionary properties (5' + 3' = 5' and 3' noncoding regions; indels = insertions and deletions of nucleotides; f-RNA = functional RNA; ti = transition substitutions; tr = transversion substitutions). Gray, black, and white circles mark the general qualities of each DNA segment.

alt *et al.* (1995), and PCR products were cloned and sequenced as above. Sequences for protamine P1 include 5' and 3' noncoding regions, two exons, and an intron.

12 S and 16 S rRNAs are structural RNAs that make up part of the mt ribosome. A contiguous stretch of 12 S and 16 S mt rDNA was amplified with primers 12SA-850 and 16SB-2860 (Kocher *et al.*, 1989; Palumbi *et al.*, 1991; Gatesy *et al.*, 1997). PCR products were cloned as above and sequenced with primers 12SA-850, 12SB-1270, 16SA-2290, and 16SB-2860 (Kocher *et al.*, 1989; Palumbi *et al.*, 1991; Gatesy *et al.*, 1997).

κ-and β-casein are distantly related nutritional milk protein genes that are tightly linked in *Bos taurus* (Threadgill and Womack, 1990). κ-Casein exon 4 and β-casein exon 7 were PCR amplified, cloned, and sequenced as in Gatesy *et al.* (1996) or PCR amplified and directly sequenced as in Gatesy *et al.* (submitted). β-Casein intron 7 was PCR amplified as in Gatesy *et al.* (submitted) using primers CASBR2INT-5' GCTG-TACCAGGAGCCTGTAC 3' and CASBR3-5' TGAAATCYTCTTAGACCTT 3'. PCR products were cloned and sequenced as above.

Cytochrome *b* is a mt protein that functions in complex III of the oxidative phosphorylation system (Irwin *et al.*, 1991). Published sequences for mt cytochrome *b* were downloaded from NCBI (e.g., Arnason *et al.*, 1991; Irwin *et al.*, 1991; Ma *et al.*, 1993; Irwin and Arnason, 1994; Stanley *et al.*, 1994; Krettek *et al.*, 1995; Arnason and Gullberg, 1996; Ledje and Arnason, 1996; Randi *et al.*, 1996; Xu *et al.*, 1996; Montgelard *et al.*, 1997).

3.1.2. Sequence Alignment

Mitochondrial 12 S + 16 S rDNA sequences were aligned with MALIGN, a multiple sequence alignment program that uses parsimony as the basis for alignment choice (Wheeler and Gladstein, 1994). Two sets of alignment parameters were tested (alignment A: internal gap cost = 4, external gap cost = 3, nucleotide mismatch cost = 2; alignment B: internal gap cost = 3, external gap cost = 2, nucleotide mismatch cost = 1). Unstable alignment

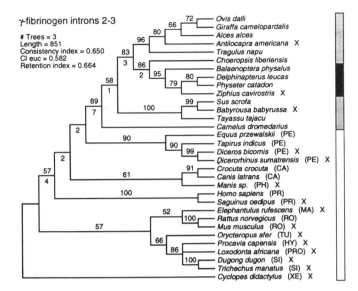

γ-fibrinogen introns 2-3

# Trees = 3
Length = 851
Consistency index = 0.650
CI euc = 0.582
Retention index = 0.664

Artiodactyla monophyly = 7 extra steps

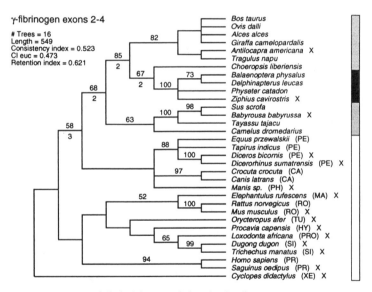

γ-fibrinogen exons 2-4

# Trees = 16
Length = 549
Consistency index = 0.523
CI euc = 0.473
Retention index = 0.621

Artiodactyla monophyly = 4 extra steps

FIGURE 9. Strict consensus trees of minimum length topologies for γ-fibrinogen introns, γ-fibrinogen exons, and γ-fibrinogen exons + introns. The number of minimum length trees (# trees), tree lengths, ensemble consistency indices (Kluge and Farris, 1969), consistency indices excluding uninformative characters (CI euc), retention indices (Farris, 1989), and the number of extra steps required to make Artiodactyla monophyletic are shown. Bootstrap scores greater than 50% are above internal branches, and branch support is below internal branches for nodes that define the relationship of Cetacea to other mammals. Gray bars are to the right of artiodactyl species, black bars are to the right of cetacean species, and white bars are to the right of other eutherians (PE = Perissodactyla, CA = Carnivora, PH = Pholidota, MA = Macroscelidea, RO = Rodentia, TU = Tubulidentata, HY = Hyracoidea, PRO = Proboscidea, SI = Sirenia, PR = Primates, and XE = Xenarthra/Edentata). X's mark taxa sequenced for this report. Branch lengths are not proportional to the number of nucleotide substitutions.

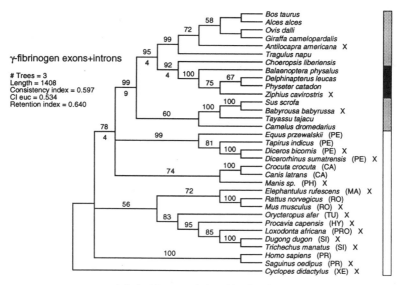

FIGURE 9. (*Continued*).

regions that differed between the two multiple alignments were discarded (Gatesy *et al.*, 1993).

Orthologous sequences for protamine P1, β-casein intron 7, and γ-fibrinogen were initially aligned with MALIGN. Adjustments were made to the algorithmic alignments by eye using SeqApp 1.9a (Gilbert, 1992). These changes were mainly the consolidation of adjacent gaps and substantially decreased the overall cost of each alignment. Intron 1 was excluded from the final alignment of protamine P1 because this region is hypervariable and composed of simple sequence repeats that do not align consistently.

The new sequences for κ-casein exon 4 and β-casein exon 7 were easily incorporated into previously published alignments for these genes (Gatesy *et al.*, 1996). Direct repeats at the 3′ ends of κ-caseins from *Cavia* (Hall, 1990) and *Dolichotis* were aligned, and ambiguity codings for these units were used in phylogenetic analysis as in Gatesy *et al.* (1996). The mt cytochrome *b* sequences were aligned as in Irwin *et al.* (1991). These data are characterized by one three-base-pair insertion in the *Loxodonta* sequence.

The combined DNA data base includes over 250 kilobases of sequence. Final alignments for all new sequences are presented in the Appendix.

3.1.3. Phylogenetic Analysis

The various sequence alignments were analyzed cladistically using PAUP 3.1.1. (Swofford, 1993) and PAUP* 4.0d55 (Swofford, in progress). Gaps were scored as missing data, and natural polymorphisms/PCR errors among clones were coded as ambiguities. All nucleotide substitutions were equally weighted. Within a parsimony framework, this

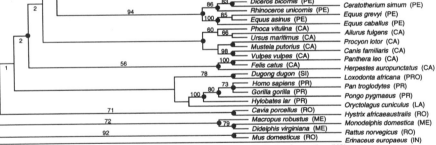

cytochrome *b*

# Trees = 8 (1944)
Length = 9306 (4530)
Consistency index = 0.149 (0.176)
CI euc = 0.141 (0.157)
Retention index = 0.492 (0.640)

Artiodactyla monophyly = 13 extra steps

FIGURE 10. Strict consensus tree of minimum length topologies for mt cytochrome *b*. Tree statistics are as in Fig. 9. In the cytochrome *b* of mammals, transition substitutions at third codon positions accumulate at a rapid rate and may not be especially reliable systematic characters (Irwin *et al.*, 1991; Milinkovitch *et al.*, 1995). Black dots mark nodes that are stable to the removal of these data. Tree statistics for the cytochrome *b* data disregarding transitions at third positions are shown in parentheses to the right of statistics for all of the cytochrome *b* data. Gray bars are to the right of artiodactyl species, the black bar is to the right of cetacean species, and the white bar is to the right of other eutherians (PE = Perissodactyla, CA = Carnivora, PRO = Proboscidea, SI = Sirenia, PR = Primates, LA = Lagomorpha, RO = Rodentia, ME = Metatheria, IN = Insectivora, and MO = Monotremata). Branch lengths are not proportional to the number of nucleotide substitutions.

12S+16S rDNA

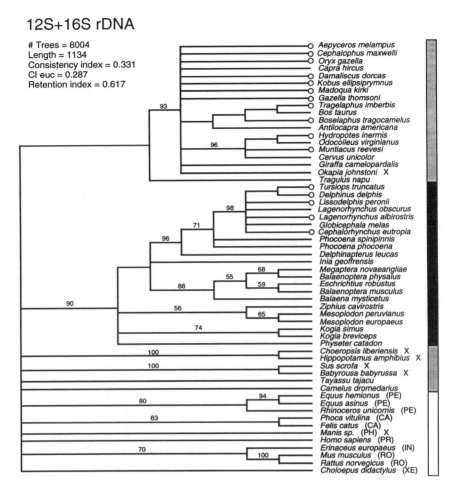

# Trees = 8004
Length = 1134
Consistency index = 0.331
CI euc = 0.287
Retention index = 0.617

Artiodactyla monophyly = 6 extra steps

FIGURE 11. Strict consensus tree of minimum length topologies for 12 S + 16 S mt rDNA. Tree statistics are as in Fig. 9. Circles at the tips of branches mark taxa excluded in bootstrap analyses and estimates of branch support. Gray bars are to the right of artiodactyl species, the black bar is to the right of cetacean species, and the white bar is to the right of other eutherians (PE = Perissodactyla, CA = Carnivora, PH = Pholidota, PR = Primates, IN = Insectivora, RO = Rodentia, and XE = Xenarthra/Edentata). X's mark taxa sequenced for this report. Branch lengths are not proportional to the number of nucleotide substitutions.

weighting scheme favors the best fit between initial and final estimates of homology (Farris, 1983). PAUP searches were branch and bound or heuristic with 10–100 random taxon addition replicates and TBR branch swapping.

The following data sets were analyzed: γ-fibrinogen exons—33 taxa [233 aligned base pairs (a.b.p.)], γ-fibrinogen introns—32 taxa (431 a.b.p.), γ-fibrinogen exons and introns—33 taxa (664 a.b.p.), mt cytochrome *b*—99 taxa (1143 a.b.p.), 12 S + 16 S rDNA—58 taxa (720 a.b.p.), κ-casein exon 4—66 taxa (519 a.b.p.), β-casein exon 7—68 taxa (520 a.b.p.), β-casein intron 7—16 taxa (677 a.b.p.), protamine P1—27 taxa (380 a.b.p.).

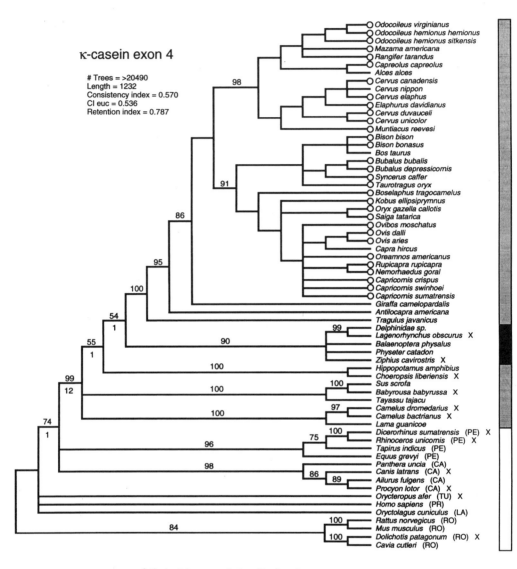

κ-casein exon 4

# Trees = >20490
Length = 1232
Consistency index = 0.570
CI euc = 0.536
Retention index = 0.787

Artiodactyla monophyly = 3 extra steps

FIGURE 12. Strict consensus tree of 20,490 minimum length topologies for κ-casein exon 4. Tree statistics are as in Fig. 9. Circles at the tips of branches mark taxa excluded in bootstrap analyses and estimates of branch support. Gray bars are to the right of artiodactyl species, the black bar is to the right of cetacean species, and the white bar is to the right of other eutherians (PE = Perissodactyla, CA = Carnivora, TU = Tubulidentata, PR = Primates, LA = Lagomorpha, and RO = Rodentia). Branch lengths are not proportional to the number of nucleotide substitutions. X's mark taxa sequenced for this report. The κ-casein sequence for *Camelus dromedarius* is the ambiguity coding for a new sequence and the published sequence, Y10082, of Kappeler and Farah.

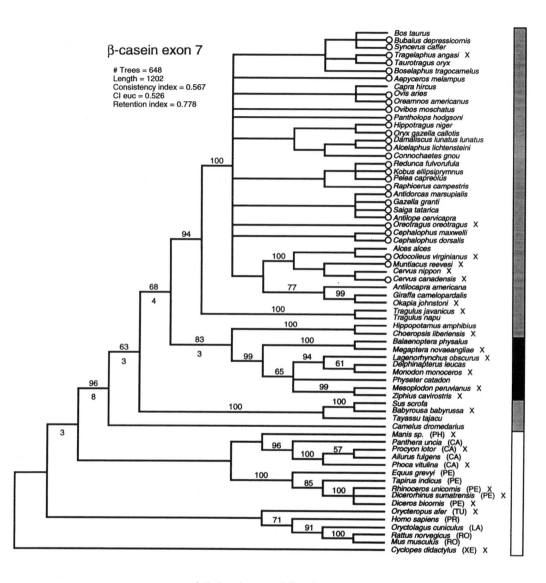

β-casein exon 7

# Trees = 648
Length = 1202
Consistency index = 0.567
CI euc = 0.526
Retention index = 0.778

Artiodactyla monophyly = 9 extra steps

FIGURE 13. Strict consensus tree of minimum length topologies for β-casein exon 7. Tree statistics are as in Fig. 9. Circles at the tips of branches mark taxa excluded in bootstrap analyses and estimates of branch support. Gray bars are to the right of artiodactyl species, the black bar is to the right of cetacean species, and the white bar is to the right of other eutherians (PH = Pholidota, CA = Carnivora, PE = Perissodactyla, TU = Tubulidentata, PR = Primates, LA = Lagomorpha, RO = Rodentia, and XE = Xenarthra/Edentata). X's mark taxa sequenced for this report. Branch lengths are not proportional to the number of nucleotide substitutions.

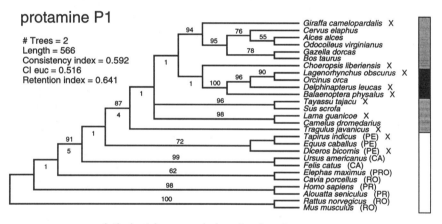

FIGURE 14. Strict consensus trees of minimum length topologies for β-casein intron 7 and protamine P1. Tree statistics are as in Fig. 9. Gray bars are to the right of artiodactyl species, black bars are to the right of cetacean species, and white bars are to the right of other eutherians (PH = Pholidota, PE = Perissodactyla, CA = Carnivora, PRO = Proboscidea, RO = Rodentia, and PR = Primates). X's mark taxa sequenced for this report. Branch lengths are not proportional to the number of nucleotide substitutions.

Depending on taxonomic representations, cladograms were rooted with Monotremata (cytochrome *b*), Edentata (γ-fibrinogen, 12 S + 16 S rDNA, β-casein exon 7), or Rodentia (κ-casein exon 4 and protamine P1). Monotremata is the sister group of Eutheria + Metatheria (Novacek, 1989), and Edentata has been implicated as the extant sister group to other eutherians (Miyamoto and Goodman, 1986). A basal positioning of Rodentia within

Eutheria is more controversial (McKenna, 1975). Therefore, the rootings for κ-casein exon 4 and protamine P1 are tentative. For β-casein intron 7, Pholidota and Perissodactyla were used to root the putative Cetacea–Artiodactyla clade. This assumes that Cetacea + Artiodactyla is a valid grouping.

At least six of the seven genes were sequenced for representatives of 19 taxa (Bovidae, Cervidae, Giraffidae, Tragulidae, Delphinoidea, Physeteridae, Balaenopteridae, Ziphiidae, Hippopotamidae, Camelidae, Suidae, Tayassuidae, Ceratomorpha, Equidae, Caniformia, Feloidea, *Homo, Mus*, and *Rattus*). For these 19 taxa, the different genic regions were analyzed separately and combined as above. Combined data sets were: the three mt genes, the four nu genes, and all seven genes. Species for these analyses are shown in Fig. 15. Because taxonomic sampling was not identical for each DNA region, higher-level terminal taxa were often hybrids of various species; all hybrid terminals were assumed to be monophyletic. Cladograms were rooted with Rodentia or Perissodactyla.

The three combined data sets were also weighted according to character fit with $k = 0$ (Goloboff, 1993). This procedure assigns higher weights to characters with fewer extra steps (i.e., less homoplasy); the constant k determines the concavity of the weighting curve (Goloboff, 1993). Searches were as above using PAUP* 4.0d55 (Swofford, in progress).

The stability of clades in minimum length trees was assessed through branch support (BS) estimates (Bremer, 1988, 1994). BS, the number of additional character transformations necessary to collapse an internal branch, was calculated for selected nodes using the "constraints" command in PAUP with 20–100 random taxon addition replicates and TBR branch swapping. In some analyses, BS was estimated by saving sets of trees that exceed minimum length (50–100 random taxon addition replicates and TBR branch swapping).

Bootstrap percentages (Felsenstein, 1985) were derived as an independent measure of clade stability. Each bootstrap analysis included 200–1000 replications. PAUP searches were heuristic with simple taxon addition and TBR branch swapping. For data sets with many equally parsimonious trees, some species were deleted before bootstrap analyses (see Figs. 11–13).

It is important to note that in all bootstrap analyses, there were missing data from incomplete taxonomic sampling, alternative PCR priming sites for divergent taxa, or insertion/deletion events (Appendix). Given the assumptions of the bootstrap (Felsenstein, 1985; Kluge and Wolf, 1993) and the amount of missing data in some cases, bootstrap percentages in this report should be interpreted cautiously.

Partitioned branch support (PBS) was calculated for the combined analysis of all seven genes as in Baker and DeSalle (1997). PBS summarizes the contribution of each genic region to BS estimates for the total DNA tree. PBS is similar to BS but differs in recognizing hidden support that emerges in combined analysis. For a particular combined data set, a particular node, and a particular data partition, PBS is the minimum number of character steps for that partition on the shortest topologies for the combined data set that do not contain that node, minus the minimum number of character steps for that partition on the shortest topologies for the combined data set that do contain that node. If there are multiple equally short topologies, tree lengths are averaged. For any node, the sum of PBS values for the different genic regions equals the BS value of that node in the combined analysis of all seven genes. Individual values for PBS can be positive or negative. A positive PBS value indicates that, within the combined analysis framework, a given data set provides at least some support for that particular node over the alternative relationships in the shortest tree(s) with-

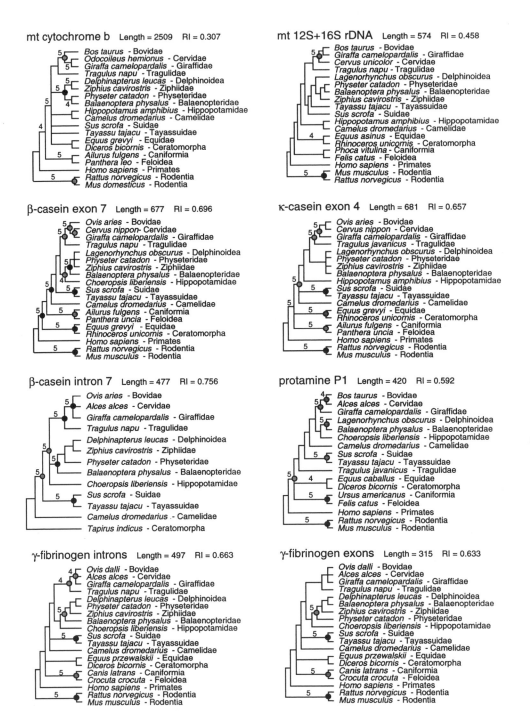

FIGURE 15. Cladograms of taxa sequenced for at least six of the seven genes in this report: minimum length topology for 12 S + 16 S mt rDNA and strict consensus trees of minimum length topologies for mt cytochrome *b*, β-casein exon 7, κ-casein exon 4, β-casein intron 7, protamine P1, γ-fibrinogen introns, and γ-fibrinogen exons. Nodes with branch support of four (4) and greater than or equal to five (5) are indicated. Bootstrap percentages (≥90% = gray dots and 100% = black dots), tree lengths, and ensemble retention indices (RI) are shown.

out the given node. A negative PBS value shows that, within this same combined analysis framework, a data set favors the shortest tree(s) without the given node over the minimum length solution(s) (Baker and DeSalle, 1997).

The incongruence length difference (ILD) test (Farris *et al.*, 1994) was used to assess the null hypothesis of congruence between data sets. This procedure employs the ILD, the number of extra character steps gained by combining data sets in simultaneous analysis (Mickevich and Farris, 1981), as a test statistic. ILDs derived from random partitions of the combined data set determine the extremity of a particular empirical ILD (Farris *et al.*, 1994). The following alignments were tested in pairwise comparisons: cytochrome *b*, 12 S + 16 S rDNA, γ-fibrinogen exons, γ-fibrinogen introns, β-casein exon 7, β-casein intron 7, κ-casein exon 4, and protamine P1. Additionally, nu and mt partitions of the total data set were compared. Species were as in Fig. 15, and all uninformative nucleotide positions were excluded. To establish a null distribution for each pair of data sets, 999 random data partitions were generated, and ILDs were calculated for each replicate with PAUP* 4.0d55 (Swofford, in progress). Searches were heuristic with simple taxon addition and TBR branch swapping.

3.2. Results and Discussion

3.2.1. Phylogenetic Trees

Phylogenetic trees derived from separate and combined analyses of the various sequence alignments are shown in Figs. 9–16. The κ-casein search was terminated at 20,490 equally parsimonious trees (Fig. 12). Most of these topologies account for different equally costly resolutions within Caprinae (Bovidae).

The combined nu DNA and total DNA data sets are for the most part stable to Goloboff weighting at $k = 0$ (Fig. 16). However, within Odontoceti, the number of equally parsimonious topologies is reduced for the nu partition and relationships are rearranged for the total DNA data set. The weighted nu and total DNA data sets each support ((Delphinoidea + Physeteridae) Ziphiidae).

For the combined mt DNA data set, few groupings within Cetacea + Artiodactyla are stable to Goloboff weighting (Fig. 16). Contrary to the equally weighted analysis of the mt data, Hippopotamidae is resolved as the sister group of Cetacea, and within Cetacea, (((Ziphiidae + Balaenopteridae) Physeteridae) Delphinoidea) is favored.

3.2.2. Incongruence among Data Sets

There is no consensus on whether to combine or separate heterogeneous data partitions in phylogenetic analysis. The various sides of this debate are divided on whether characters (e.g., Carpenter, 1992; Nixon and Carpenter, 1996), sets of characters (Miyamoto and Fitch, 1995), or generalized models of evolution (e.g., Bull *et al.*, 1993; Huelsenbeck and Bull, 1996) are of primary importance in phylogenetic analysis.

Nixon and Carpenter (1996) argued that "homogeneity among characters or sets of characters (partitions) is neither the null hypothesis nor an assumption of parsimony analysis of any dataset" and pointed out that simultaneous analysis of all data partitions maxi-

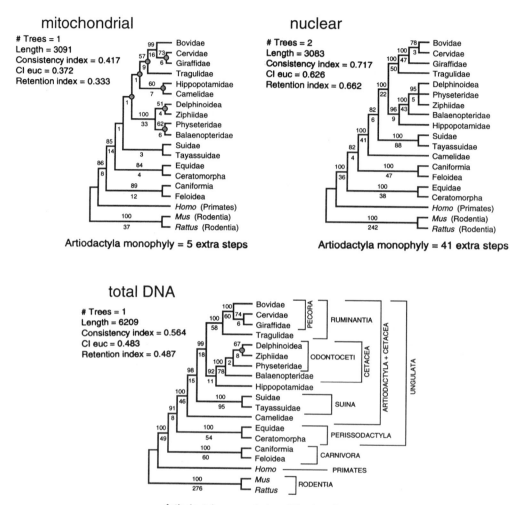

FIGURE 16. Cladograms of taxa sequenced for minimally six of the seven genes in this report: topologies for the three mt genes, the four nu genes, and all seven genes. Species for each genic region are as in Fig. 15. The number of minimum length trees (# trees), tree lengths, consistency indices, consistency indices excluding uninformative characters (CI euc), and retention indices are shown. The number of extra steps required to make Artiodactyla monophyletic is below each topology. Bootstrap scores greater than 50% are above internodes and branch support is shown below internodes. Nodes that are not stable to Goloboff weighting at $k = 0$ are marked by gray dots. Higher-level taxa are delimited by parentheses to the right of the total DNA topology. The numbers of extra nucleotide substitutions required to fit individual DNA data sets to the total DNA topology are: cytochrome b = 16, 12 S + 16 S rDNA = 20, β-casein exon 7 = 1, β-casein intron 7 = 0, κ-casein exon 4 = 4, γ-fibrinogen introns = 2, γ-fibrinogen exons = 4, protamine P1 = 12.

mizes explanatory power and allows secondary phylogenetic signals to emerge (see also Miyamoto, 1985, and Kluge, 1989, for similar viewpoints). In contrast, Bull *et al.* (1993) suggested that heterogeneous data sets should be separated. They noted that "a combined analysis of potentially diverse data is inappropriate unless it is shown that the different data sets are not significantly heterogeneous with respect to the reconstruction model." Miyamoto and Fitch (1995) stressed the importance of taxonomic congruence among topologies derived from independently evolving, unlinked, genetic loci.

For the DNA sequences in the present analysis, there is significant character incongruence between some data partitions. A division of the data into mt and nu subsets results in an extreme ILD (ILD = 35, p = 0.001). p values for other ILD tests are shown in Fig. 17. In comparisons of κ-casein exon 4 with each of the seven other data sets, ILDs are not significant. For all other genic regions, ILDs are significant in at least one pairwise comparison. Protamine P1 and mt 12 S + 16 S rDNA are incongruent with four and five of the other data sets, respectively. Because the ILD test results are not transitive, decisive boundaries between "congruent" and "incongruent" character sets do not exist. For example, protamine P1 is congruent with cytochrome *b*, which is congruent with γ-fibrinogen introns, which in turn are incongruent with protamine P1 (Fig. 17).

Patterns and rates of nucleotide substitution vary widely among the various genes in this report (Figs. 8 and 18). These differences in evolutionary tempo and mode may account for some of the character incongruence between data sets (Fig. 17). Clearly, some DNA regions are more appropriate for resolving cetacean phylogeny as evidenced by differences in BS and PBS values (Figs. 15, 16, and 19). For example, on the total DNA tree (Fig. 16), there are minimally 3119 nucleotide substitutions for the mt data and 3090 nucleotide substitutions for the nu data. However, total PBS for the nu data is 694.75 while total PBS for the mt partition is only 149.25 (Fig. 19). The mt data are not surprisingly characterized by low ensemble consistency and retention indices relative to the nu data (Figs. 15 and 16).

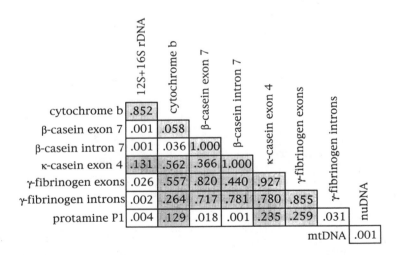

FIGURE 17. Results of incongruence length difference (ILD) tests between various data sets. p values are shown for each test. White boxes mark significant results at $p \leq 0.05$.

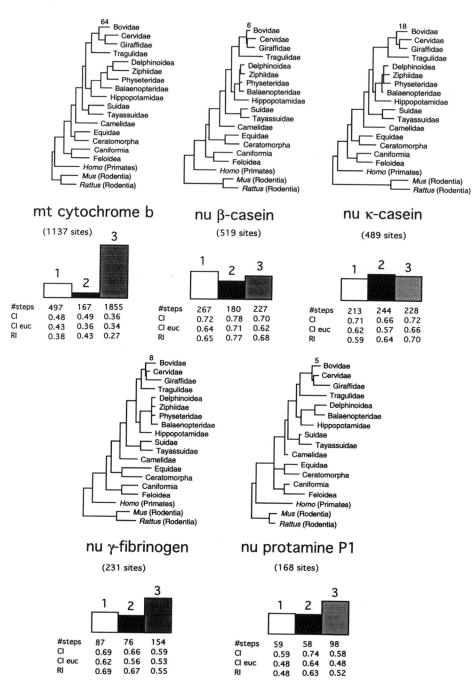

FIGURE 18. The percentage of nucleotide substitutions (vertical bars) at each of the three codon positions in cytochrome *b*, β-casein exon 7, κ-casein exon 4, γ-fibrinogen exons 2–4, and protamine P1 exons 1 and 2 for the taxa in Fig. 15. Using PAUP 3.11, the minimum number of nucleotide substitutions at each position (#steps) was determined for the total DNA topology (Fig. 16). The ensemble consistency index (CI), consistency index ex-

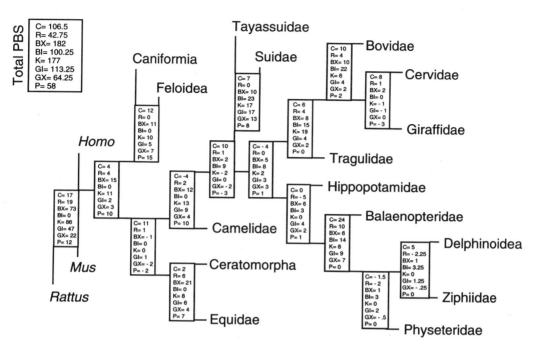

FIGURE 19. Partitioned branch support (PBS) for eight DNA regions on the total DNA topology (Fig. 16). Taxa are as in Fig. 15. C = cytochrome *b*, R = 12 S + 16 S rDNA, BX = β-casein exon 7, BI = β-casein intron 7, K = κ-casein exon 4, GI = γ-fibrinogen introns 2–3, GX = γ-fibrinogen exons 2–4, P = protamine P1. Some PBS values of zero are the result of limited taxonomic sampling for protamine P1 and β-casein intron 7 (see Fig. 15). The sum of PBS values for each gene, total PBS, is shown.

Twenty-eight extra steps are required to fit the mt data to the total DNA topology, whereas only seven extra steps are necessary for the nu data. Furthermore, the mt DNA data are not internally consistent. Goloboff weighting of the combined mtDNA data set radically re-arranges topological relationships within Artiodactyla + Cetacea (Fig. 16).

I have provisionally chosen the combined analysis of all seven genes (Fig. 16) as a best working hypothesis and use this topology to assess character incongruence and the support rendered by different data sets. The total DNA cladogram incorporates information from both the mt and nu genomes, and can be defended as a summary of the present DNA data base. Character weighting schemes that improve congruence among DNA regions will be

---

cluding uninformative positions (CI euc), and retention index (RI) for each codon position are shown. The over-all average rate of nucleotide substitution is greatest for mt cytochrome *b*, but much of this change is restricted to the rapidly evolving third codon positions. CI, CI euc, and RI values are low even for "conservative" second codon positions of cytochrome *b*. Branches in phylograms are proportional to the number of nucleotide substitutions under AccTrans optimization, and the lengths of branches that lead to Bovidae are indicated. The number of nucleotide sites for each gene is in parentheses. Note the disparities in rates of nucleotide substitution between some sister taxa.

incorporated in future studies of these data. These analyses will test the stability of the general parsimony framework presented here.

3.2.3. The Relationship of Cetacea to Other Extant Ungulate Orders

The γ-fibrinogen data set is composed of sequences from 13 eutherian orders. These data clearly support a Cetacea–Artiodactyla clade (BS = 9, Fig. 9). Evidence for this grouping comes from both the protein-coding exons (BS = 2) and noncoding introns (BS = 7) of γ-fibrinogen (Fig. 9). The other nu data sets do not have the broad taxonomic coverage of the γ-fibrinogen data set but do support a Cetacea + Artiodactyla clade with high to moderate support (BS from 4 to 13, Figs. 12–14). The mt cytochrome *b* data also weakly suggest this grouping (BS = 3, Fig. 10) as does the combined mtDNA data set (BS = 1, Fig. 16). Character support for Artiodactyla + Cetacea is extensive in the combined nu DNA tree (BS = 41, Fig. 16) and in the total DNA tree (BS = 46, Fig. 16).

The γ-fibrinogen data suggest that Perissodactyla is the next closest ungulate relative of Cetacea + Artiodactyla (Fig. 9). As in many previous molecular studies, hoofed mammals are diphyletic. Cetacea, Artiodactyla, and Perissodactyla are aligned with Carnivora + Pholidota, and Paenungulata is linked to Tubulidentata (Fig. 9). Protamine P1 and mt cytochrome *b* also favor a close relationship between Perissodactyla and Artiodactyla + Cetacea (Figs. 10 and 14). In cladograms derived from these genes, Cetacea, Artiodactyla, and Perissodactyla cluster closer to carnivorans than to paenungulate exemplars (*Loxodonta*, *Elephas*, and *Dugong*, Figs. 10 and 14).

Accepting that *Orycteropus* (Tubulidentata) is a member of Paenungulata (Miyamoto and Goodman, 1986; Shoshani, 1986; Czelusniak *et al.*, 1990; Porter *et al.*, 1996; Stanhope *et al.*, 1996), β-casein exon 7 and κ-casein exon 4 also support a diphyletic Ungulata. In both casein trees, Artiodactyla, Cetacea, and Perissodactyla join with Carnivora (+Pholidota) to the exclusion of Tubulidentata and presumably other paenungulates (Figs. 12 and 13).

3.2.4. Artiodactyl Paraphyly

In this large sample of DNA sequences from artiodactyls, there is little support for a monophyletic Artiodactyla. None of the 12 DNA data sets resolve this clade (Figs. 9–14 and 16). For the total DNA topology, 55 additional nucleotide substitutions are required to make Artiodactyla monophyletic (Fig. 16).

Most analyses place Cetacea within "Artiodactyla." However, the artiodactyl lineage that is closest to Cetacea is not identical in topologies derived from different data sets (Figs. 9–14). γ-Fibrinogen, mt cytochrome *b*, protamine P1, and β-casein exon 7 each weakly support a Cetacea + Hippopotamidae sister-group relationship. BS ranges from one to five for this clade. κ-casein exon 4 places Cetacea closest to Ruminantia with Hippopotamidae further out, and β-casein intron 7 aligns Cetacea with Ruminantia and Hippopotamidae in an unresolved trichotomy. The consensus tree for mt 12 S + 16 S rDNA is poorly resolved. No groupings of artiodactyls with cetaceans are well supported by the combined mtDNA data set (Fig. 16), but Cetacea + Hippopotamidae is favored when the mtDNA are weighted by character fit (Goloboff, 1993).

The branching order at the base of Artiodactyla + Cetacea is identical in the combined nu DNA tree and the total DNA tree (Fig. 16). Three critical clades are supported at various levels (nu BS, total DNA BS): Hippopotamidae + Cetacea (9, 11), Hippopotamidae + Cetacea + Ruminantia (22, 18), and Hippopotamidae + Cetacea + Ruminantia + Suina (6, 15). These groupings are consistent with the independent results of Graur and Higgins (1994) for five nu proteins (Fig. 4) as well as previous combined analyses of mt and nu DNA sequences (Gatesy *et al.*, 1996; Gatesy, 1997; Figs. 6 and 7).

BS is substantial in the total DNA topology (Fig. 16), but some genes have negative PBS values at various nodes (Fig. 19). For example, Artiodactyla + Cetacea + Perissodactyla is supported by only three of the eight DNA data sets. Evidence from mt cytochrome *b* accounts for almost all of the support at this node. PBS is negative for β-casein exon 7, γ-fibrinogen exons, and protamine P1, and PBS is zero for κ-casein exon 4 and β-casein intron 7 (Fig. 19). Similarly, in the combined analysis, controversial nodes within Artiodactyla + Cetacea are not generally supported by all eight DNA regions. For the Cetacea–Hippopotamidae clade, there is at least some positive PBS in five nu data sets. Ruminantia + Hippopotamidae + Cetacea has positive PBS values for all six nu data partitions, and Ruminantia + Hippopotamidae + Cetacea + Suina has positive PBS values for the mt data sets and nu β-casein (Fig. 19).

The weight of DNA evidence favors a Hippopotamidae + Cetacea clade, a Hippopotamidae + Cetacea + Ruminantia clade, and a Hippopotamidae + Cetacea + Ruminantia + Suina clade with Camelidae more distantly related. But, as mentioned previously, support is not partitioned equally among different DNA regions (Fig. 19), ILD tests are significant for several comparisons between data partitions (Fig. 17), and the shortest topologies for individual genes are not fully concordant (Figs. 9–15).

Despite the incongruence among trees derived from different data sets, only one node with BS ≥ 4 directly contradicts nodes with BS ≥ 4 in other trees (Fig. 15). When the individual molecular topologies do not match, specific disagreements usually are not solidly supported. Furthermore, in these trees (Fig. 15), 52 of 54 components with BS ≥ 4 are consistent with the total DNA topology (Fig. 16). Analyses that substitute bootstrap scores for BS as a measure of clade stability give similar results. Only one node supported by greater than 90% of bootstrap replicates, the Bovidae–Cervidae clade in the protamine P1 topology (Fig. 15), is inconsistent with the total DNA tree (Fig. 16).

Much of the topological incongruence among different published molecular trees (Figs. 2–7) is probably of the type described above. Weakly supported, spurious clades are likely to disagree with spurious clades and well-supported clades in other trees. This effect is exaggerated by the limited number of phylogenetically informative sites in many molecular data sets (see Wyss *et al.*, 1987). For example, amino acid sequences of α-crystallin A are identical in the following pairs of taxa: hippo/giraffe, elephant shrew/hyrax, rat/tree shrew, and dog/cat (De Jong *et al.*, 1993). Given a limited amount of sequence variation and extensive taxonomic sampling, spurious clades are bound to be resolved by parsimony analysis. These weakly supported, inconsistent clades should not be overinterpreted.

3.2.5. Morphological Evidence versus Molecular Evidence

The total DNA topology (Fig. 16) does not match current morphological estimates of cetacean phylogeny (Fig. 1A–C). Skeletal and dental characters suggest Cetacea is most closely related to extinct mesonychian ungulates with a monophyletic Artiodactyla more distantly related. The combined molecular data set places Cetacea as a subclade of a paraphyletic Artiodactyla. Fifty-five additional nucleotide substitutions are required to make Artiodactyla monophyletic, and 11 extra steps are necessary to break up the Cetacea–Hippopotamidae clade (Fig. 16).

From the morphological perspective, Cetacea + Mesonychia (= Cete) is a well-supported clade. Thewissen (1994) and Zhou *et al.* (1995) described the anatomical features that link Cetacea to successive clades of extinct "mesonychians." For the data matrix of Thewissen (1994), six extra character transformations are required to remove Cetacea from within Cete and group Cetacea with Artiodactyla.

Prothero *et al.* (1988), Gentry and Hooker (1988), and Prothero (1993) offered the following synapomorphies for Artiodactyla: a trochleated distal astragalus, a partial double mesocylix in distal deciduous premolars, an enlarged facial portion of the lacrimal, an expanded orbitosphenoid, narrow lower molar trigonids, loss of the alisphenoid canal, and a paraxonic foot in which the central axis runs between the third and fourth digits. Among these, narrow trigonids, enlarged lacrimals, absence of alisphenoid canal, and a paraxonic foot are also characteristic of the earliest cetaceans (Gingerich *et al.*, 1990; Thewissen, 1994; Thewissen *et al.*, 1996; Thewissen and Hussain, 1998).

The classic artiodactyl ordinal character, a trochleated distal astragalus (Schaeffer, 1948), has not been observed in cetaceans. However, the search for a cetacean astragalus has been frustrating. This bone is fused to the other tarsals in the vestigial hind limb of *Basilosaurus* (Gingerich *et al.*, 1990), the distal end of the astragalus is missing in *Ambulocetus* (Thewissen, 1994; Thewissen *et al.*, 1994), the astragalus is unknown from all other extinct whales, and it has been lost along with most of the hind limb in all extant whales. Because the entire tarsus has been lost within Cetacea, it is illogical to assume that a trochleated distal astragalus could not have been lost within "Artiodactyla."

To my knowledge, there is no morphological character matrix coded for each of the primary clades of Artiodactyla, as well as Cetacea, Perissodactyla, and outgroups. So, the relative absurdity of the molecular results is not clear. In terms of morphological change, how costly is a Cetacea + Hippopotamidae grouping or a Cetacea + Hippopotamidae + Ruminantia clade? At this juncture, the exact magnitude of these costs is unknown.

However, even ignoring the odd placement of Cetacea, the molecular data contradict morphological evidence. Within Artiodactyla, the DNA data strongly support the following arrangement ((((Pecora + Tragulidae) Hippopotamidae) (Suidae + Tayassuidae)) Camelidae). In contrast, the morphological data matrix of Gentry and Hooker (1988) favors (((Leptomerycidae + Tragulidae) Camelidae) ((Suidae + Tayassuidae) Hippopotamidae)). If the extinct Leptomerycidae is equated with Pecora, forcing the morphological characters onto the molecular topology requires ten additional character transformations.

Some morphological and behavioral characters do support a Cetacea + Hippopotamidae sister-group relationship. For example, both whales and hippos are nearly hairless, lack scrotal testes, communicate underwater, and have lost sebaceous glands (Gatesy, 1997, and references within). These and other aquatic specializations could be interpreted as adaptive convergences and dismissed as phylogenetic evidence. However, it would be difficult to argue that dental similarities between early cetaceans and mesonychians are any less adaptive than aquatic features of cetaceans and hippos. None of these characters should be ignored *a priori*. Perhaps exhaustive morphological comparisons of extant cetaceans and hippopotamids will reveal additional, more convincing, anatomical similarities between these aquatic taxa.

4. Conclusions

4.1. Summary

1. Past molecular hypotheses of cetacean phylogeny are diverse, but four common features of these studies stand out:
   A. Cetacea often groups with Artiodactyla.
   B. Perissodactyla is closely related to Artiodactyla + Cetacea.
   C. Ungulata is often diphyletic.
   D. Artiodactyla is often paraphyletic.
2. These four common features of previous molecular studies are supported by phylogenetic analyses of DNA sequences from four nu and three mt genes. A cladogram based on over 4500 aligned nucleotide positions suggests Cetacea is nested three nodes within a paraphyletic Artiodactyla. In sum, the DNA evidence supports a sister-group relationship between Cetacea and Hippopotamidae.
3. Trees derived from subsets of the total DNA data base are contradictory, and there are several examples of significant character incongruence among data partitions. However, specific topological discrepancies are not well supported in most instances.

4.2. Prospectus

Presently, there are several points of disagreement between morphology and molecules. These conflicts must be explored in tests of character congruence and subsequent reassessments of problematic characters. For the molecular evidence, several factors potentially encourage misleading results and need to be thoroughly examined in future studies. These include:

1. Ambiguities in sequence alignment (Appendix) and the treatment of gaps as characters in phylogenetic analysis
2. Nucleotide base composition biases (e.g., Irwin *et al.*, 1991)
3. Consistent rate differences among evolutionary lineages (Fig. 18)
4. Differences in evolutionary rates among genes and codons (Fig. 18)
5. Uneven and incomplete taxonomic coverage for different genes (Figs. 9–14)
6. The assumed monophyly of supraspecific hybrid taxa (Fig. 16)
7. Undetected gene duplications (e.g., Goodman *et al.*, 1985)

The above complications can be addressed through differential character weighting (e.g., Goloboff, 1993; Allard and Carpenter, 1996; Milinkovitch *et al.*, 1996), the exclusion of problematic characters or data sets (e.g., Bull *et al.*, 1993; Gatesy *et al.*, 1993), and more thorough taxonomic sampling (e.g., Philippe and Douzery, 1994).

From the morphological perspective, explicit quantification of the evidence for artiodactyl monophyly is necessary. When the morphological data are clearly characterized, molecular hypotheses can be criticized according to general criteria for homology (Patterson, 1982; DePinna, 1991).

Acknowledgments

O. Ryder, K. Hecker, M. Stanhope, M. Milinkovitch, P. Palbersol, K. Pederson, P. Arctander, G. Amato, E. Avery Stevens, M. Cronin, L. Bischoff, and P. Vrana provided DNA samples that were critical for this study. M. Allard and D. Irwin provided aligned mtDNA sequences on disk. C. Hayashi, P. O'Grady, A. Berta, S. Gatesy, M. Whiting, M. Allard, H. Thewissen, A. de Querioz, and an anonymous reviewer commented on various stages of this manuscript. R. Lewis donated his computer for several weeks. D. Swofford allowed use of PAUP*. C. Hayashi helped in entering data not stored in GenBank. The University of Arizona automated sequencing facility and LMSE significantly aided in data production. This work was supported by NSF grant DEB-9509551.

Appendix: Final Alignments for γ-Fibrinogen, 12 S rDNA, 16 S rDNA, κ-Casein Exon 4, β-Casein Exon 7, β-Casein Intron 7, and Protamine P1

Periods represent nucleotide identity to the reference sequences at the top of each alignment. Dashes are gaps introduced into the alignments. IUPAC ambiguities are natural polymorphism or TA cloning/PCR errors (also, E = A or gap, I = G or gap, J = T or gap). In the γ-fibrinogen alignment, ** mark putative intron–exon boundaries. In the 12 S + 16 S rDNA alignment, −− for all taxa show where ambiguous alignment regions were deleted; numbers above −− indicate the position of excluded nucleotides in the mtDNA sequence for *Bos taurus* (Anderson *et al.*, 1982). *** mark where intron 1 (putative) was deleted from protamine P1 sequences. Start and stop codons are also shown in the protamine P1 alignment. H's above the alignments indicate nucleotide positions that unambiguously support the Cetacea–Hippopotamidae clade in the total DNA topology (Fig. 16). If additional taxa are considered, these positions do not necessarily support the hippo–whale clade.

γ-fibrinogen

Alces alces
Bos taurus
Ovis dalli
Giraffa camelopardalis
Antilocapra americana
Tragulus napu
Choeropsis liberiensis
Balaenoptera physalus
Delphinapterus leucas
Physeter catodon
Ziphius cavirostris
Sus scrofa
Babyrousa babyrussa
Tayassu tajacu
Camelus dromedarius
Equus przewalskii
Tapirus indicus
Diceros bicornis
Dicerorhinus sumatrensis
Crocuta crocuta
Canis latrans
Manis sp.
Elephantulus rufescens
Orycteropus afer
Procavia capensis
Loxodonta africana
Dugong dugon
Trichechus manatus
Homo sapiens
Saguinus oedipus
Rattus norvegicus
Mus musculus
Cyclopes didactylus

γ-fibrinogen (continued)

Alces alces
Bos taurus
Ovis dalli
Giraffa camelopardalis
Antilocapra americana
Tragulus napu
Choeropsis liberiensis
Balaenoptera physalus
Delphinapterus leucas
Physeter catadon
Ziphius cavirostris
Sus scrofa
Babyrousa babyrussa
Tayassu tajacu
Camelus dromedarius
Equus przewalskii
Tapirus indicus
Diceros bicornis
Dicerorhinus sumatrensis
Crocuta crocuta
Canis latrans
Manis sp.
Elephantulus rufescens
Orycteropus afer
Procavia capensis
Loxodonta africana
Dugong dugon
Trichechus manatus
Homo sapiens
Saguinus oedipus
Rattus norvegicus
Mus musculus
Cyclopes didactylus

12S+16S rDNA

888-898   965-970   1018-1034   1061-1075   1086-1094

12S begin
GCTCAGCCCTAAACACAAAT--CAAGATTATTCGGCCAGAGTACTACCGGCACGGCCCAAAAACTCAAAGACTTGGCGGTGCTTTATA--TAGAGGAGCCGTTCTATAATCGATAAACCCGATAAACCTCACCAT--TATATACCGCCATCTTCAGCAAACCC--GTAAGCCACAA--ATAAAAACGTTAGGTCAAGGT

| Aepyceros melampus |
| Cephalophus maxwelli |
| Oryx gazella |
| Capra hircus |
| Damaliscus dorcas |
| Kobus ellipsiprymnus |
| Madoqua kirki |
| Gazella thomsoni |
| Tragelaphus imberbis |
| Bos taurus |
| Boselaphus tragocamelus |
| Hydropotes inermis |
| Odocoileus virginianus |
| Muntiacus reevesi |
| Cervus unicolor |
| Antilocapra americana |
| Giraffa camelopardalis |
| Okapia johnstoni |
| Tragulus napu |
| Tursiops truncatus |
| Lissodelphis peroni |
| Lagenorhynchus obscurus |
| Lagenorhynchus albirostris |
| Globicephala melas |
| Delphinus delphis |
| Cephalorhynchus eutropia |
| Phocoena spinipinnis |
| Phocoena phocoena |
| Delphinapterus leucas |
| Physeter catodon |
| Kogia simus |
| Kogia breviceps |
| Ziphius cavirostris |
| Mesoplodon peruvianus |
| Mesoplodon europaeus |
| Inia geoffrensis |
| Megaptera novaeangliae |
| Eschrichtius robustus |
| Balaenoptera physalus |
| Balaenoptera musculus |
| Balaenoptera mysticetus |
| Choeropsis liberiensis |
| Hippopotamus amphibius |
| Sus scrofa |
| Babyrousa babyrussa |
| Tayassu tajacu |
| Camelus dromedarius |
| Equus asinus |
| Rhinoceros unicornis |
| Phoca vitulina |
| Felis catus |
| Manis sp. |
| Homo sapiens |
| Erinaceus europaeus |
| Mus musculus |
| Rattus norvegicus |
| Choloepus didactylus |

12S+16S rDNA (continued)

```
                                                                       1153-1201          1220-1221                           12S end  16S begin  2323-2332            2351-2355
                              GTAACCTATGGAATGGAAAGAAATGGGCTACATTTTC--GGAGGATTTAGTAGTAAA--AAGAATAGAGTGCTTAGTTGAATTAGGCCATGAAG--CAGCATTTCCA--CACTGCCTGCCCAGTGAC--TTAAACGGCCGCGGTATCCTGACCGTGCAAAGGTAGCATAATCATTTGTTCTCTAAATAAGGACTTGTATG
Aepyceros melampus
Cephalophus maxwelli          ..........A.....G....................--..................--...................................--......CCT..--..................--.......................................................................
Oryx gazella                  .....................................--.................G--.................C.....C...........--.......C...--..................--.......................................................................
Capra hircus                  ......C.........G....................--..................--...................................--...........--..................--.................T.....................................................
Damaliscus dorcas             ......C.........G....................--..................--G.A...........C....................--......CC...--..................--.................T.....................................................
Kobus ellipsiprymnus          ...G............G....................--..................--.......................C...........--......C..T.--..................--...................................................C.......G...........
Madoqua kirki                 ................G....................--..................--...................................--.......C...--..................--.................T..................................T..................
Gazella thomsoni              .....................................--..................--.......................C...........--.......C...--..................--....................................................T..................
Tragelaphus imberbis          ..........A.....G....................--...........C......--.....C................CC...........--......CA...--..................--..T....................................................................
Bos taurus                    ..........A.....G.................C..--...........C......--...................................--.......C...--..T...............--..T....................................................................
Boselaphus tragocamelus       ..........A..........................--...........C......--.G....................C............--......CA...--..................--...........................................................G...........
Hydropotes inermis            ............G...G....................--...........C......--......................C.....A......--......AA.T.--..................--............................................C....................A.....
Odocoileus virginianus        ............G........................--...........C......--...................................--......AA.T.--..T...............--.......-...............................................................
Muntiacus reevesi             ...........G....G....................--...........C......--...................................--T......A.T.--..................--............................................C..........................
Cervus unicolor               .............C.......................--...........C......--...................................--......AA.T.--..................--............................................C..............G...........
Antilocapra americana         ...........GT........................--..................--..........C.................T......--......AC...--..................--..C................................................................C...
Giraffa camelopardalis        ................G....................--..................--...................................--.......C-T.--..................--.......-....................................C..........................
Okapia johnstoni              ................G....................--..................--................................G..--.......A.T.--..................--............................................C..........................
Tragulus napu                 ...G..A....GG...G....................--.................G--.G...............C....C.....T......--T......A.T.--............G.....--.................T.....................................................
Tursiops truncatus            ............T...G....................--...........C......--.........A......A......A......-------------.-AC-T.--..................--..................T.........................C..........................
Lissodelphis peronii          ............T...G....................--...........C......--.........A......A......A......-------------.-AC-T.--..................--..................T.........................C..........................
Lagenorhynchus obscurus       ............T...G....................--...........C......--.........A......A......A......--------------------..................--..................T.........................C..........................
Lagenorhynchus albirostris    .........A..T...G....................--...........C......--.........A......A.....AA.....--------------.-NC-..--..................--..................T.........................C..........................
Globicephala melas            ............T...G....................--...........C......--.........A......A......A......-------------.-AC-T.--..................--..................T.........................C..........................
Delphinus delphis             ............T...G....................--...........C......--.........A......A......G......-------------.-AC-T.--..................--..................T.........................C..........................
Cephalorhynchus eutropia      ............T...G....................--...........C......--.........A......A......A......-------------------.--..................--..................T.........................C..........................
Phocoena spinipinnis          ......C....GT...G....................--...........C......--.........A......A......A......-------------AG...--..................--..................T.........................C..........................
Phocoena phocoena             ...........GT...G....................--...........C......--.........A......A......A......-------------AG...--..................--..................T.........................C..........................
Delphinapterus leucas         ...........GT...G....................--...........C......--................A......A.....--------------AA.T.--..................--..................T.........................C.................G........
Physeter catadon              ......C....G....G....................--...........C......--.....C..........A......A......-------------AA...--..................--............................................C..........................
Kogia simus                   .....TC....GC...G....................--..................--....G...........A......A......-------------AA.T.--..................--............................................C..........................
Kogia breviceps               .....TC....GG...G....................--..................--................A......A......-------------AA...--..................--............................................C..........................
Ziphius cavirostris           ...........GG...G....................--...........C......--....GC..........A......A......-------------AA.T.--............G.....--............................................C.......T..................
Mesoplodon peruvianus         .....T.....G....G....................--...........C......--.......................A......-------------AG...--..................--....................................................T..................
Mesoplodon europaeus          ...........GG...G....................--...........C......--.......................A.....--------------AAT..--..................--................CT..................................T..................
Inia geoffrensis              .....T....AGT...G....................--..................--....G...........T......A......-------------AA...--..................--............................................C.......T......G...........
Megaptera novaeangliae        ...........GT...G....................--..................--....GC.........GA......A......-------------AA...--............G.....--............................................C...........T..............
Eschrichtius robustus         ...........GT...G....................--..................--....GC.........GA......A......-------------AA...--............G.....--............................................C...........T..G.......C...
Balaenoptera physalus         ......C....GT...G...T................--..................--....GC.........GA......A.........-.--T.....AA...--............G.....--............................................C...........T..G...........
Balaenoptera musculus         ......A....GT...G....................--..................--....GC.........GA......A...........--T.....AA...--............G.....--............................................C...........T..............
Balaena mysticetus            ...........GT...G....................--..................--....GC.........GA.....CA......-------------AA...--............G.....--............................................C...........T..............
Choeropsis liberiensis        ...........GC...G....................--..................--.R..G..........GA.....CA...........--T.....CA.T.--..................--...................................................CT......G...........
Hippopotamus amphibius        ...........GC...G....................--..................--....G..........GA.....CA...........--T.....CA.T.--..................--...................................................CT......G...........
Sus scrofa                    ...G.T.....GT.........G..............--...........C......--...............GA......A...........--T......A.T.--..A...............--..T..............T..........................C........C.................
Babyrousa babyrussa           ...G.T....AGT........................--...........C......--...............GA......A...........--T......A.T.--..A...............--..T..............T..........................C........C..R..............
Tayassu tajacu                ...G........T...G....................--...........C......--................A......A......-------------AA.T.--..................--..C.........................................C..........................
Camelus dromedarius           ......G....G....G....................--..................--.....C..........G.....C.......-------------.A.T.--..................--...........................................................G...........
Equus hemionus                ...G.......G.....G...................--...........C......--..........A.....A...-----------------------...T.--..................--....................A.......................C..............G...........
Equus asinus                  ...G..C....G.....G...................--..................--..........A.....A......C...........--T......C...--..................--....................A.......................C..............G...........
Rhinoceros unicornis          ...G.T.....G.....G...................--...........C......--..........A.....A.....CC........A..--T.....AC...--..................--..C.................A.......................C..........................
Phoca vitulina                ................G....................--..T..............G--..........A......C....CCG..........--.......A.T.--..................--....................................................A..................
Felis catus                   ...G.TC....AG...G...C................--..................--G.........A...C.A.....CG...........--T..........--............G.....--...................................................C.......G...........
Manis sp.                     ...G.T......T...G....G...............--.................G--..........A....GAC...........C.....--T.....GA.T.--.R..........G.....--.................T..........................C....................A.....
Homo sapiens                  ...G..C...AGG...C....................--..T........C......--G...G.................CAG....C.....--T.....CA...--...C..............--..T.............C...A.......................C......CT......G....C......
Erinaceus europaeus           .C.G.....A..T.....T...............C.T--...........C......--....G...........A.......T...A......--T.....C..A.--..................--....CG.-.....-..............................C.......T...T........C.C...
Mus musculus                  ...G..A...A.....G....................--..................--..........A.....A.......GA..A......--T......A.A.--..................--....CG.-....................................C......CT...T..G.....A.C...
Rattus norvegicus             ...G..A...A.GC.......................--..................--..........A.....A......AGA..A......--T.....AA.A.--..TC..............--.CC.CG.-...........C................G.......C......CT...T..G.....A.A...
Choloepus didactylus          ...G..A...AGG........G.....T.........--...........A......--..........A....TA.....CC.-------------------------..................--....CG.-........C.......................................C..........C...
```

12S+16S rDNA (continued)

Sequence alignment figure with position markers: H 2548-2595 2646-2659 2674-2699

Reference sequence (top):
AATGGCCACACGAGGGTTTTACTGTCTCTTACTTCCCGTGAAGAGGCGGGAATAAATAAAATGACCTTCCCGTGAAGAGACGAAGAACGACCCTATGGAGCTTAACTAACTAGTCC--TAGCAGTTTTGGTTGGGGTGACTCGGAGAACAAAAAATCTCCGGACGA--CTC-ACCAGTCAAAT--TTTGATCA

Taxa (rows):
Aepyceros melampus
Cephalophus maxwelli
Oryx gazella
Capra hircus
Damaliscus dorcas
Kobus ellipsiprymnus
Madoqua kirki
Gazella thomsoni
Tragelaphus imberbis
Bos taurus
Boselaphus tragocamelus
Hydropotes inermis
Odocoileus virginianus
Muntiacus reevesi
Cervus unicolor
Antilocapra americana
Giraffa camelopardalis
Okapia johnstoni
Tragulus napu
Tursiops truncatus
Lissodelphis peronii
Lagenorhynchus obscurus
Lagenorhynchus albirostris
Globicephala melas
Delphinus delphis
Cephalorhynchus eutropia
Phocoena spinipinnis
Phocoena phocoena
Delphinapterus leucas
Physeter catodon
Kogia simus
Kogia breviceps
Ziphius cavirostris
Mesoplodon peruvianus
Mesoplodon europoeus
Inia geoffrensis
Megaptera novaeangliae
Eschrichtius robustus
Balaenoptera physalus
Balaenoptera musculus
Balaena mysticetus
Choeropsis liberiensis
Hippopotamus amphibius
Sus scrofa
Babyrousa babyrussa
Tayassu tajacu
Camelus dromedarius
Equus hemionus
Equus asinus
Rhinoceros unicornis
Phoca vitulina
Felis catus
Manis sp.
Homo sapiens
Erinaceus europaeus
Mus musculus
Rattus norvegicus
Choloepus didactylus

12S+16S rDNA (continued)

ACGGAACAAGTTACCCTAGGGATAACAGCCAATCCTATTCAAGAGTCCTTATTCGACAA-TAGGGTTTACGACCTCGATGTTGGATCAGGACACCCGATGGTGCAACCGCTATCAAAGTTCGTTTGTTCAACGATTAAAGTCCT   16Send

Aepyceros melampus
Cephalophus maxwelli
Oryx gazella
Capra hircus
Damaliscus dorcas
Kobus ellipsiprymnus
Madoqua kirkii
Gazella thomsoni
Tragelaphus imberbis
Bos taurus
Boselaphus tragocamelus
Hydropotes inermis
Odocoileus virginianus
Muntiacus reevesi
Cervus unicolor
Antilocapra americana
Giraffa camelopardalis
Okapia johnstoni
Tragulus napu
Tursiops truncatus
Lissodelphis peronii
Lagenorhynchus obscurus
Lagenorhynchus albirostris
Globicephala melas
Delphinus delphis
Cephalorhynchus eutropia
Phocoena spinipinnis
Phocoena phocoena
Delphinapterus leucas
Physeter catodon
Kogia simus
Kogia breviceps
Ziphius cavirostris
Mesoplodon peruvianus
Mesoplodon europaeus
Inia geoffrensis
Megaptera novaeangliae
Eschrichtius robustus
Balaenoptera physalus
Balaenoptera musculus
Balaena mysticetus
Choeropsis liberiensis
Hippopotamus amphibius
Sus scrofa
Babyrousa babyrussa
Tayassu tajacu
Camelus dromedarius
Equus hemionus
Equus asinus
Rhinoceros unicornis
Phoca vitulina
Felis catus
Manis sp.
Homo sapiens
Erinaceus europaeus
Mus musculus
Rattus norvegicus
Choloepus didactylus

κ-casein exon 4

Sequence alignment of κ-casein exon 4 across species:

Cervus nippon
Odocoileus virginianus
Odocoileus hemionus hemionus
Odocoileus hemionus sitkensis
Mazama americana
Capreolus capreolus
Alces alces
Rangifer tarandus
Cervus canadensis
Cervus elaphus
Elaphurus davidianus
Cervus duvauceli
Cervus unicolor
Muntiacus reevesi
Bison bison
Bison bonasus
Bos taurus
Bubalus bubalis
Bubalus depressicornis
Syncerus caffer
Taurotragus oryx
Boselaphus tragocamelus
Kobus ellipsiprymnus
Oryx gazella callotis
Ovibos moschatus
Ovis dalli
Ovis aries
Capra hircus
Oreamnos americanus
Rupicapra rupicapra
Capricornis crispus
Capricornis swinhoei
Capricornis sumatrensis
Nemorhaedus goral
Saiga tatarica
Antilocapra americana
Giraffa camelopardalis
Tragulus javanicus
Hippopotamus amphibius
Tayassu tajacu
Choeropsis liberiensis
Delphinidae sp.
Lagenorhynchus obscurus
Balaenoptera physalus
Physeter catadon
Ziphius cavirostris
Sus scrofa
Babyrousa babyrussa
Toyassu tajacu
Camelus dromedarius
Camelus bactrianus
Lama guanicoe
Diceorhinus sumatrensis
Rhinoceros unicornis
Tapirus indicus
Equus greyi
Panthera uncia
Canis latrans
Procyon lotor
Ailurus fulgens
Orycteropus afer
Homo sapiens
Oryctolagus cuniculus
Dolichotis patagonum
Cavia cutleri
Rattus norvegicus
Mus musculus

k-casein exon 4 (continued)

κ-casein exon 4 (continued)

Cervus nippon
Odocoileus virginianus
Odocoileus hemionus hemionus
Odocoileus hemionus sitkensis
Mazama americana
Capreolus capreolus
Alces alces
Rangifer tarandus
Cervus canadensis
Cervus elaphus
Elaphurus davidianus
Cervus duvauceli
Cervus unicolor
Muntiacus reevesi
Bison bison
Bison bonasus
Bos taurus
Bubalus bubalis
Bubalus depressicornis
Syncerus caffer
Taurotragus oryx
Boselaphus tragocamelus
Kobus ellipsiprymnus
Oryx gazella callotis
Ovibos moschatus
Ovis dalli
Ovis aries
Capra hircus
Oreamnos americanus
Rupicapra rupicapra
Capricornis crispus
Capricornis swinhoei
Capricornis sumatrensis
Nemorhaedus goral
Saiga tatarica
Antilocapra americana
Giraffa camelopardalis
Tragulus javanicus
Hippopotamus amphibius
Choeropsis liberiensis
Delphinidae sp.
Lagenorhynchus obscurus
Balaenoptera physalus
Physeter catodon
Ziphius cavirostris
Sus scrofa
Babyrousa babyrussa
Tayassu tajacu
Camelus dromedarius
Camelus bactrianus
Lama guanicoe
Diceronhinus sumatrensis
Rhinoceros unicornis
Tapirus indicus
Equus greyvi
Panthera uncia
Canis latrans
Procyon lotor
Ailurus fulgens
Orycteropus afer
Homo sapiens
Oryctolagus cuniculus
Dolichotis patagonum
Cavia cutleri
Rattus norvegicus
Mus musculus

β-casein exon 7

Bos taurus
Bubalus depressicornis
Syncerus caffer
Tragelaphus angasi
Taurotragus oryx
Boselaphus tragocamelus
Aepyceros melampus
Capra hircus
Ovis aries
Oreamnos americanus
Ovibos moschatus
Pantholops hodgsoni
Hippotragus niger
Oryx gazella callotis
Damaliscus lunatus lunatus
Alcelaphus lichtensteini
Connochaetes gnou
Redunca fulvorufula
Kobus ellipsiprymnus
Pelea capreolus
Antidorcas marsupialis
Gazella granti
Saiga tatarica
Oreotragus oreotragus
Raphicerus campestris
Antilope cervicapra
Cephalophus maxwelli
Cephalophus dorsalis
Alces alces
Odocoileus virginianus
Muntiacus reevesi
Cervus nippon
Cervus canadensis
Antilocapra americana
Giraffa camelopardalis
Okapia johnstoni
Tragulus javanicus
Tragulus napu
Hippopotamus amphibius
Choeropsis liberiensis
Balaenoptera physalus
Megaptera novaeangliae
Lagenorhynchus obscurus
Delphinapterus leucas
Monodon monoceros
Physeter catodon
Mesoplodon peruvianus
Ziphius cavirostris
Sus scrofa
Babyrousa babyrussa
Tayassu tajacu
Camelus dromedarius
Manis sp.
Panthera uncia
Procyon lotor
Phoca vitulina
Ailurus fulgens
Equus grevyi
Tapirus indicus
Rhinoceros unicornis
Dicerorhinus sumatrensis
Diceros bicornis
Orycteropus afer
Homo sapiens
Oryctolagus cuniculus
Rattus norvegicus
Mus musculus
Cyclopes didactylus

β-casein exon 7 (continued)

β-casein exon 7 (continued)

Bos taurus
Bubolus depressicornis
Syncerus caffer
Tragelaphus angasi
Taurotragus oryx
Boselaphus tragocamelus
Aepyceros melampus
Capra hircus
Ovis aries
Oreamnos americanus
Ovibos moschatus
Pantholops hodgsoni
Hippotragus niger
Oryx gazella callotis
Damaliscus lunatus lunatus
Alcelaphus lichtensteini
Connochaetes gnou
Redunca fulvorufula
Kobus ellipsiprymnus
Pelea capreolus
Antidorcas marsupialis
Gazella granti
Saiga tatarica
Oreotragus oreotragus
Raphicerus campestris
Antilope cervicapra
Cephalophus maxwelli
Cephalophus dorsalis
Alces alces
Odocoileus virginianus
Muntiacus reevesi
Cervus nippon
Cervus canadensis
Antilocapra americana
Giraffa camelopardalis
Okapia johnstoni
Tragulus javanicus
Tragulus napu
Hippopotamus amphibius
Choeropsis liberiensis
Balaenoptera physalus
Megaptera novaeangliae
Lagenorhynchus obscurus
Delphinapterus leucas
Monodon monoceros
Physeter catadon
Mesoplodon peruvianus
Ziphius cavirostris
Sus scrofa
Babyrousa babyrussa
Tayassu tajacu
Camelus dromedarius
Manis sp.
Panthera uncia
Procyon lotor
Phoca vitulina
Ailurus fulgens
Equus greyyi
Tapirus indicus
Rhinoceros unicornis
Dicerorhinus sumatrensis
Diceros bicornis
Orycteropus afer
Homo sapiens
Oryctolagus cuniculus
Rattus norvegicus
Mus musculus
Cyclopes didactylus

β-casein intron 7

```
Bos taurus                 TCGGTCCTGTCCGGGGACCCTTCCCTA----------------------------------TTATTGTAAGTCTAAATTTACTAACTGTG-----CCTGTTTAACTTCTGATGTTTGTATGATATTCGAGTAATT----AAGAGTCCTATAAAAAAA--------TGAATAATGAATGGTTCCAAAATAAGCATAGCTGAG
Ovis aries                 .T........................----------------------------------..C...........................-----.......................T........----.....C...C.......--------.C.............................
Giraffa camelopardalis     ---.......................----------------------------------...............G.............-----...................C.......A........----.....C...C.......--------.............................A
Alces alces                -----.....................----------------------------------..............................-----....................---....A........----.....C...C.......---------............................
Tragulus napu              AT......A...A..AG...AA....----------------------------------..CC.......C.....C..........-----......C.....A............T....A..A....----G...A.TTC.G.......C-------........A.................A.....T.C..
Ziphius cavirostris        CT......A.......GTT..A....G---------------------------------...C........C..........T...-----......C....A........C......G..A........----G.A..CT...G..T....GAAAAAA-........A..............GCAT....T.T..
Delphinapterus leucas      .T......A.......G.T..A....G---------------------------------..........C..........T...-----......C...CA........C......G..A........----G.A..CT...G..T....GAAAAAA-......................GC.T....T.T..
Physeter catadon           .T......A.......G.T..A....G---------------------------------..........C..........T...-----......C....A........C......G..A........----G.A..CT...G..T....GAAAAAA-......................GCAT.C..T.T..
Balaenoptera physalus      AT......A.......G.TM.A....G---------------------------------..........C..........T...-----......C....A....-----------...A........----G.A..CY...G..T....GAAAAAA-......................GCAT....T.T..
Choeropsis liberiensis     .T......A.G.....C.T..A.....---------------------------------..........C................-----......C.T..A..T.....A......G..A........----G....CT...G..T....TAAACAAA..........G..............GCA.....T.T..
Babyrousa babyrussa        .Y..C...C....T..GTT..A....GTGCCTCAACCAGTTGCCCCAGTTTACAACCC------........C..G....T.C.........-----......C....A.........AGT..C.G.WA.....R..AAGC....ACT...G..T....GAGACAA-........A.G..............GCA.....T.T..
Sus scrofa                 .T..C...C...RA..GTT..A....GTGCCTCAACCAGTTGCCCCAGTTTACAACCC------........C..G....T.C.........-----......C....A.........AGT..C.G..A.........AAGC....ACT...G..T....GAGACAA-........A.G...Y.........GCA.....T.T..
Tayassu tajacu             .T..C...C.......G.T..A....GTGCCTCAACCAGTTGCCCCCGTTTACAACCCTC.C.........TC..G.........C...-----......C....AG........GT..C.G..A.........AAGC....ACT...G..TG...GCAACAA-.........................CGCA.T...T.T..
Camelus dromedarius        CT.AC..C........G.T.CA....GTGCCTCAACCACTTGTCCCTGTG------------.........GC.G.............-----......C...CAC.G...G.A.......G..A..A....C.AAGTG..G.CT...G..TG...GAAAAAAA...GC....G...A........C.CA.....T.C..
Tapirus indicus            .T..C...CCTG.A.AG.T..A....GTGACTCAACAAATTGTCCCAGTTTACAACCCGG............C.............C..------...G..C.TGCA......G.A..C...G..A..A...G.AAGTG....CT...G..TG...GGACAAA-......G...G.A.A........TCA....CT.T..
Manis sp.                  .G.ACT.CAC..A..AGTT..A....GTGACTCAACCACATGCTCCAGTTTCCAACCCCA..T.........C.T..A.........C..TTGTTG...-..C....-----------.CA..G..A..A---AAAGTG....GT...G..CG...GAATAAA-.............CAA..A.....GCA....AT.-..

                                    H
Bos taurus                 ATTAATGATTGTCAGCA-TTAGTTATAAA-TAGAATAA-------GCTGGAGAACCTTCACCTCCCCTCCACCACCAGATCTCAATGTCTAGGCTTACCCGTGGAGATTCTGATGTAATTGTTCTTTC-----TATGTAGAAGAAACT---TATTGGGAAGAAATAATATAATGGACTATGATTTAATTGGTCTGTTGAG
Ovis aries                 ..........C......-..........-........-------.....A..........................................A............T...C........-----.............G..---.................C..............C..........
Giraffa camelopardalis     ..........C......-...C......-........-------...........................................C........T..A........................T-----...............GAG..................G.C...............C..........
Alces alces                G....A..C........-...C......-........-------.................-..................T...........A.......................C.-----...............GAG....................A......................
Tragulus napu              ..C......CT......-C...C......-.......-------................TC..A....G.....T..A....A..............TG.A................G.......-----...............GA---............TG.TAT.G............TG....T.......
Ziphius cavirostris        ..C.G..G..C...A..-C...C..C...-.......-------C........T.C.C..A....--....--..A..................G.A...........A..........-----...............GAA...............CT.C.G...........................T
Delphinapterus leucas      ..C.G..G..C...A..-C...C..C...-..A....-------C........T.C.CA.A.A.ACA....--T.A....................A...........A..........-----...............GAA................T.C.G.............C..........T
Physeter catadon           ..C.G..G..C...A..-C...C..C...-.......-------CG........TCC.C..C..A.CA.-----..A....................A...........A..........-----.............T.GAA................T.C.G.....G....................T
Balaenoptera physalus      ..C.G..G..C...A..-C...C..CG..-.......-------C........T.C.C.AC....CA..---..A....................A...........AA.........-----...............GAA................T.C.G.........................T
Choeropsis liberiensis     T.C.G..G..C...A..-C..CC...C...-......----TAAC........T..GC..C..A.--------A.A..................T.A...........A---.......C..-----...............GAG........G.......T.C.G..........................
Babyrousa babyrussa        .G-.G.....C...A..-....C..C...-........CCAATAA-....GAG.GACATTTTC.T..--------..A.....T.W..MC...A......ACAA.........CA.......A...A.CTCTA.......G.....GAA..........G.....T.C.G...T.....................
Sus scrofa                 ..-.G.....C...A..-..T.C..C...-........CCAATAA-....GAG.-----TTTC.T..--------..A.......A...C...A......ACAA......-..CA.......A...A.CTCTA.......GG....T.GAA..........G.....T.C.G...T.....................
Tayassu tajacu             .G-.G.....C...A..-....C..C...-........-------C....-AG.GACATTTTCT...--------..A..T....A...C...A......ACAA.........CACCC....A..GW.TTCTT.............T.GAA....................T.C.G...T............A.........
Camelus dromedarius        ..C.G.....C...A..-...TC..C...-C.......-------C.....A...A..TTTTATTT.-------.A.A...C............T......AC............A...........GATTTTC-.................GAG..A..A.C..........T.C.G...T.........G.........G...
Tapirus indicus            ..C.G.....C...A..AC...C..C.C.A.G......-------CTC........T..TTTT....C------.A.A...C.......A.A..C.....ACA.............A.......A-.GATTTTC-.G..............GAG..C...........T..T.CTG.AGT.T.....C...A..........
Manis sp.                  ..C.G.....C...A..-C.CAC..C.CTT........-------T...T.....T..TTTT.T.T--------TG.A...............AATC...T.A.A....------.A....G.....GATTTTC-.G.....W...T...TAG..........T....T..T.C.GA..T.C............A.........

Bos taurus                 AACCAATTAAATTAGATGA----AAGCGATT-AAG-TACAATAAAGCCAAAATTGAATTTGATAATCTCATTTGGCTAAGAATAACAAACCTAAGAAGGTTTGCTATTTTCTACAATTTTGAAGTTCTCCTTATGCACAATTATTTCACCACATGACTCATTTCACA--TCGTGTTTTTGATATATGAGCATATGAGGGA
Ovis aries                 ...--------........----.G.G...-...T.............G......C........................................................T.........................................--..T...........................C
Giraffa camelopardalis     ...--------........----.G.G...-...G.............G..............G......A..............TG..........................T...............A.........................--..T.C..............A.......A..C
Alces alces                ...--------........----.G.T...-...GA............G...................A.......................C.........----------.....T..........----------...............--..T.............................C
Tragulus napu              ..G--------........----.GAG...-...G....C........G.G..A.........G.............G...-------------.........T..........G.........T.....T........G...A.T.TT.T...CT............--..T.C.......C................AC
Ziphius cavirostris        ..T--------......A.----.GAG...-...G.GT.C........G.C..A..........C.........T.GG................A...C.T...............C......T.G...G...............C.......C............--..T.........G...................C
Delphinapterus leucas      ..T--------......A.----.GAG...-...G.GT.C........G...A..........C.........T.GG................A.....T.G....................T.G...G...............C.......C............--..T.............................C
Physeter catadon           ..T--------......A.----.GAG...-...G.GT.C........G...A..........C.........T.GG................A.....T......................T.G...G...............C.......C............--..T.........G...................C
Balaenoptera physalus      ..T--------......A.----.GAG...-...G.GT.C........G...A....................T.GG................A.....T......................T...................C.......C............--..T.........G...............C..C
Choeropsis liberiensis     ..T--------......A.----.GAGC..-...G..T.C........G...A.....................GGA......T..........C...T..............C.......T..........G.......C...................--..T...............A..C.C..A..C
Babyrousa babyrussa        ..T--------.G....A.AGAG.GAG....T...G..T.G.......G...A...G.A.C...T....G...Y.TG..................T...........T...A...G..........T.........TG.........C.......C............--..T.....C-.........A...C..R...C
Sus scrofa                 ..T--------.G....A.AGAG.GAG....T...G..T.C.......G...A...G.A.C.........G....TG..................Y...........Y...A...G..........T.........TG.........C.......C............--..T.....-..........A...Y.......C
Tayassu tajacu             .GT--------.G....A.----GGAG....Y...G..TGT...........A...G.A...........G....TG.R......G........GA.....T...GA.....T...A...G..........T.........................T.......C............--..T.....C-A..G.....A...T......C
Camelus dromedarius        ..T--------......A.----.GAG...-..TG.TT.C.......G...A................R.......TG..................A.....A..............T.....G.....T..T..........T.....G.T...G....CT...........--..T..........CG...ATG.........C
Tapirus indicus            ..T--------......A.----.GAG...-...GCTT.C....A...G...A.C..............G...G....TG....--------..T.........T.T........T.......T..........T.....G....TGTC.....-.C.TG....C..T......T.--..T.........C.G...TA..........C
Manis sp.                  ..T--------.C....A.----.GAG...-...ACTTWC........G...A..Y..........G..CA....TG....--------....G....---.........G.........T.G.........G.T.T...G.....T.....-.C.--....TGAT.........CAC.T.........G..C...CA..C.......C

                                         H
Bos taurus                 AAAATACTGAGATGCTTATTTCAATACTCA-----------GGGAAAATTT-ATTGCCAAAAGGCAAGAAATGTATA
Ovis aries                 ..............................-----------..........TC.................T......
Giraffa camelopardalis     ..................A...........-----------..........TC.................T...G..
Alces alces                ........A.......G...........TG-----------..........TC................G.T......
Tragulus napu              .....GT....G..................TGAAATATTTA.......C..TC.........-..T--.G.C......
Ziphius cavirostris        ...........G..T...............TGAAATTCTTA......G.G.TC...........T--...T......
Delphinapterus leucas      ...........G..T...............TGAAATTCTTA......G.G.TC...........T--...T......
Physeter catadon           ...........GC.T..........T....TGAAATTCTTA......G.G.TC...........T--...T......
Balaenoptera physalus      ...........G.TT...............TGAAATTCTTA......G.G.TC...........T--...T......
Choeropsis liberiensis     ........G.....T...............TGAAATTCTTA..T..C..A.TCC..........T--...T......
Babyrousa babyrussa        ..G...........C...............TGGAATACTCA......T.G.TC..........AT--...T......
Sus scrofa                 ..G...........C...............TGGAATACTCN......T.G.TC..........AT--...C......
Tayassu tajacu             ..G.........C.................CGGAATACTCG......T.G.TG............T--.G.T.A....
Camelus dromedarius        ......T...AT..............T...TGGACTTCTTA......G...TC..AT.....A-T--...TW.....
Tapirus indicus            ...TG.T.................T.....CAGAATTTTTA......G.G.TT..........A.T--...T......
Manis sp.                  ...CA.T.T...............T.....CAGAATTCTTA.....-G.G.TT....T...GA.T--...T......
```

protamine P1

Giraffa camelopardalis
Cervus elaphus
Alces alces
Odocoileus virginianus
Gazella dorcas
Bos taurus
Tragulus javanicus
Choeropsis liberiensis
Lagenorhynchus obscurus
Orcinus orca
Delphinapterus leucas
Balaenoptera physalus
Tayassu tajacu
Sus scrofa
Lama guanicoe
Camelus dromedarius
Tapirus indicus
Diceros bicornis
Equus caballus
Ursus americanus
Felis catus
Homo sapiens
Alouatta seniculus
Elephas maximus
Mus musculus
Rattus norvegicus
Cavia porcellus

References

Adachi, J., and Hasegawa, M. 1996. Instability of quartet analyses of molecular sequence data by the maximum likelihood method: the Cetacea/Artiodactyla relationships. *Mol. Phylogenet. Evol.* **6**(1):72–76.

Allard, M., and Carpenter, J. 1996. On weighting and congruence. *Cladistics* **12**:183–198.

Anderson, S., DeBruijn, M., Coulson, A., Eperon, I., Sanger, F., and Young, I. 1982. Complete sequence of bovine mitochondrial DNA: conserved features of the mammalian mitochondrial genome. *J. Mol. Biol.* **156**:683–717.

Arnason, U., and Gullberg, A. 1996. Cytochrome *b* nucleotide sequences and the identification of five primary lineages of extant cetaceans. *Mol. Biol. Evol.* **13**(2):407–417.

Arnason, U., Gullberg, A., and Widegren, B. 1991. The complete nucleotide sequence of the mitochondrial DNA of the fin whale, *Balaenoptera physalis. J. Mol. Evol.* **33**:556–568.

Baba, M., Darga L., Goodman, M., and Czelusniak, J. 1981. Evolution of cytochrome c investigated by the maximum parsimony method. *J. Mol. Evol.* **17**:197–213.

Baker, R., and DeSalle, R. 1997. Multiple sources of character information and the phylogeny of Hawaiian drosophilids. *Syst. Biol.* **46**:654–673.

Beintema, J., Fitch, W., and Carsana, A. 1986. Molecular evolution of pancreatic-type ribonucleases. *Mol. Biol. Evol.* **3**:262–275.

Boyden, A., and Gemeroy, D. 1950. The relative position of the Cetacea among the orders of Mammalia as indicated by precipitin tests. *Zoologica* **35**:145–151.

Bremer, K. 1988. The limits of amino acid sequence data in angiosperm phylogenetic reconstruction. *Evolution* **42**:795–803.

Bremer, K. 1994. Branch support and tree stability. *Cladistics* **10**:295–304.

Bull, J., Huelsenbeck, J., Cunningham, C., Swofford, D., and Waddell, P. 1993. Partitioning and combining data in phylogenetic analysis. *Syst. Biol.* **42**:384–397.

Buntjer, J., Hoff, I., and Lenstra, J. 1997. Artiodactyl interspersed DNA repeats in cetacean genomes. *J. Mol. Evol.* **45**:66–69.

Cao, Y., Adachi, J., Janke, A., Pääbo, S., and Hasegawa, M. 1994. Phylogenetic relationships among eutherian orders estimated from inferred sequences of mitochondrial proteins: instability of a tree based on a single gene. *J. Mol. Evol.* **39**:519–527.

Carpenter, J. 1992. Random cladistics. *Cladistics* **8**:147–153.

Czelusniak, J., Goodman, M., Koop, B., Tagle, D., Shoshani, J., Braunitzer, G., Kleinschmidt, T., De Jong, W., and Matsuda, G. 1990. Perspectives from amino acid and nucleotide sequences on cladistic relationships among higher taxa of Eutheria, in: H. Genoways (ed.), *Current Mammology*, Volume 2, pp. 545–572. Plenum Press, New York.

De Jong, W., Gleaves, J., and Boulter, D. 1977. Evolutionary changes of α-crystallin and the phylogeny of mammalian orders. *J. Mol. Evol.* **10**:123–135.

De Jong, W., Leunissen, J., and Wistow, G. 1993. Eye lens crystallins and the phylogeny of placental orders: evidence for a macroscelid-paenungulate clade? in: F. Szalay, M. Novacek, and M. McKenna (eds.), *Mammal Phylogeny*, Volume II, pp. 5–12. Springer-Verlag, Berlin.

DePinna, M. 1991. Concepts and tests of homology in the cladistic paradigm. *Cladistics* **7**:367–394.

Douzery, E., and Catzeflis, F. M. 1995. Molecular evolution of the mitochondrial 12S rRNA in Ungulata (Mammalia). *J. Mol. Evol.* **41**:622–636.

Farris, J. S. 1983. The logical basis of phylogenetic analysis, in: N. Platnick and V. Funk (eds.), *Advances in Cladistics*, Volume 2, pp. 7–36. Columbia University Press, New York.

Farris, J. S. 1989. The retention index and the rescaled consistency index. *Cladistics* **5**:417–419.

Farris, J. S., Kallersjo, M., Kluge, A. G., and Bult, C. 1994. Testing significance of incongruence. *Cladistics* **10**:315–319.

Felsenstein, J. 1985. Confidence limits on phylogenies: an approach using the bootstrap. *Evolution* **39**:783–791.

Fitch, W., and Beintema, J. 1990. Correcting parsimonious trees for unseen nucleotide substitutions: the effect of dense branching as exemplified by ribonuclease. *Mol. Biol. Evol.* **7**:438–443.

Gatesy, J. 1997. More support for a Cetacea/Hippopotamidae clade: the blood-clotting protein gene γ-fibrinogen. *Mol. Biol. Evol.* **14**(5):537–543.

Gatesy, J., DeSalle, R., and Wheeler, W. 1993. Alignment-ambiguous nucleotide sites and the exclusion of systematic data. *Mol. Phylogenet. Evol.* **2**(2):152–157.

Gatesy, J., Hayashi, C., Cronin, M., and Arctander, P. 1996. Evidence from milk casein genes that cetaceans are close relatives of hippopotamid artiodactyls. *Mol. Biol. Evol.* **13**(7):954–963.

Gatesy, J., Amato, G., Vrba, E., Schaller, G., and DeSalle, R. 1997. A cladistic analysis of mitochondrial ribosomal DNA from the Bovidae. *Mol. Phylogenet. Evol.* **7**(3):303–319.

Gatesy, J., Arctander, P., and Friis, P. Submitted. Phylogenetic placement of the saola (*Pseudoryx nghetinhensis*, Bovidae, Artiodactyla) based on DNA sequences from five genes. *Syst. Biol.*

Gentry, A. W., and Hooker, J. J. 1988. The phylogeny of the Artiodactyla, in: M. Benton (ed.), *The Phylogeny and Classification of the Tetrapods*, Volume 2, pp. 235–272. Clarendon Press, Oxford.

Gilbert, D. 1992. SeqApp version 1.9a. Indiana University, Bloomington.

Gingerich, P. D., Wells, N. A., Russell, D. E., and Shah, S. M. I. 1983. Origin of whales in epicontinental remnant seas: new evidence from the early Eocene of Pakistan. *Science* **220**:403–406.

Gingerich, P. D., Smith, B. H., and Simons, E. L. 1990. Hind limbs of Eocene *Basilosaurus*: evidence of feet in whales. *Science* **249**:154–157.

Gingerich, P. D., Raza, S. M., Arif, M., Anwar, M., and Zhou, X. 1994. New whale from the Eocene of Pakistan and the origin of cetacean swimming. *Nature* **368**:844–847.

Goloboff, P. 1993. Estimating character weights during tree search. *Cladistics* **9**:83–91.

Goodman, M., Czelusniak, J., and Beeber, J. E. 1985. Phylogeny of Primates and other eutherian orders: a cladistic analysis using amino acid and nucleotide sequence data. *Cladistics* **1**:171–185.

Graur, D., and Higgins, D. G. 1994. Molecular evidence for the inclusion of cetaceans within the order Artiodactyla. *Mol. Biol. Evol.* **11**:357–364.

Hall, L. 1990. Nucleotide sequence of guinea pig κ-casein cDNA. *Nucleic Acids Res.* **18**:6129.

Hasegawa, M., and Adachi, J. 1996. Phylogenetic position of cetaceans relative to artiodactyls: reanalysis of mitochondrial and nuclear sequences. *Mol. Biol. Evol.* **13**(5):710–717.

Hasegawa, M., Adachi, J., and Milinkovitch, M. 1997. Novel phylogeny of whales supported by total molecular evidence. *J. Mol. Evol.* **44**:S117–S120.

Honeycutt, R., Nedbal, M., Adkins, R., and Janecek, L. 1995. Mammalian mitochondrial DNA evolution: a comparison of the cytochrome *b* and cytochrome *c* oxidase II genes. *J. Mol. Evol.* **40**:260–272.

Huelsenbeck, J., and Bull, J. 1996. A likelihood ratio test to detect conflicting phylogenetic signal. *Syst. Biol.* **45**(1):92–98.

Irwin, D. M., and Arnason, U. 1994. Cytochrome *b* gene of marine mammals: phylogeny and evolution. *J. Mamm. Evol.* **2**:37–55.

Irwin, D. M., Kocher, T. D., and Wilson, A. C. 1991. Evolution of the cytochrome *b* gene of mammals. *J. Mol. Evol.* **32**:128–144.

Kluge, A. G. 1989. A concern for evidence and a phylogenetic hypothesis of relationships among *Epicrates* (Boidae, Serpentes). *Syst. Zool.* **38**:7–25.

Kluge, A. G., and Farris, J. S. 1969. Quantitative phyletics and the evolution of anurans. *Syst. Zool.* **18**:1–32.

Kluge, A. G., and Wolf, A. 1993. Cladistics: what's in a word? *Cladistics* **9**:183–199.

Kocher, T., Thomas, W., Meyer, A., Edwards, S., Pääbo, S., Villablanca, F., and Wilson, A. 1989. Dynamics of mitochondrial DNA evolution in animals: amplification and sequencing with conserved primers. *Proc. Natl. Acad. Sci. USA* **86**:6196–6200.

Krettek, A., Gullberg, A., and Arnason, U. 1995. Sequence analysis of the complete mitochondrial DNA molecule of the hedgehog, *Erinaceus europaeus*, and the phylogenetic position of the Lipotyphla. *J. Mol. Evol.* **41**:952–957.

Lavergne, A., Douzery, E., Stichler, T., Catzeflis, F., and Springer, M. 1996. Interordinal mammalian relationships: evidence for paenungulate monophyly is provided by complete mitochondrial 12S rRNA sequences. *Mol. Phylogenet. Evol.* **6**:245–258.

Ledje, C., and Arnason, U. 1996. Phylogenetic analyses of complete cytochrome *b* genes of the order Carnivora with particular emphasis on the Caniformia. *J. Mol. Evol.* **42**:135–144.

Ma, D., Zharkikh, A., Graur, D., Vandeberg, J., and Li, W. 1993. Structure and evolution of opossum, guinea pig, and porcupine cytochrome *b* genes. *J. Mol. Evol.* **36**:327–334.

Maxson, L., and Maxson, R. 1990. Proteins II: immunological techniques, in: D. Hillis and C. Moritz (eds.), *Molecular Systematics*, pp. 127–155. Sinauer, Sunderland, MA.

McKenna, M. C. 1975. Toward a phylogenetic classification of the Mammalia, in: W. D. Luckett and F. S. Szalay (eds.), *Phylogeny of the Primates*, pp. 21–46. Plenum Press, New York.

McKenna, M. C. 1987. Molecular and morphological analysis of high-level mammalian interrelationships, in:

C. Patterson (ed.), *Molecules and Morphology in Evolution: Conflict or Compromise?* pp. 55–93. Cambridge University Press, London.

Mickevich, M. F., and Farris, J. S. 1981. The implications of congruence in *Menidia*. *Syst. Zool.* **30**:351–370.

Milinkovitch, M. C. 1992. DNA–DNA hybridizations support ungulate ancestry of Cetacea. *J. Evol. Biol.* **5**:149–160.

Milinkovitch, M. C., Orti, G., and Meyer, A. 1993. Revised phylogeny of whales suggested by mitochondrial ribosomal DNA sequences. *Nature* **361**:346–348.

Milinkovitch, M. C., Meyer, A., and Powell, J. 1994. Phylogeny of all major groups of cetaceans based on DNA sequences from three mitochondrial genes. *Mol. Biol. Evol.* **11**:939–948.

Milinkovitch, M. C., Orti, G., and Meyer, A. 1995. Novel phylogeny of whales revisited but not revised. *Mol. Biol. Evol.* **12**:518–520.

Milinkovitch, M. C., LeDuc, R., Adachi, J., Farnir, F., Georges, M., and Hasegawa, M. 1996. Effects of character weighting and species sampling on phylogeny reconstruction: a case study based on DNA sequence data in cetaceans. *Genetics* **144**:1817–1833.

Miyamoto, M. 1985. Consensus cladograms and general classifications. *Cladistics* **1**:186–189.

Miyamoto, M., and Fitch, W. 1995. Testing species phylogenies and phylogenetic methods with congruence. *Syst. Biol.* **44**:64–76.

Miyamoto, M., and Goodman, M. 1986. Biomolecular systematics of eutherian mammals: phylogenetic patterns and classification. *Syst. Zool.* **35**:230–240.

Montgelard, C., Catzeflis, F. M., and Douzery, E. 1997. Phylogenetic relationships of artiodactyls and cetaceans as deduced from the comparison of cytochrome *b* and 12S rRNA mitochondrial sequences. *Mol. Biol. Evol.* **14**(5):550–559.

Nixon, K., and Carpenter, J. 1996. On simultaneous analysis. *Cladistics* **12**:221–241.

Novacek, M. J. 1989. Higher mammal phylogeny: the morphological–molecular synthesis, in: B. Fernholm, K. Bremer, and H. Jornvall (eds.), *The Hierarchy of Life*, pp. 421–435. Elsevier, Amsterdam.

Palumbi, S., Martin, A., Romano, S., McMillan, W., Stice, L., and Grabowski, G. 1991. *The Simple Fool's Guide to PCR*. University of Hawaii, Honolulu.

Patterson, C. 1982. Morphological characters and homology, in: K. A. Joysey and A. E. Friday (eds.), *Problems of Phylogenetic Reconstruction*, pp. 21–74. Academic Press, New York.

Philippe, H., and Douzery, E. 1994. The pitfalls of molecular phylogeny based on four species, as illustrated by the Cetacea/Artiodactyla relationships. *J. Mamm. Evol.* **2**:133–152.

Porter, C., Goodman, M., and Stanhope, M. 1996. Evidence on mammalian phylogeny from sequences of exon 28 of the von Willebrand factor gene. *Mol. Phylogenet. Evol.* **5**:89–101.

Prothero, D. R. 1993. Ungulate phylogeny: molecular versus morphological evidence, in: F. S. Szalay, M. J. Novacek, and M. C. McKenna (eds.), *Mammal Phylogeny*, Volume II, pp. 173–181. Springer-Verlag, Berlin.

Prothero, D. R., Manning, E. M., and Fischer, M. 1988. The phylogeny of the ungulates, in: M. J. Benton (ed.), *The Phylogeny and Classification of the Tetrapods*, Volume 2, pp. 201–234. Clarendon Press, Oxford.

Queralt, R., Adroer, R., Oliva, R., Winkfein, R., Retief, J., and Dixon, G. 1995. Evolution of protamine P1 genes in mammals. *J. Mol. Evol.* **40**:601–607.

Randi, E., Lucchini, V., and Hoong Diong, C. 1996. Evolutionary genetics of the Suiformes as reconstructed using mtDNA sequencing. *J. Mamm. Evol.* **3**:163–194.

Rixon, M., Chung, D., and Davie, E. 1985. Nucleotide sequence of the gene for the γ chain of human fibrinogen. *Biochemistry* **24**:2077–2086.

Sarich, V. M. 1985. Rodent macromolecular systematics, in: W. P. Luckett and J. Hartenberger (eds.), *Evolutionary Relationships among Rodents: A Multidisciplinary Approach*, pp. 423–452. Plenum Press, New York.

Sarich, V. M. 1993. Mammalian systematics: twenty-five years among their albumins and transferrins, in: F. S. Szalay, M. J. Novacek, and M. C. McKenna (eds.), *Mammal Phylogeny*, Volume II, pp. 103–114. Springer-Verlag, Berlin.

Schaeffer, B. 1948. The origin of a mammalian ordinal character. *Evolution* **2**:164–175.

Shimamura, M., Yasue, H., Ohshima, K., Abe, H., Kato, H., Kishiro, T., Goto, M., Munechika, I., and Okada, N. 1997. Molecular evidence from retroposons that whales form a clade within even-toed ungulates. *Nature* **388**:666–670.

Shoshani, J. 1986. Mammalian phylogeny: comparison of morphological and molecular results. *Mol. Biol. Evol.* **3**(3):222–242.

Slijper, E. J. 1962. *Whales*. Hutchinson, London.

Smith, M., Shivji, M., Waddell, V., and Stanhope, M. 1996. Phylogenetic evidence from the IRBP gene for the paraphyly of toothed whales, with mixed support for Cetacea as a suborder of Artiodactyla. *Mol. Biol. Evol.* **13**:918–922.

Springer, M., and Kirsch, J. A. W. 1993. A molecular perspective on the phylogeny of placental mammals based on mitochondrial 12S rDNA sequences, with special reference to the problem of the Paenungulata. *J. Mamm. Evol.* **1**:149–166.

Springer, M., Hollar, L., and Burk, A. 1995. Compensatory substitutions and the evolution of the mitochondrial 12S rRNA gene in mammals. *Mol. Biol. Evol.* **12**:1138–1150.

Stanhope, M., Bailey, W. J., Czelusniak, J., Goodman, M., Si, J., Nickerson, J., Sgouros, J., Singer, G., and Kleinschmidt, T. 1993. A molecular view of primate supraordinal relationships from the analysis of both nucleotide and amino acid sequences, in: R. MacPhee (ed.), *Primates and Their Relatives in Phylogenetic Perspective*, pp. 251–292. Plenum Press, New York.

Stanhope, M., Smith, M., Waddell, V., Porter, C., Shivji, M., and Goodman, M. 1996. Mammalian evolution and the interphotoreceptor retinoid binding protein (IRBP) gene: convincing evidence for several superordinal clades. *J. Mol. Evol.* **43**:83–92.

Stanley, H., Kadwell, M., and Wheeler, J. 1994. Molecular evolution of the family Camelidae: a mitochondrial DNA study. *Proc. R. Soc. London Ser. B* **256**:1–6.

Swofford, D. 1993. PAUP: Phylogenetic Analysis Using Parsimony, Version 3.1.1. Illinois Natural History Survey, Champaign.

Thewissen, J. G. M. 1994. Phylogenetic aspects of cetacean origins: a morphological perspective. *J. Mamm. Evol.* **2**:157–184.

Thewissen, J. G. M., and Hussain, S. 1998. Systematic review of the Pakicetidae, early and middle Eocene Cetacea (Mammalia) from Pakistan and India. *Bull. Carnegie Mus. Nat. Hist.* **34**:220–238.

Thewissen, J. G. M., Hussain, S. T., and Arif, M. 1994. Fossil evidence for the origin of aquatic locomotion in archaeocete whales. *Science* **263**:210–212.

Thewissen, J. G. M., Madar, S., and Hussain, S. 1996. *Ambulocetus natans*, an Eocene cetacean (Mammalia) from Pakistan. *Courier Forsch.-Inst. Senckenberg* **191**:1–86.

Threadgill, D., and Womack, J. 1990. Genomic analysis of the major bovine milk protein genes. *Nucleic Acids Res.* **18**:6935–6942.

Van Valen, L. 1966. Deltatheridia, a new order of mammals. *Am. Mus. Nat. Hist. Bull.* **132**:1–126.

Werman, S., Springer, M., and Britten, R. 1990. Nucleic acids I: DNA–DNA hybridization, in: D. Hillis and C. Moritz (eds.), *Molecular Systematics*, pp. 204–249. Sinauer, Sunderland, MA.

Wheeler, M., and Gladstein, D. 1994. MALIGN, Version 2.1. American Museum of Natural History, New York.

Wyss, A. R., Novacek, M. J., and McKenna, M. C. 1987. Amino acid sequence versus morphological data and the interordinal relationships of mammals. *Mol. Biol. Evol.* **4**:99–116.

Xu, X., Janke, A., and Arnason, U. 1996. The complete mitochondrial DNA sequence of the greater Indian rhinoceros, *Rhinoceros unicornis*, and the phylogenetic relationship among Carnivora, Perissodactyla, and Artiodactyla (+ Cetacea). *Mol. Biol. Evol.* **13**(9):1167–1173.

Zardoya, R., and Meyer, A. 1996. Phylogenetic performance of mitochondrial protein-coding genes in resolving relationships among vertebrates. *Mol. Biol. Evol.* **13**(7):933–942.

Zhou, X., Zhai, R., Gingerich, P. D., and Chen, L. 1995. Skull of a new mesonychid (Mammalia, Mesonychia) from the late Paleocene of China. *J. Vertebr. Paleontol.* **15**(2):387–400.

CHAPTER 4

Cetaceans Are Highly Derived Artiodactyls

MICHEL C. MILINKOVITCH, MARTINE BÉRUBÉ,
and PER J. PALSBØLL

1. Introduction

Cetaceans (whales, dolphins, and porpoises) form one of the most dramatically derived group of mammals and modern representatives are easily recognized by the telescoping of the skull, posterior movement of the narial openings, isolation of the earbones, shortening of the neck, loss of external hind limbs, reduction of the pelvic girdle, and addition of vertebrae (e.g., Barnes, 1984). These skeletal character states are among the most conspicuous features within a suite of transformations that cetaceans experienced in basically all of their biological systems during their adaptation to the aquatic environment.

Although the highly modified and specialized morphology of all living cetaceans makes homology statements and polarization of the morphological characters difficult, analyses of neontological data have long suggested that cetaceans are closely related to ungulates (e.g., Flower, 1883; Prothero et al., 1988). These results were corroborated by paleontological data suggesting that artiodactyls is the sister group to Cete, i.e., a clade comprising cetaceans and a group of presumed carnivorous hoofed mammals: the mesonychians (Gingerich et al., 1990; Thewissen and Hussain, 1993; Thewissen, 1994). The close relationship between cetaceans and ungulates (and especially artiodactyls) was strongly supported by basically all molecular data published to date (reviewed in Gatesy, this volume): from immunological (e.g., Boyden and Gemeroy, 1950; Shoshani, 1986) and DNA–DNA hybridization (Milinkovitch, 1992) studies, to analyses of mitochondrial and nuclear amino-acid and DNA sequences (e.g., Gatesy, 1997).

One surprising result from the molecular analyses was the suggestion that cetaceans are nested within the artiodactyl phylogenetic tree, making artiodactyls paraphyletic (e.g.,

MICHEL C. MILINKOVITCH • Evolutionary Genetics, Free University of Brussels (ULB), 1050 Brussels, Belgium. MARTINE BÉRUBÉ • Department of Population Biology, Copenhagen University, Copenhagen Ø, DK-2100, Denmark. PER J. PALSBØLL • Department of Ecology and Evolutionary Biology, University of California, Irvine, Irvine, California 92697-2525. *Present address for MB and PJP*: Evolutionary Genetics, Free University of Brussels (ULB), 1050 Brussels, Belgium.

The Emergence of Whales, edited by Thewissen. Plenum Press, New York, 1998.

Goodman *et al.*, 1985; Czelusniak *et al.*, 1990; Irwin *et al.*, 1991; Graur and Higgins, 1994; Gatesy *et al.*, 1996; Smith *et al.*, 1996; Gatesy, 1997; Montgelard *et al.*, 1997). However, most of these studies (Czelusniak *et al.*, 1990; Irwin *et al.*, 1991; Smith *et al.*, 1996; Montgelard *et al.*, 1997) provided only weak support for this hypothesis and the analyses by Graur and Higgins (1994) have been strongly criticized because of sensitivity to taxon sampling (Philippe and Douzery, 1994; Hasegawa and Adachi, 1996). Nevertheless, the hypothesis of artiodactyl paraphyly has subsequently been supported by: (1) extensive cladistic analyses of DNA sequence data from seven nuclear and mitochondrial genes (e.g., Gatesy, this volume) and (2) by phylogenetic interpretation of retropositional events that lead to the insertion of short interspersed elements (SINEs) at particular loci in the nuclear genome of various artiodactyl and cetacean ancestors (Shimamura *et al.*, 1997). Gatesy (this volume) demonstrated that the data he analyzed indicate (1) a sister relationship between cetaceans and hippopotamuses (hippo) and (2) the grouping of the hippo/whale clade with ruminants (Tragulidae, Bovidae, Cervidae, and Giraffidae), whereas tylopods (camels and llamas), Suidae (pigs), and Tayassuidae (peccaries) were more basal in the (artiodactyls + cetaceans) phylogenetic tree (Fig. 1). The analysis by Shimamura *et al.* (1997) provided a striking support for the grouping of cetaceans, hippos, and ruminants in a monophyletic group (i.e., hypothesis 2 above), but did not resolve the placement of hippos within that clade. It is the nature of the molecular data analyzed by Shimamura *et al.* that makes their results so remarkable. Indeed, the likelihood of SINEs being independently inserted at the same locus in different lineages, or precisely excised in different lineages, seems virtually nil. Hence, these markers can reasonably be considered to be essentially free of noise (Milinkovitch and Thewissen, 1997).

In summary, the evidence available today from molecular data unequivocally indicates the paraphyly of artiodactyls, and more specifically a sister-group relationship between hippos and cetaceans, as well as a sister-group relationship between ruminants and the hippo/cetacean clade. These results are in striking contradiction to the common interpretation of the available morphological data, namely, supporting artiodactyl monophyly (e.g.,

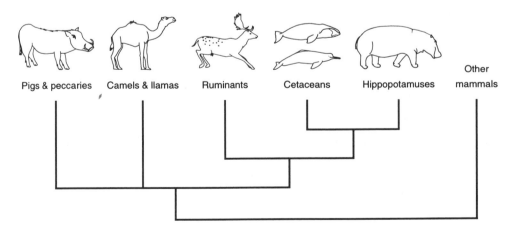

FIGURE 1. Molecular phylogenetic hypothesis of artiodactyl paraphyly (e.g., Graur and Higgins, 1994; Gatesy *et al.*, 1996; Gatesy, 1997, this volume; Milinkovitch *et al.*, this analysis.

Geisler and Luo, this volume). In the present study we analyzed phylogenetically cetacean and artiodactyl α-lactalbumin sequences to test the artiodactyl paraphyly hypothesis. The α-lactalbumin is a secretory protein present in the mammary tissue and, among other secondary functions, is one of the two components of lactose synthetase (Ebner and Schanbacher, 1974; Vilotte *et al.*, 1987). We discuss the implications of our results on the interpretation of some morphological characters.

2. An Analysis of Cetacean Origin Based on Lactalbumin DNA Sequences

2.1. Materials and Methods

2.1.1. Data Collection and Sequence Alignments

Genomic DNA from various cetacean and artiodactyl taxa was extracted from tissue samples using standard protocols of cell lysis with 1% SDS (sodium dodecyl sulfate) followed by overnight digestion with Proteinase K (100 μg/ml) at 65°C, multiple phenol–chloroform extractions, and, finally, ethanol precipitation (Sambrook *et al.*, 1989). The α-lactalbumin nucleotide sequences were obtained by direct sequencing of asymmetrically amplified DNA. Initial symmetrical amplifications were performed in a 10-μl volume with 67 mM Tris-HCl (pH 8.8), 2 mM $MgCl_2$, 16.6 mM $(NH_4)_2SO_4$, 10 mM β-mercaptoethanol, 200 μM dNTP, 1 μM of each primer, and 0.4 unit of Taq™ DNA polymerase. An initial denaturation step of 2 min at 94°C was followed by 32 cycles of 15 s at 94°C, 15 s at 54°C, 30 s at 72°C on a Genemachine Junior™ Thermal Cycler (USA/Scientific Plastics). Asymmetrical amplifications were conducted under similar conditions except that the concentration of one primer was reduced to 0.01 μM and the reaction volume was 50 μl. The amplifications were carried out on a Techne Thermocycler™ for 30 cycles: 60 s at 94°C, 60 s at 55°C, and 90 s at 72°C. The oligonucleotide primer sequences were: LacI.R (5′-CTC ACT GTC ACA GGA GAT GT-3′) and LacII.F (5′-CCA AAA TGA TGT CCT TTG TC-3′), LacIII.F2 (5′-GGG TCT GTA CCG TAT TTC ATA-3′), and LacIV.R (5′-GAC TCA CCA GTA GGT AAT TC-3′). All primers were used for symmetrical and asymmetrical amplifications as well as for subsequent sequencing reactions. Sequencing (Sequenase™ Version 2.0, US Biochemicals Inc.) was conducted using ^{35}S-labeled dCTP, following the manufacturer's instructions. The oligonucleotides Lac1011.F (5′-ATT ATC AAA CAA TTC TCT TAT-3′) and Lac1179.R (5′-AGA TAT CAC AGG GAT GTC CAC-3′) were used as internal sequencing primers. The sequencing products were separated by vertical electrophoresis on a 5% denaturing polyacrylamide gel (Long Ranger™) and visualized by autoradiography. Both strands, amplified from independent asymmetric PCR reactions, were sequenced to ensure accurate results. We analyzed our cetacean and artiodactyl sequences together with published sequences in order to test the artiodactyl paraphyly hypothesis (e.g., Gatesy, 1997). Ingroup sequences used were from (*binomial name* [common name, EMBL or GenBank accession number]): *Balaena mysticetus* [bowhead, AJ007809], *Megaptera novaeangliae* [humpback whale, AJ007810], *Phocoena phocoena* [harbor porpoise, AJ007811], and *Monodon monoceros* [narwhal, AJ007812] belonging to the order Cetacea; *Bos taurus* [bovine, 06366] and *Capra hircus* [domestic goat, M63868] belonging to the artiodactyl suborder Ruminantia; *Lama guanacoe* [llama, AJ007814] belonging to the artiodactyl suborder Tylopoda; and *Hippopotamus amphibius* [hippopotamus, AJ007813] belonging to the artiodactyl suborder Suiformes. Outgroup lactalbumin intronic sequences were available in databanks only for *Mus musculus* [mouse, M87863], *Rattus norvegicus* [rat, X00461], and *Cavia cutleri* [guinea pig, Y00726] from the order Rodentia; and for *Homo sapiens* [human, X05153] from the order Primates.

Given that the human intronic sequences were difficult to align with the ingroup taxa, we used that taxon only for the analyses of the exonic sequences (see below). The remaining sequences were aligned with Clustal W (Thompson *et al.*, 1994). Given that different alignments (obtained with different sets of alignment parameters) can yield different phylogenetic results, we used two different sets of alignment parameters (alignment 1: weighted matrix, gap penalty = 10, extension penalty = 5; alignment 2: weighted matrix, gap penalty = 8, extension penalty = 4) and positions at which the two alignments differed were excluded in the subsequent analyses (Gatesy *et al.*, 1993). All gaps introduced during alignment were located in introns except for a single three-nucleotide gap in exon 2 of the pig sequence. Gaps were encoded as single characters irrespective of their length, and added to the nucleotide data set. When gaps of different lengths overlapped, each size class was considered a different character state.

Given that mRNA sequences were also available for *Sus scrofa* [pig, M80520] and *Ovis aries* [sheep, X06367], we performed analyses using only exonic sequences together with these taxa.

2.1.2. Phylogenetic Analyses

Using different sets of characters (exons only, or introns + exons) and different outgroup and ingroup samplings (see results) all maximum parsimony (MP) analyses were performed with PAUP* (versions 4.0d56, d57, and d59; Swofford, in progress) with exact branch-and-bound searches. All characters were weighted equally. However, it is now well known that different types of changes can occur at different evolutionary rates, which may, in specific cases, justify differential weighting or encoding (e.g., Swofford *et al.*, 1996; Milinkovitch *et al.*, 1996). We therefore checked for possible saturation of nucleotide substitution types by plotting the number of transitions (Ti) and transversions (Tv) against the uncorrected pairwise distances. We also compared the outcome of our unweighted analyses with that of parsimony analyses where all Ti were excluded. We estimated the reliability of the various inferred clades by bootstrapping (1000 replicates, exact branch-and-bound searches), although we realize that bootstrap values may be misleading estimates of accuracy under specific conditions (e.g., Milinkovitch *et al.*, 1996). For selected branches and using the "constraints" command in PAUP* with exact branch-and-bound searches, we calculated Bremer branch support (BS)—the number of additional character transformations necessary to collapse an internal branch (Bremer, 1994)—as an alternative to bootstrap values (BV) as estimate of clade stability.

The stability of the cladogram obtained with unweighted parsimony was also tested using the Goboloff fit criterion (Goboloff, 1993) allowing the weighting of individual characters according to their implied homoplasy (less homoplasious characters are given more weight than more homoplasious characters).

Because maximum likelihood (ML) may outperform unweighted parsimony methods under some models of evolution (e.g., Swofford *et al.*, 1996), we used PAUP* (versions 4.0d56, d57, and d59; Swofford, in progress) heuristic searches to estimate ML trees. The settings used were: nucleotide frequencies computed from the data, some sites assumed to be invariable with proportion estimated via maximum likelihood, rates (for variable sites) assumed to follow a gamma distribution with shape parameter estimated by maximum likelihood (number of rate categories = 4, 6, 8, and 12; average rate for each category represented by mean), Hasegawa, Kishino, and Yano (1985) (HKY85) model with rate heterogeneity, transition/transversion ratio (Ti/Tv) estimated by maximum likelihood, starting branch lengths obtained using Rogers–Swofford approximation method, molecular clock not enforced, starting tree obtained by stepwise addition (as-is), and tree-bisection-reconnection (TBR) branch-swapping. We did not attempt to estimate bootstrap supports for the various nodes on the ML trees because of the high computational intensity of ML estimations.

We also generated ML trees with the program PUZZLE 3.1 (Strimmer and von Haeseler, 1997a), which implements a fast tree search heuristic called "quartet puzzling" (Strimmer and von Haeseler, 1996). The method works by reconstructing all possible quartet ML trees, then combining the quartet trees to an overall *n*-taxon tree. The procedure is typically repeated 1000 times (with randomization of the *n*-taxon input order), and the majority-rule consensus tree is computed from the 1000 *n*-taxon trees. This method is very fast and therefore especially well adapted to the ML analysis of data sets that include a large number of taxa. We did not use PUZZLE because of the speed of the analyses—our data set was small enough to allow estimation of ML trees in reasonable time with the more exhaustive searches implemented in PAUP* (versions 4.0d56, d57, and d59; Swofford, in progress)—but because it provides quartet puzzling branch support values that probably have the same practical meaning as bootstrap values (Strimmer and von Haeseler, 1996; von Haeseler, personal communication). We used the following settings for the PUZZLE analyses: nucleotide frequencies computed from the data, rates assume to follow a gamma distribution with shape parameter estimated via maximum likelihood (number of rate categories = 4 and 8), HKY85 model (Hasegawa *et al.*, 1985), Ti/Tv estimated via ML.

We also used PUZZLE 3.1 (Strimmer and von Haeseler, 1997a) with the JTT model (Jones *et al.*, 1992) and four gamma rate categories, and ProtML 2.3b3 (Adachi and Hasegawa, 1996) with the JTT model (Jones *et al.*, 1992), to estimate ML trees from the amino-acid translation of the exonic nucleotide sequences. Bootstrap proportions were estimated in ProtML 2.3b3 (Adachi and Hasegawa, 1996) by the RELL method (Kishino *et al.*, 1990) with 10^4 replications. Lastly, we used PAUP* (version d59; Swofford, in progress) to perform neighbor joining (NJ) and bootstrap NJ analyses after LogDet transformation (Steel, 1994; Lockhart *et al.*, 1994) of the data.

2.1.3. RASA and Likelihood Mapping Analyses

Optimal trees can be inferred from random (i.e., uninformative) data under MP, ML, or any other criterion. However, a recently developed method of data exploration termed RASA (relative apparent synapomorphy analysis; Lyons-Weiler *et al.*, 1996) may provide an *a priori* deterministic statistical measure of phylogenetic signal (i.e., natural cladistic hierarchy) in any (molecular, morphological, combined) character state matrix. RASA works by comparing the rate of increase of cladistic hierarchy among taxon pairs as phenetic similarity increases to a null rate of increase. The method is tree-independent, requires no randomization, and yields solutions in polynomial time. This is a practical advantage over other methods such as PTP (Archie, 1989; Faith and Cranston, 1991), which also test for nonrandom hierarchical structure but require permutations and tree building, making the tests difficult (or impossible) to apply with large numbers of taxa.

Testing for hierarchical structure was also performed with likelihood mapping (Strimmer and von Haeseler, 1997b) implemented in PUZZLE 3.1 (Strimmer and von Haeseler, 1997a). The method works by comparing, for each possible quartet of taxa, the maximum likelihood of each of the three possible fully resolved topologies among these four taxa. The three likelihoods are represented as a single point inside an equilateral triangle where seven regions have been defined: the central region represents starlike evolution (i.e., the three topologies have very similar likelihoods), the three areas corresponding to the triangle corners represent well-resolved topologies, and the three areas along the edges of the triangle correspond to situations where two of the three topologies have similarly high likelihoods. When phylogenetic signal is high, all or most data points get distributed among the three corner areas.

Because multiple substitutions can accumulate along long branches, the likelihood that molecular character states shared by one taxon and an outgroup will be based on random similarity rather than on history increases with increasing divergence between the outgroup and the ingroup taxa (e.g., Wheeler, 1990; Milinkovitch, 1995). Such random phenetic similarity between the outgroup and the ingroup increases the likelihood of spurious rooting of the ingroup (e.g., Wheeler, 1990; Milinkovitch *et al.*, 1996). However, RASA test statistics (*tRASA*) can be recalculated with the character states of an outgroup taxon (or of a combination of outgroup taxa) assumed to represent plesiomorphies for the ingroup (optimal outgroup analysis; Lyons-Weiler *et al.*, in press). The criterion required to accept any outgroup taxon or combination of outgroup taxa as a reasonable estimate of ingroup plesiomorphy is an increase of *tRASA* from unrooted to rooted analyses. Using different outgroup combinations, comparison among the test statistics might permit an objective identification of the combination that yields maximum phylogenetic signal in the ingroup. This allows the user to define the outgroup taxon (taxa) that will provide the best estimate of plesiomorphy for the ingroup in a global (i.e., outgroup + ingroup) phylogenetic analysis. Importantly, it is not necessarily the sister taxon to the ingroup that will provide the best estimate of plesiomorphy, even if evolutionary rates are equal (Lyons-Weiler *et al.*, in press). This does not contradict the fact that the probability of keeping more plesiomorphy information increases with closer outgroup.

One major concern in phylogeny estimation is that the loss of signal related to long branches (because of multiple substitutions on the same character) can create an artifact, well known as the "branch attraction" phenomenon (Felsenstein, 1978), which can yield erroneous phylogenetic estimates. Importantly, adding more characters causes the attraction to be stronger (Felsenstein, 1978). RASA methodology might provide an objective approach by which long branches can be identified. Indeed, taxa defining long branches will appear above the 1:1 line in graphs where cladistic variance is plotted versus phenetic variance for each taxon ("taxon variance plots"; Lyons-Weiler and Hoelzer, 1997). Note that the problematic branch(es) may not appear long in the phylogeny estimates (especially in parsimony analyses) because multiple hits will go undetected and, thus, the actual branch length underestimated. All RASA analyses were performed with RASA 2.1 (Lyons-Weiler, 1997).

2.2. Results and Discussion

2.2.1. DNA Sequences

We collected between 1071 and 1089 nucleotides (depending on the species) of DNA sequence spanning exon 2, introns 1 and 2, and partial exons 1 and 3 of the α-lactalbumin gene from various cetacean and artiodactyl taxa. The observed mean base frequencies were A: 25.3%, C: 22.4%, G: 21.2%, T: 31.1%. Uncorrected pairwise sequence divergences range from 0.9 to 40.4%. All new sequences were deposited at EMBL. The final alignment (after coding of gaps and removal of ambiguously aligned positions) included 994 characters and is available, on request, from the authors.

2.2.2. Phylogenetic Trees

The unweighted parsimony analysis yielded a single MP tree shown in Fig. 2. Bootstrap values (BV) for all branches and Bremer support (BS) for selected branches are indicated above and below the branches, respectively. That cladogram was stable to Goboloff weighting (for $k = 0, 2, 4,$ and 8). These MP results strongly support the paraphyly of artiodactyls and, more specifically, a sister-group relationship between hippos and cetaceans, as well as a sister-group relationship between ruminants and the hippo/cetacean clade. Constraining artiodactyl monophyly required a minimum of 19 additional evolutionary events and this traditional grouping was supported only in 0.2% of the bootstrap replicates.

Figure 3 shows that there was no obvious sign of saturation of Ti or Tv substitutions within the whole range of taxon comparisons. This strongly suggested keeping Ti and Tv

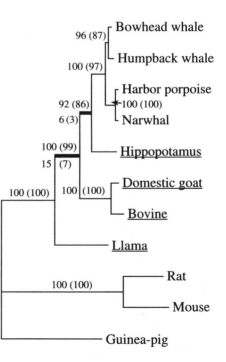

FIGURE 2. The unweighted parsimony analysis (exact branch-and-bound search) of the final alignment yielded a single tree whose phylogram is shown here (tree length = 895, CI = 0.8570, CI excluding uninformative characters = 0.7838, RI = 0.7797, RC = 0.6682). BV for all branches and BS for selected branches (thick) are indicated above and below the branches, respectively. The BV and BS in parentheses are those obtained in the transversion parsimony analysis (exact branch-and-bound). The ML analysis yielded a tree whose topology is identical to that presented here [−ln likelihood = 4735.22339, estimated Ti/Tv ratio = 2.251605 (κ = 4.542785), 14.6039% of sites estimated invariable, estimated value of gamma shape parameter = infinity; these results were stable when the number of rate categories for discrete approximation of gamma distribution was set to 4, 6, 8, or 12]. Artiodactyl taxa are underlined.

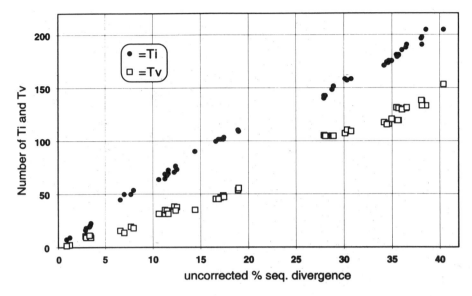

FIGURE 3. Accumulation of Ti and Tv with increasing sequence divergence (uncorrected) in the lactalbumin alignment including introns and exons. Ti and Tv seem to accumulate linearly within the whole range of taxon comparisons (outgroups are included). This result suggested keeping Ti and Tv unweighted in our MP analyses.

unweighted in our parsimony analyses. However, this obviously does not imply that the data set is void of undetected multiple substitutions. In Fig. 3, the lower-left group of data points (below 20% uncorrected sequence divergence) all correspond to pairwise comparisons among cetaceans and artiodactyl taxa—with the exception of the rat versus mouse comparison that also falls into this group—while the upper-right group of data points (above 25% uncorrected sequence divergence) all correspond to pairwise comparisons between one of the outgroup and one of the ingroup taxa. Even in the absence of extensive saturation, and even if one assumes that the lactalbumin sequences evolve in a clocklike fashion, lower Ti/Tv ratios for higher levels of divergence should be observed. This was exactly what we observed as the linear regression of Ti versus Tv across the whole range of taxon comparisons yielded a slope of 1.2578 ($R^2 = 0.9674$), whereas linear regression of Ti versus Tv up to 20% sequence divergence yielded a Ti/Tv ratio of 1.9865 ($R^2 = 0.949$), which is closer to that estimated by ML (see Fig. 2 caption). Parsimony analysis of transversions alone (exact branch-and-bound search) yielded a single tree with a topology identical to that shown in Fig. 2. BV and BS obtained under this coding scheme (Tv cost = 1, Ti cost = 0) are indicated on Fig. 2 in parentheses above and below the branches, respectively. Constraining artiodactyl monophyly required 10 additional Tv's.

The ML analysis with the maximum possible (in PAUP*) number of parameters estimated from the data (including the proportion of invariable sites and the shape parameter of the gamma distribution for variable sites; see Materials and Methods) yielded a tree whose topology is identical to that presented in Fig. 2. The length of the branches defined by the most basal taxa (including the outgroup branches and the llama branch) were, as expected, slightly increased compared with those estimated in the MP analysis because the ML procedure provides a correction for multiple hits. Surprisingly, the ML estimate of the gamma shape parameter was infinity, suggesting that there is a single rate category for variable sites. This result (and the likelihood of the tree) was stable when the number of rate categories for discrete approximation of gamma distribution was set to 4, 6, 8, or 12. This is surprising because the analyzed sequences included intronic and exonic sequences, and the latter category is usually observed to accumulate substitutions much more slowly than the former. Figure 4 shows that α-lactalbumin exonic and intronic sequences have evolved at very similar rates, confirming the ML result. In addition, for exonic sequences, one would anticipate much fewer constraints at third positions, and thus a much higher substitution rate, than at first and second positions. This was not what we observed for lactalbumin sequences (see Fig. 5a); the nucleotide substitutions are distributed more evenly among the three codon positions than, e.g., in cytochrome *b* (where third positions account for 73.7% of all substitutions) with similar taxa. A comparable result was observed for κ- and β-casein sequences (Gatesy *et al.*, 1996). Panels b–e of Fig. 5 show, for each codon position, various indices of fit (lower values indicate higher noise) obtained by constraining the ingroup topology shown in Fig. 2. These results indicate that (1) surprisingly, the third positions of the analyzed exonic sequences are less noisy than the first and second positions and (2) the cytochrome *b* data set is much more noisy than the lactalbumin exonic sequences, regardless of what codon position is considered. Note that the MP analysis of the full cytochrome *b* sequences (data not shown) from the ingroup species shown in Fig. 2 yielded the same topology as that based on the lactalbumin sequences. The ML analysis with the gamma shape parameter estimated from the data (eight discrete categories) but proportion of invariable sites set to zero yielded a tree whose topology is identical and likelihood very similar to those obtained when the proportion of invariables sites was estimated from the data (see above). This supports the idea discussed above that the analyzed characters fall into two categories: (1) the invariable sites and (2) the variable sites with a single rate. The

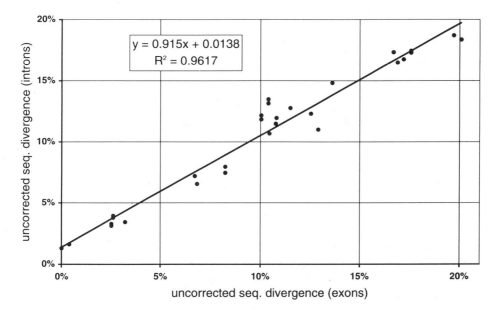

FIGURE 4. Exonic and intronic lactalbumin sequences have evolved at very similar rates. The exonic and intronic sequences include 279 and 673 characters, respectively.

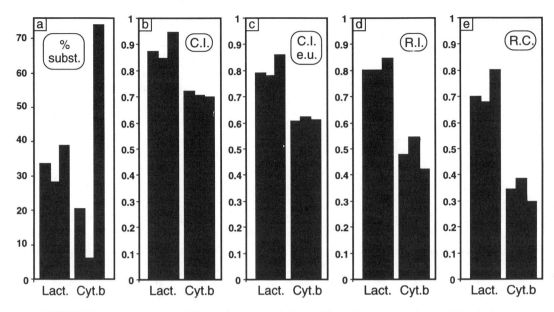

FIGURE 5. Percentage of nucleotide substitutions and indices of fit. (a) Proportions of nucleotide substitutions at each of the three condon positions in lactalbumin (present study) and full cytochrome *b* (e.g., from Milinkovitch *et al.*, 1996) sequences. For each codon position, we plotted consistency index (b), consistency index excluding uninformative characters (c), retention index (d), and rescaled consistency index (e) computed from the lactalbumin and the cytochrome *b* data sets with the ingroup topology shown in Fig. 2 constrained. Indices are consistently lower (hence, noise is consistently higher) for the cytochrome *b* than for the lactalbumin sequences. The MP analysis of the cytochrome *b* sequences from the ingroup species shown in Fig. 2 yielded the same topology (data not shown) as that obtained with lactalbumin sequences.

ML method can explain the data with very similar likelihoods either with proportion of invariable sites set to zero and a gamma distribution with a small shape parameter (here: 3.291010), or with proportion of invariable sites higher than zero (estimated to 0.146039) and the gamma shape parameter estimated to infinity. The issue of very similar likelihoods obtained over a wide range of values for invariable sites and gamma shape parameter is discussed in detail by Sullivan and Swofford (in press).

Rejecting the grouping of cetaceans with hippo in a clade did not yield artiodactyl monophyly and required a decrease in ln likelihood of 9.76931. Rejecting the grouping of cetaceans with hippo and ruminants in a clade did not yield artiodactyl monophyly and required a decrease in ln likelihood of 10.61802. Constraining artiodactyl monophyly decreased the ln likelihood by 20.25252. The two latter decreases in ln likelihood are significant at $p < 0.05$ under the test of Kishino and Hasegawa (1989).

The PUZZLE ML analyses (four or eight categories; see Materials and Methods) yielded the same topology as that shown in Fig. 2 with all nodes supported by 100% quartet puzzling support values except the baleen whale clade and the [cetaceans + hippo] clade that were supported by 99 and 94% values, respectively. Artiodactyl monophyly was supported in none of the replicates. Note that the three rodent taxa differed, on the basis of a 5% level chi-square test, in nucleotide composition from the frequency distribution of the ML model. This result is important because violation of the assumptions of stationarity and time re-versibility of the substitution probability matrix might yield systematic errors in tree esti-mation. We therefore applied to the lactalbumin data the LogDet distance transformation (Steel, 1994; Lockhart *et al.*, 1994), which is robust against base-composition heterogene-ity among taxa. We computed NJ trees and bootstrap NJ consensus trees with the propor-tion of invariable sites assumed to be zero and with the proportion of invariable sites as-sumed to be 0.146 (as estimated from the above ML analyses). Under the latter setting, we removed identical sites proportionally to base frequencies estimated from constant sites only (Waddell, 1995; Swofford *et al.*, 1996) because base composition was not homoge-neous throughout the tree. All of these analyses yielded NJ trees and bootstrap NJ consen-sus trees with the same topology as that shown in Fig. 2; the [cetaceans + hippo] and the [cetaceans + hippo + ruminants] clades were supported by 95 and 99% BV when the pro-portion of invariable sites was assumed to be zero, and 88 and 94% BV when the propor-tion of invariable sites was assumed to be 0.146. All other nodes were supported by BV higher than 95%.

2.2.3. RASA, Likelihood Mapping Analyses, and Additional Phylogenetic Trees

Results of rooted (with all possible combinations of outgroup taxa) and unrooted RASA are shown in Table I. Unrooted *tRASA* was lower when Ti were excluded (data not shown), supporting the suggestion that Ti should not be downweighted or excluded (see above). All rootings resulted in a significant increase of *tRASA*, indicating that each of these combina-tions of outgroup taxa is a reasonable estimate of ingroup plesiomorphy and, hence, is un-likely to produce spurious rooting. The presence of hierarchical structure was confirmed by likelihood mapping; 97.3% of the data points were observed in the corner areas (Fig. 6).

The taxon variance plot including only ingroup taxa (Fig. 7) indicated that the llama defines a long branch, hence, this taxon was identified as putatively problematic. Analyses of each separate intron and exon yielded a similar conclusion: Llama is located at the end of a long branch. This indicated that the long branch was not a product of an accelerated evolution at portion(s) of the analyzed sequences. RASA results without llama are also

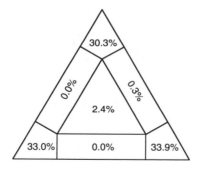

FIGURE 6. Likelihood-mapping analysis for the lactalbumin data set (exons + introns). The central triangular area of attraction represents starlike evolution, the three corner areas of attraction correspond to well-resolved topologies, and the three remaining areas of attraction correspond to situations where two topologies have simi-larly high likelihoods.

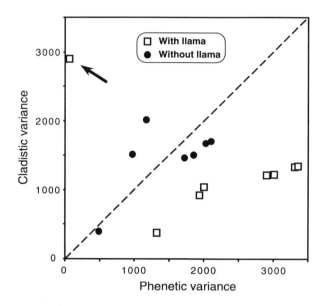

FIGURE 7. Taxon variance plot computed with all ingroup taxa included (open squares) or with llama deleted (solid circles). Putatively problematic taxa (because on long branches) are outliers above the 1/1 line, while apparently unproblematic taxa appear below that line (Lyons-Weiler and Hoelzer, 1997). This analysis recognized llama (indicated by an arrow) as defining a long branch. The taxon variance plot excluding llama (solid circles) is much less heterogeneous.

shown in Table I. Note that signal was more significant in the unrooted analysis when llama was deleted than when included (*tRASA* was higher and degrees of freedom lower). Recomputation of the cladistic and phenetic variance after removal of llama (Fig. 7) produced a plot with much less heterogeneous variance. This could indicate a lower chance of branch attraction among the remaining taxa. After removal of the llama, RASA optimum outgroup analysis revealed that the outgroup combinations [mouse + guinea pig] and [mouse + rat + guinea pig] yield the highest increase of *tRASA* (Table I), hence suggesting that these

Table I. Results of Unrooted and Rooted RASA

Outgroup	With llama			Without llama		
	β obs.	β null	*tRASA*	β obs.	β null	*tRASA*
None	1.815408	1.548353	0.9600289	2.400797	1.722297	2.439481
Rat (r)	2.200867	1.700296	1.739504	2.611616	1.846722	2.251941
Mouse (m)	2.418056	1.734929	2.073221	2.941844	1.88247	2.870231
Guinea pig (gp)	2.394627	1.824448	2.00842	2.962997	2.010626	2.908487
r + m	2.463855	1.762772	2.150757	2.834132	1.890309	2.575113
r + gp	2.563023	1.928926	2.063819	3.191096	2.174484	2.861135
m + gp	2.72218	1.93488	2.414013	3.438589	2.202041	3.439528
m + r + gp	2.739802	1.967532	2.371699	3.420066	2.234177	3.294901

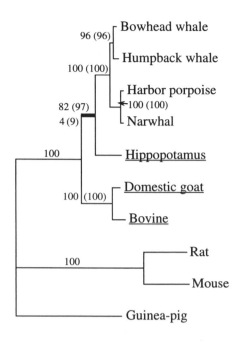

FIGURE 8. Unweighted parsimony analysis (branch-and-bound) excluding llama yields a single tree whose phylogram is shown here (tree length = 785, CI = 0.8981, CI excluding uninformative characters = 0.8403, RI = 0.8394, RC = 0.7538). BV for all branches and BS for one selected branch (thick) are indicated above and below the branches, respectively. The BV and BS in parentheses are those obtained when only the guinea pig is used as outgroup. The ML analysis yielded a tree whose topology is identical to that presented here [ln likelihood = 4268.95245, estimated Ti/Tv ratio = 2.212468 (κ = 4.469899), 15.4017% of sites estimated invariable, estimated value of gamma shape parameter = infinity]. Artiodactyl taxa are underlined.

combinations be used in the subsequent MP analyses. Note that the taxon variance plot exhibited very low variance heterogeneity when apparent autapomorphies were excluded, even when llama was included (data not shown).

Given the above RASA results, we reanalyzed our alignment with llama excluded. The resulting single MP tree is shown in Fig. 8 with BV and BS indicated above and below the branches, respectively. Among the single-taxon rooted RASA, the guinea pig yielded the highest increase of *tRASA*, suggesting that, when llama was deleted, this taxon provided a better estimate of the ingroup plesiomorphy than the two other rodent taxa. In addition, the guinea pig (1) was less of an outlier than rat or mouse in the unrooted taxon variance plot including all ingroup and outgroup taxa (data not shown) and (2) was positioned at the tip of a shorter branch than those of rat and mouse in the estimated phylograms (Figs. 2 and 8). We realize that branch lengths are underestimated in a parsimony analysis. BV and BS obtained under this outgroup choice are given in parentheses in Fig. 8 above and below the branches, respectively.

Goboloff weighting (k = 0, 2, 4, and 8) but also all ML analyses (see Materials and Methods) yielded a tree whose topology is identical to that presented in Fig. 8.

2.2.4. Analysis of the Extended Exon Data Set

We used lactalbumin exonic sequences from all available eutherian mammalian species in an extended (in terms of number of taxa) data set including all of the species analyzed above plus human, pig, and sheep. using rat, mouse, guinea pig, and human as outgroup taxa, the unweighted parsimony analysis yielded ten MP trees whose strict consensus is shown in Fig. 9. It is difficult to estimate whether the reduction in resolution (in compari-

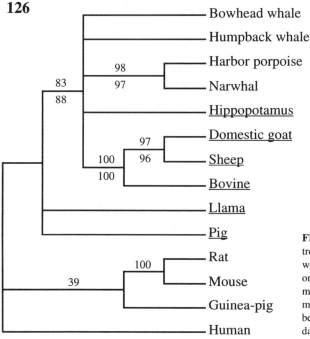

FIGURE 9. Strict consensus of the ten best trees (tree length = 261) obtained under unweighted MP analysis of all lactalbumin exonic sequences. BV obtained with [rat, mouse, guinea pig, human] and [mouse, human] as outgroup are indicated above and below the branches, respectively. Artiodactyl taxa are underlined.

son with the analyses above) is related to an increase in the number of taxa or a decrease in the number of characters, or both. BV for resolved branches are indicated above the branches. Constraining artiodactyl monophyly required a minimum of seven additional evolutionary events and this traditional grouping was supported in none of the bootstrap replicates. The values below the branches (Fig. 9) are BV obtained with [mouse + human] as outgroup. RASA analyses identified this combination as the best combination of outgroup taxa (data not shown). None of the ML analyses of the extended data set of exonic sequences supported artiodactyl monophyly (data not shown).

ML analyses of the amino-acid translation of these exonic sequences yielded topologies compatible with that shown in Fig. 9. The [cetaceans + hippo + ruminant] clade was supported by a 85% quarter puzzling value (PUZZLE 3.1; Strimmer and von Haeseler, 1997a) and a 93% RELL-bootstrap value (ProtML 2.3b3; Adachi and Hasegawa, 1996).

In summary, these analyses of exonic sequences alone again indicate artiodactyl paraphyly despite the low number of characters analyzed.

2.2.5. Morphology versus Molecules

Undoubtedly, some of the problems encountered in phylogeny estimation are specific to molecular data. For example, extinct taxa cannot be sampled for molecular studies but in exceptional cases (e.g., DeSalle *et al.*, 1992). Furthermore, the widely used mitochondrial DNA sequences might be evolving too quickly to provide reliable resolution of ancient nodes. However, it is unlikely that these two aspects are problematic regarding the artiodactyl monophyly/paraphyly issue as numerous artiodactyl, cetacean, and outgroup taxa have been sampled for various mitochondrial and nuclear genes (this study, and Gatesy, this volume). It is ironic that the results of molecular data analyses are now criticized because "functional convergence of encoded proteins is a potential problem" (Theodor, 1996). It is obvious that homoplasy (including convergences) is the major confounding problem (almost by definition) in phylogeny estimation, but it is certainly not more the case in analyses of nuclear DNA sequences than in the analyses of morphological characters.

Prothero *et al.* (1988), Gentry and Hooker (1988), and Prothero (1993) suggested only seven morphological synapomorphies for artiodactyls, and at least two of these character states (e.g., paraxony) are also present in early whales (Thewissen *et al.*, 1996). Probably referring to the remaining five characters, it is often considered that the hypothesis of artiodactyl monophyly is "strongly supported" by morphological data (Theodor, 1996). This view, which implicitly assumes that molecular data are necessarily more noisy than morphological data, is debatable (e.g., Milinkovitch and Thewissen, 1997) as some molecular characters indicating artiodactyl paraphyly seem essentially devoid of noise (Shimamura *et al.*, 1997).

If the molecular hypothesis of artiodactyl paraphyly is correct, missing data might be one of the reasons (besides, e.g., functional convergences) why morphological data yield a seemingly erroneous support for artiodactyl monophyly. For example, the presence of a trochleated astragalus is generally considered as one of the best synapomorphies for this traditional grouping because this character state is present in all artiodactyls and in no other mammal. However, the character cannot be scored in modern whales (because of the dramatic reduction of the hind limbs) and no complete functional astragalus is known for a fossil whale. In fact, the whale lineage is considered to lack a trochleated astragalus because the mesonychians (regarded, on the basis of dental characters, to be the sister group of cetaceans, Thewissen, 1994; Zhou *et al.*, 1995) have a nontrochleated astragalus. We think that the possibility either (1) that the astragalus character is homoplasious or (2) that mesonychians are not the sister group to cetaceans should be considered as reasonable alternatives to the rejection of the molecular hypothesis based on thousands of characters from several independent loci. The discovery of new early cetacean fossils could potentially help to resolve the astragalus issue.

Another character considered as a good synapormorphy for artiodactyls is the deciduous lower premolar 4 that consists of three rows of cusps only in artiodactyls (and not in early cetaceans or mesonychians). It is, however, well known that dental characters are subject to convergences because of functional constraints, and thus, dental characters are likely to covary with dietary similarities.

A general major problem in the investigation of cetacean origin is their highly modified and specialized morphology (even in early whales) that makes homology statements and polarization of the morphological characters difficult (e.g., Milinkovitch, 1995, 1997). Regarding the molecular hypothesis that cetaceans and hippopotamuses form a clade, several nonmolecular characters support this provocative grouping (Gatesy, 1997). One of these characters is the ability to vocalize underwater (Barklow, 1995). In the context of the molecular hypothesis of toothed-whale paraphyly, we first suggested (Milinkovitch *et al.*, 1993, 1994; Milinkovitch, 1994, 1995) that echolocation capabilities may have been present in the ancestor of all extant whales (making the presence of echolocation in odontocetes a shared ancestral character state). Because (1) the melon (a fatty acoustic lens located in the forehead of toothed whales) is considered an important component of the echolocation system and (2) a vestigial melon has been described in baleen whales (Heyning and Mead, 1990), we suggested that the ancestor of all extant whales may have possessed a well-developed melon and correspondingly well-developed echolocation capabilities (Milinkovitch, 1994, 1995). We tentatively suggest that the ability to use underwater sound emissions for communication and navigation might have started to develop in the common ancestor of cetaceans and hippopotamuses. Obviously, further investigations on the use and function of the sounds emitted underwater by hippopotamuses are warranted.

It is unfortunately difficult to evaluate the amount of conflict between morphological and molecular characters because no morphological character state matrix including all major artiodactyl clades is available. Such a matrix would allow identification of morphological characters that increase/decrease signal when combined with the molecular data.

We do not consider that molecular data are necessarily superior to morphological data for phylogeny estimation. We are well aware of the problems inherent to molecular sequence data such as nucleotide and codon composition biases, ambiguities in alignments, and undetected paralogies. We consider that these potential problems are unlikely to have played a significant role in the molecular analysis of artiodactyl and cetacean phylogeny because of the diversity of the genes analyzed (e.g., alignment was trivial for some of these loci). On the other hand, we consider that it has not been convincingly demonstrated that well-known problems inherent to morphological data (such as the difficulties in objectively defining morphological characters and character states) have not played a significant role in the support of the artiodactyl monophyly hypothesis.

Regarding the lactalbumin data analyzed in the present study, the identified potential problems (nucleotide composition bias for rodents, and the existence of a long branch defined by llama) will probably be easily overcome by the use of less divergent outgroup taxa (such as several representatives of the orders Perissodactyla and Carnivora) and by the use of additional ingroup taxa (especially pigs and/or peccaries but also additional tylopods). When all morphological and molecular data available to date are considered, the suggestion that cetaceans are nested within the artiodactyl phylogenetic tree (making artiodactyls paraphyletic) is the best-supported hypothesis for the origin of whales. To acknowledge that whales, dolphins, and porpoises are, most likely, highly derived artiodactyls, we consider that the taxonomic rank of cetaceans should be lowered from an ordinal level and included in the order Artiodactyla.

Acknowledgments

Dr. Dave Irwin kindly provided some of the primers, the Copenhagen Zoo kindly provided samples from the llama and hippopotamus, and Greenlandic Nature Research Institute the humpback whale, the narwhal, and the harbor porpoise samples. We thank Ralph Tiedemann and Arndt von Haeseler for helpful comments on an earlier version of the manuscript. We are particularly grateful to Dave L. Swofford for giving us the opportunity to use the successive beta versions of PAUP*.

References

Adachi, J., and Hasegawa, M. 1996. MOLPHY: programs for molecular phylogenetics, version 2.3. Institute of Statistical Mathematics, Tokyo.

Archie, J. W. 1989. A randomization test for phylogenetic information in systematic data. *Syst. Zool.* **38**:219–252.

Barklow, W. 1995. Hippo talk. *Nat. Hist.* **104**:54.

Barnes, L. G. 1984. Whales, dolphins, and porpoises: origin and evolution of the Cetacea, in: T. W. B. Broadhead (ed.), *Mammals: Notes for a Short Course organized by P. D. Gingerich and C. E. Badgley*, pp. 139–158. University of Tennessee Studies in Geology 8.

Boyden, A., and Gemeroy, D. 1950. The relative position of the Cetacea among the orders of Mammalia as indicated by precipitin tests. *Zoologica* **35**:145–151.

Bremer, K. 1994. Branch support and tree stability. *Cladistics* **10**:295–304.

Czelusniak, J., Goodman, M., Koop, B. F., Tagle, D. A., Shoshani, J., Braunitzer, G., Kleinschmidt, T. K., De Jong, W. W., and Matsuda, G. 1990. Perspectives from amino acid and nucleotide sequences on cladistic relationships among higher taxa of Eutheria, in: H. H. Genoways (ed.), *Current Mammalogy*, Volume 2, pp. 545–572. Plenum Press, New York.

DeSalle, R. J., Gatesy, J., Wheeler, W., and Grimaldi, D. 1992. DNA sequences from a fossil termite in Oligo-Miocene amber and their phylogenetic implications. *Science* **257**:1933–1936.

Ebner, K. E., and Schanbacher, F. 1974. In: B. L. Larson and V. R. Smith (eds.), *Lactation: A Comprehensive Treatise*, Volume 2, Academic Press, New York.

Faith, D. P., and Cranston, P. S. 1991. Could a cladogram this short have arisen by chance alone? On permutation test for cladistic structure. *Cladistics* **7**:1–28.

Felsenstein, J. 1978. Cases in which parsimony and compatibility methods will be positively misleading. *Syst. Zool.* **27**:401–410.

Flower, W. H. 1883. On whales, present and past and their probable origin. *Proc. Zool. Soc. London* **1883**:466–513.

Gatesy, J. 1997. More support for a Cetacea/Hippopotamidae clade: the blood-clotting protein gene γ-fibrinogen. *Mol. Biol. Evol.* **14**:537–543.

Gatesy, J., DeSalle, R., and Wheeler, W. 1993. Alignment-ambiguous nucleotide sites and the exclusion of systematic data. *Mol. Phylogenet. Evol.* **2**:152–157.

Gatesy, J., Hayashi, C., Cronin, M. A., and Arctander, P. 1996. Evidence from milk casein genes that cetaceans are close relatives of hippopotamid artiodactyls. *Mol. Biol. Evol.* **13**:954–963.

Gentry, A., and Hooker, J. 1988. The phylogeny of the Artiodactyla, in: M. Benton (ed.), *The Phylogeny and Classification of the Tetrapods,* Volume 2, pp. 25–272. Clarendon Press, Oxford.

Gingerich, P. D., Smith, B. H., and Simons, E. L. 1990. Hand limbs of Eocene *Basilosaurus isis:* evidence of feet in whales. *Science* **249**:154–157.

Goboloff, P. 1993. Estimating character weights during tree search. *Cladistics* **9**:83–91.

Goodman, M., Czelusniak, J., and Beeber, J. E. 1985. Phylogeny of primates and other eutherian orders: a cladistic analysis using amino acid and nucleotide sequence data. *Cladistics* **1**:171–185.

Graur, D., and Higgins, D. G. 1994. Molecular evidence for the inclusion of cetaceans within the order Artiodactyla. *Mol. Biol. Evol.* **11**:357–364.

Hasegawa, M., and Adachi, J. 1996. Phylogenetic position of cetaceans relative to artiodactyls: reanalysis of mitochondrial and nuclear sequences. *Mol. Biol. Evol.* **13**:710–717.

Hasegawa, M., Kishino, H., and Yano, T. 1985. Dating of the human–ape splitting by a molecular clock of mitochondrial DNA. *J. Mol. Evol.* **22**:160–174.

Heyning, J. E., and Mead, J. G. 1990. Evolution of the nasal anatomy of cetaceans, in: J. Thomas and R. Kastelein (eds.), *Sensory Abilities of Cetaceans*, pp. 67–79. Plenum Press, New York.

Irwin, D. M., Kocher, T. D., and Wilson, A. C. 1991. Evolution of the cytochrome *b* gene of mammals. *J. Mol. Evol.* **32**:128–144.

Jones, D. T., Taylor, W. R., and Thornton, J. M. 1992. The rapid generation of mutation data matrices from protein sequences. *Comput. Appl. Biosci.* **8**:275–282.

Kishino, H., and Hasegawa, M. 1989. Evaluation of the maximum likelihood estimate of the evolutionary tree topologies from DNA sequence data, and the branching order in Hominoidea. *J. Mol. Evol.* **29**:170–179.

Kishino, H., Miyata, T., and Hasegawa, M. 1990. Maximum likelihood inference of protein phylogeny, and the origin of chloroplasts. *J. Mol. Evol.* **30**:151–160.

Lockhart, P. J., Steel, M. A., Hendy, M. D., and Penny, D. 1994. Recovering evolutionary trees under a more realistic model of sequence evolution. *Mol. Biol. Evol.* **11**:605–612.

Lyons-Weiler, J. 1997. RASA 2.1. Software and documentation for Macintosh. Distributed by the author; http://loco.biology.unr.edu/archives/rasa/rasa.html.

Lyons-Weiler, J., and Hoelzer, G. A. 1997. Escaping from the Felsenstein zone by detecting long branches in phylogenetic data. *Mol. Phylogenet. Evol.* **8**(3):375–384.

Lyons-Weiler, J., Hoelzer, G. A., and Tausch, R. J. 1996. Relative apparent synapomorphy analysis (RASA) I: the statistical measurement of phylogenetic signal. *Mol. Biol. Evol.* **13**:749–757.

Lyons-Weiler, J., Hoelzer, G. A., and Tausch, R. J. In press. Relative apparent synapomorphy analysis (RASA) II: optimal outgroup analysis. *Biol. J. Linn. Soc.*

Milinkovitch, M. C. 1992. DNA–DNA hybridizations support ungulate ancestry of Cetacea. *J. Evol. Biol.* **5**:149–160.

Milinkovitch, M. C. 1994. Phylogenetic analyses of molecular data in vertebrates with special emphasis on the implications of mitochondrial DNA sequences for reevaluating morphological and behavioral evolution in cetaceans. Ph.D. thesis, Brussels Free University.

Milinkovitch, M. C. 1995. Molecular phylogeny of cetaceans prompts revision of morphological transformations. *Trends Ecol. Evol.* **10**:328–334.

Milinkovitch, M. C. 1997. The phylogeny of whales: a molecular approach, in: A. E. Dizon, S. J. Chivers, and W. F. Perrin (eds.), *Molecular Genetics of Marine Mammals*, pp. 317–338. Society for Marine Mammology, Lawrence, KS.

Milinkovitch, M. C., and Thewissen, J. G. M. 1997. Eventoed fingerprints on whale ancestry. *Nature* **388**:622–624.

Milinkovitch, M. C., Ortí, G., and Meyer, A. 1993. Revised phylogeny of whales suggested by mitochondrial ribosomal DNA sequences. *Nature* **361**:346–348.

Milinkovitch, M. C., Meyer, A., and Powell, J. R. 1994. Phylogeny of all major groups of cetaceans based on DNA sequences from three mitochondrial genes. *Mol. Biol. Evol.* **11**:939–948.

Milinkovitch, M. C., Leduc, R. G., Adachi, J., Farnir, F., Georges, M., and Hasegawa, M. 1996. Effects of character weighting and species sampling on phylogeny reconstruction: a case study based on DNA sequence data in cetaceans. *Genetics* **144**:1817–1833.

Montgelard, C., Catzeflis, F. M., and Douzery, E. 1997. Phylogenetic relationships of artiodactyls and cetaceans as deduced from the comparison of cytochrome *b* and 12S rRNA mitochondrial sequences. *Mol. Biol. Evol.* **14**:550–559.

Philippe, H., and Douzery, E. 1994. The pitfalls of molecular phylogeny based on four species, as illustrated by the Cetacea/Artiodactyla relationships. *J. Mamm. Evol.* **2**:133–152.

Prothero, D. 1993. Ungulate phylogeny: molecular *versus* morphological evidence, in: F. Szalay, M. Novacek, and M. McKenna (eds.), *Mammal Phylogeny*, Volume 2, pp. 173–181. Springer-Verlag, Berlin.

Prothero, D., Manning, E., and Fischer, M. 1988. The phylogeny of the ungulates, in: M. J. Benton (ed.), *The Phylogeny and Classification of the Tetrapods*, Volume 2, pp. 201–234. Clarendon Press, Oxford.

Sambrook, J., Fritsch, E. F., and Maniatis, T. 1989. *Molecular Cloning: A Laboratory Manual*, 2nd ed. Cold Spring Harbor Laboratory Press, Cold Spring Harbor, NY.

Shimamura, M., Yasue, H., Ohshima, K., Abe, H., Kato, H., Kishiro, T., Goto, M., Munechika, I., and Okada, N. 1997. Molecular evidence from retroposons that whales form a clade within even-toed ungulates. *Nature* **388**:666–671.

Shoshani, J. 1986. Mammalian phylogeny: comparison of morphological and molecular results. *Mol. Biol. Evol.* **3**:222–242.

Smith, M., Shivji, M., Waddell, V., and Stanhope, M. 1996. Phylogenetic evidence from the IRBP gene for the paraphyly of toothed whales, with mixed support for Cetacea as a suborder of artiodactyls. *Mol. Biol. Evol.* **13**:918–922.

Steel, M. 1994. Recovering a tree from the Markov leaf colourations it generates under a Markov model. *Appl. Math. Lett.* **7**:19–23.

Strimmer, K., and von Haeseler, A. 1996. Quartet puzzling: a quartet maximum likelihood method for reconstructing tree topologies. *Mol. Biol. Evol.* **13**:964–969.

Strimmer, K., and von Haeseler, A. 1997a. PUZZLE 3.1, maximum likelihood analysis for nucleotide and amino acid alignments. Software and documentation distributed by the authors; http://www.zi.biologie.uni-muenchen.dev́strimmer/puzzle.html.

Strimmer, K., and von Haeseler, A. 1997b. Likelihood-mapping: a simple method to visualize phylogenetic content of a sequence alignment. *Proc. Natl. Acad. Sci. USA* **94**:6815–6819.

Sullivan, J., and Swofford, D. L. In press. Uncertainty in estimating parameters of invariable-sites plus gamma models of rate heterogeneity: the effect of taxon sampling. *Mol. Biol. Evol.*

Swofford, D. L. 1997. PAUP*: Phylogenetic Analysis Using Parsimony (and other methods), Versions 4.0d56, d57, d59, in progress. Sinauer Associates, Sunderland, MA.

Swofford, D. L., Olsen, G. J., Waddell, P. J., and Hillis, D. 1996. Phylogenetic inference, in: D. M. Hillis, C. Moritz, and B. K. Mable (eds.), *Molecular Systematics*, 2nd ed, pp. 407–514. Sinauer Associates, Sunderland, MA.

Theodor, J. M. 1996. Why do molecules and morphology conflict? Examination of the Artiodactyl–Cetacea relationship. *J. Vertebr. Paleontol.* **17**(Suppl.):80A.

Thewissen, J. G. M. 1994. Phylogenetic aspects of cetacean origins: a morphological perspective. *J. Mamm. Evol.* **2**:157–184.

Thewissen, J. G. M., and Hussain, S. T. 1993. Origin of underwater hearing in whales. *Nature* **361**:444–445.

Thewissen, J. G. M., Madar, S. I., and Hussain, S. T. 1996. *Ambulocetus natans*, an Eocene cetacean (Mammalia) from Pakistan. *Cour. Forsch.-Inst. Senckenberg* **191**:1–86.

Thompson, J. D., Higgins, D. G., and Gibson, T. J. 1994. CLUSTAL W: improving the sensitivity of progressive multiple sequence alignment through sequence weighting, positions-specific gap penalties and weight matrix choice. *Nucleic Acids Res.* **22**:4673–4680.

Vilotte, J.-L., Soulier, S., Mercier, J.-C., Gaye, P., Hue-Delahaie, D., and Furet, J.-P. 1987. Complete nucleotide sequence of α-lactalbumin gene: comparison with its rat counterpart. *Biochimie* **69**:609–620.

Waddell, P. J. 1995. Statistical methods of phylogenetic analysis, including Hadamar conjugations, LogDet transforms, and maximum likelihood. Ph.D. dissertation, Massey University.

Wheeler, W. C. 1990. Nucleic acid sequence phylogeny and random outgroups. *Cladistics* **6**:363–367.

Zhou, X., Zhai, R., Gingerich, P. D., and Chen, L. 1995. Skull of a new mesonychid (Mammalia, Mesonychia) from the late Paleocene of China. *J. Vertebr. Paleontol.* **15**:387–400.

CHAPTER 5

Phylogenetic and Morphometric Reassessment of the Dental Evidence for a Mesonychian and Cetacean Clade

MAUREEN A. O'LEARY

1. Introduction

Van Valen (1966, 1968, 1969, 1978) first hypothesized that mesonychians, an aberrant group of carnivorous mammals nested within the ungulates, are closely related to cetaceans. Discoveries of new cetacean and mesonychian fossils have prompted detailed descriptions of taxa in both groups (Szalay, 1969a,b; Ting and Li, 1987; Gingerich and Russell, 1990; O'Leary and Rose, 1995a,b; Zhou *et al.*, 1995; Thewissen *et al.*, 1996), classificatory changes (McKenna, 1975; McKenna and Bell, 1997), and cladistic analyses (Prothero *et al.*, 1988; Thewissen, 1994; Zhou *et al.*, 1995) (Fig. 1).

 Mesonychia has included Mesonychidae (Mesonychinae and Andrewsarchinae) and Hapalodectidae (Hapalodectinae) (Szalay and Gould, 1966; Zhou *et al.*, 1995). Others argue for the inclusion of Didymoconidae (Gingerich, 1981), but this arrangement is not widely accepted (Meng *et al.*, 1994; McKenna and Bell, 1997). It is very likely that inclusion of all four families noted above in Mesonychia makes the order paraphyletic with respect to Cetacea.

 McKenna (1975) established the mirorder Cete for Cetacea and its sister taxon. Different authors include different taxa in Cete (Fig. 1); for example, Mesonychia, Cetacea, triisodontines, *Oxyclaenus*, and *Baioconodon* (Prothero *et al.*, 1988), or Mesonychia, Cetacea, and Didymoconidae (Thewissen, 1994).

 The living sister taxon of Cetacea has been shown by some studies to be Perissodactyla (Novacek and Wyss, 1986; Prothero *et al.*, 1988; Novacek, 1989; Thewissen, 1994). However, an increasing number of molecular (Fitch and Beintema, 1990; Graur and Higgins, 1994; Gatesy *et al.*, 1996; Xu *et al.*, 1996) and morphological (Gingerich *et al.*, 1990; Novacek, 1992; Thewissen and Hussain, 1993) studies now find that the closest extant cetacean

MAUREEN A. O'LEARY • Department of Cell Biology, New York University School of Medicine, New York, New York 10016. *Present address*: Department of Anatomical Sciences, Health Sciences Center, State University of New York at Stony Brook, Stony Brook, New York 11794-8081.

The Emergence of Whales, edited by Thewissen. Plenum Press, New York, 1998.

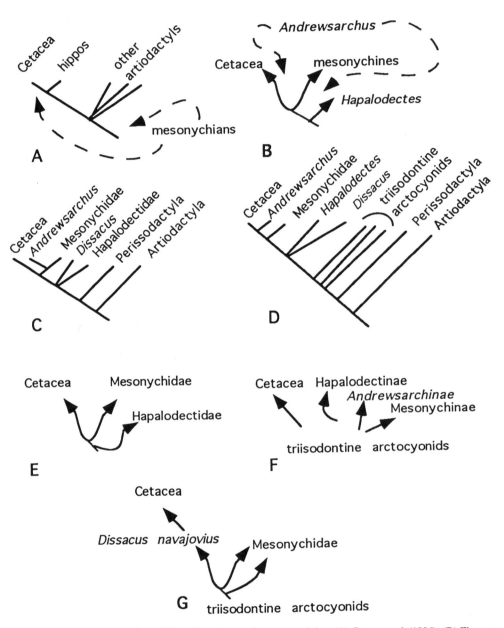

FIGURE 1. Summary of various different hypotheses of cetacean origins. (A) Gatesy *et al.* (1996); (B) Zhou *et al.* (1995); (C) Thewissen (1994); (D) Prothero *et al.* (1988); (E) Ting and Li (1987); (F) Szalay (1969a); (G) Van Valen (1966, 1978). A is based on molecular evidence, B through G on morphological evidence. Disagreement exists as to whether Mesonychia or Artiodactyla is paraphyletic with respect to Cetacea.

relative is among the Artiodactyla. Furthermore, some molecular studies find not only a close relationship between Cetacea and Artiodactyla, but also that Cetacea is nested within Artiodactyla, a result for which there has been little support from the fossil record.

Despite the growing consensus among morphologists that mesonychians are the sister taxon of Cetacea, there is no consensus as to which mesonychian(s) is the sister taxon or which characters support this relationship (Fig. 1). Most morphological studies argue that the sister taxon is Mesonychidae (sometimes excluding *Dissacus*) and/or *Andrewsarchus* on the basis of a variety of dental and cranial characters. The amount of disagreement (Fig. 1) underscores the need for detailed morphological descriptions and phylogenetic analyses of the relevant taxa, the results of which should narrow the variety of hypotheses.

Thewissen's (1994) study of cranial, dental, and postcranial data found that dental characters are among the best morphological criteria for diagnosing the clade Cete. An examination of his character taxon matrix indicates that it is dental characters that support a sister-group relationship between Cetacea and a clade including *Andrewsarchus* and the Mesonychidae (excluding *Dissacus*). The distribution of some of Thewissen's (1994) dental characters merits closer examination. For example, absence of the P^4 protocone, a supposed synapomorphy linking Mesonychidae and Cetacea (Thewissen, 1994), does not occur in a number of mesonychids.

Synapomophies used by other authors to support a close relationship between Cetacea and Mesonychia also require further examination. For example, one synapomorphy, rostrocaudally aligned incisors, used to support a sister taxon relationship between *Andrewsarchus* and Cetacea to the exclusion of other mesonychians (Prothero *et al.*, 1988), has been questioned (Thewissen *et al.*, 1983; Thewissen, 1994) because a similar condition exists in the dichobunid artiodactyl *Diacodexis*. Ting and Li (1987, p. 185) also challenged whether the presence of vascularized embrasure pits is a synapomorphy supporting a sister taxon relationship between hapalodectids and Cetacea. They emphasized that embrasure pits appear in other mammals such as *Didelphis*, and may develop convergently in mammals with tall protoconids on the lower molars. Finally, the morphology of the reentrant groove on the lower molars, another potential synapomorphy of Cetacea and Hapalodectidae, has not been well described or illustrated. A clear understanding of dental synapomorphies diagnosing both the sister taxon of Cetacea and the larger clade Cete is crucial for the accurate placement of new fossil cetans, many of which are first known only from dentitions. This paper examines and synthesizes dental information pertaining to the debate on cetacean origins.

The dental evidence for a cetacean–mesonychian relationship is examined here with particular focus on recently described archaic cetaceans from Pakistan: *Nalacetus, Pakicetus, Ambulocetus*, and *Attockicetus* (Thewissen *et al.*, 1996; Thewissen and Hussain, 1998, in press). A new mesonychid from Pakistan is described and a cladogram based on dental characters assembled. The shape of the occlusal outlines of P^4 and M^1 of mesonychians, archaic cetaceans, arctocyonids, and a primitive artiodactyl are then examined using eigenshape analysis. Eigenshape analysis describes the occlusal outline shapes of the P^4's and M^1's of early cetaceans to determine whether or not archaic cetaceans most closely resemble mesonychians or other mammals such as *Chriacus, Arctocyon*, and *Diacodexis*. P^4 is of particular interest because it assumes very distinctive morphologies early in cetacean evolution with an enlarged paracone and a reduced protocone (Thewissen, 1994; Thewissen and Hussain, 1998; Thewissen *et al.*, 1996), and M^1 because it has been argued that mesonychids and primitive cetaceans may have similarly narrow trigon basins (Thewissen, 1994). Finally, a general comparison of toothwear patterns in mesonychians and archaic cetaceans is presented.

Dental terminology follows Van Valen (1966, pp. 7–9). Institutional abbreviations are listed in Section 3.1.1.

2. Systematic Paleontology

Mirorder CETE McKenna, 1975
Order MESONYCHIA Van Valen, 1969
Family MESONYCHIDAE Cope, 1875
Mesonychid sp. A
Figure 2A and B

Referred Material.—H-GSP 96134, a single tooth, either left P^4 or M^2, collected by M. Arif, 1996 and ONG/K/19 (Oil and Natural Gas Commission, Dehra Dun, India; Ranga Rao, 1973).

Localities.—H-GSP 96134 from H-GSP Locality 9205, Upper Kuldana Formation, Ganda Kas Area, Kala Chitta Hills, Pakistan. Locality 9205 is 500 m west of H-GSP Locality 58 (West and Lukacs, 1979). ONG/K/19 is from the Subathu Formation of northern India.

Age and Distribution.—Eocene of Pakistan and India.

Diagnosis.—Differs from all other mesonychids, except *Dissacus willwoodensis*, in having well-developed pre- and postcingula. Approximately three times larger than *D. willwoodensis.*

Description.—The three-rooted tooth (H-GSP 96134) has a well-developed protocone lobe, similar in shape to the P^4's of *Ankalagon saurognathus, Pachyaena gracilis*, and *Mesonyx obtusidens* and to the M^2's of *Harpagolestes orientalis*. The protocone is not preserved, but based on the occlusal outline shape compared with other mesonychids, it was a well-developed cusp subequal to the paracone. There is no indication that a paraconule was present, but there is an incipient cuspule at the distal aspect of the trigon basin, either a highly reduced metaconule or evidence of slight scalloping along the postprotocrista as seen in other mesonychids (O'Leary and Rose, 1995a). The tooth has a small parastyle; the metastylar region is not preserved, but the occlusal outline suggests that it was not well developed.

Despite poor preservation, the presence of cingula is apparent. A distinct and complete ectocingulum extends over the summit of the parastyle, becoming faint along the poorly preserved mesial border of the tooth and pronounced again as it reaches the lingual extreme. The strong postcingulum apparently connected with both the precingulum and the ectocingulum but poor preservation prevents confirmation of this.

Comparison.—The tooth is most likely from a member of the Mesonychidae as it is larger than the teeth of any hapalodectid, including *Hapalorestes* (Gunnell and Gingerich, 1996) and is less than 50% the size of the molars and premolars of *Andrewsarchus*. The precise locus of the tooth is difficult to establish because it resembles the P^4's and M^2's of different mesonychids.

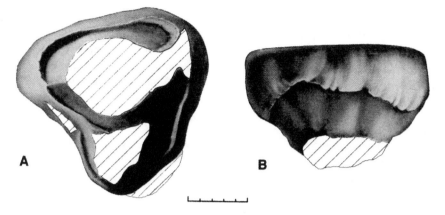

FIGURE 2. Occlusal view of H-GSP 96134, P^4 or M^2 of a mesonychid, which most likely represents a new genus and species. Scale = 5mm.

The P^4's of certain mesonychids (e.g., *Sinonyx, Mesonyx,* and *Pachyaena gigantea*) have paracones that are relatively central in position (directly labial to the protocone) and have highly reduced metacones. The condition of H-GSP 96134 is similar except that it has pre- and postcingula. The M^2's of *Mongolonyx, Harpagolestes,* and *Mongolestes* also lack metacones making them similar to H-GSP 96134, particularly *Harpagolestes,* which is also similar in size. M^2's of *Harpagolestes,* however, also differ from H-GSP 96134 in lacking pre-, post-, and ectocingula.

H-GSP 96134 is most likely not a tooth of an archiac cetacean. It differs from the P^4 of *Pakicetus* and also the more poorly preserved *Nalacetus,* in lacking the hypertrophied paracone present in these basal cetaceans (Thewissen and Hussain, 1998, Fig. 2), even though both *Nalacetus* and *Pakicetus* have pre- and postcingula on P^4. The M^2 of *Nalacetus* is not known, but H-GSP 96134 differs from the M^2 of *Pakicetus* in being less labiolingually elongate and in having a shorter paracone. At present it is not possible to make meaningful comparisons between the new mesonychid and the ambulocetid *Gandakasia* (Dehm and Oettingen-Spielberg, 1958; Thewissen *et al.,* 1996) as *Gandakasia* is known only from isolated lower molars. *Gandakasia* appears to be somewhat larger than H-GSP 96134.

The mesiodistal length of H-GSP 96134 is 18.3 mm, and the buccolingual width is 17.5 mm.

Discussion.—H-GSP 96134 and ONG/K/19 (Ranga Rao, 1973) are the only mesonychids to be described from Indo-Pakistan. Both are isolated teeth and they are of similar size. Like H-GSP 96134, the locus of ONG/K/19 is uncertain; it has been identified as either an upper molar or P^4 (Ranga Rao, 1973). Regardless of their loci, these isolated teeth most likely represent a new genus and species of mesonychid for one of two reasons. If these teeth are M^2's, then the Indo-Pakistani specimens differ from other mesonychids in their size range in having lingual cingula on the upper molars. If these teeth are P^4's, then the Indo-Pakistani specimens are unique among mesonychids in having lingual cingula on the P^4.

Another character that is important for placing these specimens among the mesonychids is the presence of continuous ectocingula. Ectocingula are present in more primitive mesonychids (e.g., *Dissacus*) and lost in more derived ones (e.g., *Harpagolestes*) (Zhou *et al.*, 1995). Although Zhou *et al.* (1995) score the ectocingulum in mesonychids as occurring on P^4 as well as upper molars, I have found that even in the most primitive mesonychids, the ectocingulum is extremely faint or absent on P^4; one possible exception to this is *Sinonyx*, which I have not examined.

Pre- and postcingula rarely occur on the upper molars and premolars of mesonychids with the exception of the M^1 and M^2 of *Dissacus willwoodensis* (O'Leary and Rose, 1995a). They do occur on the molars of *Triisodon, Microclaenodon, Arctocyon, Chriacus, Diacodexis, Pakicetus*, and *Nalacetus*; the latter two lack precingula on M^1. Of these, *Diacodexis, Microclaenodon, Pakicetus*, and *Nalacetus* have pre- and postcingula on P^4 as well. Presence of the ecto-, pre-, and postcingula appears to be plesiomorphic as they are present in a number of taxa outside of Mesonychia (see also below) and suggests that the Indo-Pakistani mesonychids fall close to the base of the mesonychian clade.

Generic attribution of the Indo-Pakistani mesonychids at this time is problematic because the material is so fragmentary despite the fact that it differs from known forms. Ranga Rao (1973) referred ONG/K/19 to cf. *Honanodon* based largely on comparisons with the poorly known *H. hebetis* from China (Chow, 1965), citing similarities of "size and structure" (Rango Rao, 1973, p. 5). I think it unlikely, however, that either ONG/K/19 or H-GSP 96134 represents *Honanodon*.

Ranga Rao (1973) was more correct in stating that ONG/K/19 was "clearly distinct from all the North American and Mongolian mesonychids" (p. 5). The morphology of the P^4 metacone of *H. hebetis* is unclear as the only described tooth of this species is of uncertain locus; it could be either an M^1 or M^2 (Chow, 1965) or a P^4 (Szalay and Gould, 1966). Its metacone is larger than those of the P^4's of most mesonychids with the exception of *Harpagolestes*. The holotype of *H. hebetis* also differs from the two Indo-Pakistani specimens in lacking an ectocingulum and pre- and postcingula.

Although the Indo-Pakistani material most likely represents a new genus and species, I refrain from naming it here because of the fragmentary nature of the material. I think the tooth is most likely an M^2 rather than a P^4 as the presence of cingula occurs with rarity on mesonychid molars but has never been reported for mesonychid premolars.

These specimens merit attention because their provenance matches that of the earliest cetaceans. H-GSP 96134 comes from beds yielding *Gandakasia, Pakicetus* occurs below these sediments, and *Ambulocetus* above them.

3. Phylogenetic and Morphometric Analyses

3.1. Phylogenetic Analysis

3.1.1. Materials

The phylogenetic analysis included archaic cetaceans, mesonychians, and triisodontines, as well as *Diacodexis*, the oldest artiodactyl (Rose, 1982). Artiodactyla is considered to be the extant order most closely related to cetaceans based on the results of a number of recent morphological and molecular studies (Fitch and Beintema, 1990; Gingerich *et al.*, 1990; Novacek, 1992; Graur and Higgins, 1994; Xu *et al.*, 1996). Specimens scored in the analysis (Appendix) were predominantly from the collection in the American Museum of Natural History, New York (AMNH). Casts of specimens from the following collections were also used: Howard University–Geological Survey of Pakistan, Washington, D.C. (H-GSP, specimens studied at Northeastern Ohio Universities College of Medicine); Institute of Vertebrate Paleontology and Paleoanthropology, Beijing, China (IVPP); Johns Hopkins University–United States Geological Survey, now at the Smithsonian Institution, Washington, D.C. (USGS); New Mexico Museum of Natural History, Albuquerque, New Mexico (NMMNH); Princeton University, now at the Yale Peabody Museum, New Haven (YPM-PU); and the University of Michigan, Ann Arbor (UM). Where possible, several individuals of the same species were examined.

A character taxon matrix was assembled from the following groups. Basal cetaceans were represented by the families Pakicetidae (*Nalacetus* and *Pakicetus*) and Ambulocetidae (*Ambulocetus*). More advanced cetaceans (including *Attockicetus*) were not included in the phylogenetic analysis. They are assumed to be nested within the cetacean clade, following Thewissen (1994). Mesonychia was represented by members of Mesonychidae (*Sinonyx, Dissacus, Ankalagon, Pachyaena, Harpagolestes, Mongolestes, Mongolonyx*, and *Mesonyx*) and Hapalodectidae (*Hapalodectes*). Other taxa included were Triisodontinae (*Andrewsarchus, Triisodon, Microclaenodon*, and *Eoconodon*) and Artiodactyla (*Diacodexis*). The outgroups were the arctocyonids, *Arctocyon* and *Chriacus* (contra Prothero *et al.*, 1988). Didymoconids were not included as their relationship to Cete may be quite remote (Meng *et al.*, 1994). Characters and character states are listed in Table I. Information on dental characters for which specimens were not available came from the literature (Appendix); 23 characters were scored for 19 taxa (Table II).

I will explain here why *Andrewsarchus* has been included with the triisodontine arctocyonids rather than with the mesoncyhids. The taxonomic position of *Andrewsarchus* from the late Eocene Irdin Manha Formation of Mongolia has long been controversial primarily because of the fragmentary nature of the holotype of *A. mongoliensis* (Osborn, 1924), a poorly preserved skull and upper dentition. Osborn (1924) named *A. mongoliensis* as a member of the Mesonychidae but did not identify derived features uniting it with other mesonychids. Despite some descriptions of why *Andrewsarchus* might be related to cetaceans (Prothero *et al.*,1988; Thewissen, 1994) the characters that make *Andrewsarchus* a mesonychid are not particularly clear. Those cited by Zhou *et al.* (1995) may simply be related to large size.

Dental characters form the basis of much of the taxonomy of the Mesonychia (Matthew, 1937; O'Leary and Rose, 1995a). Derived characters supporting the monophyly of *Andrewsarchus* and other mesonychians, however, are difficult to specify in part because the upper dentition of the holotype of *A. mongoliensis* is extremely worn but also because they do not seem to exist.

Chow (1959) named *Paratriisodon henanensis* from the "upper Eocene beds of Lushih, Honan Province" (p. 133) of China. These beds can be correlated to those of the Irdin Manha formation that yielded *Andrewsarchus* (Chow, 1959). *P. henanensis* was described on the basis of upper and lower teeth and attributed to the Arctocyonidae. Chow *et al.* (1973) then named *Paratriisodon gigas* for an animal with molars "twice as larger [sic]

Table I. Character States for Phylogenetic Analysis

1. Hypocone (0) present, (1) reduced, (2) absent
2. P^4 metacone (0) highly reduced/absent, (1) present
3. P^4 protocone (0) present, (1) highly reduced/absent
4. M^3 area (0) large, subequal to M^2 or larger, (1) approximately half the size of M^2, (2) small or absent
5. M_3 hypoconulid (0) long, protrudes as separate distal lobe, (1) reduced, does not protrude substantially beyond rest of talonid, (2) absent
6. M^2 metacone (0) distinct cusp, subequal to paracone, (1) distinct cusp, approximately half the size of the paracone, (2) highly reduced, indistinct from paracone
7. Ectocingula on upper molars (0) present, (1) absent
8. M_1 metaconid (0) present, (1) highly reduced/absent
9. M_2 metaconid (0) present, (1) highly reduced/absent. The exact loci of the molars of *Nalacetus* (H-GSP 91045) and *Pakicetus* (H-GSP 18410) are unknown, although it is unlikely that they are M_1's as the M_1 of H-GSP 96505 does not have a reentrant grove on the anterior margin and both of these teeth have reentrant grooves and a small interstitial facet at the distal extreme
10. M_3 metaconid (0) present, (1) highly reduced/absent
11. Occlusal surface of trigon basin (0) broad, (1) somewhat narrow, (2) very narrow
12. P^4 paracone size (0) equal or subequal to height of paracone of M^1, (1) twice height of paracone of M^1
13. P^1 (0) absent, *(1) present, one root*, (2) present, two roots
14. P^3 roots (0) three, (1) two
15. P_1 (0) present, (1) absent
16. Lingual cingulid on molars (0) poorly defined or absent, (1) continuous from mesial to distal extremes
17. M_2 talonids (0) broad, entoconid present, (1) laterally compressed, hypoconid only (and possibly accessory cuspules, O'Leary and Rose, 1995a)
18. Lower molar paraconid (0) directly anterior, paralophid (if present) directed mesiodistally (Fig. 5B), *(1) lingual cusp well-developed, paralophid winds lingually* (Fig. 5A), (2) lingual, cusp reduced, paralophid winds lingually (Fig. 5C), (3) highly reduced (no cusp) paralophid winds lingually (Fig. 5D), (4) no cusp, paralophid is straight crest on lingual margin
19. Reentrant grooves on lower molars (0) proximal (usually well developed), *(1) absent,* (2) distal (usually poorly developed)
20. Conules on upper molars (0) well developed, (1) small, (2) extremely small or absent
21. Molar protoconid (0) subequal to height of talonid, (1) approximately twice height of talonid or greater
22. Pre- and postcingula on M^2 (0) present, (1) absent
23. Parastyle on M^1 (0) absent, (1) weak, (2) well developed

^{a}All characters were treated as unordered except for three that changed as linear transformation series (Wiley et al., 1991) as determined on the basis of morphological criteria. For each of these three ordered characters, the character state present in the outgroup is italicized.

as those of the type [*P. henanensis*]" (p. 181). *P. gigas* is very close in size to *A. mongoliensis*. Other species of *Andrewsarchus* [e.g., Ding *et al.*, 1977, *A. crassum* (two lower premolars)] are fragmentary and have unfortunately aided little in establishing the broader affinities of this genus.

Van Valen (1978) synonymized *Paratriisodon* with *Andrewsarchus, P. gigas* with *A. mongoliensis*, and moved *Andrewsarchus* to the Triisodontinae, a subfamily of the Arctocyonidae. Unfortunately, Van Valen provided no explanation for this synonymy and family-level change for *Andrewsarchus*. However, based on examination of the holotype of *Andrewsarchus* and of the published illustrations and descriptions of *Paratriisodon*, the synonymy appears to be valid and his reasons may be as follows. The outline shape of the molars and premolars of *Andrewsarchus* and *Paratriisodon* resemble each other and the molars of *P. gigas* are of comparable size to those of *Andrewsarchus* and much larger than those

Table II.   Character Taxon Matrix with Percent Completeness for Each Taxon

	5	10	15	20	23	% complete
Andrewsarchus	20?00	80???	1?100	????0	?00	56%
Ankalagon	21012	10100	20?00	01022	112	100%
Dissacus	21012	18000	20100	01022	182	100%
Hapalodectes	01012	11000	10100	01002	112	100%
Harpagolestes	21022	21111	20111	01022	112	100%
Mongolestes	2102?	2??1?	200?1	01012	112	78%
Mesonyx	20022	11111	20110	01022	112	100%
Mongolonyx	2102?	21??1	2?110	010??	???	61%
Pachyaena	21012	18111	20180	01022	112	100%
Sinonyx	20012	10101	20100	01022	112	100%
Triisodon	20011	00000	10?00	00111	101	96%
Eoconodon	10010	00000	10100	00111	100	100%
Microclaenodon	200?1	00000	20??0	00211	102	87%
Ambulocetus	20102	2??11	21???	1140?	100	74%
Nalacetus	2?1??	?0?0?	21???	11202	1?0	56%
Pakicetus	2010?	20?8?	212?0	11302	100	83%
Diacodexis	20000	00000	00100	00110	000	100%
Chriacus	00080	00000	00100	00110	001	100%
Arctocyon	00080	00000	00100	00110	000	100%

of other mesonychids (Chow, 1959; Qi, 1980). It is unlikely that two extremely large and morphologically similar omnivore/carnivore grade mammals existed in the same middle Eocene fauna, suggesting strongly that *Paratriisodon* and *Andrewsarchus* are the same taxon. *Andrewsarchus* and *Paratriisodon* both have unreduced M³'s; mesonychids all show some reduction of M³. McKenna and Bell (1997) uphold this synonymy.

More characters can be scored for the genus *Andrewsarchus* once the synonymy is accepted because lower dentitions are known for species initially attributed to *Paratriisodon* (Chow, 1959; Qi, 1980). In this context the mesonychian affinities of *Andrewsarchus* become increasingly remote, reflecting the fact that certain specimens have well-developed conules on the upper molars and a fully developed talonid on M₃ with a distinct and well-separated hypoconulid and entoconid (Chow, 1959, Fig. 3). These two characters make *Andrewsarchus* (sensu lato) too primitive to be included in a monophyletic Mesonychia (see below). Mesonychia includes only Hapalodectidae and Mesonychidae (McKenna and Bell, 1997).

3.1.2. Methods

The character taxon matrix was assembled in MacClade (Maddison and Maddison, 1992) and a parsimony analysis performed using the branch and bound algorithm in PAUP 3.1.1 (Swofford, 1993). All characters were unordered except three, which were determined, on the basis of morphological criteria, to change along a linear transformation series (Wiley *et al.*, 1991) (Table I). The tree was rooted using the two outgroups specified above. After the initial analysis, two constraint trees were run to find the number of additional steps required to achieve certain alternative scenarios.

3.2. Eigenshape Analysis of P⁴ and M² Outlines

3.2.1. Materials

Outline data were collected from the following specimens (see Section 3.1.1 for abbreviations), for P⁴: *Dissacus* (AMNH 3360), *Pachyaena* (YPM 50000), *Mesonyx* (AMNH 12643), *Harpagolestes* (AMNH 26301), *Hapalodectes* (AMNH 80802), *Ankalagon* (AMNH 776), *Andrewsarchus* (AMNH 20135), *Microclaenodon* (AMNH 102160), *Triisodon* (AMNH 3175), *Eoconodon* (AMNH 764), *Chriacus* (AMNH 3100), *Diacodexis* (AMNH 27787), *Arctocyon* (AMNH 2456), *Mongolestes* (AMNH 26064), *Mongolonyx* (AMNH 26662), *Pakicetus* (H-GSP 18470), *Attockicetus* (H-GSP 96232), *Nalacetus* (H-GSP 91045), Mesonychidae, sp. A (H-GSP 96134); and for M¹: same specimens as for P⁴ except: *Harpagolestes* (AMNH 26301), *Ambulocetus* (H-GSP 18507), *Mesonyx* (USNM 251548). The M¹ of *Mongolonyx* was too poorly preserved to include. *Sinonyx* was not included because it was unavailable for study.

3.2.2. Methods

Outline shape was traced once for each tooth using a Nikon binocular dissecting microscope with camera lucida. Consistency of orientation was maintained by positioning each tooth under the scope such that the line of sight was perpendicular to the occlusal surface; therefore, no particular side of the tooth was visible preferentially. Left teeth were used for outline tracings when possible. If only right teeth were available, the traced right outline was reversed before digitization to achieve orientational consistency. Outlines were digitized with an Acecat II 5″ × 5″ graphics tablet. The subsequent procedure follows methods outlined in MacLeod and Rose (1993).

Each outline was digitized three times from the ectoflexus, considered to be a homologous initial landmark. The digitized outlines consisted of 128 evenly spaced x, y coordinates. Following MacLeod and Rose (1993), outlines were interpolated back to 100 evenly spaced points. The outlines were then converted to the unstandardized ϕ^* format (Zahn and Roskies, 1972) by subtraction of the net angular bend associated with a circle (Lohmann, 1983; Lohmann and Schweitzer, 1990; MacLeod and Rose, 1993).

The unstandardized ϕ^* shape function data sets were assembled into two pooled samples, one for P⁴ and one for M¹, and each pooled sample was subjected separately to eigenshape analysis. Eigenshape analysis, as executed here, involves the singular value decomposition of the covariance matrix calculated between the individual ϕ^* functions in the pooled sample (following MacLeod and Rose, 1993; see also Jöreskog *et al.*, 1976; Lohmann, 1983; Rohlf, 1986; Lohmann and Schweitzer, 1990).

Eigenshape analysis results in a series of mutually orthogonal, latent (eigen) shape functions that describe different components of shape variation in the sample (MacLeod and Rose, 1993, p. 309). The eigenshape functions are ordered hierarchically, the first eigenshape axis accounting for the largest percentage of the shape variation in the sample and all others accounting for progressively less, until all of the shape variation in the sample is described. The eigenshape axes define a shape space within which the distribution of individual taxa in the sample can be studied. Graphic representations, models, of the position of individual specimens within the defined shape space are generated following MacLeod and Rose (1993, pp. 309, 313–314).

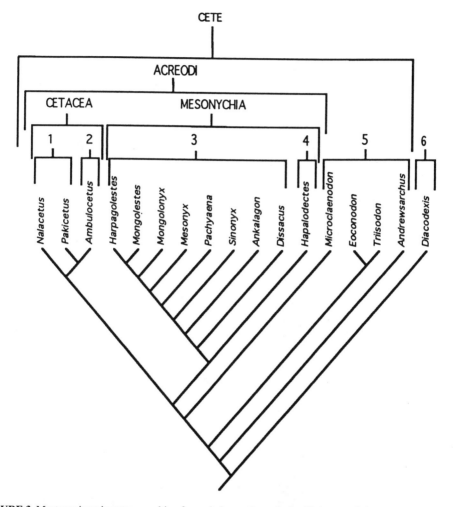

FIGURE 3. Most parsimonious tree resulting from phylogenetic analysis. *Chriacus* and *Arctocyon* were used as outgroups. Numbers refer to families or subfamilies to which these genera are currently allocated: (1) Pakiceti-dae, (2) Ambulocetidae, (3) Mesonychidae, (4) Hapalodectidae, (5) Triisodontinae, (6) Dichobunidae.

4. Results

4.1. Results of Phylogenetic Analysis

The phylogenetic analysis yielded one most parsimonious tree (Fig. 3) of 49 steps with CI = 0.714, RI = 0.873, rescaled CI = 0.623. Many of the character states (Table I) are il-lustrated in Figs. 4 and 5. A constraint tree with *Andrewsarchus* forming a monophyletic group with Mesonychia results in one most parsimonius tree with 55 steps. One that forces monophyly of *Diacodexis* and Cetacea results in three trees with 51 steps each.

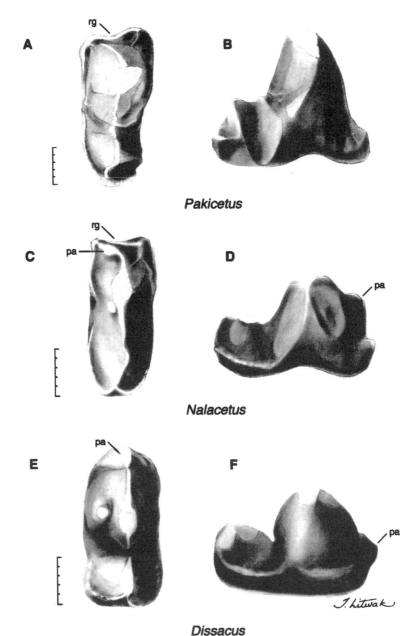

FIGURE 4. Right lower molars of cetans, occlusal views on left, labial views on right. (A, B), *Pakicetus attocki* (H-GSP 18410), M_x; (C, D), *Nalacetus ratimitus* (H-GSP 91045), M_x; (E, F), *Dissacus praenuntius* (YPM-PU 19597), M_1; (G, H) *Hapalodectes compressus* (AMNH 14748), M_2 (reversed); (I, J) *Microclaenodon assurgens* (AMNH 2473a), M_2 (reversed); (K, L), *Triisodon quivirensis* (AMNH 3352), M_1. "pa" = paraconid (absent in *Pakicetus*) and "rg" = proximal molar reentrant groove. Although the exact locus of the two cetacean molars is unknown, it is unlikely that they are M_1's because of the relatively well-developed reentrant grooves, not present in the M_1 of archaic cetacean specimens where the molars are *in situ* (e.g., H-GSP 96505). Scale = 5mm.

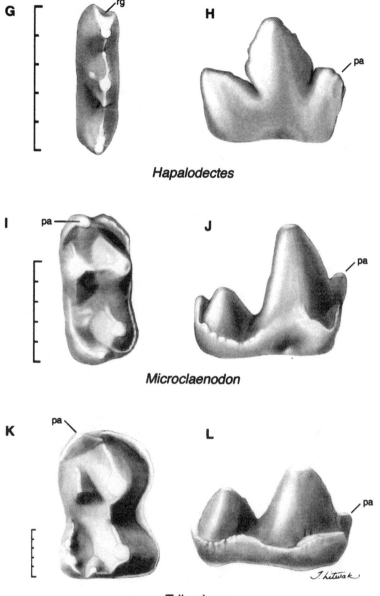

Hapalodectes

Microclaenodon

Triisodon

FIGURE 4. (*Continued*).

4.1.1. Dental Apomorphies for Mesonychia

Dental evidence favors uniting Hapalodectidae and Mesonychidae, but not *Andrewsarchus*, in a clade, Mesonychia (Fig. 3). The monophyly of this group is supported by one dental character: lower molar paraconids oriented directly anterior to the protoconids (character 18) (Fig. 4E,G). Zhou *et al.* (1995) characterized Mesonychidae in part by having an M^{1-2} with a moderate to strong parastyle (character 23). This character describes not only Mesonychidae (Fig. 5C), but also Hapalodectidae, *Microclaenodon* (Fig. 5D), and to some degree *Triisodon* (Fig. 5E).

Within Mesonychia, presence of a hypocone (character 1) appears to be an autapomorphy of Hapalodectidae. Characters that give structure to the mesonychid clade are largely those described by Zhou *et al.* (1995). A stepwise reduction of the metaconid (characters 8–10) on the lower molars occurs in mesonychids from *Dissacus*, which has metaconids on all molars, through *Sinonyx* and *Ankalagon*, which have lost metaconids on some of the molars, and finally to *Pachyaena*, *Mesonyx*, *Mongolestes*, and *Harpagolestes*, which have no metaconids on their molars. Loss of ectocingula on the upper molars (character 7), present in primitive mesonychids, occurs in middle Eocene forms such as *Mesonyx*, *Mongolonyx*, *Mongolestes*, and *Harpagolestes*.

Two types of reentrant groove (character 19) appear in the mesonychian clade. Proximal molar reentrant grooves occur in hapalodectids (Fig. 4G), and Zhou *et al.* (1995) described reentrant grooves in *Sinonyx* but at the distal end of the tooth and these are often much more poorly developed than proximal ones. Reentrant grooves, therefore, developed convergently at least once within Mesonychia alone.

4.1.2. Dental Apomorphies for Cetacea

Three dental apomorphies diagnose Cetacea: (1) lack of a protocone on P^4 (character 3), (2) P^4 paracone twice the height of the paracone of M^1 (character 12), (3) presence of a well-developed lingual cingulid on the lower molars (character 16). Cetaceans have an M^3 that is subequal to the M^2 in area (character 4), a condition also seen in artiodactyls and *Andrewsarchus*, but which differs from mesonychids and from triisodontines. Cetaceans have the plesiomorphic condition of an ectocingulum on the upper molars (character 7). Some of the later occurring and more derived mesonychians, by contrast, have lost this feature.

Cetaceans differ from mesonychians in the position of the paraconid on the lower molars. The morphology of the paraconid in archaic cetaceans was not clear prior to the discovery of *Nalacetus*. *Nalacetus* and *Pakicetus* (Fig. 4A–D) represent two different stages in the loss of the paraconid. The paraconid is small and lingual in *Nalacetus* (Fig. 4C,D) and has disappeared entirely in some individuals of *Pakicetus* (Fig. 4A; see also Thewissen and Hussain, 1998), but the position of the mesial end of the paralophid curves lingually indicating that *Pakicetus* differs from *Nalacetus* only in the absence of a well-defined cusp. This morphology is not homologous with or derived from a mesonychian paraconid condition (Fig. 4E–H). The archaic cetacean *Nalacetus* more closely resembles triisodontines (Fig. 4I–L), artiodactyls, and the outgroups (*Chriacus* and *Arctocyon*) in the relatively lingual position of the paraconid.

4.1.3. Dental Apomorphies for Acreodi

Acreodi, a name used by Prothero *et al.* (1988) for Cetacea and Mesonychia, is used in a similar sense here with the exception that Mesonychia does not include *Andrewsarchus*. The clade Acreodi is diagnosed by three synapomorphies: loss of M_3 hypoconulid (character 5), laterally compressed M_2 talonid (character 17), and loss of conules on the upper molars (character 20). Some of these characters, such as the molar conules (Fig. 5D), and the M_3 hypoconulid apparently have been lost in a stepwise fashion, as both features are present but reduced in *Microclaenodon*, which falls outside the clade Acreodi (Fig. 3). Contra Thewissen (1994), Cetacea and its sister group do not share absence of the protocone on P^4 (character 3). Also, absence of a P^4 metacone (character 2), which characterizes archaic cetaceans, triisodontines, and *Diacodexis*, is not present in all mesonychians. Primitive mesonychians apparently had this cusp, but the condition reversed in *Sinonyx* and *Mesonyx*.

Proximal reentrant grooves on the molars (character 19) are found in both archaic cetaceans and *Hapalodectes*. On the basis of the evidence here, it is equally parsimonious to conclude that proximal reentrant grooves arose at the base of the clade Acreodi and reversed at the base of the clade Mesonychidae (Fig. 3) as it is to argue that they arose independently in *Hapalodectes* and again at the base of the clade Cetacea.

The metaconid (characters 8–10) also appears to have been lost independently in Cetacea, Mesonychidae, and Hapalodectidae, as the most primitive members of each of these groups have a metaconid on the lower molars. Reduced metaconids do not appear to be a synapomorphy of Acreodi because they are also reduced in *Microclaenodon* (Fig. 4I), which falls outside of Acreodi. Subtle morphological differences also exist in the position of the metaconid in archaic cetaceans and in mesonychians. *Nalacetus* (Fig. 4C) has a more open trigonid than mesonychians, and the extremely small metaconid is positioned slightly distal to the protoconid. In mesonychians (Fig. 4E,G), the metaconid is lingual and slightly mesial to the protoconid.

Compression of the trigon basin of M^3 (Thewissen, 1994) (character 11) does not appear to support an exclusive clade of mesonychians and Cetacea as it also occurs to varying degrees in the Triisodontinae. In *Microclaenodon*, especially, trigon compression approximates the condition seen in mesonychids. Various cetans, therefore, not just Acreodi, are characterized by the reduction of the trigon basin (Fig. 5).

One implication of this phylogenetic analysis is that whereas cetaceans and mesonychians have a number of shared derived features in common, mesonychians are autapomorphic in the position of the paraconid. This derived feature is present and retained throughout the entire mesonychian clade and apparently did not undergo any reversals. Thus, although mesonychians (Mesonychidae and Hapalodectidae) may be the sister taxon of the cetaceans, they are too derived dentally to have been ancestral to cetaceans. This conclusion runs contrary to earlier suggestions that *Dissacus* (Van Valen, 1966) or Mesonychidae (Zhou *et al.*, 1995) were ancestral to cetaceans.

4.1.4. Dental Apomorphies for Cete

Cete is here comprised of Cetacea, Mesonychidae, Hapalodectidae, and Triisodontinae. One dental character unites Cete: protoconids twice the height of the talonid (charac-

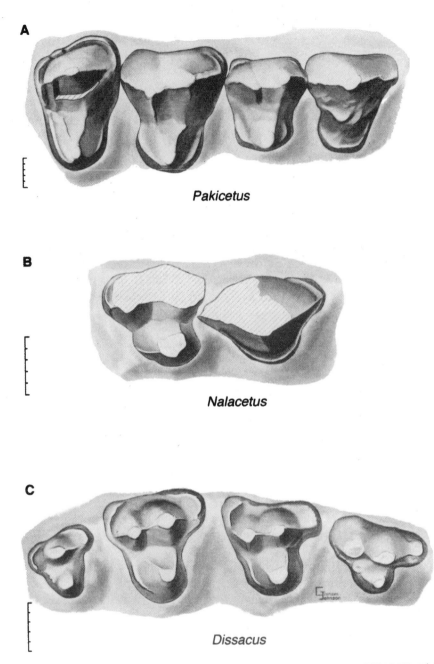

FIGURE 5. Occlusal views of right upper dentitions of cetans. (A) *Pakicetus attocki* (H-GSP 18470), P^4–M^3; (B) *Nalacetus ratimitus* (H-GSP 18521), P^4–M^1; (C) *Dissacus navajovius* (AMNH 3360), P^4–M^3 (reversed); (D) *Microclaenodon assurgens*; (E) *Triidoson antiquus* (AMNH 3175), P^4–M^3 (reversed). Note reduction of the M^3 area in *Dissacus* and *Triisodon* in contrast to *Pakicetus*, and the relative development of pre- and postcingula, least developed in *Dissacus*. Scale = 5mm.

D

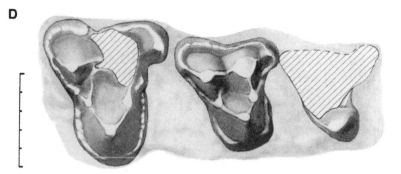

Microclaenodon

E

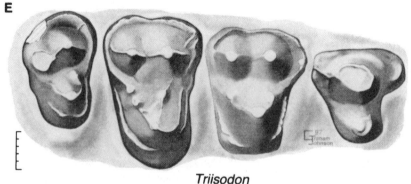

Triisodon

FIGURE 5. (*Continued*).

ter 21). Cete here contains Mesonychia, Cetacea, *Microclaenodon, Triisodon, Eoconodon*, and *Andrewsarchus*. The condition of the protoconids in *Andrewsarchus* is inferred to be similar to other cetans but could not be coded because none of the specimens, including those of *Paratriisodon*, preserve this feature.

A number of cetans, *Eoconodon* and *Andrewsarchus*, have a well-developed hypo-conulid that protrudes distally from the talonid (character 5) (see also Chow, 1959, Fig. 3). In *Triisodon* and *Microclaenodon*, this cusp is reduced to varying degrees.

4.2. Results of Outline Shape Analysis

4.2.1. P⁴

The first eigenshape axis of the P^4 data set accounts for 61.4% of the shape variation in the sample. As seen in Fig. 6, the modeled shape along eigenshape axis one (Es-1) that most closely resembles a circle (e.g., 0.13) has a covariance closest to zero. P^4's with low covariances are relatively circular because they have indistinct protocone lobes. Higher co-

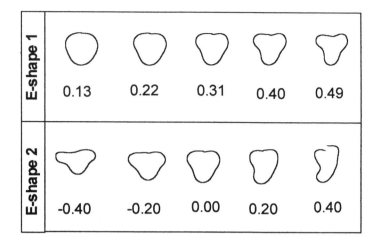

FIGURE 6. Along axis shape models for occlusal views of left P⁴'s, eigenshapes one (Es-1) and two (Es-2), together accounting for 82.9% of the total variation in the sample. All modeled shapes pass through the mean shape (Es-1 = 0.31; Es-2 = 0.00) for the P⁴ sample. The P⁴'s of dentitions in Fig. 4 indicate the general orientation of the outlines as traced from original specimens. The number below each model is its covariance with the particular eigenshape axis. A covariance of 0.40 on Es-2 results in a biologically impossible shape (see MacLeod and Rose, 1993, for further explanation).

variances on Es-1 (e.g., 0.49) describe P⁴ outlines that have a more sharply defined ectoflexus and a well-defined protocone lobe that is narrowed in a mesiodistal direction (Fig. 6). The mean shape for the sample has a covariance of 0.31 on Es-1. No specimens in the sample have covariances with Es-1 greater than 0.49.

Variation along Es-2 accounts for 21.5% of the total variation in the P⁴ sample. For Es-2, the mean shape of the sample has a covariance of 0.0 and sample variation extends in both positive and negative directions from the mean (Fig. 6). Negative covariances with Es-2 correspond to P⁴'s that have mesiodistally elongate labial borders and poorly defined protocone lobes that are reduced in a labiolingual direction (Fig. 6). Positive covariances with Es-2 represent P⁴'s with mesiodistally short labial borders, but with protocone lobes that are relatively elongate in a labiolingual direction. The protocone lobes in individuals with high values on Es-2 also tend to be angled in a mesial direction such that the tooth is convex along the distal margin (Fig. 6).

A scatterplot of Es-1 versus Es-2 (Fig. 7A) shows that the sample is generally divided between two corners of the graph (numbers following the names of taxa below refer to outlines in Fig. 7A). *Nalacetus* (18), *Attockicetus* (15), and *Ambulocetus* (16) cluster together with low values on both eigenshape axes. These cetaceans have P⁴'s with mesiodistally elongate labial borders and relatively reduced protocone lobes, doubtless reflecting loss of the P⁴ protocone, a synapomorphy of cetaceans. The clade Cetacea (Fig. 7A,B), however, is not entirely characterized by this morphology. *Pakicetus* (17) does not cluster with this group but instead with a group consisting of both triisodontines and many mesonychids (Fig. 7A). *Andrewsarchus* (19) also clusters with the cetaceans even though the phylogenetic analysis does not show that it is particularly closely related to cetaceans (Fig. 7B). The outline shape of P[4] of *Diacodexis* (9) is not similar to any of the cetaceans.

Mesonychians, with the exception of *Mongolestes* (5), *Dissacus* (2), and *Hapalodectes* (8), generally have higher values on both eigenshape axes than cetaceans do. In other words, they have well-defined, labiolingually elongate protocone lobes and less elongate labial borders. *Dissacus* (2) and *Hapalodectes* (8), the two most primitive mesonychians (Fig. 7B), differ from other mesonychians and resemble cetaceans [except *Pakicetus* (17)] in the relative mesiodistal elongation of the labial borders of their P[4]'s. Mesiodistal elongation of the labial border of the P[4] in cetaceans and mesonychians, however, does not appear to be homologous. The P[4]'s of *Pakicetus* and *Nalacetus* (Fig. 5A,B) have broad labial borders but virtually no development of stylar cusps. The labial borders of the P[4]'s of these archaic cetaceans are broad because the hypertrophied paracone has a wide base compared with the paracones of other mammals, e.g., *Dissacus* (Fig. 5C) or *Triisodon* (Fig. 5E). *Dissacus* (Fig. 5C) and *Hapalodectes* (not shown, but see Szalay, 1969a, Figs. 7 and 12), by contrast, have broad labial borders because they have well-developed parastyles (Table I, character 23). Thus, although the eigenshape analysis has detected similarities in outline shape of the occlusal surface of the P[4]'s of *Dissacus, Hapalodectes,* and certain archiac cetaceans, on closer inspection this similarity does not appear to be homologous.

Andrewsarchus (19) has P[4] shape that closely resembles that of cetaceans, with both a relatively elongate labial border and a reduced protocone lobe (negative values on both eigenshape axes). Unfortunately, the holotype of *Andrewsarchus* is so poorly preserved that it is difficult to assess what this similarity of outline shape means in terms of the gross morphology of the P[4]. On the basis the phylogenetic and morphometric information available here, the similarity in shape between *Andrewsarchus* and a number of the archaic cetaceans appears to be convergent.

The outline shape of the P[4] of *Pakicetus* (17) does not group with the cetaceans but instead with a variety of mesonychids, triisodontines, and the two outgroups. The taxa with which *Pakicetus* groups have P[4]'s with inelongate labial borders and well-developed protocone lobes that support fully developed protocones. *Pakicetus* differs from the taxa with which it clusters, however, in having lost its protocone and in having a greatly hypertrophied paracone. Its outline shape, therefore, must represent a reversal to the condition more closely resembling the shape of other taxa at lower nodes on the tree (Fig. 7B).

Mongolestes (5) differs from all other taxa in having an outline that is relatively circular with low values on Es-1 and high values on Es-2. The distinctive shape of *Mongolestes*, which consists of a poorly defined protocone lobe with a relatively convex distal border, appears to be autapomorphic within the mesonychian clade.

The outgroups *Arctocyon* (11) and *Chriacus* (10) occupy different positions on the graph: *Arctocyon* (11) clusters with a group containing mesonychids, triisodontines, and *Pakicetus* (17). *Chriacus* (10), which is at the lower right corner of the graph, has a relatively deep ectoflexus and an elongate, well-defined protocone lobe (pinched mesiodistally). These shapes are indicative of the primitive outline shape for the group of mammals under study here.

If the phylogenetic hypothesis of Fig. 7B is correct, then the relatively inelongate protocone lobe [low values on Es-1 (Figs. 6 and 7)] originated three times independently. It originated once at the base of the cetacean clade, once in *Mongolestes* (5), and once in *Andrewsarchus* (19). It also reversed in *Pakicetus* (17). This scenario is one step shorter than

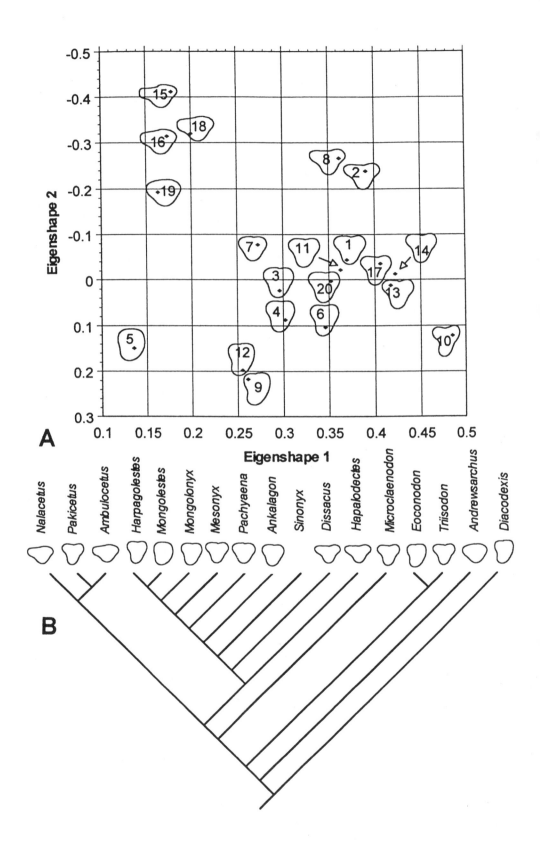

positing its independent origin in each of the five taxa with this feature or its origin at the base of Cete and its reversal in a number of mesonychians and triisodontines that have more well-developed protocone lobes.

4.2.2. M^1

Pure shape variation along eigenshape axes one (Es-1) and two (Es-2) in the M^1 sample (Fig. 8) resembles that seen in the P^4 sample (Fig. 6). Es-1 for the M^1 sample accounts for 72.8% of the sample variation. Covariances closest to zero on Es-1 represent M^1's with a poorly defined ectoflexus and a protocone lobe that is relatively wide in a mesiodistal direction. Higher covariances along Es-1 describe outlines that have a more pinched protocone lobe. Es-2, which accounts for 13.6% of the variation in the sample, describes changes in the relative elongation of the labial border of the M^1. Covariances toward the negative extreme of the sample variation represent M^1's with elongate labial borders and a protocone lobe that has a somewhat asymmetrical shape. At the positive end of sample variation along Es-2, the M^1's have a deeper ectoflexus and a broad protocone lobe that is symmetrical. The mean shape has a covariance of 0.31 with Es-1 and 0.00 with Es-2. Numbers following names of different taxa below refer to outlines in Fig. 9A.

A scatterplot of Es-1 versus Es-2 (Fig. 9A) reveals that the trend of shape variation in the sample forms a rough diagonal. *Eoconodon* (12), *Arctocyon* (11), *Chriacus* (10), and *Diacodexis* (9) cluster together at one end of the graph. These taxa have relatively short labial borders (positive Es-2 values). This group intergrades with various mesonychids, as well as *Andrewsarchus* (19) and *Microclaenodon* (14), which have slightly more elongate labial borders. *Hapalodectes* (8) has the most extreme elongation of the labial border, which reflects the relative development of stylar cusps on the M^1's of this genus. *Ankalagon* (3) is close to the mean shape for the sample.

The archaic cetaceans [Fig. 9A (16, 17, 18)] have a varied distribution in this shape space. *Pakicetus* (17) and *Nalacetus* (18) fall close to the mean shape on both axes, indicating that both show some degree of elongation of the labial border of M^1, but that neither is highly derived in this regard. *Nalacetus* (18) has a slightly more pinched protocone lobe (mesiodistally narrow, higher Es-1 value) than *Pakicetus* (17). Importantly, *Pakicetus* (17), does not resemble mesonychids any more than it does *Eoconodon* (12), *Diacodexis* (9), and *Arctocyon* (11) on Es-1. *Ambulocetus* (16), a more derived cetacean (Fig. 9B), has the most pronounced elongation of the labial border of the M^1.

Relative elongation of the labial border of the M^1 is most pronounced (Es-2 values greater than -0.1) in the following members of Acreodi: *Mesonyx* (1), *Pachyaena* (7), *Hapalodectes* (8), and *Ambulocetus* (16). It appears that this high degree of elongation devel-

◀───

FIGURE 7. (A) Two-dimensional shape models of left P^4 occlusal outlines in a plane defined by Es-1 versus Es-2. Diamonds represent the locations of the occlusal P^4 outlines of different taxa in the shape space. Shape models are superimposed over the diamonds to show the relative shape differences among taxa in the shape space. (1) *Mesonyx*, (2) *Dissacus*, (3) *Ankalagon*, (4) *Mongolonyx*, (5) *Mongolestes*, (6) *Harpagolestes*, (7) Pachyaena, (8) Hapalodectes, (9) *Diacodexis*, (10) *Chriacus*, (11) *Arctocyon*, (12) *Eoconodon*, (13) Triisodon, (14) *Microclaenodon*, (15) *Attockicetus*, (16) *Ambulocetus*, (17) *Pakicetus*, (18) *Nalacetus*, (19) *Andrewsarchus*, (20) Pakistan mesonychid (H-GSP 96134), new genus and species. (B) Modeled left P^4 occlusal outlines mapped onto the cladogram. *Sinonyx* was not available for study.

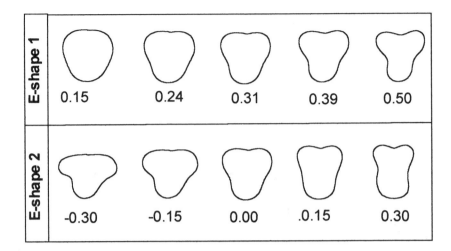

FIGURE 8. Along axis shape models for occlusal views of left M¹'s, eigenshapes one (Es-1) and two (Es-2), which together account for 86.4% of sample variation. All modeled shapes pass through the mean shape (Es-1 = 0.31; Es-2 = 0.00) for the M¹ sample. The M¹'s of dentitions in Fig. 4 indicate the general orientation of the outlines as traced from original specimens. The number below each model is its covariance with the particular eigenshape axis.

oped convergently in these four genera, as it would be less parsimonious to hypothesize that it was present in the common ancestor of Acreodi.

As with P⁴, elongation of the labial border of the M¹ in cetaceans and mesonychians reflects different gross morphological features. Cetacean M¹'s (Fig. 5A,B; see also Thewissen *et al.*, 1996, for *Ambulocetus*) do not have well-developed stylar cusps but instead a paracone with a broad base. Mesonychians that cluster with taxa such as *Ambulocetus* in the eigenshape analysis have broad labial borders because they have well-developed stylar cusps.

5. Toothwear

Subtle differences in molar morphology among mesonychians and archaic cetaceans are reflected in toothwear differences. In early stages of wear, mesonychid teeth have blunt cusps and small, oval shearing facets irregularly positioned on the labial sides of the protoconid, paraconid, and talonid (Fig. 4F; Szalay and Gould, 1966, Fig. 11). These facets

FIGURE 9. (A) Two-dimensional shape models of left M¹ occlusal outlines in a plane defined by Es-1 versus Es-2. Diamonds represent the locations in the shape space of the occlusal M¹ outlines of different taxa. Shape models are superimposed over the diamonds to show the relative shape differences among taxa in the shape space. (1) *Mesonyx*, (2) *Dissacus*, (3) *Ankalagon*, (5) *Mongolestes*, (6) *Harpagolestes*, (7) *Pachyaena*, (8) *Hapalodectes*, (9) *Diacodexis*, (10) *Chriacus*, (11) *Arctocyon*, (12) *Eoconodon*, (13) *Triisodon*, (14) *Microclaenodon*, (16) *Ambulocetus*, (17) *Pakicetus*, (18) *Nalacetus*, (19) *Andrewsarchus*. (B) Modeled left M¹ occlusal outlines mapped onto cladogram of cetans. *Sinonyx* was not available for study and *Mongolestes* was too poorly preserved to include.

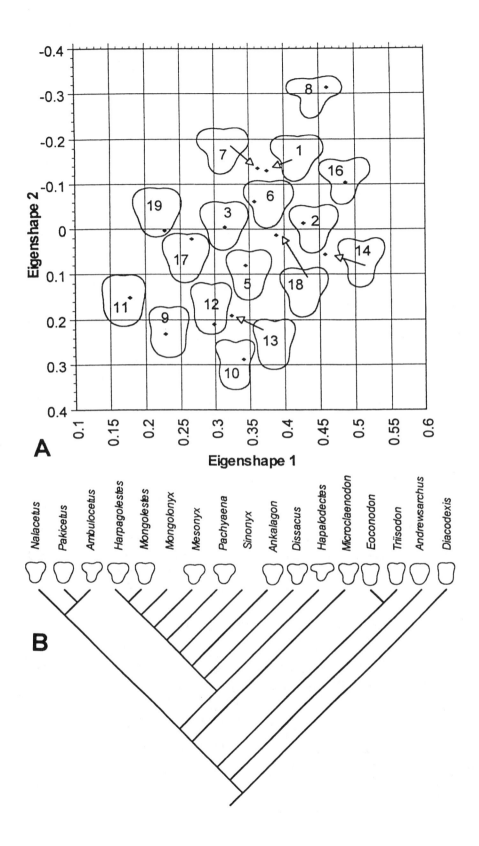

A

B

Nalacetus *Pakicetus* *Ambulocetus* *Harpagolestes* *Mongolestes* *Mongolonyx* *Mesonyx* *Pachyaena* *Sinonyx* *Ankalagon* *Dissacus* *Hapalodectes* *Microclaenodon* *Eoconodon* *Triisodon* *Andrewsarchus* *Diacodexis*

tend not to extend to the gum line and, as noted by Szalay and Gould (1966, p. 161), "it is doubtful that shear between upper and lower molars of mesonychids played an important part in mastication." Heavily worn mesonychid teeth are often flattened in a horizontal plane with the enamel completely worn through and dentine exposed. Heavily worn mesonychid teeth retain little of their original morphology.

The teeth of primitive cetaceans such as *Pakicetus, Nalacetus*, and *Ambulocetus*, when relatively unworn, exhibit some apical cusp wear like mesonychids, but instead of developing small, oval shearing facets, cetaceans develop long shearing facets that often extend to the gum line (Fig. 4B,D). These facets appear in three places, most clearly seen in *Nalacetus* (Fig. 4D): on the anterior and distal aspects of the protoconid, and on the labial aspect of the talonid.

Heavily worn molars of archaic cetaceans exhibit markedly different toothwear from mesonychids. Archaic cetacean molars do not become flattened in the occlusal plane like a mesonychid molar. Instead the long, relatively vertical shearing facets in cetaceans are gouged into the labial side of the tooth resulting in exposed dentin, especially between the protoconid and the talonid (Fig. 4B,D). *Ambulocetus*, for example, has deep, vertical wear facets, gouged until concave, extending from the crown of the tooth to the base. Extensive shearing capability never present in mesonychid dentitions is retained even at very advanced stages of wear in these early cetaceans.

Contrary to Van Valen's (1966) argument that cetaceans and mesonychians have identical toothwear, knowledge of more primitive cetaceans, particularly *Ambulocetus* and *Nalacetus*, reveals that the earliest cetaceans have quite different toothwear from mesonychians especially in late stages of wear. The mesonychian chewing mechanism involved more crushing and possibly a component of mediolateral movement especially in late stages of wear. The chewing mechanism of archaic cetaceans involved heavy shearing and almost exclusively orthal motions, even in advanced stages of wear.

6. Discussion

The dental evidence presented here supports a union of Hapalodectidae and Mesonychidae, but not *Andrewsarchus*, in a monophyletic Mesonychia. Mesonychians share an anteriorly positioned paraconid on the lower molars, a feature derived over the condition in more generalized mammals where this cusp is positioned lingually. This study finds no support for a clade Hapalodectini that groups *Hapalodectes* and *Dissacus* to the exclusion of other mesonychids (Prothero *et al.*, 1988).

On the basis of dental characters, *Nalacetus* is the most primitive known cetacean. Cetacea share three dental synapomorphies: a lingual cingulid on the lower molars, absence of a P^4 protocone, and a P^4 paracone twice the height of the paracone of M^1. Cetacea and Mesonychia form a clade, Acreodi, whose sister taxon is *Microclaenodon*. The members of Acreodi share three dental synapomorphies: a laterally compressed M_2 talonid, absence of molar conules, and loss of the hypoconulid on M_3. Neither *Andrewsarchus* nor *Diacodexis* fall particularly close to the clade Acreodi on the phylogenetic tree. A constraint tree forcing monophyly of artiodactyls and cetaceans, however, only changes the total number of steps on the tree from 49 to 51.

Importantly, mesonychian lower molar morphology does not represent the ancestral condition for the cetacean clade as has been suggested (Van Valen, 1966; Zhou *et al.*, 1995). When Van Valen (1966) first proposed a close cetacean–mesonychid relationship, knowledge of primitive cetaceans consisted of *Protocetus* and *Pappocetus* (Van Valen, 1966, pp. 90–93). In that context it was relatively parsimonious to derive a *Protocetus* molar from a mesonychid or hapalodectid molar. Such a transition would simply have involved loss of the anteriorly positioned paraconid and loss of the metaconid in a mesonychid ancestor. Discovery of *Pakicetus*, and particularly *Nalacetus*, have, however, broadened our understanding of primitive cetacean dental morphology. These fossils indicate that derivation of an archaic cetacean molar from a mesonychid one was not so obvious as it would have involved reversal of the derived mesonychian paraconid position and then reduction and ultimate loss of the paraconid. This scenario seems particularly unlikely as mesonychians and cetaceans have distinctive differences in toothwear on their molars. This phylogenetic analysis predicts that the common ancestor of cetaceans and mesonychians had a relatively primitive molar trigonid with a lingually positioned paraconid, similar to the triisodontine condition.

The distribution of various other dental characters within this clade has been clarified. Loss of the metaconid occurred independently in the clades Mesonychia and Cetacea, as revealed by its presence in the basal members of both clades. Reduction of this cusp such that it is twinned with the protoconid also occurred to some degree in *Microclaenodon*; therefore, twinning of this cusp is not a synapomorphy of the clade Acreodi. Absence of the metacone on P^4 (Thewissen, 1994), which occurs in archaic cetaceans, certain more derived mesonychians like *Mesonyx*, as well as *Diacodexis*, appears to be convergent. Finally, it is equally parsimonious to conclude that proximal molar reentrant grooves found in hapalodectids and archaic cetaceans are convergent as it is to conclude that they are shared derived features. These structures also exhibit morphological differences between hapalodectids and archaic cetaceans.

Eigenshape analyses of the outline shape of the P^4 and M^1 of a number of archaic cetaceans, mesonychians, triisodontines, and an artiodactyl reveal that convergent evolution has resulted in similarly shaped teeth in certain cases. For example, the P^4's and M^1's of certain archaic cetaceans and certain primitive mesonychians are alike in having relatively elongate labial borders. In the archaic cetaceans, this elongation is the result of the presence of a relatively hypertrophied paracone with a wide base. In the mesonychians, however, the elongation is caused by the presence of a well-developed parastyle cusp. Mesonychians also do not show any tendency toward loss of the protocone on P^4, a synapomorphy of cetaceans that is reflected in the occlusal outline shape of a cetacean P^4. *Pakicetus* is one exception to this trend.

The morphological differences in the dentitions of mesonychids and archaic cetaceans translate into differences in dental function as interpreted from differences in molar toothwear. Although mesonychid lower molars resemble those of archaic cetaceans in having tall protoconids and laterally compressed talonids, gross examination of toothwear in these two groups indicates that similar morphology does not translate directly into similar toothwear patterns. Cetaceans and mesonychians both achieved predominantly orthal crushing mechanisms, but the primitive cetacean condition involved more precise occlusion and the retention of strong shearing facets even at very advanced stages of wear. These facets became heavily gouged and often reached the gum line in the most worn specimens. Mesonychids, by contrast, engaged in less shearing and more crushing, possibly incorporating more mediolateral movement into their chewing cycle. Mesonychids generated much more apical cusp wear than cetaceans and wore their teeth in a horizontal plane such that they were virtually incapable of even the crudest shearing in advanced stages of wear. The anterior position of the paraconid, the dental apomorphy uniting mesonychians, contributed to the surface area available for crushing in the molars of these mammals.

Described here also is the second, albeit highly fragmentary, mesonychian tooth from cetacean-bearing beds in Indo-Pakistan. This specimen has significance for the problem of cetacean origins because of its stratigraphic and geographic position. Information related to provenance is often given less consideration than shared derived morphological features determined by phylogenetic analysis. Simpson (1945, p. 7) emphasized, however, that "animals cannot have common ancestry without also having common geographic origin. These and other principles are highly pertinent in interpreting phylogeny, and they deserve equal weight with morphological data." The Pakistani material differs from other mesonychids in possessing well-developed pre- and postcingula, a character found in cetaceans, triisodontines, and *Diacodexis*. Retrieval of more material of this intriguing form, in particular, will undoubtedly contribute greatly to our understanding of the internal structure of the clade Cete.

Acknowledgments

For access to specimens I would like to thank P. D. Gingerich, G. F. Gunnell, and W. J. Sanders, University of Michigan; S. G. Lucas, University of New Mexico at Albuquerque; M. C. McKenna, R. H. Tedford, and J. P. Alexander, American Museum of Natural History; J. G. M. Thewissen, S. I. Madar, and E. M. Williams, Northeastern Ohio Universities College of Medicine; K. C. Beard and A. Tabrum, Carnegie Museum of Natural History; and K. D. Rose, Johns Hopkins University School of Medicine. I am grateful to M. Arif for finding the Pakistani mesonychid and to J. G. M. Thewissen for granting me permission to study several undescribed H-GSP specimens. The program for digitizing outline shapes was written by R. E. Chapman, Smithsonian Institution, and all programs for eigenshape analysis by N. MacLeod, Natural History Museum, London. For access to a dissecting scope I would like to thank E. W. Heck. An earlier version of the manuscript was improved by comments from J. H. Geisler, W. P. Luckett, J. G. M. Thewissen, and an anonymous reviewer. The following individuals provided helpful discussion: W. P. Luckett, S. I. Madar, M. C. McKenna, M. B. Meers, K. D. Rose, J. G. M. Thewissen, M. D. Uhen. T. Litwak prepared Fig. 4 (part), G. T. Johnson prepared Figs. 2, 4 (part), and 5, and M. J. Ellison assisted with Figs. 7 and 9. The Pakistani mesonychid and cetacean specimens were collected on an NSF grant to J. G. M. Thewissen.

Appendix: Specimens and References Used for Phylogenetic Analysis

(Institutional abbreviations are defined in Section 3.1.1.)

Andrewsarchus; AMNH 20135, Chow (1959)
Ankalagon; AMNH 776, 777, 2454, NMMNH 16309

Dissacus; AMNH 3356, 3357, 3360, 3361, 15996, 39276 (cast), 55925 (cast), IVPP 4266 (cast), 5478 (cast), 5479 (cast), USGS 27612, 27635, YPM-PU 13295, 16135, 16137, Zhou *et al.* (1995)

Hapalodectes; AMNH 78, 12781, 14748 (cast), 20172, 80802 (cast), USGS 275 (cast), 9628, 10293 (cast), Guthrie (1967) Szalay (1969a), Ting and Li (1987), Zhou and Gingerich (1991)

Harpagolestes; AMNH 1878, 1892, 26300, 26301, Peterson (1931), Zhou *et al.* (1995)

Mongolestes; AMNH 26064, 26065.

Mesonyx; AMNH 1716, 11552, 12641, 12643 (cast), 122122

Mongolonyx; AMNH 26661, 26662

Pachyaena; AMNH 72, 2959, 4262, 15222, 15224, 15228, 15728, USGS 7185, YPM 50000

Sinonyx; IVPP V10760 (cast of lower dentition), Zhou *et al.* (1995).

Triisodon; AMNH 3172, 3174, 3175, 3176, 3178, 3225, 3352, 16559

Eoconodon; AMNH 764, 774, 3181, 3187, 16329, 16341

Microclaenodon; AMNH 2473a, 3215, 10571 (cast), 10858, 15999, 102160, Matthew (1937), Taylor (1981)

Ambulocetus; H-GSP 18507 (cast), Thewissen (1994), Thewissen *et al.* (1996)

Nalacetus; H-GSP 18521, 91045

Pakicetus; H-GSP 18410, 18470, 96505, Gingerich and Russell (1990)

Diacodexis; AMNH 15527, 27787, Thewissen (1994)

Chriacus; AMNH 2382, 3100, 16586, 48699, Matthew (1897), Rose (1996)

Arctocyon; AMNH 2456, 2459, 55902, Russell (1964)

References

Chow, M. 1959. A new arctocyonid from the upper Eocene of Lushih, Honan. *Vertebr. PalAsiat.* **3**:133–138.

Chow, M. 1965. Mesonychids from the Eocene of Honan. *Vertebr. PalAsiat.* **9**:286–291.

Chow, M. M., Li, C.-K., and Chang, Y.-P. 1973. Late Eocene mammalian faunas of Honan and Shansi with notes on some vertebrate fossils collected therefrom. *Vertebr. PalAsiat.* **14**:12–34.

Dehm, R., and Oettingen-Spielberg, T. zu. 1958. Paläontologische und geologische Untersuchungen im Tertiär von Pakistan. 2. Die mitteleocänen Säugetiere von Ganda Kas bei Basal in Nordwest-Pakistan. *Abh. Bayer. Akad. Wiss. Math. Naturwiss. Kl. N. F.* **91**:1–54.

Ding, S.-Y., Zheng, J.-J., Zhang, Y.-P., and Tong, Y.-S. 1977. The age and characteristics of the vertebrate fauna from Liuniu and Dongjun Formations of the Bose Basin, Zhuang autonomous region. *Vertebr. PalAsiat.* **15**:35–44.

Fitch, W. M., and Beintema, J. J. (1990). Correcting parsimonious trees for unseen nucleotide substitutions: the effect of dense branching as exemplified by ribonuclease. *Mol. Biol. Evol.* **7**:438–443.

Gatesy, J., Hayashi, C., Cronin, M. A., and Arctander, P. 1996. Evidence from milk casein genes that cetaceans are close relatives of hippopotamid artiodactyls. *Mol. Biol. Evol.* **13**:954–963.

Gingerich, P. D. 1981. Radiation of early Cenozoic Didymoconidae (Condylarthra, Mesonychia) in Asia, with a new genus from the early Eocene of western North America. *J. Mammal.* **62**:526–538.

Gingerich, P. D., and Russell, D. E. 1990. Dentition of early Eocene *Pakicetus* (Mammalia, Cetacea). *Contrib. Mus. Paleontol. Univ. Michigan* **28**:1–20.

Gingerich, P. D., Smith, H. B., and Simons, E. L. 1990. Hind limbs of Eocene *Basilosaurus:* evidence of feet in whales. *Science* **249**:154–157.

Graur, D., and Higgins, D. G. 1994. Molecular evidence for the inclusion of cetaceans within the order Artiodactyla. *Mol. Biol. Evol.* **11**:357–364.

Gunnell, G. F., and Gingerich, P. D. 1996. New hapalodectid *Hapalorestes lovei* (Mammalia, Mesonychia) from the early middle Eocene of northwestern Wyoming. *Contrib. Mus. Paleontol. Univ. Michigan* **29**:413–418.

Guthrie, D. A. 1967. The mammalian fauna of the Lysite Member, Wind River Formation, (early Eocene) of Wyoming. *Mem. South. Calif. Acad. Sci.* **5**:1–53.

Jöreskog, K. G., Klovan, J. E., and Reyment, R. A. 1976. *Geological Factor Analysis*. Elsevier, Amsterdam.

Lohmann, G. P. 1983. Eigenshape analysis of microfossils: a general morphometric procedure for describing changes in shape. *Math. Geol.* **15**:659–672.

Lohmann, G. P., and Schweitzer, P. N. 1990. On eigenshape analysis, in: F. J. Rohlf and F. L. Bookstein (eds.), *Proceedings of the Michigan Morphometrics Workshop, Special Publication 2: Ann Arbor, Michigan, The University of Michigan Museum of Zoology*, pp. 147–166.

MacLeod, N., and Rose, K. D. 1993. Inferring locomotor behavior in Paleogene mammals via eigenshape analysis. *Am. J. Sci.* **293-A**:300–355.

Maddison, W. P., and Maddison, D. R. 1992. *MacClade Program (Version 3.01)*. Sinauer Associates, Sunderland, MA.

Matthew, W. D. 1897. A revision of the Puerco Fauna. *Bull. Am. Mus. Nat. Hist.* **9**:59–110.

Matthew, W. D. 1937. Paleocene faunas of the San Juan Basin, New Mexico. *Trans. Am. Philos. Soc.* **30**:1–510.

McKenna, M. C. 1975. Toward a phylogenetic classification of the Mammalia, in: W. P. Luckett and F. S. Szalay (eds.), *Phylogeny of the Primates*, pp. 21–46. Plenum Press, New York.

McKenna, M. C., and Bell, S. K. 1997. *Classification of Mammals above the Species Level*. Columbia University Press, New York.

Meng, J., Suyin, T., and Schiebout, J. A. 1994. The cranial morphology of an early Eocene didymoconid (Mammalia, Insectivora). *J. Vertebr. Paleontol.* **14**:534–551.

Novacek, M. J. 1989. Higher mammal phylogeny: the morphological-molecular synthesis, in: B. Fernholm, K. Bremer, and H. Jörnvall (eds.), *The Hierarchy of Life*, pp. 421–435. Elsevier, Amsterdam.

Novacek, M. J. 1992. Mammalian phylogeny: shaking the tree. *Nature* **356**:121–125.

Novacek, M. J., and Wyss, A. R. 1986. Higher-level relationships of the recent eutherian orders: morphological evidence. *Cladistics* **2**:257–287.

O'Leary, M. A., and Rose, K. D. 1995a. New mesonychian dentitions from the Paleocene and Eocene of the Bighorn Basin, Wyoming. *Ann. Carnegie Mus.* **64**:147–172.

O'Leary, M. A., and Rose, K. D. 1995b. Postcranial skeleton of the early Eocene mesonychid *Pachyaena* (Mammalia: Mesonychia). *J. Vertebr. Paleontol.* **15**:401–430.

Osborn, H. F. 1924. *Andrewsarchus*, giant mesonychid of Mongolia. *Am. Mus. Novit.* **146**:1–5.

Peterson, O. A. 1931. New mesonychids from the Uinta. *Ann. Carnegie Mus.* **20**:333–339.

Prothero, D. R., Manning, E. M., and Fischer, M. 1988. The phylogeny of the ungulates, in: M. J. Benton (ed.), *The Phylogeny and Classification of the Tetrapods*, Volume 2, pp. 201–234. Clarendon Press, Oxford.

Qi, T. 1980. Irdin Manha upper Eocene and its mammalian fauna at Huhebolhe Cliff in central Inner Mongolia. *Vertebr. PalAsiat.* **18**:28–32.

Ranga Rao, A. 1973. Notices of two new mammals from the upper Eocene Kalakot Beds, India. *Directorate Geol. Oil Nat. Gas Comm. Dehra Dun India Spec. Pap.* **2**:1–6.

Rohlf, F. J. 1986. Relationships among eigenshape analysis, Fourier analysis and analysis of coordinates. *Math. Geol.* **18**:845–857.

Rose, K. D. 1982. Skeleton of *Diacodexis*, oldest known artiodactyl. *Science* **216**:621–623.

Rose, K. D. 1996. On the origin of the order Artiodactyla. *Proc. Natl. Acad. Sci. USA* **93**:1705–1709.

Russell, D. E. 1964. Les Mammifères Paléocènes d'Europe. *Mem. Mus. Natl. Hist. Nat. Paris Ser. C* **13**:1–324.

Simpson, G. G. 1945. The principles of classification and a classification of mammals. *Bull. Am. Mus. Nat. Hist.* **85**:1–350.

Swofford, D. L. 1993. *PAUP: Phylogenetic Analysis Using Parsimony, Version 3.1.1*. Illinois Natural History Survey, Champaign.

Szalay, F. S. 1969a. The Hapalodectinae and a phylogeny of the Mesonychidae (Mammalia, Condylarthra). *Am. Mus. Novit.* **2361**:1–26.

Szalay, F. S. 1969b. Origin and evolution of function of the mesonychid condylarth feeding mechanism. *Evolution* **23**:703–720.

Szalay, F. S., and Gould, S. J. 1966. Asiatic Mesonychidae (Mammalia, Condylarthra). *Bull. Am. Mus. Nat. Hist.* **132**:129–173.

Taylor, L. H. 1981. The Kutz Canyon local fauna, Torrejonian (Middle Paleocene) of the San Juan Basin, New Mexico, in: S. G. Lucas, J. K. Rigby, Jr., and B. S. Kues (eds.), *Advances in San Juan Basin Paleontology*, pp. 242–263. University of New Mexico Press, Albuquerque.

Thewissen, J. G. M. 1994. Phylogenetic aspects of cetacean origins: a morphological perspective. *J. Mamm. Evol.* **2**:157–184.

Thewissen, J. G. M., and Hussain, S. T. 1993. Origin of underwater hearing in whales. *Nature* **361**:444–445.

Thewissen, J. G. M., and Hussain, S. T. In press. *Attockicetus praecursor*, a new remingtonocetid from marine Eocene sediments of Pakistan, in: L. G. Barnes (ed.), *The Origin and Early Evolution of Cetacea*. Natural History Museum of Los Angeles County Science Series.

Thewissen, J. G. M., and Hussain, S. T. 1998. Systematic review of the Pakicetidae, early and middle Eocene Cetacea (Mammalia) from Pakistan and India. *Bull. Carnegie Mus. Nat. Hist.* **34**:220–238.

Thewissen, J. G. M., Russell, D. E., Gingerich, P. D., and Hussain, S. T. 1983. A new artiodactyl (Mammalia) from the Eocene of north-west Pakistan. Dentition and classification. *Proc. K. Ned. Akad. Wet. Ser. B* **86**:153–180.

Thewissen, J. G. M., Madar, S. I., and Hussain, S. T. 1996. *Ambulocetus natans*, an Eocene cetacean (Mammalia) from Pakistan. *Cour. Forsch.-Inst. Senckenberg* **191**:1–86.

Ting, S., and Li, C. 1987. The skull of *Hapalodectes* (?Acreodi, Mammalia), with notes on some Chinese Paleocene mesonychids. *Vertebr. PalAsiat.* **25**:161–186.

Van Valen, L. 1966. Deltatheridia, a new order of mammals. *Bull. Am. Mus. Nat. Hist.* **132**:1–126.

Van Valen, L. 1968. Monophyly or diphyly in the origin of whales. *Evolution* **22**:37–41.

Van Valen, L. 1969. The multiple origins of the placental carnivores. *Evolution* **23**:118–130.

Van Valen, L. 1978. The beginning of the age of mammals. *Evol. Theory* **4**:45–80.

West, R. M., and Lukacs, J. R. 1979. Geology and vertebrate-fossil localities, Tertiary continental rocks, Kala Chitta Hills, Attock District, Pakistan. *Contrib. Biol. Geol. Milwaukee Public Mus.* **26**:1–20.

Wiley, E. O., Siegel-Causey, D., Brooks, D. R., and Funk, V. A. 1991. The compleat cladist: a primer of phylogenetic techniques. *Univ. Kans. Mus. Nat. Hist. Spec. Publ. No. 19.*

Xu, X., Janke, A., and Arnason, U. 1996. The complete mitochondrial DNA sequence of the greater Indian rhinoceros, Rhinoceros unicornis, and the phylogenetic relationship among Carnivora, Perissodactyla, and Artiodactyla (+ Cetacea). *Mol. Biol. Evol.* **13**:1167–1173.

Zahn, C. T., and Roskies, R. Z. 1972. Fourier descriptors for closed plane curves. *IEEE Trans. Comput.* **C-21**:269–281.

Zhou, X., and Gingerich, P. D. 1991. New species of *Hapalodectes* (Mammalia, Mesonychia) from the early Wasatchian, early Eocene, of northwestern Wyoming. *Contrib. Mus. Paleontol. Univ. Michigan* **28**:215–220.

Zhou, X., Zhai, R., Gingerich, P. D., and Chen, L. 1995. Skull of a new mesonychid (Mammalia, Mesonychia) from the late Paleocene of China. *J. Vertebr. Paleontol.* **15**:387–400.

CHAPTER 6

Relationships of Cetacea to Terrestrial Ungulates and the Evolution of Cranial Vasculature in Cete

JONATHAN H. GEISLER and ZHEXI LUO

1. Introduction

1.1. Cranial Vasculature of Extant Cetaceans

Concomitant with the evolution of numerous aquatic adaptations, extant cetaceans have developed a cranial vascular system that is very different from those of terrestrial mammals. Terrestrial mammals rely on the internal carotid, external carotid, and vertebral arteries for blood supply to the brain. In extant cetaceans, however, these cranial vessels are either lost or extremely reduced early in ontogeny (such as the internal carotid and its branch the stapedial artery), or ramify along part of their course into anastomotic networks of small arteries (such as the vertebral artery and branches of the external carotid). These networks of arteries, and their morphologically similar complexes of veins, are termed *retia mirabile*, "wonderful nets." In contrast to the typical ungulate cranial circulation in which a few large arteries make up the main supply channels to the brain, the cetacean cranial circulation is characterized by a series of the retia mirabile. The cranial vascular patterns of cetaceans (Boenninghaus, 1904; Slijper, 1936; Walmsley, 1938; Vogl and Fisher, 1981a) are so different from those of terrestrial eutherian mammals (Sisson, 1921; Miller *et al.*, 1964; Bugge, 1974; Gray, 1974; Hunt, 1974; Presley, 1979; MacPhee, 1981; Wible, 1987) that they have attracted the attention of morphologists since the first half of the nineteenth century (Breschet, 1836; Stannius, 1841).

The abundance and position of the retia mirabile in cetaceans may be an aquatic adaptation. Nagel *et al.* (1968) hypothesized that one function of the retia mirabile in cetaceans is to modulate and dampen fluctuations in the flow of blood to the central nervous system. A rapid rise of blood pressure during diving, which is potentially damaging to surrounding

JONATHAN H. GEISLER • Department of Vertebrate Paleontology, American Museum of Natural History, New York, New York 10024-5192.   ZHEXI LUO • Section of Vertebrate Paleontology, Carnegie Museum of Natural History, Pittsburgh, Pennsylvania 15213-4080.
The Emergence of Whales, edited by Thewissen. Plenum Press, New York, 1998.

tissues, could be prevented by the retia mirabile (Dormer *et al.*, 1977). The retia mirabile may also act as a reservoir of oxygenated blood for the central nervous system during long dives. Compression of the lungs under high pressure at great depth would allow the retia within the thoracic cavity, which does not collapse, to expand into the space formerly occupied by the lungs, by filling with blood (Walmsley, 1938). The retia in the basicranial region may regulate the blood supply to the venous plexus of the pneumatic ("air") sinuses, which function as a barrier to interaural acoustic interference in the skull. The pneumatic sinuses are foamlike, highly vascularized structures that are continuous with the middle ear space and separate the petrosal from the skull (Fraser and Purves, 1960). This acoustic barrier is critical for directional hearing under water (Gingerich *et al.*, 1983).

The generalized cranial vascular pattern of extant cetaceans is unusual, relative to other eutherians, in the following aspects: 1, involution or extreme reduction of the internal carotid artery (Boenninghaus, 1904; Slijper, 1936; Walmsley, 1938; Vogl and Fisher, 1981b); 2, loss of the stapedial artery and all of its branches (Slijper, 1936; Galliano *et al.*, 1966; Vogl and Fisher, 1981b, Fig. 6; Wible, 1984, 1987); 3, absence of the capsuloparietal emissary vein (Slijper, 1936); 4, an extensive rostral endocranial arterial rete (Breschet, 1836; Boenninghaus, 1904; Slijper, 1936; Walmsley, 1938; McFarland *et al.*, 1979; Vogl and Fisher, 1981a); 5, development of a caudal endocranial arterial and venous retia mirabile (Breschet, 1836; Breathnach, 1955; Pilleri, 1991; Bajpai *et al.*, 1996; Melnikov, 1997); 6, development of a rete along the spinal cord and within the vertebral canal (Boenninghaus, 1904; Slijper, 1936; Vogl and Fisher, 1981a,b); 7, development of arterial retia ventral to the basicranium, lateral to the cervical vertebrae, and lateral to the thoracic vertebrae (Breschet, 1836; Slijper, 1936; Walmsley, 1938; Galliano *et al.*, 1966); 8, vascularization of the pneumatic sinuses and the middle ear by various retia in the basicranium especially around the petrosal and ectotympanic bulla (Fraser and Purves, 1960).

The rostral endocranial rete is located in the anteroventral (or sphenoidal) part of the cranial cavity. The rete is contained within the cavernous sinus and surrounds the hypophysis (Slijper, 1936; Walmsley, 1938; Vogl and Fisher, 1981a, Fig. 4). The rostral rete is supplied by the spinal meningeal arteries or a homologous arterial rete, which extends posteriorly from the posteroventral corners of the cerebral hemispheres to the dorsolateral sides of the cerebellum before exiting the cranial cavity through the foramen magnum (Slijper, 1936; Walmsley, 1938; Galliano *et al.*, 1966; McFarland *et al.*, 1979; Vogl and Fisher, 1981a; Melnikov, 1997). Four vessels originate from the rostral rete and directly supply the brain, a rostral and a caudal pair. Subdural branches of the four vessels are homologous to parts of the circle of Willis. However, unlike the interconnected vessels in the circle of Willis, these vessels are not connected to each other resulting in subdural independence of blood supply to each cerebral hemisphere (Vogl and Fisher, 1981a).

The caudal endocranial rete mirabile is located in the posterodorsal part (parietal region) of the cranial cavity. This massive rete is well developed in mysticetes and absent in almost all odontocetes (Slijper, 1936; Walmsley, 1938; Pilleri, 1991; Melnikov, 1997). It covers the posteroventral corners of the cerebral hemispheres, as well as the entire dorsal and lateral aspects of the cerebellum. Posteriorly, the caudal endocranial rete exits the cranial cavity through the foramen magnum to connect with the spinal rete in the vertebral canal (Boenninghaus, 1904; Walmsley, 1938; McFarland *et al.*, 1979; Melnikov, 1997).

Cetaceans also have a series of extracranial retia mirabile. Within the vertebral canal, the spinal meningeal arteries are extensively anastomosed with and supplied by an epidural spinal rete, the rete arteriales columnae vertebralis of Slijper (Slijper, 1936; Walmsley, 1938; Galliano *et al.*, 1966; McFarland *et al.*, 1979; Vogl and Fisher, 1981a). Through the intervertebral foramina, the spinal rete is extensively anastomosed with a series of retia lateral to the cervical and thoracic parts of the vertebral column (Breschet, 1836; Slijper, 1936; Walmsley, 1938; Galliano *et al.*, 1966; Melnikov, 1997). The arterial rete external to the cervical vertebrae is partially supplied by the omooccipital or occipital artery (Slijper, 1936; Galliano *et al.*, 1966; Wible, 1984). The rete lateral to the thoracic vertebrae is supplied by intercostal branches of the thoracic aorta, the subclavian artery and branches, or the supreme intercostal branch [possibly homologous to the subcostal branch of *Equus* as illustrated by Sisson (1921)] of the costocervical artery (Slijper, 1936; Galliano *et al.*, 1966; Melnikov, 1997). The right costocervical artery originates from one of the following: the trunk of the brachiocephalic artery (arteria anonyma of older anatomical literature), right subclavian artery, or from one of the 7–10 segmental arteries. The left costocervical artery arises from either the left subclavian, external carotid, or the tenth segmental arteries (Slijper, 1936). In summary, the brain of cetaceans receives blood from a series of connected retia that are ultimately supplied by the costocervical (or subclavian) and occipital (or omooccipital) arteries, but not the maxillary branch of the external carotid, internal carotid, and vertebral arteries as in most mammals (Sisson, 1921; Miller *et al.*, 1964; Bugge, 1974; Gray, 1974; Presley, 1979; MacPhee, 1981; Wible, 1987).

Numerous cranial osteological features are related to various vessels by function and by the development of bone shape during ossification. When preserved in fossils, these bony characters can be used to infer the circulatory patterns in extinct taxa. When viewed in a phylogenetic context, vascular reconstructions in different taxa can be linked to develop a hypothesis for the morphological changes of different vessels. This chapter describes the bony structures associated with vessels in extant taxa, in early cetaceans, and their extinct relatives. By comparison with the vasculature of extant whales, artiodactyls, and perissodactyls, we describe the transformation of arterial and venous cranial circulation patterns through the terrestrial ungulate–cetacean transition.

1.2. Unresolved Phylogenetic Issues

Van Valen (1966) advocated the ancestry of Cetacea within the Mesonychidae and supported this hypothesis with a list of shared resemblances between mesonychids and *Protocetus atavus*, the oldest and most plesiomorphic cetacean known at that time. Although he did not separate synapomorphies from symplesiomorphies, the mesonychid–cetacean relationship has been widely accepted (Szalay 1969a,b; Barnes and Mitchell, 1978; Barnes *et al.*, 1985; Gaskin, 1985; many others). McKenna (1975) recognized the close phylogenetic affinity of both taxa in his classification of mammals by placing mesonychids (as order Acreodi) and Cetacea within the mirorder Cete to the exclusion of other mammalian taxa. Prothero *et al.* (1988) found a mesonychid and cetacean clade that was diagnosed by at least two synapomorphies (Fig. 1D).

Prothero *et al.* (1988) specifically suggested that *Andrewsarchus* was the sister taxon to Cetacea and described three synapomorphies as supporting this node: "Medial portion of lambdoid crest high; I^{2,3} aligned with cheek teeth; elongated premaxilla" (Prothero *et al.*, 1988, p. 213). The shape of the lambdoid crests in *Andrewsarchus* does not resemble the

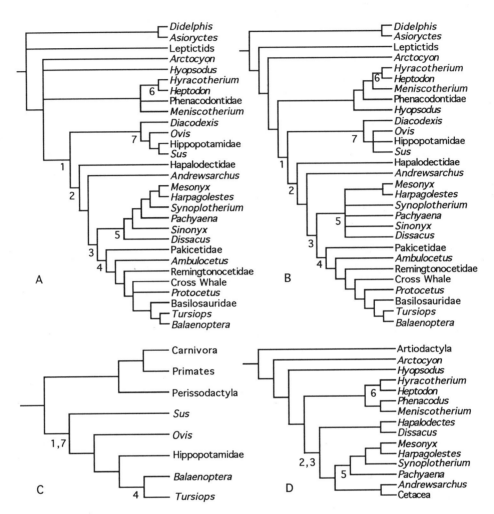

FIGURE 1. Phylogenetic hypotheses of this study based on the character/taxon matrix in Appendix 2, and a comparison of the phylogenies of other studies. (A) Strict consensus of 12 most parsimonious trees (from 100 repetitions of a heuristic search using PAUP 3.1.1., all multistate characters unordered). (B) Strict consensus of 15 most parsimonious trees (from 100 repetitions of a heuristic search using PAUP 3.1.1., all multistate characters ordered). (C) One hypothesis of ungulate phylogeny based on molecular sequence data from Fig. 4 of Gatesy (1997); some taxa are omitted to facilitate comparison. (D) Phylogenetic hypothesis presented by Prothero *et al.* (1988) that is based on morphological data; some taxa are omitted to facilitate comparison. Clade names: 1, Paraxonia; 2, Cete; 3, Acreodi; 4, Cetacea; 5, Mesonychidae; 6, Perissodactyla; 7, Artiodactyla.

morphology in the earliest cetaceans (Barnes and Mitchell, 1978; Gingerich *et al.*, 1983, Fig. 1). Only I³ in *Andrewsarchus mongoliensis* (AMNH 20135) is clearly aligned with the posterior premolars and molars, but not I². The premaxilla elongation is an order of magnitude less than the condition in Cetacea and the morphologies in both taxa should be treated as two distinct character states. Thus, it is questionable if *Andrewsarchus* can be regarded as the sister taxon to cetaceans to the exclusion of other "mesonychian" taxa. A recent parsimony analysis by Thewissen (1994) found a mesonychid and *Andrewsarchus* clade to be the sister group to Cetacea. Zhou *et al.* (1995) also grouped *Andrewsarchus* with the mesonychids and suggested that this group is closely related to Cetacea, although the exact phylogenetic relationships of these taxa were not described.

The ungulate affinities of mesonychids were recognized by Van Valen (1966) by moving them from the Carnivora (*sensu* Simpson, 1945) into the Condylarthra, a group of archaic ungulates that are now considered to be paraphyletic. This was based on the presence of hooflike terminal phalanges in mesonychids. In an effort to reduce paraphyletic grades, such as the "condylartha," McKenna (1975) removed mesonychids from the "condylarthra" and placed them with cetaceans and all other ungulates in the grandorder Ungulata. McKenna and Bell (1997) reduced Cetacea to subordinal level and placed Cetacea with Mesonychidae, Triisodontidae, and Hapalodectidae in the Order Cete. Acreodi was redefined as including all of the above groups except Cetacea. Since the classification of McKenna and Bell (1997) was developed to minimize paraphyletic groups, these taxa are understood as hypotheses of monophyletic groups.

Most molecular evidence indicates that the sister group of Cetacea is within the Ungulata. Early in the application of immunological and molecular data to systematic problems, precipitin tests suggested the artiodactyl affinities of cetaceans (Boyden and Gemeroy, 1950); however, the relative recency of their common ancestry was not clear because other ungulate groups were not included in the study. Recent phylogenies based on DNA and amino acid sequence data have produced more explicit, albeit somewhat conflicting, results. A clade often termed Cetungulata, which includes Cetacea, Artiodactyla, and Perissodactyla to the exclusion of other mammalian orders, is supported by several molecular studies including Czelusniak *et al.* (1990), Irwin *et al.* (1991), McKenna (1992), and Stanhope *et al.* (1996). The vast majority of molecular-based phylogenies support a closer relationship of artiodactyls to cetaceans than either is to perissodactyls (Fitch and Beintema, 1990; Czelusniak *et al.*, 1990; Irwin *et al.*, 1991; de Jong *et al.*, 1993; Irwin and Arnason, 1994; Janke *et al.*, 1994; Queralt *et al.*, 1995; Honeycutt *et al.*, 1995; Stanhope *et al.*, 1996; Xu *et al.*, 1996; Smith *et al.*, 1996; Gatesy *et al.*, 1996, Montgelard *et al.*, 1997; Gatesy, 1997). In fact, the preponderance of molecular data show that Artiodactyla, as currently defined, is paraphyletic and that Cetacea is nested within this order (Fig. 1C) (Irwin *et al.*, 1991; Irwin and Arnason, 1994; Graur and Higgins, 1994; Queralt *et al.*, 1995; Honeycutt *et al.*, 1995; Smith *et al.*, 1996; Gatesy *et al.*, 1996; Montgelard *et al.*, 1997; Gatesy, 1997; Shimamura *et al.*, 1997). By contrast, de Jong (1985) and McKenna (1992) suggested a clade of Cetacea and Perissodactyla to the exclusion of the Artiodactyla.

The possibility of Cetacea originating within Artiodactyla has been viewed with skepticism by most paleontologists because of the clear morphological evidence for the monophyly of Artiodactyla to the exclusion of Cetacea (see Prothero *et al.*, 1988; Gentry and Hooker, 1988). It should be noted that most of these characters occur on the astragalus. Although they appear to be independent, it is possible that they are related and may be overweighted by morphologists.

Even though most molecular analyses produce topologies with a paraphyletic Artiodactyla, the subclade of artiodactyls that is closest to Cetacea is variable (Smith *et al.*, 1996). The different possibilities include camels (Irwin *et al.*, 1991), ruminants (Graur and Higgins, 1994; Honeycutt *et al.*, 1995; Smith *et al.*, 1996), or the Hippopotamidae (Irwin and

Arnason, 1994; Gatesy *et al.*, 1996; Gatesy, 1997, this volume; Montgelard *et al.*, 1997). There also remains a question of the validity of phylogenies of family-level groups based on molecular data if only a small number of species are sampled within each family, and if the number of nucleotide base pairs sampled in most studies are sufficient to resolve the questions they address (Hervé and Douzery, 1994). The Hippopotamidae and Cetacea sister-group relationship generally occurs in the studies with the most diverse sampling of sequences and taxa (Fig. 1C).

Previous ordinal-level eutherian phylogenies based on morphological data have consistently shown cetaceans to be closer to perissodactyls than to artiodactyls. Novacek and Wyss (1986a) had Cetacea forming a trichotomy with Perissodactyla, and a clade including the Hyracoidea, Sirenia, and Proboscidea. Artiodactyla was the sister group to this trichotomy. Novacek (1986) also showed (with a similar data set) that artiodactyls are outside a clade including perissodactyls, cetaceans, and other ungulates. Two synapomorphies were cited to support a clade including Cetacea and other ungulates, except Artiodactyla: (1) absence of a postglenoid foramen and (2) posterior lacerate foramen coalesced with basicochlear fissure (= basicapsular fissure of this study; Novacek, 1986, character states 45d and 67c). This node is also supported by the absence of the proximal stapedial artery (Wible, 1987).

Prothero *et al.* (1988) also found Cetacea to be more closely related to Perissodactyla than to Artiodactyla; however, this phylogeny was not generated with a computer algorithm and therefore may not be the most globally parsimonious topology. Thewissen (1994) published a phylogeny that reaffirmed the close affinity of cetaceans and perissodactyls, and Shoshani (1993) found that myological characters support Ungulata monophyly, including Cetacea, but were unable to resolve the position of Cetacea within Ungulata.

2. Materials and Methods

2.1. Materials

Vascular reconstructions were based on taxa represented by specimens with well-preserved basicrania from several different collections (Tables I, II). Early artiodactyls are represented by an excellent basicranium of *Diacodexis* sp. initially described by Coombs and Coombs (1982; AMNH 16141; for institutional abbreviations see Table I) and *Diacodexis pakistanensis* (Russell *et al.*, 1983). Early Cete, here defined as all taxa more closely related to the mesonychid/cetacean clade than to artiodactyls, are best represented by a juvenile skull of *Hapalodectes hetangensis*, which was described by Ting and Li (1987). The vascular reconstruction for mesonychids is based on *Dissacus* sp. (UM 75501; Luo and Gingerich, in press) and *Mesonyx obtusidens* (AMNH 12643). Well preserved early cetacean basicrania include several pakicetids (cast of GSP-UM 085, HGSP 964312, 96386, 96231) and an unnamed cetacean of the protocetid grade, which is referred to as the Cross Whale (ChM PV5401; Geisler *et al.*, 1996). Endocranial vascular grooves and foramina are also available for *Dalanistes ahmedi* (GSP-UM 3106, 3052) and *Indocetus* (Bajpai *et al.*, 1996). Basilosaurid basicrania are particularly well represented in the University of Michigan Museum of Paleontology collections including but not limited to: *Dorudon atrox* (GSP-UM 93220, 97516, 94797), *Basilosaurus isis* (GSP-UM 97507), and *Saghacetus osiris* (GSP-UM 83905, 97550) (reviewed by Uhen, 1996).

Table I. Institutional Abbreviations

AMNH	American Museum of Natural History (New York, New York)
ChM PV	Vertebrate Fossil Collections, the Charleston Museum (Charleston, South Carolina)
GSP-UM	Geological Survey of Pakistan (Islamabad, Pakistan)–University of Michigan project. Specimens are deposited in GSP collections
H-GSP	Howard University (Washington, DC) and Geological Survey of Pakistan (Islamabad, Pakistan). Specimens are deposited in GSP collections
IVPP	Institute of Vertebrate Paleontology and Paleoanthropology, Chinese Academy of Sciences (Beijing, China)
NR	Staatliches Museum für Naturkunde (Stuttgart, Germany)
UM	University of Michigan Museum of Paleontology (Ann Arbor, Michigan)
USNM	United States National Museum of Natural History, Smithsonian Institution (Washington, DC)
YPM-PU	Princeton University Collection at Yale Peabody Museum (New Haven, Connecticut)

Taxa (Table II) were selected to address two questions: (1) Are mesonychids the sister group to cetaceans? (2) Are artiodactyls or perissodactyls more closely related to cetaceans? The second question is crucial for the vascular reconstructions in fossil taxa because the distribution of vascular characters in related extant taxa, such as perissodactyls and artiodactyls, is the key to the reconstruction of these vessels in extinct taxa. Only taxa that could be scored for the majority of the characters were included in the phylogenetic analysis, with one exception (see below). The following relatively complete fossil taxa were included to restrict the misidentification of convergences between extant taxa as synapomorphies (see Gauthier *et al.*, 1988; Novacek *et al.*, 1988; Gould, 1995): *Arctocyon, Meniscotherium*, Phenacodontidae, and *Hyopsodus*. The analysis of Williamson and Lucas (1992) placed *Meniscotherium* within the Phenacodontidae; however, their selection of taxa allowed them to determine the relationships within Phenacodontidae but not whether it is monophyletic. For several characters, *Meniscotherium* is more plesiomorphic than other phenacodontids (see Appendixes 1 and 2, characters 6, 7, 51, 71); therefore, it is coded separately. *Andrewsarchus* can be coded for only 20% of the 78 characters. It was nevertheless included because of its potentially close phylogenetic relationship with Cetacea (Prothero *et al.*, 1988; Thewissen, 1994). Didymoconids were not considered because basicranial evidence supports a closer relationship with Lipotyphla than with Ungulata (Meng *et al.*, 1994).

Resolving cetacean phylogeny is not the primary goal of this study; therefore, only taxa that document the main morphological transitions were included: pakicetids, *Ambulocetus*, Remingtonocetidae, the Cross Whale, *Protocetus*, and Basilosauridae. The extant cetacean genera *Balaenoptera* and *Tursiops* were used as representatives of the Mysticeti and Odontoceti, respectively. Monophyly of the Perissodactyla was assumed allowing fossil taxa, *Hyracotherium* and *Heptodon*, to be scored for soft morphological characters preserved only in extant taxa. The following taxa were included to sample some of the diversity of Artiodactyla and as a preliminary test of the monophyly of this order: *Diacodexis*, Hippopotamidae (including both *Hippopotamus amphibius* and *Choeropsis liberiensis*), *Sus scrofa*, and *Ovis aries*.

Although the closest possible outgroup is often used to root trees, it is not a requirement for phylogenetic reconstruction (Nixon and Carpenter, 1993). By contrast, it is important not to violate the assumption of ingroup monophyly. If the outgroup is within the ingroup, then all shortest trees will be incorrectly rooted. Since eutherian phylogeny above

Table II.   Specimens Examined in This Study and Used to Code 29 Taxa
for 80 Morphological Characters[a]

Taxa	Specimens
Cetacea	
Tursiops truncatus	AMNH 120920*, 184930*, 212554*
Balaenoptera physalus	AMNH 28274*, 84870*, 148407*
Basilosauridae	
Basilosaurus isis	GSP-UM 97507
Dorudon atrox	GSP-UM 93220, 94797, 94812, 97516
Saghacetus osiris	GSP-UM 83905, 97550
Zygorhiza kochi	USNM 11692
Protocetidae	
Protocetus atavus	NR 11084
"Cross Whale"	ChM PV5401
Remingtonocetidae	
Remingtonocetus sp.	GSP-UM 3009, 3054, 3057
Dalanistes ahmedi	GSP-UM 3052, 3106
Ambulocetidae	
Ambulocetus natans	H-GSP 18507
Pakicetidae	GSP-UM 084 (cast), H-GSP 96231, 96386, 96431
Archaic Acreodi	
Mesonychidae	
Mesonyx obtusidens	AMNH 12643
Harpagolestes uintensis	AMNH 1878, 1945
Harpagolestes orientalis	AMNH 26300
Pachyaena gigantea	AMNH 2823
Pachyaena ossifraga	AMNH 75, 1522, 4262, 15730, YPM-PU 14708
Sinonyx jiashanensis	IVPP V10760
Dissacus navajovius	AMNH 3359, 3360, 3361
Dissacus praenuntius	AMNH 131919 (cast)
Archaic Cete	
Hapalodectes hetangensis	IVPP V5253
Andrewsarchus mongoliensis	AMNH 20135
Artiodactyla	
Ovis aries	AMNH 53598*, 100072*, 146547*
Sus scrofa	AMNH 235190*, 235192*, 236144*, 694422*
Hippopotamus amphibius	AMNH 24289*, 130247*
Choeropsis liberiensis	AMNH 89626*, 146848*, 202423*
Diacodexis sp.	AMNH 16141
Perissodactyla	
Hyracotherium vasacciense	AMNH 55267, 55269, 96274
Hyracotherium tapirinum	AMNH 96277, 96298
Heptodon calciculus	AMNH 294, 4858, 14884
Archaic ungulates	
Arctocyon primaevus	AMNH 55900 (cast), 55901 (cast)
Hyopsodus paulus	AMNH 11350, 11415, 11899
Hyopsodus sp.	AMNH 39
Meniscotherium terraerubrae	AMNH 4414, 4426, 4434, 4447
Meniscotherium chamense	AMNH 2560, 48083, 48126, 48129
Phenacodontidae	
Phenacodus sp.	AMNH 1582, 15262, 15266, 15268, 15271
Phenacodus trilobatus	AMNH 15275
Phenacodus vortmani	AMNH 4378
Tetraclaenodon puercensis	AMNH 93000

Table II. *(Continued)*

Taxa	Specimens
Outgroups	
Didelphis virginiana	AMNH 219212*, 219220*, 242660*
Leptictidae	
Frictops sp.	AMNH 76745
Leptictis sp.	AMNH 2144481

^aAll AMNH specimens are from the Department of Vertebrate Paleontology except those marked with an asterisk, which are from the Department of Mammalogy. For abbreviations see Table 1.

the ordinal level is unresolved (Novacek, 1992), the extinct taxa Leptictidae and *Asioryctes* were chosen as outgroups. The marsupial *Didelphis* was added as a third outgroup to increase the phylogenetic information of the three soft tissue characters (characters 78, 79, 80), which cannot be scored for leptictids and *Asioryctes*.

2.2. Phylogenetic Methods

The computer program PAUP 3.1.1 (Swofford, 1993) was used to find the most parsimonious trees. Data entry, character optimizations, and tree manipulations were performed on MacClade 3.01 (Maddison and Maddison, 1992). Twenty-six ingroup and three outgroup taxa were scored for 80 characters. Among these characters, 27 are basicranial, 18 other cranial, 14 dental, 18 postcranial, and 3 soft morphological (Appendixes 1 and 2). The characters used in this phylogenetic analysis are incorporated from several previous parsimony analyses of cetaceans and/or other ungulates (Thewissen and Domning, 1992; Thewissen, 1994; Geisler and Luo, 1996; Geisler *et al.*, 1996; Luo and Gingerich, in press; and many others, see Appendix 1). Several characters used by major ordinal-level cladistic studies such as Novacek (1986) and Novacek and Wyss (1986a) were also included. The taxa were scored for soft tissue morphological characters that have a bearing on the relationships of the ungulate taxa examined (Wible, 1987; Thewissen, 1994). Most of the characters in Prothero *et al.* (1988), relevant to the affinities of Cetacea and Mesonychidae, were included. Our intention was to reevaluate these characters in the context of all characters available. However, several characters used by Prothero *et al.* (1988) were excluded for the following reasons: no clear morphological distinction between the taxa examined (position of tympanic aperture of facial nerve canal relative to the oval window), high levels of homoplasy within the operational taxonomic units [position and size of third trochanter of the femur (see also Rose, 1996), size of the deltoid crest of the humerus, size of the coracoid process of the scapula], and unresolved character distribution (presence or absence of a clavicle, see Thewissen, 1990; Williamson and Lucas, 1992).

Two separate phylogenetic analyses were undertaken, one with all characters unordered and the other with all multistate characters ordered. When ordering transformation series, character states were aligned to minimize the morphological gap between adjacent states. Characters were polarized *a posteriori* during rooting. Limited by our computer capacity, a branch-and-bound search proved unfeasible; therefore, in both the ordered and unordered runs a heuristic search with the random addition option was repeated for 100 replicates. All trees were constrained for monophyly of the ingroup.

2.3. Soft-Tissue Reconstruction

We distinguish three categories of osteological correlates, osseous structures or morphologies that are closely associated with a soft tissue feature. For vessels, these correlates are grooves, canals, and foramina. Although the types and application of osteological correlates deserve a more thorough treatment than is given here, they are briefly discussed because it is a crucial component in our vascular reconstructions. The three types of osteological correlates are: two-way, one-way positive, and one-way negative correlates. A two-way osteological correlate is always present whenever its corresponding soft tissue structure is present, as observed in extant taxa. In turn, the soft tissue structure is only present in conjunction with its osteological correlate; therefore, the presence or absence of either can be used to infer the presence or absence of the other. A soft tissue feature in a fossil taxon can be reconstructed on the presence of its osteological correlate; however, in this study, reconstructions are not extended to other taxa with morphologically similar but non-homologous osteological correlates. The homology of potential osteological correlates in extinct taxa to known correlates in extant taxa is determined by optimizing the correlate's distribution on a phylogenetic tree.

One-way positive osteological correlates are partially correlated with their corresponding soft tissue features. In extant taxa, the soft tissue structure is always present if its one-way positive correlate is present; however, the soft tissue structure may still be present in the absence of its correlate. The distribution of a one-way positive osteological correlate among fossil taxa may be an underestimate of the distribution of its corresponding soft tissue feature, but never an overestimate. By contrast, a one-way negative correlate is occasionally present in the absence of its supposedly correlative soft tissue feature yet is never absent in the presence of its soft tissue structure, as observed in extant taxa. The distribution of a one-way negative osteological correlate among fossil taxa may be an overestimate of the distribution of its corresponding soft tissue. There is equal confidence in reconstructing a vessel's presence based on the presence of its two-way or one-way positive correlate while equal confidence is given to reconstructing a vessel's absence based on the absence of its two-way or one-way negative correlate. The distinction between two-way and one-way positive correlates becomes important in the development of vascular evolutionary hypotheses. Whereas the timing of vascular changes associated with a two-way correlate are highly constrained on a phylogeny, a vascular feature with a one-way positive correlate may either predate the appearance or postdate the disappearance of its corresponding osteological feature.

The most conservative method of soft-tissue reconstruction in extinct taxa is the extant phylogenetic bracket. As initially outlined by Bryant and Russell (1992) and Witmer (1992), the extant phylogenetic bracket uses a known phylogeny of extinct and extant taxa to reconstruct soft tissues in extinct taxa. Extant taxa are first surveyed to identify osteological structures that always co-occur with a specific soft tissue feature. A vascular or soft tissue structure can be reconstructed in an extinct taxon with a high degree of confidence if: (1) its unique osteological correlate is present and (2) its two closest extant relatives have the soft tissue structure and osteological correlate (Witmer, 1995). In cases where only the closest extant relative has the soft tissue structure, the extant phylogenetic bracket must be expanded, by adding more outgroups, to determine the most parsimonious optimization for the extinct taxon (Maddison *et al.*, 1984; Witmer, 1995).

We use two-way and one-way positive osteological correlates as sufficient evidence to reconstruct vessels in extinct taxa; however, optimization on our phylogeny and on the phylogenies of Novacek (1992) is used to ascertain whether the osteological features in fossils are homologous to correlates in extant taxa. The extant phylogenetic bracket is viewed here as too conservative, especially because most evidence suggests that some vessels of the eutherian morphotype have been lost multiple times in different lineages (Wible, 1987). We have included assessments of the extant phylogenetic bracket to demonstrate instances where reconstructions based on phylogeny and just morphology agree.

For vessels that do not have osteological correlates, the extant phylogenetic bracket is the only available method for reconstructing soft tissues in fossils (Witmer, 1995). We have used the extant phylogenetic bracket in this manner to view vascular patterns of extinct taxa as a fully functional complex. The extant phylogenetic bracket is also used to reconstruct vessels that have one-way negative correlates. One-way negative correlates can be positively misleading; therefore, we have chosen a conservative stance with regard to correlates of this kind.

Where possible, cetacean vessels are named after homologues in terrestrial eutherian mammals following Schaeffer (1953), Miller *et al.* (1964), Gray (1974) and Getty (1975). We believe that this approach will help with comparisons of cetacean vasculature with those of noncetacean ungulates, and will facilitate the placement of cetaceans with regard to other ungulates in eutherian phylogeny.

3. Results of Phylogenetic Analysis

The phylogenetic analysis is based on 80 morphological characters (listed in Appendix 1), and their distributions among 29 taxa (matrix presented in Appendix 2). The analysis with unordered multistate characters produced 12 most parsimonious trees of 259 steps with CI = 0.459, RI = 0.708 (Fig. 1A). The analysis with ordered multistate characters produced 15 most parsimonious trees of 271 steps with CI = 0.439, RI = 0.714 (Fig. 1B). In describing the results of our analysis, we use stem-based definitions for most taxa (see de Queiroz and Gauthier, 1990). Stems have been chosen such that the taxic content of these groups follows, as closely as possible, the most widely accepted usage. Monophyly of Cetacea, which is defined here as all taxa more closely related to the cetacean crown group than to *Mesonyx obtusidens*, is strongly supported. It is diagnosed by 12 unequivocal synapomorphies. Cete, here defined as all taxa more closely related to Cetacea than to Artiodactyla, occurred in all most parsimonious trees and is diagnosed by 8 unequivocal synapomorphies: 1, presence of the alisphenoid canal (character 28); 2, presence of foramen rotundum (31); 3, absence of mastoid foramen (35); 4, presence of preglenoid process (38); 5, strong parastyle on M^1 and M^2 (47); 6, M_3 metaconid absent or forms lingual swelling on protoconid (55); 7, talonid basin rectangular with cristid oblique shifted lingually (57); 8, absence of M_3 hypoconulid (58). We suggest defining the Mesonychidae as all taxa that are more closely related to *Mesonyx obtusidens* than to the crown group of cetaceans. This would exclude *Andrewsarchus mongoliensis* and the Hapalodectidae from the Mesonychidae based on our cladogram. The Mesonychidae is diagnosed by one unequivocal synapomorphy, maximum mesodistal length of M^3 less than 60% of the maximum length of M^2 (50). Both Andrewsarchinae and Hapalodectinae were listed by Szalay and Gould (1966) as two subfamilies of the Mesonychidae. Our phylogenetic analysis supports the removal of *Andrewsarchus* and *Hapalodectes* from Mesonychidae.

The clade Acreodi is defined here as the most recent common ancestor of *Mesonyx obtusidens* and extant cetaceans, plus all of its descendants. As far as taxic content, this follows the use of Acreodi by Prothero *et al.* (1988), except for the exclusion of *Andrewsarchus*, but differs considerably from McKenna and Bell (1997). The Acreodi contains two monophyletic groups, the Mesonychidae and the Cetacea. It is diagnosed by at least one, the foramen ovale is situated anterior to the glenoid fossa, but potentially seven unequivocal synapomorphies. Ambiguity in the number of synapomorphies exists because *Andrewsarchus*, the sister group to Acreodi, is only 20% complete, in terms of number of characters scored. If *Andrewsarchus* is removed from the analysis, then Acreodi is supported by seven unequivocal synapomorphies regardless of the optimization option used including: 1, anterior process of petrosal present (character 3); 2, facial nerve sulcus present on the mastoid process of petrosal (9); 3, elongate and deep external auditory meatus (20); 4, presence of paired ridges (one on each side of the sagittal plane) on basioccipital (21); 5, absence of postglenoid foramen in squamosal (24); 6, M^1–M^3 hypocone absent (48).

Crucial for the vascular reconstruction in basal members of Cete is the presence of a Cetacea and Artiodactyla clade, here termed Paraxonia (see Thewissen, 1994). This is similar to the use of this name by Kalandadze and Rautian (1992). Paraxonia is defined here as the most recent common ancestor of Artiodactyla and Cetacea and all of its descendants. Paraxonia is diagnosed by nine unequivocal synapomorphies: 1, stylomastoid foramen encircled ventromedially by tympanic bulla (character 8); 2, presence of sigmoid process (14); 3, absence of a lacrimal tubercle on the orbital rim (41); 4, entepicondyle of humerus 25% or less than the width of the combined ulnar and radial articulation surface (64); 5, paraxonic foot (75); 6, four functional digits of the foot (76); 7, elongate blastocyst (78); 8, penis erection via relaxation of retractor muscle with little cavernous tissue (79); and 9, three primary bronchi (80). Paraxonia, including both Cetacea and Artiodactyla to the exclusion of other ungulates, is congruent with most molecular-based phylogenies but is at odds with previous morphological-based phylogenies (Fig. 1A,B). However, unlike most molecular studies, we found Artiodactyla to be monophyletic and supported by four unequivocal synapomorphies: 1, absence of posttemporal canal (37); 2, absence of astragalar canal (71); 3, wide sustentacular facet of the astragalus (73); 4, broad contact of distal articulating surface of astragalus with cuboid (74).

4. Vasculature

4.1. Arteries

4.1.1. Proximal Stapedial Artery

The proximal stapedial artery is the stem of the stapedial artery between its origin from the internal carotid artery and its bifurcation into the inferior and superior rami (Fig. 2C: sa). The artery, if present, is located on the ventral surface of the promontorium (MacPhee, 1981; Wible, 1987). The only known exception to this position in mammals is found in the monotreme *Ornithorhynchus* in which the proximal stapedial artery is lateral to the promon-

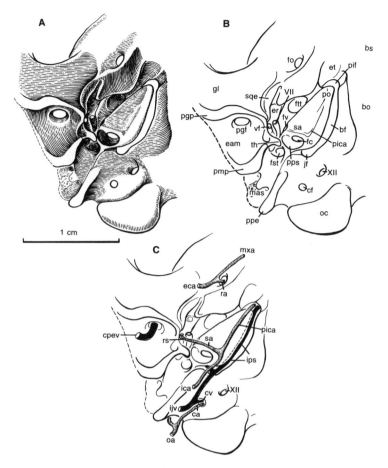

FIGURE 2. *Diacodexis* (Artiodactyla): compository reconstruction of basicranial anatomy and vasculature. (A) Reconstruction of the right basicranium of *Diacodexis* (primarily based on AMNH 16141). The exoccipital features are after *Homacodon* (AMNH 12685, Coombs and Coombs, 1982). The squamosal features are after *Diacodexis pakistanensis* (Russell *et al.,* 1983). (B) Outline of basicranium with labels to identify the structure. (C) Vascular restoration illustrating our interpretation of the pathways and connections, but not the size, of the original vessels. In several cases, vessels are reconstructed based on their distribution in extant taxa, in a phylogenetic context, without the presence of a two-way or one-way positive osteological correlate (denoted with an asterisk in the following). Abbreviations: bf, basicapsular fissure; bo, basioccipital; bs, basisphenoid; ca, condyloid artery*, a branch of the occipital artery; cf, condyloid foramen for the condyloid artery and vein; cpev, capsuloparietal emissary vein, or "postglenoid vein," a tributary of the external jugular vein; cv, condyloid vein*; eam, external auditory meatus; eca, external carotid artery*; er, epitympanic recess; et, trough for Eustachian tube; fc, fenestra cochleae; fo, foramen ovale; fst, stapedial muscle fossa; ftt, fossa for tensor tympani muscle; fv, fenestra vestibuli; gl, glenoid fossa; ica, internal carotid artery; ijv, internal jugular vein; ips, inferior or medial petrosal sinus/vein; jf, jugular foramen (posterior lacerate foramen); mas, mastoid process of petrosal; mxa, maxillary artery*; oa, occipital artery*, a branch from the external carotid artery; oc, occipital condyle; pgf, postglenoid foramen; pgp, postglenoid process; pica, promontorial branch of the internal carotid artery and its related groove; pif, piriform fenestra; pmp, postmeatal process of squamosal; po, pole of the promontorium; ppe, paroccipital process of exoccipital; pps, postpromontorial sinus; ra, ramus anastomoticus*; rs, superior ramus of stapedial artery; sa, proximal stapedial artery or its sulcus; sqe, entoglenoid process of the squamosal; th, attachment site for tympanohyal; vf, vascular foramen for the ramus superior of the stapedial artery; VII, facial nerve foramen; XII, foramen for hypoglossal nerve.

torium, within the facial nerve sulcus (Wible and Hopson, 1995). The bifurcation of the proximal stapedial artery from the internal carotid artery is within the middle ear cavity in most mammals, except for muroid rodents in which the origin is posterior to the fibrous membrane (Wible, 1987). In all eutherian mammals with a proximal stapedial artery, the artery passes through the stapedial foramen, an aperture within the stapes just distal to the footplate. After passing laterally through the stapedial foramen, the proximal stapedial artery bifurcates into a superior ramus and an inferior ramus (MacPhee, 1981; Wible, 1987).

Within the middle ear cavity, the proximal stapedial artery often travels in a groove or canal on the surface of the promontorium of the petrosal (Fig. 2: sa; MacPhee, 1981). The bifurcation between the internal carotid and stapedial arteries typically occurs in the posterior part of the middle ear cavity; therefore, the stapedial groove or canal usually extends from the anteromedial edge of the fenestra rotunda to the edge of the fenestra vestibuli, which receives the footplate of the stapes (Wible, 1987). Although a proximal stapedial artery is present in microchiropteran bats (Wible, 1987; Wible and Novacek, 1988), a distinct groove for the stapedial artery is not always present (personal observation). The same holds true for *Ornithorhynchus* (Rougier *et al.*, 1992). Therefore, a groove on the promontorium for the proximal stapedial artery is a one-way positive correlate. In several groups of eutherians, such as primates (Szalay, 1975; MacPhee, 1981), the proximal stapedial artery is completely enclosed by an osseous canal. We know of no example among extant mammals in which the stapedial canal is present when the proximal stapedial artery is absent; therefore, we consider the stapedial canal a two-way correlate of the proximal stapedial artery. The stapedial foramen, which perforates the stapes, is a one-way negative correlate because it may be present in the absence of the proximal stapedial artery.

Among adults of the extant taxa studied, the proximal stapedial artery and its corresponding groove or canal on the promontorium of the petrosal are absent in the following: *Didelphis* (Wible, 1984, 1987, 1990; Novacek and Wyss, 1986b), Perissodactyla (Kampen, 1905; Sisson, 1921; Wible, 1987), Artiodactyla (Kampen, 1905; Sisson, 1921; du Boulay and Verity, 1973; Ghoshal, 1975a,b; Wible, 1984, 1987; Smut and Bezuidenhout, 1987), and Cetacea (Boenninghaus, 1904; Slijper, 1936; Vogl and Fisher, 1981b; Wible, 1984, 1987; Melnikov, 1997). Therefore, the extant phylogenetic bracket is decisively negative for all of the fossil taxa considered in the analysis. In some extant cetaceans, hypertrophy of the stapedial crura coincides with the disappearance or extreme reduction of the stapedial foramen (Oelschläger, 1986; Novacek and Wyss, 1986b); however, this condition is restricted to some members of the cetacean crown group. A patent stapedial foramen occurs in basilosaurid archaeocetes (Lancaster, 1990), the sister group to the cetacean crown group (Fig. 1A,B). Therefore, reduction of the stapedial foramen in extant cetaceans is interpreted as an apomorphy of these taxa and not the primitive condition of Cetacea.

A groove for the proximal stapedial artery occurs in leptictids (Novacek, 1986), *Arctocyon* (Russell, 1964), *Hyopsodus* (Cifelli, 1982), *Meniscotherium* (AMNH 48555, 2560; Cifelli, 1982), *Diacodexis* (AMNH 16141; Fig. 2B: sa), *Hapalodectes hetangensis* (IVPP 5253; Ting and Li, 1987; Fig. 3A: sa), *Dissacus* (UM 75501, Luo and Gingerich, in press), and *Mesonyx obtusidens* (AMNH 12643; Figs., 4, 5: sa). By contrast, a proximal stapedial groove is absent in extant and extinct cetaceans, perissodactyls (Kampen, 1905; Radinsky, 1965), and in two of the three outgroups: *Didelphis* (Wible, 1987, 1990) and *Asioryctes* (Kielan-Jaworowska, 1981).

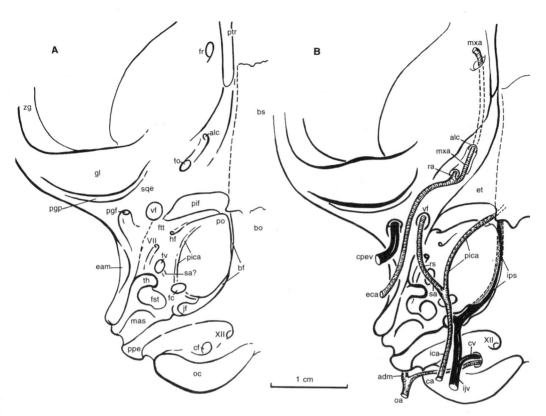

FIGURE 3. *Hapalodectes* (an archaic ungulate of the Cete clade). (A) Right basicranium (based on IVPP 5253; modified from Luo and Gingerich, in press). (B) Vascular reconstruction (modified from Ting and Li, 1987). Abbreviations (see also Fig. 2 caption): adm, arteria diploetica magna; alc, alisphenoid canal; fr, foramen rotundum; hf, hiatus Fallopi for the great petrosal nerve; ptr, pterygoid ridge, forming the base of pterygoid hamulus; sa?, proximal stapedial artery sulcus?; th, tympanohyal; vf, vascular foramen, for the superior ramus of the stapedial artery; zg, zygoma.

The presence of a proximal stapedial artery is plesiomorphic for Eutheria based on ontogenetic evidence (Wible, 1987) and was present in the putative ungulate ancestor based on optimization of a groove for the proximal stapedial artery on the phylogeny generated here. The proximal stapedial groove in extinct Ungulata is homologous with the same structure in extant taxa such as the Erinaceidae or Soricomorpha (MacPhee, 1981) unless the Ungulata is the sister group to the Edentata and Pholidota clade, as suggested by some optimal trees of Novacek (1992). Under these topologies, the homology of the stapedial groove is equivocal.

Based on the distribution of its osteological correlates, the proximal stapedial artery was ubiquitous in early taxa of Paraxonia including mesonychids, *Hapalodectes*, and probably *Andrewsarchus*. It was also present in *Diacodexis* (Fig. 2: sa) but lost within Artiodactyla and all Cetacea (including *Pakicetus*). Ontogenetic evidence does offer some sup-

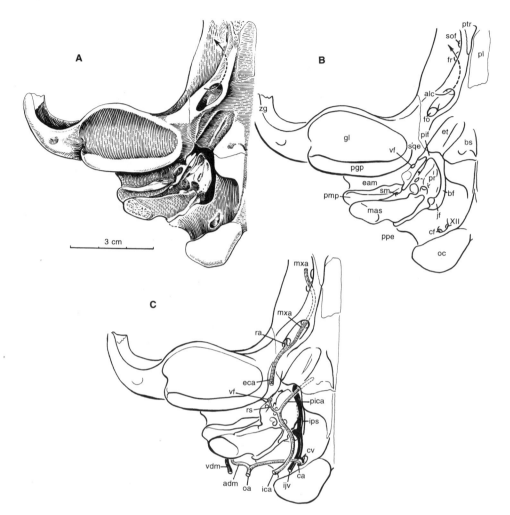

FIGURE 4. *Mesonyx obtusidens* (Mesonychidae). (A) Restoration of right basicranium (AMNH 12643). (B) Outline of basicranium with labels of structures. (C) Vascular reconstruction. Abbreviations (see also Fig. 2 caption): adm, arteria diploetica magna; alc, alisphenoid canal; fr, foramen rotundum; pl, pterygoid lamina; pr, promontorium; ptr, pterygoid ridge, forming the base for pterygoid hamulus; sm, stylomastoid foramen; sof, sphenorbital fissure; vdm, vena diploetica magna*.

port for the hypothesized wide distribution of the proximal stapedial artery in extinct members of Paraxonia. A proximal stapedial artery has been reported in the early ontogenetic stages of two artiodactyls, *Sus scrofa* (Hofmann, 1914) and *Bos taurus* (Hammond, 1937), and of two odontocete cetaceans, *Globicephala melaena* (Schreiber, 1916) and *Phocoena phocoena* (Slijper, 1936). The proximal stapedial artery was absent in the common ancestor of the perissodactyl crown group but was present in the stem groups of Perissodactyla: *Meniscotherium* and phenacodontids.

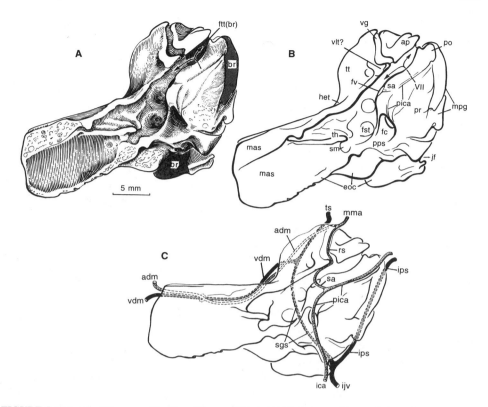

FIGURE 5. Petrosal of *Mesonyx obtusidens* (Mesonychidae). (A) Right petrosal (ventral view of AMNH 12643). (B) Outline with labels to identify the structures. (C) Reconstruction of vasculature. Abbreviations (see also Fig. 2 caption): adm, arteria diploetica magna; ap, anterior process = anterior extension of tegmen tympani; br, breakage and/or matrix; eoc, contact for the exoccipital on the petrosal; ftt (br), fossa for tensor tympani muscle, partially broken; het, hiatus epitympanicus; mas, occipital exposure of the mastoid process; mma, middle meningeal artery; mpg, medial promontorial groove for inferior petrosal sinus; pr, promontorium; sgs, sigmoid sinus; ts, transverse sinus; tt, tegmen tympani = superior process; vdm, vena diploetica magna*; vg, vascular groove for the superior ramus of the stapedial artery (the groove is enclosed into a foramen when the petrosal is articulated with the squamosal); vlt?, ventrolateral tuberosity; VII, pathways of branches of the facial nerve.

4.1.2. Superior Ramus of Stapedial Artery

The superior ramus is one of two major branches of the proximal stapedial artery. In many mammals that have the superior ramus, it branches from the proximal stapedial artery either within the epitympanic recess or slightly anterior to it. On its entry into the cranial cavity, the superior ramus may join the middle meningeal artery. The anterior branch of the superior ramus exits the cranial cavity into the temporal fossa or orbit either with the ophthalmic nerve through the sphenorbital fissure or through a separate foramen (Kielan-Jaworowska *et al.*, 1986; Wible, 1987; Rougier *et al.*, 1992). If the stem of the superior ramus involutes during development, then its anterior branch may reverse its direction of blood flow and join the external carotid artery (Daniel *et al.*, 1954; Wible, 1987). The proximal stapedial artery ends as a bifurcation into the superior and inferior rami; therefore, the division between the superior ramus and the proximal stapedial artery is arbitrary if the inferior ramus is absent.

A potentially confounding condition of the superior ramus occurs in Lagomorpha, Rodentia, Macroscelidea, Chiroptera, and some Scandentia. These forms share a derived state in which the proximal stapedial artery enters the cranial cavity through an apparent superior ramus foramen before bifurcating into superior and inferior rami (Wible, 1987; Wible and Novacek, 1988). The stem of the superior ramus is confined inside the cranial cavity as the result of an endocranial bifurcation of the proximal stapedial artery. An endocranial bifurcation does not occur in any extant ungulates and cannot be inferred from the osteology of the skull [unless the bifurcation occurs within the cranial wall (intracranial) as in Scandentia; MacPhee, 1981; Wible and Zeller, 1994]; therefore, it is not considered in reconstructing the vasculature of fossil ungulates.

If the superior ramus of the stapedial artery originates within the tympanic cavity, it exits the tympanic cavity to enter the cranial cavity through: (1) a canal or foramen in the petrosal/squamosal suture (see Figs. 4, 5: vg, rs), (2) a canal or foramen through the tegmen tympani, or (3) through the piriform fenestra (MacPhee, 1981; Wible, 1987; Rougier et al., 1992; Wible and Hopson, 1995). Both the canal in the petrosal/squamosal suture and the canal within the tegmen tympani are one-way positive correlates for the superior ramus. If both canals and the piriform fenestra are absent, then the superior ramus is absent as well. The superior ramus may course in a groove on the ventral surface of the tegmen tympani and/or adjacent to the dorsolateral rim of the fenestra vestibuli; however, depending on where the proximal stapedial bifurcates into the superior and inferior rami, these grooves may carry the proximal stapedial artery instead. Thus, both grooves are one-way positive correlates for the stapedial artery, but it is ambiguous as to which specific portion of the artery they are correlated to.

A superior ramus of the stapedial artery and its corresponding foramen in the tympanic roof are absent in the following extant taxa: *Didelphis* (Wible, 1984, 1987, 1990), artiodactyls (personal observation; Kampen, 1905; Sisson, 1921; du Boulay and Verity, 1973; Ghoshal, 1975a,b; Wible, 1984, 1987; Smut and Bezuidenhout, 1987), perissodactyls (Kampen, 1905; Sisson, 1921; Wible, 1987), and extant cetaceans (Boenninghaus, 1904; Slijper, 1936; Vogl and Fisher, 1981b; Wible, 1984, 1987). A superior ramus of the stapedial artery has been reported early in the ontogeny of *Sus scrofa* (Hofmann, 1914) and *Bos taurus* (Hammond, 1937). However, the extant phylogenetic bracket is decisively negative for the superior ramus in the adults of the extinct ungulates studied here, although it potentially supports its presence at some early stage in ontogeny in extinct Paraxonia.

A foramen for the superior ramus has been observed in several extinct taxa: leptictids (Novacek, 1986), *Arctocyon* (Russell, 1964), *Diacodexis* (AMNH 16141: Fig. 2: vf), potentially *Hapalodectes hetangensis* (IVPP 5253), *Mesonyx obtusidens* (AMNH 12643; Fig. 4: vf; Fig. 5B: vg), and *Dissacus* (UM 75501). A nearly continuous sulcus, which wraps around the ventrolateral tuberosity (Fig. 5: vlt), connects the proximal stapedial groove to the vascular foramen situated on the petrosal/squamosal suture in *Mesonyx*. The only break in the groove is a 1.5-mm swath across the facial nerve sulcus. The portion of the stapedial sulcus lateral to the fenestra vestibuli is also indicative of a ramus superior in mesonychids (Fig. 5: vg). When present, the position of the foramen for the superior ramus exhibits two basic states. In *Arctocyon* (Russell, 1964) and *Diacodexis* (Fig. 2: vf, er) the foramen is lateral to the epitympanic recess whereas in *Hapalodectes* (Fig. 3: vf) and mesonychids it is anterior to the epitympanic recess. The anterior position, as in mesonychids, required a bend in either the superior ramus or the proximal stapedial artery. A foramen for the superior ramus is absent in *Asioryctes* (Kielan-Jaworowska, 1981), and Phenacodontidae (AMNH 15268; Cifelli, 1982).

In *Hapalodectes hetangensis* (IVPP 5253), a canal for the superior ramus of the stapedial artery through either the tegmen tympani or the petrosal/squamosal suture is apparently absent, but both the piriform fenestra and an unidentified foramen in the squamosal, medial to the postglenoid foramen, are present (Fig. 3: vf). The unidentified foramen is completely formed on the right side but only partially separated from the piriform fenestra on the left (Fig. 3: pif). As the exact position of the petrosal/squamosal suture in this specimen is unclear, the posterior edge of the foramen may be formed by the petrosal, thus placing it on the petrosal/squamosal suture like the superior ramus foramen in mesonychids. There are two possible occupants of this foramen, based on vessels that enter or exit the cranial cavity in this region of the skull in extant taxa: (1) a branch of the capsuloparietal emissary vein or (2) the superior ramus of the stapedial artery. Although multiple foramina for the capsuloparietal emissary vein are not unusual in the squamosal of extant mammals, they are usually fairly close together and within a larger recess. Instead of being associated with the postglenoid foramen, the unidentified foramen is apparently a subdivision of the piriform fenestra, which does transmit the superior ramus in some extant taxa (Macroscelidea; MacPhee, 1981). Therefore, we tentatively identify this as a foramen for the superior ramus, otherwise this vessel has to enter the skull through the piriform fenestra.

The presence of the superior ramus of the stapedial artery is plesiomorphic not only for Eutheria but probably for more inclusive clades, depending on whether it is homologous to similar vessels in sauropsids (Wible, 1987). Its presence is also plesiomorphic for Ungulata, based on the cladogram produced in this study. The foramen for the superior ramus in the tympanic roof of extinct ungulates is homologous with the same structure in extant taxa, such as primates and lipotyphlans (MacPhee, 1981; Wible, 1987), unless Ungulata turns out to be the sister group of an Edentata and Pholidota clade (see Novacek, 1992). In this unlikely case, the homology of the foramen is equivocal. The absence of the superior ramus in the clade including Perissodactyla and Phenacodontidae is best interpreted as a secondary loss that these two groups share.

The superior ramus of the stapedial artery is also absent in all extant cetaceans. It is hypothesized that the absence of the proximal stapedial groove in all extant and extinct cetaceans is indicative of the absence of the proximal stapedial artery and all of its branches. However, as the sulcus for the proximal stapedial artery is a one-way positive osteological correlate, then the stapedial artery and the superior ramus could have been lost at any time between the origination of the cetacean crown group and the first appearance of the earliest cetaceans. Character optimization on the phylogeny advocated here suggests that the absence of the superior ramus of the stapedial artery may diagnose Pantomesaxonia (including perissodactyls, hyracoids, desmostylians, sirenians, and proboscideans, *sensu* Thewissen and Domning, 1992) and not Ungulata (if Tubulidentata is excluded from Ungulata) as suggested by different phylogenies (Wible, 1987; Prothero *et al.*, 1988).

4.1.3. Ramus Anastomoticus

The ramus anastomoticus is a vessel that connects the vascular network of the external carotid artery with the inferior ramus of the stapedial artery, a branch of the internal carotid artery (Fig. 2: ra; Wible, 1987). If present, it originates from the maxillary branch of the external carotid, then enters the cranial cavity through the foramen ovale (Fig. 2: ra, fo; Wible, 1987). Its endocranial branches cannot be separated from the anterior branch of the superior ramus of the stapedial artery in most adult mammals (Daniel *et al.*, 1954). The middle meningeal artery of *Bos taurus* (Sisson, 1921) and *Camelus dromedarius* (Smut and Bezuidenhout, 1987) is probably homologous to the ramus anastomoticus of the maxillary artery. The ramus anastomoticus is present in almost all artiodactyls (du Boulay and Verity, 1973; Wible, 1984).

A ramus anastomoticus has been reported in *Monodon monoceros* (Wible, 1984) and possibly other odontocetes (Fraser and Purves, 1960). A vascular plexus is near the foramen ovale in an embryonic stage of *Balaenoptera acutorostrata* and accompanies the mandibular branch of the trigeminal nerve into the skull where it anastomoses with the rostral rete (de Burlet, 1915a, Figs. 20, 22, 23). We interpret this vascular network as a homologous ramification of the ramus anastomoticus. We further hypothesize that the presence of a ramus anastomoticus is plesiomorphic for Cetacea, pending verification of its presence in other extant cetaceans that have not been studied. If our interpretation is correct, then the extant phylogenetic bracket indicates that all Cete had a ramus anastomoticus. The occurrence of this vessel in at least some odontocetes, possibly in mysticetes, plus its widespread occurrence in the next two extant outgroups suggests that this vessel was present in some if not all archaic cetaceans.

In mesonychids, such as *Mesonyx obtusidens* (AMNH 12643), *Dissacus* (UM 75501), and *Sinonyx jiashanensis* (Zhou *et al.*, 1995), a deep vascular groove is present between the foramen ovale and the posterior opening of the alisphenoid canal (Fig. 4A: fo, alc), possibly indicating that the foramen ovale had a vascular occupant in addition to the mandibular branch of the trigeminal nerve. This observation is consistent with the presence of the ramus anastomoticus in these taxa as based on the extant phylogenetic bracket; however, the groove cannot be viewed as an osteological correlate because it is not homologous to any structure in extant taxa.

4.1.4. Spinal Meningeal Arteries

The spinal meningeal arteries are a group of neomorphic vessels of Cetacea that are embedded within the rete mirabile of the vertebral canal. Before entering the cranial cavity, the spinal meningeal arteries split into bilateral groups (McFarland *et al.*, 1979). Within the cranial cavity, the vessels become intradural, course ventrolaterally around the cerebellum, and connect with the rostral arterial rete to become the only channel supplying blood to the brain (Boenninghaus, 1904; Slijper, 1936; McFarland *et al.*, 1979). They occur in some odontocetes including *Phocoena* (Boenninghaus, 1904; Slijper, 1936) and *Tursiops* (Galliano *et al.*, 1966; McFarland *et al.*, 1979).

The endocranial portions of the spinal meningeal arteries may have corresponding transverse grooves on the endocranial surface of the exoccipital, as in *Phocoena* (Breathnach, 1955, Fig. 1). By contrast, the endocranial parts of the spinal meningeal arteries in *Tursiops* are not apparent when the superficial layer of the dura mater is in place (McFarland *et al.*, 1979), and they do not have correlative sulci on the exoccipital (personal observation). Therefore, the transverse grooves on the endocranial surface of the exoccipital are one-way positive osteological correlates for the endocranial portion of the spinal meningeal arteries. The limited number of observations regarding this correlate in extant taxa warrants caution in applying this correlate to vascular reconstructions of extinct taxa.

We suggest that the lateral portion of the caudal endocranial arterial rete mirabile of mysticetes and the odontocete *Physeter* is homologous to the endocranial portions of the spinal meningeal arteries of odontocetes. This hypothesis is based on the following observations: Both structures occupy the same position in the skull relative to the cerebellum and the foramen magnum, both contain vessels that are primarily aligned anteroposteriorly, and both have fairly large lumina relative to vessels of the rostral rete (Boenninghaus, 1904; Walmsley, 1938; McFarland *et al.*, 1979; Melnikov, 1997). The main difference between the odontocete spinal meningeal arteries and the caudal arterial rete is the greater number of vessels in the latter. Ontogenetic data can be used to bridge this morphological difference. The caudal rete lateral to the cerebellum in an embryonic *Balaenoptera acutorostrata* is composed of less than ten vessels (Slijper, 1936, Figs. 8, 9), which is in conspicuous contrast to the multitude of vessels in the rete of adult mysticetes (Walmsley, 1938), yet is much closer to the number of spinal meningeal arteries in odontocetes. We suggest that the caudal rete of mysticetes and *Physeter* ramifies from the spinal meningeal arteries during development. This is consistent with observations, in other taxa, of rete development from a few simple vessels during ontogeny (Hammond, 1937; Wible, 1984). It is unlikely that the rete in adult mysticetes is a retention of vascular plexuses formed in early development (for an example of the embryonic plexuses, see Padget, 1948, 1957). If the endocranial portion of the spinal meningeal arteries is homologous to the lateral portions of the caudal arterial rete and the latter develops from the former through ontogeny, as suggested here, then the spinal meningeal arteries in adult odontocetes probably formed by paedomorphosis.

In the Cross Whale (ChM PV5401) there is lateral groove on the endocranial surface of the exoccipital possibly for the spinal meningeal arteriès. It originates at the dorsolateral corner of the foramen magnum and travels laterally until ending at the medial border of an air sinus fossa between the pars cochlearis of the petrosal and the exoccipital. This groove is morphologically similar to the groove for the spinal meningeal artery in some odontocetes; however, both grooves are not homologous because of its absence in most odontocetes, mysticetes, and basilosaurids (Dart, 1923, Figs. 4, 6, 10, 14, 17). Therefore, the reconstruction of the spinal meningeal arteries in the Cross Whale is highly speculative.

4.1.5. Branches of Occipital Artery

The occipital artery is usually a branch of the external carotid artery (Figs. 2–4: oa). The occipital artery of some eutherian mammals and many noneutherian mammals gives rise to the arteria diploetica magna, which enters the cranial cavity to join the superior ramus of the stapedial artery (Figs. 6, 7: adm, rs) (Kielan-Jaworowska *et al.*, 1986; Wible, 1987). In ungulates the occipital artery may give rise to the posterior meningeal or posterior auricular arteries which in some cases perforate the back of the skull to supply the meninges (Sisson, 1921). Some or all of the posterior meningeal arteries may be homologous to the arteria diploetica magna (see below). Two conditions of the posterior meningeal arteries are known with regard to the bones of the cranium among extant ungulates. (1) It enters the cranial cavity through the mastoid foramen (*sensu* Cope, 1880) between the exoccipital and the mastoid portion of the petrosal. This condition occurs in the artiodactyls *Camelus* (Smut and Bezuidenhout, 1987), *Capra, Ovis* (Ghoshal, 1975a), and *Hippopota-*

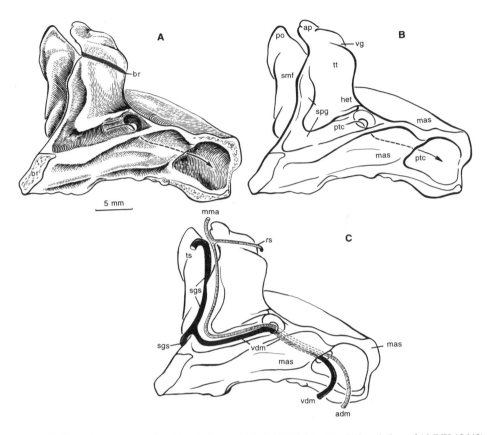

FIGURE 6. Petrosal of *Mesonyx obtusidens* (Mesonychidae). (A) Right petrosal (dorsal view of AMNH 12643). (B) Outline with labels to identify the structure. (C). Reconstruction of vasculature. Abbreviations (see also Fig. 2 caption): adm, arteria diploetica magna; ap, anterior process = anterior extension of tegmen tympani; br, breakage and/or matrix; het, hiatus epitympanicus; mma, middle meningeal artery, the anterior–dorsal continuation of arteria diploetica magna; ptc, posttemporal canal; sgs, sigmoid sinus; smf, suprameatal fossa; spg, superior petrosal sulcus or groove; ts, transverse sinus*; tt, tegmen tympani = superior process; vdm, vena diploetica magna*; vg, vascular groove.

mus amphibius (du Boulay and Verity, 1973), in the carnivore *Canis familiaris* (Miller *et al.*, 1964), and in the primate *Homo sapiens* (Schaeffer, 1953; Gray, 1974). (2) The posterior meningeal artery in *Equus* courses through a posttemporal canal between the pars canicularis portion (the mastoid process) of the petrosal and the temporal fossa portion of the squamosal to enter the cranial cavity (Sisson, 1921).

The posterior meningeal artery in *Equus* is probably homologous with the arteria diploetica magna in noneutherian and eutherian mammals (Kielan-Jaworowska *et al.*, 1986; Wible, 1987), because this vessel enters the cranial cavity in an intraosseous canal between the mastoid process of the petrosal and the squamosal, which is very similar to the posttemporal canal in some marsupials that contains the arteria diploetica magna (Wible, 1987). The evidence for the homology of the artiodactyl posterior meningeal artery, which enters the cranium through the mastoid foramen, with the arteria diploetica magna is not as clear.

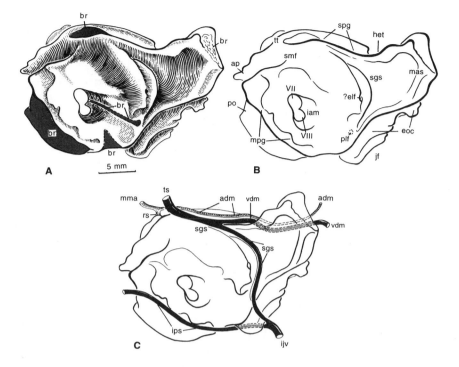

FIGURE 7. *Mesonyx obtusidens* (Mesonychidae). (A) Right petrosal (medial or endocranial view of AMNH 12643)). (B) Outline with labels to identify the structures. (C) Reconstruction of vasculature. Abbreviations (see also Fig. 2 caption): adm, arteria diploetica magna; ap, anterior process = anterior extension of tegmen tympani; br, breakage and/or matrix; ?elf, endolymphatic foramen; eoc, contact for the exoccipital on the petrosal; het, hiatus epitympanicus; iam, internal acoustic meatus; mma, middle meningeal artery = anterior continuation of arteria diploetica magna and the superior ramus; mpg, medial promontorial groove for inferior petrosal sinus; plf, perilymphatic foramen; po, pole of the promontorium; sgs, sigmoid sinus; smf, suprameatal fossa; spg, superior petrosal groove; ts, transverse sinus*; tt, tegmen tympani = superior process; vdm, vena diploetica magna*; VII, facial foramen; VIII, foramina for vestibulocochlear nerves.

Two different applications of parsimony yield two different conclusions regarding the homology of the posterior meningeal artery of artiodactyls and the arteria diploetica magna. The most widespread method of identifying prospective hypotheses of homology, which can later be tested by optimization of its distribution on a phylogenetic tree, is shared similarity (Patterson, 1982). With vessels, accompanying structures are often used as a measure of similarity, particularly nerves (for examples with the internal carotid artery, see Wible, 1984), but also surrounding osseous structures. Similarity, in the sense of spatial relationships to other structures, minimizes the vessel's positional variation to other structures. Applying this logic to the posterior meningeal artery of artiodactyls means that it is not homologous to the arteria diploetica magna because each vessel has a different course relative to the bones of the occiput. Thus, in taxa with a posttemporal canal and mastoid foramen, the former would contain the arteria diploetica magna and possibly the vena diploetica magna while the latter would transmit the mastoid emissary vein and occasionally a posterior meningeal artery.

A second method for dealing with the posterior meningeal artery/arteria diploetica magna homology question is by minimizing the number of vessels, and thus assuming greater variation in the vessel's positional relationships. This method uses parsimony by not assuming the presence of multiple arteries, for which there is no direct evidence. If the number of branches of the occipital artery is minimized, the posterior meningeal artery of artiodactyls is homologous to the arteria diploetica magna. Under this hypothesis, the posterior meningeal artery and the arteria diploetica magna still pass a crude test of similarity because they both: (1) branch from the occipital artery and (2) enter the skull toward the lateral side of the occiput. Therefore, in taxa with only a mastoid foramen the arteria diploetica magna and the mastoid emissary vein would both pass through the mastoid foramen whereas in taxa with two foramina, the arteria diploetica magna would pass through the posttemporal canal and the mastoid emissary vein would enter through the mastoid foramen. The arteria diploetica magna and mastoid emissary vein would be adjacent in some cases and not others (separated by the mastoid process of the petrosal). Postulating homology between the posterior meningeal arteries and the arteria diploetica magna is testable by a conjunction test (Patterson, 1982), specifically by dissecting animals that have skulls with two foramina on the occiput and observing whether both the arteria diploetica magna and posterior meningeal artery are present. We are unaware of any reports of both vessels occurring in the same individual; therefore, we consider this a valid hypothesis. The current evidence, in our view, is equivocal with regard to the homology of the artiodactyl posterior meningeal artery and the arteria diploetica magna.

The posttemporal canal is a two-way osteological correlate of the arteria diploetica magna for all extant mammals. However, the mastoid foramen does not always contain a posterior meningeal artery, as is the case in *Bos taurus* (Ghoshal, 1975a). Therefore, the mastoid foramen is a one-way negative osteological correlate.

A mastoid foramen, or canal, is present in many extant and extinct ungulates including: the artiodactyls *Bos taurus, Ovis aries, Capra hircus* (Ghoshal, 1975a), *Camelus dromedarius* (Smut and Bezuidenhout, 1987), *Ovibos moschatus* (AMNH 2868), *Rangifer tarandus* (AMNH uncataloged), *Homacodon vagans* (AMNH 1265), and *Diacodexis pakistanensis* (Russell *et al.*, 1983); the perissodactyl *Tapirus terrestris* (AMNH 1403); and the archaic ungulates *Hyopsodus* (Gazin, 1968) and *Arctocyon* (Russell, 1964). It is also present in the insectivore *Leptictis dakotensis* (Novacek, 1986). The mastoid foramen is absent in: *Hyracotherium* (personal observation), *Phenacodus* sp. (AMNH 15268), *Meniscotherium* (Gazin, 1965), all cetaceans and mesonychids with the exception of *Synoplotherium vorax* (Wortman, 1901), and *Didelphis virginiana* (AMNH 242660).

The posttemporal canal occurs in the following taxa: *Didelphis virginiana* (AMNH 242660); the archaic ungulates *Arctocyon* (Russell, 1964), *Hyopsodus* (Gazin, 1968; Cifelli, 1982; personal observation), *Phenacodus* sp. (AMNH 15268), and *Meniscotherium* (AMNH 48555, 2560); the perissodactyls *Deperetella cristata* (AMNH 20290) and *Hyracotherium*; and all mesonychids including *Dissacus* (UM 75501), *Mesonyx obtusidens* (AMNH 12643; Fig. 6: ptc), *Harpagolestes uintensis* (AMNH 1892), *Synoplotherium vorax* (Wortman, 1901), and *Pachyaena ossifraga* (AMNH 15730). In mesonychids the morphology of the posttemporal canal is slightly different from other mammals because its distal (external) end is completely enclosed by the mastoid process of the petrosal (Fig. 6: ptc, mas). We feel confident with our identification of this structure in mesonychids because its proximal (internal) end is lateral to the mastoid process of the petrosal and medial to the squamosal. The posttemporal canal is present in *Equus*, but absent in all artiodactyls including *Diacodexis pakistanensis* (Russell *et al.*, 1983) and all cetaceans with the possible exception of pakicetids.

Pakicetids have a large foramen near the junction of the exoccipital, supraoccipital, and squamosal (cast of GSP-UM 084, H-GSP 96386, 96231) that leads directly into the cranial cavity. A smaller foramen in the same position occurs in the artiodactyl *Choeropsis liberiensis* (AMNH 89626) and transmits a branch of the occipital artery. In contrast to the morphology in mesonychids, the mastoid process of the petrosal of cetaceans has been excluded from the occiput and confined to a ventral and sometimes lateral exposure (Geisler and Luo, 1996; Luo and Gingerich, in press). The foramen of pakicetids is near the occipital/squamosal suture; therefore, it is different from both the mastoid foramen and the posttemporal canal.

If the posterior meningeal arteries of ungulates and the arteria diploetica magna of other mammals are homologous, then the mastoid foramen in artiodactyls, the posttemporal canal of *Equus*, and the unnamed foramen of *Pakicetus* are three different locations where homologous vessels enter the cranial cavity. Alternatively, the posterior meningeal artery of artiodactyls and the arteria diploetica magna may not be homologous, and the foramen in pakicetids transmitted one of the two vessels depending on the homology of the pakicetid foramen to either the posttemporal canal or the mastoid foramen. We hypothesize that the large foramen in the squamosal/occipital suture of pakicetids is homologous to the posttemporal canal of mesonychids. This choice is based on our phylogeny where the transformation of the pakicetid foramen from the posttemporal canal is one step shorter than the transformation from the mastoid foramen. Given this homology, the position of the posttemporal canal shifted medially in cetaceans to the junction of the squamosal, supraoccipital, and exoccipital. This shift is consistent with a reduction in the occipital exposure of the mastoid process of the petrosal.

The arteria diploetica magna is potentially a supplier of blood to the brain, and could have been an evolutionary intermediate between the plesiomorphic condition of cranial supply (via the internal carotid and vertebral arteries) and the derived condition in cetaceans (via a pair of spinal meningeal arteries or spinal rete). Based on the girth of the posttemporal canal and its postulated homologue in pakicetids, the arteria diploetica magna was large in mesonychids and the earliest cetaceans. The endocranial portion of the arteria diploetica magna in these taxa could have reached the rostral rete via a short anastomosis and become a supplier of blood to the brain. Under this hypothesis, the anastomosis between the arteria diploetica magna and the rostral rete would be homologous to the extreme anterior end of the spinal meningeal arteries of extant cetaceans. Such a hypothetical vascular pattern is comparable to that of the artiodactyl *Hippopotamus amphibius* in which a meningeal branch of the occipital artery, a possible homologue of the arteria diploetica magna, does anastomose with the rostral arterial rete (du Boulay and Verity, 1973).

A second anastomosis could have linked the endocranial portion of the arteria diploetica magna, as described above, with the spinal meningeal arteries, thereby establishing an arterial route from the spinal canal to the cranial cavity. Therefore, the development of two short anastomoses could have transformed the endocranial part of the arteria diploetica magna into the spinal and endocranial retial connection that is unique to cetaceans. The external and intracranial (within the skull wall) portions of the arteria diploetica magna had to have been lost to make this hypothesis conform to observation because no branch of the occipital artery enters the skull through the occiput in extant cetaceans. A possible test of this morphological scenario would be to document the ontogenetic development of the spinal meningeal and occipital arteries in cetaceans.

4.2. Veins

The capsuloparietal emissary vein (Figs. 2, 3, 8: cpev) is a major venous drainage of the cranial cavity in many eutherians (Gelderen, 1924). It is also known as the temporal sinus (Sisson, 1921; Miller *et al.*, 1964), postglenoid vein (Butler, 1967), or petrosquamous sinus (Padget, 1957). The cerebral venous system early in mammalian development consists of a longitudinal lateral head vein, ventrolateral to the developing brain, and three primary dorsal tributaries. The middle of the three, the middle cerebral vein, is immediately anterior to the otic capsule and becomes the prootic sinus (Padget, 1957; Butler, 1967; Wible and Hopson, 1995). Later in ontogeny, the three dorsal tributaries, the middle one being the prootic sinus, are joined by anastomotic vessels dorsal to the otic capsule and trigeminal ganglion. These anastomoses, plus the dorsal ends of the three primary dorsal tributaries that they connect, form a vessel that roughly trends anterodorsally to posteroventrally. It is the embryonic precursor to the transverse and sigmoid sinuses (Padget, 1957; Butler, 1967). The junction of the prootic sinus with the dorsal anastomoses marks the division between the transverse sinus anteriorly and the sigmoid sinus posteriorly. The capsuloparietal emissary vein originates posterior to the prootic sinus from the sigmoid sinus and joins the ventral branch(es) of the prootic sinus (Padget, 1957, Pl. 2, Figs. 36, 37). This is concomitant with the degeneration of the prootic sinus. In adult eutherian mammals, only the dorsal end of the prootic sinus remains as the stem of the superior petrosal sinus (Padget, 1957; Butler, 1967). While the prootic sinus or its branch, the sphenoparietal emissary vein, exits the chondrocranium through the sphenoparietal fenestra (opening in the chondrocranium between the orbital plate and the otic capsule), the capsuloparietal emissary vein exits through the capsuloparietal foramen, a small opening between the otic capsule and parietal plate (Gelderen, 1924; Padget, 1957; Wible, 1990; Wible and Hopson, 1995). An increase in size of the capsuloparietal emissary vein at the expense of the prootic sinus appears to be the predominant pattern in eutherian mammals (Gelderen, 1924; Wible and Hopson, 1995), although there are possible exceptions (see Butler, 1967). Vascular definitions including ontogenetic information, although crucial in separating superficially similar but nonhomologous vessels, cannot be directly applied to extinct organisms.

We hypothesize that the vein exiting the adult skull in the postglenoid region of all ungulates is the capsuloparietal emissary vein, instead of the prootic sinus or the sphenoparietal emissary vein (see Wible, 1990; Wible and Hopson, 1995). We base this on the development of the dural sinuses in Artiodactyla, Perissodactyla (Gelderen, 1924), and Cetacea (Gelderen, 1924; Slijper, 1936) where the capsuloparietal emissary vein, identified by its exit from the chondrocranium through the capsuloparietal foramen, can be traced through ontogeny to the vein in the postglenoid foramen of adult ungulates.

Osteological correlates for the capsuloparietal emissary vein include: (1) the temporal canal, which travels within either the squamosal or the petrosal/squamosal suture, and (2) its ventral aperture, the postglenoid foramen (*sensu* Cope, 1880; Figs. 2, 3: pgf; Gelderen, 1924; Wible, 1990). The foramen and its corresponding canal are two-way osteological cor-

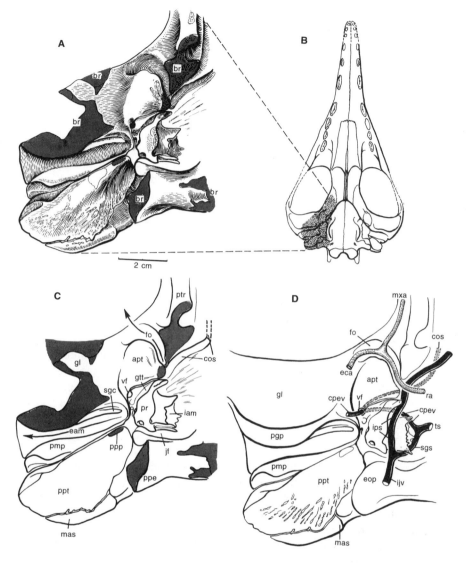

FIGURE 8. "Cross Whale" (an unnamed taxon of the protocetid grade, Cetacea). Ventral view of the right basicranium based on ChM PV5401, a partial skull and postcranial skeleton from the Cross Quarry near Charleston, South Carolina, mid to late Eocene (see Geisler *et al.*, 1996). (A) The partial right basicranium (ChM PV 5401). (B) The approximate position and orientation of the partial basicranium in relation to the whole skull (the skull outline is modified from *Protocetus*, a related taxon of the same protocetid grade, modified from Fraas, 1904, and Kellogg, 1936). (C) Outline of basicranium with labels to identify structures. (D) Reconstruction of the basicranium and its vascular pattern (outline proportion based on *Gaviacetus*, Gingerich *et al.*, 1995; Luo and Gingerich, in press; and *Protocetus*, Fraas, 1904; Kellogg, 1936). Abbreviations (see also Fig. 2 caption): apt, contacting site on the squamosal for the anterior process of tympanic; br, broken or matrix-filled areas; cos, cranio-orbital sinus, anterior continuation of the capsuloparietal emissary vein; eop, bullar process of the exoccipital, incomplete on ChM PV5401; gtt, tensor tympani groove; iam, internal acoustic meatus; ppp, posterior pedicle of tympanic bulla; ppt, posterior process of the tympanic; pr, promontorium; ptr, pterygoid ridge; sgc, contact site on squamosal for the sigmoid process of the bulla; sgs, sigmoid sinus*; ts, transverse sinus; vf, vascular foramen for the capsuloparietal emissary vein, also known as the petrosquamous vein.

relates. A canal in the petrosal/squamosal suture for the superior ramus of the stapedial artery can be differentiated from a morphologically similar structure for the capsuloparietal emissary vein by its position relative to the middle ear cavity, specifically the fibrous membrane or its osseous ontogenetic homologue the tympanic bulla. When the temporal canal is in the petrosal/squamosal suture, the postglenoid foramen opens just lateral to the tympanic bulla such as in *Equus* (Sisson, 1921) or in some rodents (personal observation). By contrast, the foramen for the superior ramus of the stapedial artery is within the middle ear cavity; therefore, it is medial to the fibrous membrane or tympanic bulla.

A capsuloparietal emissary vein with a temporal canal and postglenoid foramen is present in all extant artiodactyls studied including: *Camelus dromedarius* (Smut and Bezuidenhout, 1987), some individuals of *Sus scrofa*, although highly reduced or vestigial in the adult (Dennstedt, 1904; Butler, 1967), *Ovis aries* (Gelderen, 1924), and *Bos taurus* (Dennstedt, 1904; Gelderen, 1924). The capsuloparietal emissary vein and correlates are also present in the perissodactyl *Equus* (Dennstedt, 1904; Sisson, 1921; Gelderen, 1924). By contrast, a capsuloparietal emissary vein is absent in the limited number of extant adult cetaceans studied (Boenninghaus, 1904; Slijper, 1936). It is possible that the extensive vascular plexus associated with pneumatic sinuses in the same region of the skull in extant cetaceans may have annexed the capsuloparietal vein, or rendered the vein indistinguishable.

The capsuloparietal foramen is present in mysticetes (de Burlet, 1915a; Gelderen, 1924), but absent in odontocetes (de Burlet, 1915b). Slijper (1936) reported a small capsuloparietal emissary vein in a 105-mm embryo of *Balaenoptera acutorostrata*. If Slijper's observation is representative of most mysticetes, (1) its absence in odontocetes is an autapomorphy and (2) the occurrence of a capsuloparietal emissary vein in stem Cetacea and in mesonychids at some stage of development is unequivocal based on the extant phylogenetic bracket.

Whereas the postglenoid foramen is usually in the squamosal (such as in *Diacodexis*, Fig. 2: pgf), it is situated in the petrosal/squamosal suture in early cetaceans (*Protocetus*, NR 11084; Cross Whale, ChM PV5401, and basilosaurids: UM 94812, 83905, 97550, USNM 11962) (Fig. 8: vf). Although the postglenoid foramen of archaeocete cetaceans is obscured in ventral view by the ectotympanic bulla, suggesting it opens within the tympanic cavity, matching and adjacent grooves on the base of the postglenoid process of the squamosal and the ventrolateral edge of bulla demonstrate that the vessel of this canal was immediately lateral to the edge of the ectotympanic bulla. Unlike the canal for the superior ramus, the temporal canal of archaic cetaceans has an anterior osseous branch for the cranio-orbital sinus (Fig. 8: cos). The temporal canal is absent in all mesonychids (*Sinonyx jiashanensis* IVPP V10760, *Pachyaena ossifraga* AMNH 4262, and *Harpagolestes uintensis* AMNH 1878) and in *Andrewsarchus*. A temporal canal and postglenoid foramen are present in all other taxa examined in this study (Table II).

The temporal canal loses its morphological identity and becomes incorporated into the peribullar air sinus cavity in extant cetaceans, especially in odontocetes. In addition, the ectotympanic bulla and petrosal become progressively isolated by the dorsal wall of the peribullar cavity from the cranial cavity (Fraser and Purves, 1960; Fordyce, 1994). The disappearance of the capsuloparietal emissary vein in odontocetes and adult mysticetes may be related to increases in its length and tortuosity (defined as actual length of vessel divided by shortest distance between two points). In odontocetes and mysticetes, the parietal and squamosal have developed secondary internal laminae (or septa) that exclude the petrotympanic complex from the cranial cavity (Fraser and Purves, 1960; Fordyce, 1994). Progressive isolation of the petrosal from the cranial cavity would have displaced the exit of the capsuloparietal emissary vein ventrolaterally, increasing the length of the vessel. Its connection to the transverse sinus was shifted posteriorly by a marked increase in the size of the cerebral hemispheres in extant cetaceans relative to basilosaurids (Dart, 1923). Again, this would have increased the length of the vessel. In both odontocetes and mysticetes, the temporal sinus becomes incorporated into the fossa for the peribullar air sinus and can no loner be differentiated.

A medial shift in the capsuloparietal emissary vein, such that the postglenoid foramen opens within the petrosal/squamosal foramen, is concomitant with a reduction in the size of this vessel. The capsuloparietal emissary vein was greatly reduced in size in stem cetaceans (and possibly the most recent common ancestor of mesonychids and cetaceans) based on the size of the postglenoid foramen and the temporal canal, and was finally involuted in the cetacean crown group and in all mesonychids, based on the loss of the temporal canal.

The presence of a capsuloparietal emissary vein is plesiomorphic for Eutheria and for Ungulata based on the optimization of its distribution (from Gelderen, 1924) both on a cladogram of Novacek (1992, Fig. 4) and on the phylogeny of this study. Ontogenetic evidence for the primitive state of Mammalia is complex. It suggests that a prootic sinus is the most plesiomorphic, followed by a sphenoparietal emissary vein, and finally the capsuloparietal emissary vein (Gelderen, 1924; Padget, 1957; Wible, 1990). This developmental pattern is mirrored by the phylogeny of the major groups of Mammalia and the adult character states within these groups: Monotremes have a prootic sinus, marsupials retain only the stem of the prootic sinus connected to a sphenoparietal emissary vein, and eutherians have a capsuloparietal emissary vein (Wible, 1990; Wible and Hopson, 1995). The absence of any major sinus anterior to the otic capsule is derived, based on ontogenetic evidence (Butler, 1967).

4.3. Endocranial Retia Mirabile and Related Vessels

4.3.1. Character Distribution in Extant Taxa

Retia mirabile are complex nets of small anastomosing vessels, either arteries or veins or a combination of both. Endocranial retia are intradural, situated between the inner and outer layers of the dura mater, and within the cranial cavity, although they may be continuous with extracranial retia. The homology of individual vessels within retia of different taxa is uncertain (Daniel *et al.*, 1954). The homology of various retia, but not their constituent vessels, can be established by their position relative to other structures and by the embryonic precursor vessels from which the adult retia have ramified.

The cavernous sinus of some eutherians contains the rostral arterial rete mirabile; this is the same as the internal carotid rete of Slijper (1936) and Wible (1984) (Sisson, 1921; Daniel *et al.*, 1954; Ghoshal, 1975a,b; Wible, 1984, 1987). In development, the rostral rete is primarily formed by ramifications of the endocranial parts of the internal carotid artery, but occasionally the ramus anastomoticus of the maxillary artery and by branches of the maxillary artery within the sphenorbital fissure as well (Daniel *et al.*, 1954; Wible, 1984).

The rostral rete mirabile occurs in all extant cetaceans (Boenninghaus, 1904; Slijper,

1936; Walmsley, 1938; Galliano *et al.*, 1966; McFarland *et al.*, 1979; Vogl and Fisher, 1981b; Wible, 1984, 1987; Melnikov, 1997) and all extant artiodactyls (Sisson, 1921; Daniel *et al.*, 1954; du Boulay and Verity, 1973; Ghoshal, 1975a,b; Wible, 1984, 1987; Smut and Bezuidenhout, 1987) yet is absent in *Equus* (Sisson, 1921). The rostral rete in cetaceans has also been termed the internal carotid rete (Slijper, 1936; McFarland *et al.*, 1979; Vogl and Fisher, 1981b), carotid rete (Melnikov, 1997), or the internal ophthalmic rete (Galliano *et al.*, 1966).

The caudal arterial rete is present in mysticetes (Slijper, 1936; Walmsley, 1938) but is absent and is functionally replaced by the spinal meningeal arteries in odontocetes (e.g., *Phocoena*, Boenninghaus, 1904; Slijper, 1936; and *Tursiops*, Galliano *et al.*, 1966; McFarland *et al.*, 1979). One exception is the odontocete *Physeter catodon*, which like mysticetes has an extensive arterial and venous caudal rete that has components lateral and dorsal to the cerebellum and ventral to the posterior part of the medulla oblongata (Walmsley, 1938; Pilleri, 1991; Melnikov, 1997). The dura mater adhering to the anterior two-thirds of the basioccipital is avascular in mysticetes indicating the absence of the rete in this region (Walmsley, 1938). The caudal rete or its equivalent, the spinal meningeal arteries, is continuous anteriorly with the rostral rete and posteriorly with the extradural spinal retia mirabilia (Walmsley, 1938; McFarland *et al.*, 1979; Vogl and Fisher, 1981a). The portion of the rete dorsal to the cerebellum in primarily venous (Walmsley, 1938).

Among other ungulates, a simple and ventrally confined caudal arterial rete also occurs in the artiodactyls *Tragelaphus strepsiceros*, *Hippopotamus amphibius* (du Boulay and Verity, 1973), *Sus scrofa* (Ghoshal, 1975b), and *Bos taurus* (Sisson, 1921). With the exception of *Bos*, the caudal rete in artiodactyls is not complex and is confined either to the floor of the cranial cavity immediately anterior to the foramen magnum or to the spinal canal of the atlas vertebra. In all artiodactyls, the caudal rete is supplied by branches of the vertebral and occipital arteries.

4.3.2. Character Distribution among Fossil Taxa

A hypertrophied endocranial rete can be identified in an endocast or skull by the presence of extraneous endocranial space that cannot be accounted for by the brain, nerve ganglia, or the spinal cord (Breathnach, 1955). The identification of a rete mirabile in a fossil skull depends on the conservative assumption that parts of the central nervous system were not significantly larger in extinct taxa than in closely related extant taxa. Extraneous endocranial space is a one-way positive correlate of a rete mirabile because a small rete may not be recognizable in an endocast. Based on the extant phylogenetic bracket, extinct members of Cete had a rostral rete. However, because the rostral rete mirabile in extant taxa does not cause a noticeable expansion of the cranial cavity, its presence in extinct taxa cannot be corroborated by any osteological correlates.

The caudal rete mirabile is one of the most obvious autapomorphies in the vasculature of some extant cetaceans. Its evolutionary origin can be traced to cetaceans of protocetid grade, based on the shape of the cranial cavity. In basilosaurids, the rete was already developed to the same extent as in mysticetes (Figs. 9, 10: crm). A hypertrophied caudal rete is present in the basilosaurid *Dorudon atrox* (Breathnach, 1955). This part of the endocast was misinterpreted as a hypertrophied lobe of the cerebellum by Dart (1923) and Edinger (1955). Breathnach (1955) reinterpreted this part of the endocast as being evidence for a

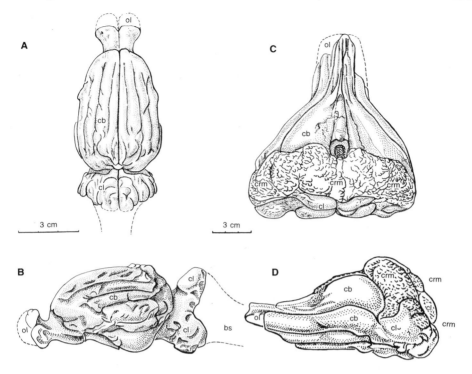

FIGURE 9. The cranial endocasts and endocranial vascular retia in mesonychids and basilosaurids. (A) *Synoplotherium* (dorsal view of cranial endocast; modified from Radinsky, 1976b). (B) *Synoplotherium* (lateral view of the same endocast). (C) *Dorudon atrox* (dorsal view of the cranial endocast). Based on a cast in UM collection, courtesy Dr. P. D. Gingerich, of a specimen described by Dart (1923). (D) Lateral view of the brain endocast of *Dorudon atrox*. The stippled area represents the surface of the brain. The rippled area represents the endocranial vascular rete external to the cerebrum and cerebellum. The brain in *Synoplotherium* has little or no endocranial rete. By contrast, the endocranial vascular rete (rete mirabile) in the basilosaurid is hypertrophied. It fills the entire space between the cerebrum and the cerebellum and covers the entire dorsal aspect of the cerebellum. Abbreviations: bs, brain stem; cb, cerebrum; cl, cerebellum; crm, caudal rete mirabile; ol, olfactory lobe.

caudal rete mirabile. This view has been accepted by subsequent authors (Pilleri, 1991; Bajpai *et al.*, 1996). Extraneous endocranial space indicative of a caudal rete mirabile is also present medial and dorsomedial to the petrosal of *Indocetus* (Bajpai *et al.*, 1996, Fig. 1A). The rete was present medial to the petrosal and to a lesser degree beneath the vertex of the cranial cavity in the Cross Whale (ChM PV5401). Numerous diploic foramina on the endocranial surface of the parietal in the Cross Whale and especially in basilosaurid archaeocetes (Pilleri, 1991; GSP-UM 97516, 94797) indicate that a venous rete was situated dorsal to the cerebellum, as in extant mysticetes (Walmsley, 1938). In the remingtonocetid *Dalanistes ahmedi* (Gingerich *et al.*, 1995), a rete probably filled the region dorsomedial to the petrosals (GSP-UM 3106, 3052), although there is little evidence for a caudal rete dorsal to the cerebellum (Gingerich *et al.*, 1995).

By contrast, a hypertrophied caudal rete mirabile is clearly absent in *Synoplotherium* (Fig. 9A,B; see Radinsky, 1976b) and *Mesonyx obtusidens* (AMNH 12643), both members

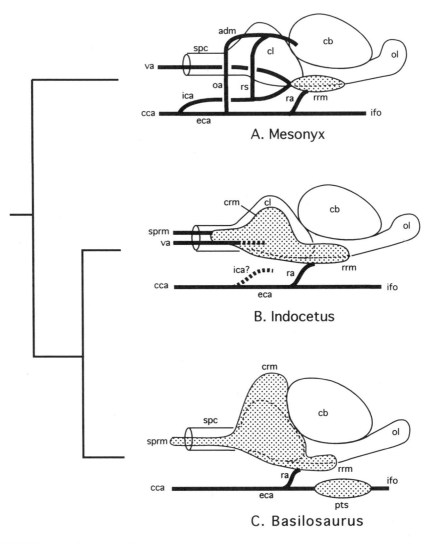

FIGURE 10. Evolutionary origin of the endocranial vascular retia and the reorganization of arterial channels in Cete. (A) *Mesonyx:* schematic pattern of brain endocast, and major cranial arteries. Proportions of endocast are based on Radinsky (1976b). (B) *Indocetus:* schematic pattern of brain endocast, and major cranial arteries. The proportions of the endocranial retia are based on observations from the Cross Whale (ChM PV5401), which is a whale of the same evolutionary grade. (C) *Durodon atrox* (Dart, 1923; Pilleri, 1981). The endocast is the same as in Fig. 9. Abbreviations: adm, arteria diploetica magna, a branch from the occipital artery; cb, cerebrum; cca, common carotid artery; cl, cerebellum; crm, caudal rete mirabile; eca, external carotid artery; ica, internal carotid artery; ifo, infraorbital (maxillary artery); mma, middle meningeal artery; oa, occipital artery; ol, olfactory lobe; pts, extracranial pterygoid sinus; ra, ramus anastomoticus; rrm, rostra rete mirabile; rs, superior ramus of the stapedial artery; spc, spinal cord; sprm, spinal rete mirabile; va, vertebral artery.

of the sister group to Cetacea. There is also no evidence for this structure in other early ungulates, such as *Arctocyon* (Russell and Sigogneau, 1965), *Hyopsodus* (Gazin, 1968), *Meniscotherium* (Gazin, 1965), *Phenacodus* (Simpson, 1933), *Hyracotherium* (Radinsky, 1976a), and early perissodactyls (Edinger, 1948, 1949, 1964; Radinsky, 1976a). Our examination of pakicetids (H-GSP 96231) suggests that they probably lacked a hypertrophied endocranial caudal rete. We cannot rule out a small caudal rete in any of the fossil taxa because extraneous endocranial space is a one-way positive correlate.

Given that a large caudal rete is absent in mesonychids and pakicetids, an extensive caudal rete is a synapomorphy of more derived cetaceans. The caudal rete was a sizable structure in post-*Pakicetus* cetaceans and increased in size, becoming the largest feature in the cranial cavity of basilosaurid archaeocetes. The absence of the caudal rete in most odontocetes is interpreted as a secondary loss.

A simple caudal arterial rete is present in a few extant artiodactyls. A small rete is present in some taxa of the bunodont and selenodont clades of extant artiodactyls, based on the taxic definition of these clades by Gentry and Hooker, (1988). It is possible that a small rete mirabile was present in the common ancestor of Paraxonia, or that the small rete in extant artiodactyls is a convergence with cetaceans. To choose between these two alternative interpretations, it is necessary to establish the distribution of the caudal rete in extant and extinct artiodactyls. The distribution of a caudal rete in extinct artiodactyls is unknown because the presence of space for a small rete in a fossil skull cannot be identified.

5. Evolution of Cranial Vasculature

5.1. Morphotype of Paraxonia and Cete

The hypothetical ancestral cranial vascular pattern for Paraxonia is based on the extant phylogenetic bracket and optimization of vascular osteological correlates on the cladogram generated in the phylogenetic analysis (Fig. 1A). In the common ancestor of Paraxonia, the internal carotid artery was patent and a major supplier of blood to the brain, as inferred by the width and depth of the corresponding sulcus (Fig. 2: ica, pica). It lies in the transpromontorial position (*sensu* Wible, 1986). Immediately anterior to the fenestra cochleae, the proximal stapedial artery branched from the internal carotid artery and continued dorsolaterally from the stapedial foramen, in the stapes, as the superior ramus (Fig. 2: rs). The superior ramus entered the cranial cavity through a foramen lateral to the epitympanic recess positioned on the petrosal/squamosal suture (Fig. 2: vf). The inferior ramus of the stapedial artery was absent in the hypothetical ancestor of Paraxonia based on the extant phylogenetic bracket and the absence of this vessel in the next extant outgroup, Perissodactyla (Sisson, 1921; du Boulay and Verity, 1973; Wible, 1984). The ramus anastomoticus, a branch of the maxillary artery, entered the cranium through the foramen ovale (Fig. 2: fo, ra). The internal carotid artery and the ramus anastomoticus supplied the rostral arterial rete. The circle of Willis, the immediate blood supply source to the brain, would have received arterial blood from two channels: (1) the endocranial part of the internal carotid artery after it reemerged from the rostral rete and (2) from the basilar artery, which was probably linked to the vertebral artery and branches of the occipital artery as in many extant mammals (du Boulay and Verity, 1973). Unfortunately, there is no osteological evidence that can either confirm or refute a connection between the basilar and/or occipital arteries. The occipital artery (Fig. 2C: oa) probably had three branches: the recurrent branch which joined the vertebral artery near the transverse foramen of the atlas vertebra; a condyloid artery which entered the cranial cavity through the condyloid foramen, and then connected endocranially with the vertebral or basilar artery; and an arteria diploetica magna in the posttemporal canal. The presence and position of these three branches are based on the extant phylogenetic bracket. A small rostral rete was also present and possibly a simple, ventrally confined caudal arterial rete mirabile.

The hypothetical cranial venous vasculature in the common ancestor of Paraxonia was similar to the primitive eutherian condition. The dorsal venous sinuses drained into the transverse sinus which exited the skull through two branches: the capsuloparietal emissary vein via the postglenoid foramen (Fig. 2: cpev) and the sigmoid sinus through the confluent posterior lacerate foramen and basicapsular fissure; both the presence and position of these vessels are based on the extant phylogenetic bracket. The cranio-orbital sinus, if present, emptied into the capsuloparietal emissary vein near the latter's connection with the sigmoid sinus. In mammals, the cranio-orbital sinus accompanies the ramus supraorbitalis, anterior branch of the superior ramus of the stapedial artery, along the endocranial surface of the braincase into the orbital region. It connects the superior ophthalmic vein anteriorly to the capsuloparietal emissary vein posteriorly (Diamond, 1992; Wible and Zeller, 1994). The cranio-orbital sinus is homologous to part of the middle meningeal sinuses of Padget (1957) and has a correlative groove on, or a canal within, the sidewall of the cranial cavity (Diamond, 1992). A groove or canal for the cranio-orbital sinus is ubiquitous among early Ungulata (Thewissen, 1985). The presence of the cranio-orbital sinus is based on endocranial grooves; however, its presence in any ungulate taxon is not supported by the extant phylogenetic bracket.

The ventral venous sinuses in the hypothetical Paraxonia ancestor drained into the cavernous sinus, which in turn emptied into the inferior petrosal sinus. The inferior petrosal sinus coursed in a groove or a fissure between the petrosal and basioccipital in an intracranial (within the cranial wall) position (Fig. 2: ips), and then joined the sigmoid sinus at the posterior lacerate foramen to form the internal jugular vein (Fig. 2: ijv). The condyloid vein may have been present. If so, it exited the cranial cavity through the condyloid foramen with the condyloid artery (Fig. 2: cv). The ventral venous sinuses reconstructed are supported by the extant phylogenetic bracket except the condyloid vein for which it is indecisive.

The vascular pattern in basal Cete was relatively unchanged with respect to the plesiomorphic Paraxonia pattern. The only exceptions concern the foramen for the superior ramus of the stapedial artery and the maxillary artery. Instead of entering the cranial cavity through a foramen lateral to the epitympanic recess, the foramen for the superior ramus (see Section 4.1.2) is anterior to the epitympanic recess and lies within the squamosal medial to the postglenoid foramen and lateral to the piriform fenestra or in the petrosal/squamosal suture (Fig. 3: vf). The maxillary artery ran through the alisphenoid canal whereas in the hypothetical ancestor of Paraxonia the alisphenoid canal was absent (Fig. 3: mxa, alc).

5.2. Morphotype of Acreodi

Notable cranial vascular changes occurred in the common ancestor of Acreodi as compared with the plesiomorphic Cete pattern. The internal carotid artery was in the transpromontorial position although its corresponding sulcus is much fainter, suggesting the vessel was decreasing in girth. The canal for the superior ramus in mesonychids and the temporal canal for the capsuloparietal emissary vein in cetaceans are identical except that the ventral opening of the former is slightly medial to the latter. We hypothesize that the dorsal ends of both canals were confluent in the common ancestor of Acreodi. Although there is no modern analogue for this condition, we suspect that the morphologies of the foramen for the superior ramus as observed in extant taxa underestimate the total variety of morphologies in both extant and extinct taxa. This is based on the apparent independent loss of the superior ramus and its foramen in many mammalian lineages. The posterior branch of the ramus superior probably joined the endocranial part of the arteria diploetica magna. The size of the posttemporal canal implies that the arteria diploetica magna was a large vessel.

A clear groove in mesonychids connects the posterior opening of the alisphenoid canal to the foramen ovale, probably marking the course of the ramus anastomoticus from the maxillary artery into the skull. It is hypothesized that a reduction in the internal carotid artery was compensated by one of or any combination of three vessels: the arteria diploetica magna via an anastomosis homologous to the anterior end of the spinal meningeal arteries (see Section 4.1.5), the ramus anastomoticus from the external carotid, and the basilar artery via the vertebral artery and branches of the occipital artery.

In contrast to the plesiomorphic Cete condition, the cranial cavity of the putative Acreodi ancestor was drained almost exclusively by the sigmoid and inferior petrosal sinuses (Figs. 5, 7: sgs, ips). The width of the postglenoid foramen, consequently the width of the capsuloparietal emissary vein, is reduced by at least 70% in basal cetaceans (Fig. 8: vf) relative to *Hapalodectes* and *Diacodexis* (Fig. 2: pfg). The postglenoid foramen is on the petrosal/squamosal suture, medial to the position in early Cete. The vena diploetica magna (see Wible and Hopson, 1995), if present, ran through the posttemporal canal (Figs. 5–7: vdm). The vena diploetica magna is a possible homologue of the mastoid emissary vein, using the same reasoning concerning the homology of the posterior meningeal arteries and the arteria diploetica magna (see Section 4.1.5).

5.3. Morphotype of Cetacea

The primitive cranial vascular pattern for Cetacea is primarily based on pakicetid cetaceans (H-GSP 96431, 96386, 96231; cast of GSP-UM 084). The size of the internal carotid artery in early cetaceans is uncertain. The groove on the promontorium of the petrosal that may have received the internal carotid (H-GSP 96386) is poorly defined in most pakicetid specimens (H-GSP 96231, GSP-UM 084) and absent in other cetaceans. The extant phylogenetic bracket is decisively positive for an internal carotid artery in all Cetacea. The timing of the extreme reduction of this artery as seen in all extant cetaceans (Boenninghaus, 1904; Slijper, 1936; Walmsley, 1938; Vogl and Fisher, 1981b) is unknown. We hypothesize that this change occurred in pakicetid-grade cetaceans based on the presence and absence of this groove in different members of this group. The proximal stapedial artery and its branches were absent, as inferred from the absence of the corresponding sulci and foramina. The sulcus for the proximal stapedial artery is a one-way positive osteological correlate; therefore, it could have been present in basal cetaceans even in the absence of its correlative groove.

The alisphenoid canal is absent in all cetaceans. Thus, the maxillary artery has an extracranial course. As discussed earlier (see Section 4.1.5), the exclusion of the mastoid process of the petrosal from the occiput of the skull plus a thickening of the lambdoid crest of the squamosal displaced the foramen of the arteria diploetica magna medially into the occipital/squamosal suture. We hypothesize that the arteria diploetica magna was anastomosed to and primarily supplied the rostral rete mirabile. The circle of Willis was supplied by the carotid rete, as in extant cetaceans (Boenninghaus, 1904; Walmsley, 1938; Vogl and Fisher, 1981a), and possibly the basilar artery. The basilar artery, and possibly a small caudal rete, may have been supplied by the vertebral and condyloid arteries, as in an early embryo of the odontocete *Stenella* (Sinclair, 1967). The reconstruction of the basilar artery and its connections to other vessels are speculative and should be considered tentative. An occipital emissary vein, homologous to the mastoid vein, the vena diploetica magna, or both, may have been present in the foramen for the arteria diploetica magna.

5.4. Morphotype of Postpakicetid Cetaceans

In postpakicetid cetaceans, the spinal arterial retia is hypothesized to have anastomosed with the endocranial portions of the arteria diploetica magna of each side. Both intraosseous (i.e., within the posttemporal canal or homologous structure) and extracranial portions of the arteria diploetica magna then involuted. The remaining endocranial portions of the arteria diploetica magna became incorporated into two groups of primarily longitudinal vessels within the cranial cavity, the spinal meningeal arteries. In this scenario, the spinal meningeal arteries are formed from two anastomoses of the arteria diploetica magna linked by the endocranial segment of this vessel.

The condyloid artery and vein were absent in cetaceans more derived than pakicetids, as inferred from the absence of the condyloid foramen. Cetaceans that postdate pakicetids but predate the cetacean crown group show a general increase in extraneous endocranial space in the posterior cranial fossa. This space is interpreted as being filled by a vascular rete mirabile (Breathnach, 1955; Pilleri, 1991; Bajpai *et al.*, 1996). In pakicetids, there is no positive evidence for a caudal rete (H-GSP 96231). By contrast, the endocast of the posterior cranial fossa in *Indocetus* (Bajpai *et al.*, 1996) and remingtonocetids (H-GSP 96232, GSP-UM 3106) is wider than the endocast of the middle cranial fossa, indicative of extensive vasculature lateral to the cerebellum and medial to the petrosal. Unlike pakicetids and mesonychids (Radinsky, 1976b), later cetaceans including remingtonocetids, *Indocetus*, the Cross Whale, and basilosaurids have a tentorium osseum that is restricted to the roof of the cranial cavity. The absence of the tentorium osseum from the sidewall of the cranial cavity may be correlated with the presence of an endocranial rete mirabile. In the Cross Whale, the dorsal surface of the posterior cranial fossa is rugose and contains numerous foramina for diploic veins that drained into the venous rete situated dorsal to the cerebellum. The venous rete in basilosaurids is hypertrophied and domed above the rest of the cranial cavity (Figs. 9, 10: crm; Dart, 1923; Breathnach, 1955; Pilleri, 1991). The arterial caudal rete was probably supplied by the spinal meningeal arteries.

The development of the arterial caudal rete in early Cetacea is followed by a decrease in the diameter of the transverse canal for the vertebral artery in the axis vertebra, as seen in basilosaurids (GSP-UM 101222; Kellogg, 1936). It is hypothesized that the anterior end of the vertebral artery was functionally replaced by an anastomosis between the rete in the spinal canal and the endocranial portions of the spinal meningeal arteries (Fig. 10: va, sprm). There have been no reports of vertebral arteries in extant cetaceans except for an early embryo of *Stenella* (Sinclair, 1967). The cervical arterial rete mirabile, situated lateral to the vertebral column, of extant cetaceans is probably homologous to the vertebral arteries, the former being extensive ramifications of the latter during development. Osteology supports this homology; the transverse foramen in noncetacean mammals transmits the vertebral arteries (Sisson, 1921; Miller *et al.*, 1964; Gray, 1974) whereas in cetaceans it contains the cervical rete (Walmsley, 1938).

The channel for the cranio-orbital sinus in archaic ungulates (Radinsky, 1976b), artiodactyls (Whitmore, 1953), and mesonychids (Radinsky, 1976b; AMNH 12643) is an open sulcus, except for its extreme anterior end where it exits the skull through a short canal. In the Cross Whale, the cranio-orbital sinus channel is enlarged and enclosed in a canal except for a short segment anterodorsal to the anterior process of the petrosal (Fig. 8: cos). The cranio-orbital sinus of basilosaurids is almost completely intracranial (*Dorudon atrox*: GSP-UM 97516, 94797), and there is no clear cranio-orbital foramen indicating that the cranio-orbital sinus does not connect with the superior ophthalmic vein of the orbit. Instead its anterior end leads into the diploe of the parietal; therefore, the cranio-orbital sinus in basilosaurids is the functional equivalent of a diploic vein (Pilleri, 1991). The cranio-orbital sinus is absent in the cetacean crown group.

It is possible that some of the vascular innovations in cetaceans were necessary aquatic adaptations. The development of retia mirabile in cetaceans is probably related to life underwater (see Section 1.1), even though their specific function(s) is not known (Walmsley, 1938; Nagel *et al.*, 1968; Dormer *et al.*, 1977). It is hypothesized that the caudal endocranial rete mirabile was a preadaptation to the basicranial air sinuses. The air sinuses initially developed in gaps within the cranial wall, specifically between the petrosal and its surrounding elements: the pterygoid, basioccipital, and exoccipital. As inferred from the basicranium of basilosaurid archaeocetes, the air sinuses intruded dorsally into the cranial cavity. Through subsequent bone resorption caused by expansion of the air sinuses, the petrosal and tympanic became isolated in a basicranial fossa, which is separated from the cranial cavity by a ridge or bony lamina (Fraser and Purves, 1960). Before the development of a separating bony ridge or lamina, the expansion and contraction of the air sinuses during diving could have been lethal, if there was no retial tissue between the petrosal and the cerebellum. The rete also provided a vascular network needed to control the size of the air sinuses via blood flow (Fraser and Purves, 1960). Separation of the petrosal and the tympanic from the skull by the air sinuses was probably an adaptation for directional hearing underwater (Gingerich *et al.*, 1983).

6. Summary and Conclusions

A phylogenetic analysis of relatively complete ungulate taxa strongly supports a hierarchy of monophyletic groups: the Paraxonia, a clade including Artiodactyla and Cete to the exclusion of Perissodactyla; a monophyletic Mesonychiadae, to the exclusion of *Hapalodectes* and *Andrewsarchus*; and Acreodi, a clade containing Cetacea and the Mesonychidae. Artiodactyla is also monophyletic and does not include Cetacea. This phylogeny, in conjunction with morphological data, was used to reconstruct the vasculature of extinct members of Paraxonia. The reconstructions were used to develop hypotheses concerning the evolution of cranial vasculature in these groups.

Grooves on the promontorium and tegmen tympani of the petrosals of mesonychids suggest that the internal carotid, proximal stapedial, and superior ramus of the stapedial arteries were patent and functional. The absence of the osteological correlates of these vessels in all cetaceans suggests that the stapedial artery was involuted and the internal carotid artery was reduced in the earliest cetaceans. The vertebral arteries, present in terrestrial ungulates and probably protocetid cetaceans, were lost in basilosaurids. The arteria diploetica magna, as inferred from the posttemporal canal and homologous foramina, was large in mesonychids and pakicetids. It is hypothesized that this vessel formed a transition between cerebral blood supply via the internal carotid artery, the plesiomorphic condition, and via the spinal meningeal arteries or homologous arterial rete mirabile as in extant cetaceans.

The capsuloparietal emissary vein was significantly reduced in size and shifted medially in Acreodi, as compared with early Cete and Artiodactyla, based on the position and size of the temporal canal and postglenoid foramen among these group. The capsuloparietal emissary vein was subsequently lost in the cetacean crown group. The inferior petrosal and sigmoid sinuses compensated for the reduction and eventual loss of the capsuloparietal emissary vein, which may be associated with the development of vascularized air sinuses surrounding the middle ear and the separation of the petrosal from the cranial cavity.

A caudal endocranial rete mirabile was absent or not enlarged in mesonychids and pakicetids. It developed lateral to the cerebellum in remingtonocetids, *Indocetus*, and the Cross Whale. The arterial component of the rete coevolved with the reduction or loss of some of the arteries that supply the brain in noncetacean ungulates: the internal carotid, vertebral, and branches of the occipital arteries (Fig. 10). The replacement of these primary supplying channels to the brain by the endocranial arterial retia began to appear in protocetids and had been completely established in basilosaurids, before the origin of extant cetaceans.

Acknowledgments

We thank Drs. P. D. Gingerich (University of Michigan Museum of Paleontology), J. G. M. Thewissen (Northeastern Ohio Universities College of Medicine), A. E. Sanders (Charleston Museum), M. C. McKenna (Department of Vertebrate Paleontology) and N. B. Simmons (Department of Mammalogy, American Museum of Natural History), E. Heizmann (Staatliches Museum für Naturkunde, Stuttgart), and R. C. Hulbert (Georgia Southern University) for allowing us to study specimens of whales and related ungulates. Both J. G. M. Thewissen and J. R. Wible made comments on an earlier draft of this chapter that improved both its content and clarity. We benefited from discussions with and suggestions from M. A. Carrasco, P. J. Makovicky, and Drs. P. D. Gingerich, R. C. Hulbert, S. G. McGehee, M. C. McKenna, M. A. O'Leary, G. W. Rougier, A. E. Sanders, J. G. M. Thewissen, M. D. Uhen, and J. R. Wible. This research was partially funded by NSF grants DEB 941898 and DEB 9527892 to Z. Luo and an NSF graduate fellowship to J. Geisler. Unpublished H-GSP material was collected under an NSF grant to J. G. M. Thewissen.

Appendix 1: Character List

The taxa were coded for 80 morphological characters based on the specimens in Table II. When specimens were unavailable or incomplete, the coding for the taxa was supplemented by the following references: *Didelphis* (Bremer, 1904), *Asioryctes* (Kielan-Jaworowska, 1977, 1981), leptictids (Novacek, 1980, 1986), *Arctocyon* (Russell, 1964), *Hyopsodus* (Gazin, 1968; Thewissen and Domning, 1992), Phenacodontidae (Cifelli, 1982; Thewissen, 1990; Thewissen and Domning, 1992), *Meniscotherium* (Gazin, 1965; Cifelli, 1982), *Heptodon* (Radinsky, 1965), *Hyracotherium* (Thewissen and Domning, 1992), *Diacodexis* (Russell *et al.*, 1983; Thewissen *et al.*, 1983; Rose, 1985; Thewissen and Hussain, 1990; Thewissen and Domning, 1992), *Hapalodectes* (Szalay, 1969b; Ting and Li, 1987), *Pachyaena* (O'Leary and Rose, 1995b; Rose and O'Leary, 1995), *Sinonyx* (Zhou *et al.*, 1995), *Synoplotheirum* (Wortman, 1901), *Mesonyx* (Scott, 1888), *Pakicetus* (Gingerich and Russell, 1981, 1990; Thewissen and Hussain, 1993; Thewissen, 1994), *Ambulocetus* (Thewissen *et al.*, 1996), Remingtonocetidae (Kumar and Sahni, 1986; Gingerich *et al.*, 1995), *Protocetus* (Fraas, 1904), and basilosaurids (Kellogg, 1936). Coding for character 76 was based on Mossman (1987) and Hartman (1916), for character 77 (McCrady, 1940; Daudt, 1898, with condition in *Tursiops* based on other odontocetes; Sisson, 1921), and for character 78 (Müller, 1898, with condition in *Tursiops* based on other odontocetes; Getty, 1975). The hypothetical most primitive state is coded "0" with progressively more derived states being coded "1," "2," and so on. In one run the characters were left unordered, in a second all multistate characters were ordered. Unqualified citations indicate that the character is worded with little or no modification from the given reference. Characters that are "modified" from references have been significantly changed, and those that are "derived" have been extracted from a diagnosis or morphological description and converted into a form ready for cladistic analysis. Citations generally indicate the first use of the character in a cladistic analysis but by no means indicate the first use of the character in any form. The following restrictions apply to this list of characters: 11–19 cannot be scored when the ectotympanic is not preserved, 12–19 cannot be scored for taxa that lack an ectotympanic bulla, 29 and 30 cannot be scored for taxa that lack an alisphenoid canal, 55 and 58 cannot be scored for taxa that lack M_3, 71–77 cannot be scored for taxa that lack or have extremely reduced hind limbs, 78–80 cannot be scored for extinct taxa (exceptions are described in Section 2.1).

1. Subarcuate fossa: (0) present, (1) absent (Novacek, 1986).
2. Tegmen tympani inflation: (0) absent, (1) present, forms barrel-shaped ossification lateral to the facial nerve canal, (2) hyperinflated, transverse width greater than or equal to width of promontorium. Usually cannot be scored for petrosals that are attached to the skull (modified from Cifelli, 1982).
3. Anterior process of petrosal: (0) absent, (1) present but small with gap between anterior edge of tegmen tympani and promontorium, (2) present and enlarged with anterior edge of tegmen tympani extending well past anterior edge of promontorium (modified from Luo and Marsh, 1996).
4. Epitympanic recess: (0) uniformly concave, (1) differentiated, with distinct fossa for the head of the malleus (derived from Luo and Eastman, 1995).

5. Fossa for tensor tympani muscle: (0) shallow anteroposteriorly elongate fossa, (1) circular pit, (2) circular pit with deep tubular anterior groove, (3) long narrow groove between tegmen tympani and promontorium (modified from Luo and Marsh, 1996).

6. Transpromontorial sulcus for the internal carotid artery: (0) present, forms anteroposterior groove on promontorium, medial to the fenestrae rotunda and ovalis, (1) absent (Cifelli, 1982; Thewissen and Domning, 1992).

7. Sulcus on promontorium for proximal stapedial artery: (0) present, forms a groove that branches from the transpromontorial sulcus anteromedial to the fenestra rotunda and extends to the medial edge of the fenestra vestibuli, (1) absent (Cifelli, 1982; Thewissen and Domning, 1992).

8. Stylomastoid foramen: (0) large notch, open medially, (1) small orifice ventromedially encircled by a tympanic bulla. Cannot be scored for taxa that lack an ectotympanic bulla.

9. Facial nerve sulcus: (0) absent, (1) on anteroventral edge of mastoid process, (2) on posteroventral edge of mastoid process (modified from Luo and Marsh, 1996; Geisler and Luo, 1996).

10. Articulation of pars cochlearis of petrosal with basisphenoid/basioccipital: (0) present, (1) absent (Thewissen and Domning, 1992).

11. Ectotympanic: (0) simple ring, (1) medial edge expanded into bulla. The expanded ventral apex of the ectotympanic ring of *Asioryctes* is not considered a bulla. Cannot be scored for taxa in which the ectotympanic is not preserved (derived from Novacek, 1977; MacPhee, 1981).

12. Articulation of ectotympanic bulla to squamosal: (0) broad articulation with medial base of postglenoid process, (1) circular facet on short entoglenoid process, (2) contact reduced to a crest of the entoglenoid process, (3) contact absent.

13. Contact between exoccipital and ectotympanic bulla: (0) absent, (1) present.

14. Sigmoid process (medial part of anterior wall of tympanic meatal tube): (0) absent, only anterior crus of tympanic ring present, (1) present, transverse plate that projects laterally from anterior crus of the ectotympanic ring or homologous structure. State (1) includes taxa with a complete ectotympanic meatal tube.

15. Morphology of sigmoid process: (0) thin and transverse plate, (1) broad and flaring, base of the sigmoid process forms dorsoventral ridge on lateral surface of ectotympanic bulla.

16. Pachyosteosclerotic involucrum of bulla: (0) absent, (1) present (Thewissen, 1994).

17. Lateral furrow of tympanic bulla: (0) absent, (1) present, forms a groove on the lateral surface of the ectotympanic bulla anterior to the base of the sigmoid process.

18. Median furrow of tympanic bulla: (0) absent, (1) median notch on posterior rim of bulla, (2) prominent anteroposteriorly oriented furrow that splits the ventral surface of the bulla into medial and lateral halves.

19. Meatal portion of tympanic: (0) absent, (1) present but short, floors medial part of the external auditory meatus, length (transverse) less than 30% the maximum transverse width of bulla, (2) present and completely floors external auditory meatus, length greater than 60% the maximum width of bulla.

20. Squamosal portion of external auditory meatus: (0) absent or shallow, depth less than 25% of its transverse length, (1) elongate deep groove, depth greater than 35% of its length (derived from Van Valen, 1966).

21. Basioccipital eminence: (0) absent, (1) present, paired anteroposterior ridges in ventral view, situated on each side of sagittal plane, (2) developed into ventrolaterally flaring basioccipital crests also known as the falcate processes (derived from Barnes, 1984; modified from Thewissen, 1994).

22. Internal carotid foramen: (0) absent or confluent with the piriform fenestra, (1) present, at basisphenoid/basioccipital suture with lateral wall of foramen formed by both of these bones and thus separated from the piriform fenestra.

23. Condyloid foramen: (0) absent, (1) present, situated posterior to hypoglossal foramen.

24. Postglenoid foramen in squamosal: (0) present, immediately posterior to postglenoid process, (1) absent (modified from Novacek, 1986).

25. Foramen in petrosal/squamosal suture in vicinity of epitympanic recess, skull is in ventral view: (0) present, (1) absent.

26. Position of foramen in petrosal/squamosal suture: (0) lateral to epitympanic recess, (1) anterolateral to epitympanic recess, adjacent to ventrally convex portion of the tegmen tympani. Cannot be scored for taxa that lack a foramen on the petrosal/squamosal suture.

27. Foramen ovale: (0) medial to glenoid fossa, (1) anterior to glenoid fossa (derived from Zhou *et al.*, 1995).

28. Alisphenoid canal (alar canal): (0) absent, (1) present (Novacek, 1986; Thewissen and Domning, 1992).

29. Anterior opening of alisphenoid canal: (0) within sphenorbital fissure, cannot be distinguished from sphenorbital fissure in lateral view, (1) opens posterior to sphenorbital fissure (Novacek, 1986; Thewissen and Domning, 1992).

30. Posterior opening of alisphenoid canal: (0) well separated from foramen ovale, (1) within a deep groove or recess with foramen ovale (derived from Zhou *et al.*, 1995).

31. Foramen rotundum: (0) absent, maxillary division of trigeminal nerve exits skull through the sphenorbital fissure, (1) present (Novacek, 1986; Thewissen and Domning, 1992).

32. Ethmoid foramen: (0) present, pierces the frontal bone near the ventral border of its orbital exposure, (1) absent (Shoshani, 1986; Thewissen and Domning, 1992).

33. Cranio-orbital sinus canal or groove: (0) present, within or on the endocranial surface of the parietal, anteriorly it exits the cranium through a foramen in the orbit, posteriorly it connects with a groove or canal for the transverse sinus or one of its branches, (1) absent (modified from Thewissen and Domning, 1992).

34. Temporal foramina: (0) present, within the squamosal or on the squamosal/parietal suture in the temporal fossa, (1) absent.

35. Mastoid foramen: (0) present, skull in posterior view, (1) absent (see MacPhee, 1994).

36. Occipital foramen: (0) absent, (1) present, occurs in suture between exoccipital and squamosal or just within the exoccipital.

37. Posttemporal canal (for arteria diploetica magna, also called percranial foramen):

(0) present, occurs at petrosal/squamosal suture with skull in posterior view, the canal continues within the petrosal/squamosal suture, (1) absent (Wible, 1990; MacPhee, 1994).

38. Preglenoid process: (0) absent, (1) present, forms transverse, ventrally projecting ridge at anterior edge of glenoid fossa (modified from Thewissen, 1994).

39. Supraorbital process: (0) absent, region over orbit does not project lateral from sagittal plane, (1) present, laterally elongate and tabular (derived from Barnes, 1984).

40. Contact of frontal and maxilla in orbit: (0) absent, (1) present (Novacek, 1986; Thewissen and Domning, 1992).

41. Lacrimal tubercle: (0) absent, (1) present, situated on anterior edge of orbit adjacent to the lacrimal foramen (Novacek, 1986).

42. Posterior margin or external nares: (0) anterior to or over the canines, (1) between P^1 and P^2, (2) posterior to P^2.

43. Rostrum: (0) premaxillae short with incisors arranged in transverse arc, (1) premaxillae elongate, incisors aligned longitudinally with intervening diastemata (modified from Prothero et al., 1988; Thewissen, 1994).

44. Embrasure pits on palate: (0) absent, (1) present, situated medial to the tooth row, accommodated the cusps of the lower dentition when the mouth was closed (modified from Thewissen, 1994).

45. P^4 protocone: (0) present, (1) absent (Thewissen, 1994).

46. P^4 paracone: (0) less than twice the height of M^1 paracone, (1) greater than twice the height of M^1 paracone (Thewissen, 1994).

47. M^1–M^2 parastyle: (0) weak, (1) moderate to strong (Zhou et al., 1995).

48. M^1–M^3 hypocone: (0) absent, (1) present (Thewissen and Domning, 1992).

49. Postprotocrista on upper molars: (0) present, (1) absent (Thewissen and Domning, 1992).

50. M^3: (0) present, (1) reduced, maximum mesodistal length less than 60% the length of M^2, (2) absent (modified from Zhou et al., 1995).

51. P_3 metaconid: (0) absent, (1) present (Thewissen and Domning, 1992).

52. P_4 metaconid: (0) absent, (1) present (Thewissen and Domning, 1992).

53. M_1 metaconid: (0) present, (1) absent (Thewissen, 1994).

54. M_2 metaconid: (0) present, forms distinct cusp, (1) absent or occasionally present as swelling on lingual side of protoconid (modified from Zhou et al., 1995).

55. M_3 metaconid: (0) present, forms distinct cusp, (1) absent or occasionally present as swelling on lingual side of protoconid (modified from Zhou et al., 1995).

56. Rostral trigonid: (0) flat anterior border, (1) concave, accommodates posterior convex border of the talonid of adjacent tooth (modified from Thewissen, 1994).

57. Talonid basins: (0) triangular, (1) compressed, with cristid obliqua displaced lingually and centered on the width of the tooth (modified from Zhou et al., 1995; description based on O'Leary and Rose, 1995a).

58. M_3 hypoconulid: (0) present, (1) absent (Thewissen, 1994).

59. Mandibular foramen: (0) small, maximum height of opening 25% or less the height of the mandible at M_3, (1) enlarged and continuous with a large posterior fossa, maximum height greater than 50% the height of the mandible at M_3 (modified from Thewissen, 1994).

60. Cervical vertebrae: (0) long, length of centrum greater than or equal to the centra

of the anterior thoracics, (1) short, length shorter than centra of anterior thoracics (derived from Gingerich *et al.*, 1995).

61. Sacral vertebrae articulation with the pelvis: (0) present, dorsoventrally and anteroposteriorly broad articulation between pelvis and S1 and occasionally S2, (1) present, small articulation between the end of the transverse process of S1, (2) articulation absent.

62. Number of sacral vertebrae: (0) one, (1) two or three, (2) four, (3) five or six. Cannot be scored for taxa that lack articulation of vertebral column to ilium (Thewissen and Domning, 1992; Gingerich *et al.*, 1995).

63. Scapular spine: (0) bears large acromion process which overhangs glenoid fossa, (1) scapular spine with acromion process reduced or absent, does not encroach on glenoid fossa (derived from O'Leary and Rose, 1995b).

64. Entepicondyle of humerus: (0) wide, width 50% or greater than the width of the ulnar and radial articulation facets, (1) narrow, 25% or less than the width of the ulnar and radial articulation facets (derived from O'Leary and Rose, 1995b).

65. Entepicondylar foramen: (0) present, (1) absent (Thewissen and Domning, 1992).

66. Length of olecranon process: (0) short, less than 10% of total ulnar length, (1) long, greater than 20% of ulnar length (derived from O'Leary and Rose, 1995b).

67. Proximal end of radius: (0) single fossa for edge of trochlea and capitulum of humerus, (1) two fossae, for the medial edge of the trochlea and the capitulum, (2) three fossae, same as state (1) but with additional fossa for the lateral lip of the humeral articulation surface.

68. Distal articulation surface of radius: (0) single concave fossa, (1) split into scaphoid and lunate fossae (derived from O'Leary and Rose, 1995b).

69. Centrale: (0) present, (1) absent (Thewissen, 1994).

70. Greater trochanter of femur: (0) below level of head of femur, (1) approximately same level as head of femur, (2) elevated well beyond head of femur (derived from O'Leary and Rose, 1995b).

71. Astragalar canal: (0) present, (1) absent (derived from Shoshani, 1986).

72. Navicular facet of astragalus: (0) convex, (1) saddle-shaped, (2) highly concave with notch V-shaped (modified from Thewissen and Domning, 1992).

73. Sustentacular facet of astragalus: (0) narrow, less than 40% the width of the astragalus, (1) wide, greater than 70% the width of the astragalus.

74. Distal end of astragalus: (0) does not contact cuboid, (1) contact present but small, width of contact less than 30% the width of the distal articulating facet, (2) contact broad, greater than 40% of articulating surface.

75. Foot: (0) 3rd digit slightly longer than other digits or roughly equal in length to 2nd and 4th digits, (1) paraxonic (derived from Gingerich *et al.*, 1990; Thewissen, 1994).

76. Digits of foot: (0) 5, (1) 4, (2) 3 (Thewissen and Domning, 1992).

77. Unguals: (0) clawlike, (1) hooflike, flattened dorsoventrally (derived from Van Valen, 1966; Thewissen, 1994).

78. Elongation of blastocyst: (0) absent, (1) present (derived from Mossman, 1987; Thewissen, 1994).

79. Penis erection: (0) via filling of cavernous tissue, (1) via relaxation of retractor penis muscle, little cavernous tissue (Thewissen, 1994).

80. Primary bronchi of lung: (0) two, (1) three (Thewissen, 1994).

Appendix 2: Character/Taxon Matrix Compiled in This Study for Addressing the Phylogenetic Relationships of Cetaceans and Closely Related Extant and Extinct Taxa

A = 0&1, B = 1&2 (these characters either are polymorphic for the taxonomic unit or are the possible character states in an ambiguous fossil taxon, the degree of preservation allows some states to be excluded). In the parsimony search, both A and B were treated as polymorphic characters.

```
Taxa               Characters
                           10         20         30         40         50         60         70         80

Didelphis          0000011000 0??0??????0 0110120022 0020120000 0000000000 0100010200 0100000002 0000000000
Asioryctes         ?????11200 0??0??????0 0200120022 0000?02102 2010001000 0000010000 ?????????? ?0??00????
Lepticids          0000000010 0??0??????0 0000000110 0000001000 1000000100 2000000020 0200000000 0002000???
Artocyon           00000000?? ?????????0 0000000002 1000000000 1000000100 0000000020 0200000000 1001000???
Hyopsodus          0101000?00 ?????????0 0200?0010? 0210000002 1000000100 1100000000 0200001120 1010000???
Phenacodontidae    0101011200 1??0??????0 02A0120110 0100120000 1000000100 1100000000 2??0001112 00A0001???
Meniscotherium     0101000?0? ?????????0 0200?20110 1??0120012 1000001110 0100000020 0200011122 1000021???
Heptodon           ??20111000 0??0??????0 0200?20110 200022001? 1000001110 1100000000 ?2?1101112 210102122?
Hyracotherium      ??20111201 ??20??????1 0?10120110 1102120010 1000000110 1100000000 0311101112 0100021000
Diacodexis         ??20100200 1??2?020A0 0010000022 0000001001 0000000100 0000000020 012110211B 0212110111
Hippopotamidae     120??2????  1311000021 0000?21022 0010?120A1 0000000100 A100000000 0311112111 0212111???
Sus                0001111101 1001000120 1001121022 0011111010 A010000110 0100010000 0211112111 0212110111
Ovis               1000101181 1301001120 100012102? 0010001010 1??0002201 ?100000000 0211102111 0112110111
Andrewsarchus      ?????????? ?????????1 1??12?0120 ?????2102? 2000?2000  ?????????? ?????????? ??????????
Hapalodectidae     ?200000200 ?????????0 0110?20100 102?001101 0001?2?120 2000111122 ?????????? ??????????
Dissacus           0110100212 1111010111 ?211011121 ?211011122 2??0001001 0000001122 ?2?00221?? 10???????
Sinonyx            ??20102110 1001000021 1?1???21?1 1020200100 1000001021 0010101102 010101220B 110111122?
Pachyaena          ????????2? ??21?21021 121???21?? ?????2001?? 1000001012 0111101120 1121011200 A101111122
Mesonyx            1110100211 1021000021 0001011101 1210200012 2000001012 1211201120 ?2??201120 0101111122
Synoplotherium     ?????????? 1?21?20021 ?201?21101 1000000102 1000???21  2???201120 ?101021101 0101111122
Harpagolestes      ??????211? 1011200021 1?0??20121 ??20200102 100A0?2022 ?221201122 ??20?2???? ??????????
Pakicetidae        ?212A1120  1111010111 121101102? 1??011102? 2???110000 000?211102 ?2??210101 1??211102?
Ambulocetus        ?12???202? 11?1?11A0? 1?210?2022 020??2?202 ?211111000 201?211012 ?2??210101 1??211122?
Remingtonocetidae  ?121211021 11?11?1101 2?01011022 0??0?21002 2011?2?220 ?????21120 02???????2 0??????????
Cross Whale        1121211021 11??11?221 2101101?20 0?00111210 2011?21210 00??211111 ?????????? ????????
Protocetus         ??21211021 1111011221 2?0101102? ??20?21002 2121111020 0???211111 102???2?2? ??????????
Basilosauridae     1221311021 1201011120 2?01011022 0000121010 0B111?2022 0011111211 2?1111212? 2?2123111
Balaenoptera       1222311021 1301111001 2101121022 0?111?1011 ?????????? ?????????? ?011100122 2?2??????1
Tursiops           1221311021 1301111200 2101120022 0?111?1011 2220??21?1 ?????????11 ?01100122 2011?????1
```

References

Bajpai, S., Thewissen, J. G. M., and Sahni, A. 1996. *Indocetus* (Cetacea, Mammalia) endocasts from Kachchh (India). *J. Vertebr. Paleontol.* **16**:582–584.

Barnes, L. G. 1984. Whales, dolphins and porpoises: origin and evolution of the Cetacea, in: T. W. Broadhead (ed.), *Mammals. Notes for a Short Course Organized by P. D. Gingerich and C. E.*, pp. 139–154. University of Tennessee Department of Geological Sciences, Studies in Geology 8 (1–4).

Barnes, L. G., and Mitchell, E. 1978. Cetacea, in: V. J. Maglio and H. B. S. Cooke (eds.), *Evolution of African Mammals*, pp. 582–602. Harvard University Press, Cambridge, MA.

Barnes, L. G., Domning, D. P., and Ray, C. E. 1985. Status of studies on fossil marine mammals. *Mar. Mamm. Sci.* **1**(1):15–53.

Boenninghaus, G. 1904. Das Ohr des Zahnwhales, zugleich ein Beitrag zur Theorie der Schalleitung. *Zool. Jahrb. Abt. Anat. Ontog. Tiere* **19**:189–360.

Boyden, A., and Gemeroy, D. 1950. The relative position of Cetacea among the orders of Mammalia as indicated by precipitin tests. *Zoologica* **35**:145–151.

Breathnach, A. S. 1955. Observations on endocranial casts of recent and fossil cetaceans. *J. Anat.* **89**:532–546.

Bremer, J. L. 1904. On the lung of the opossum. *Am. J. Anat.* **3**:67–73.

Breschet, M. G. 1836. Histoire anatomique et physiologique d'un organe de nature vasculaire découvert dans les cetacés. Bechet Jeune, Paris. Reprinted in supplement to *Invest. Cetacea* **20**:1–82.

Bryant, H. N., and Russell, A. P. 1992. The role of phylogenetic analysis in the inference of unpreserved attributes of extinct taxa. *Philos. Trans. R. Soc. London Ser. B* **337**:405–418.

Bugge, J. 1974. The cephalic arterial system in insectivores, primates, rodents, and lagomorphs, with special reference to the systematic classification. Supplement 62 to *Acta Anat.* **87**:160.

Butler, H. 1967. The development of mammalian dural venous-sinuses with especial reference to the postglenoid vein. *J. Anat.* **102**:33–56.

Cifelli, R. L. 1982. The petrosal structure of *Hyopsodus* with respect to that of some other ungulates, and its phylogenetic implications. *J. Paleontol.* **56**(3):795–805.

Coombs, M. C., and Coombs, W. P., Jr. 1982. Anatomy of the ear region of four Eocene artiodactyls: *Gobiohyus, ?Helohyus, Diacodexis*, and *Homacodon*. *J. Vertebr. Paleontol.* **2**(2):219–236.

Cope, E. D. 1880. On the foramina perforating the posterior part of the squamosal bone of the Mammalia. *Proc. Am. Philos. Soc.* **18**(105):452–461.

Czelusniak, J., Goodman, M., Koop, B. F., Tagle, D. O., Shoshani, J., Braunitzer, G., Kleinschmidt, T. K., de Jong, W. W., and Matsuda, G. 1990. Perspectives from amino acid and nucleotide sequences on cladistic relationships among higher taxa of Eutheria, in: H. H. Genoways (ed.), *Current Mammalogy*, Volume 2, pp. 545–572. Plenum Press, New York.

Daniel, P. M., Dawes, J. D. K., and Prichard, M. M. L. 1954. Studies of the carotid rete and its associated arteries. *Philos. Trans. R. Soc. London Ser. B* **237**:173–208.

Dart, R. A. 1923. The brain of the Zeuglodontidae (Cetacea). *Proc. Zool. Soc. London* **42**:615–648.

Daudt, W. 1898. Beiträge zur Kenntnis des Urogenitalapparates der Cetaceen. Jena. *Z. Naturwiss.* **32**:231–312.

De Burlet, H. M. 1915a. Zur Entwicklungsgeschichte des Walschädels 3, Das Primordialcranium eines Embryo von *Balaenoptera rostrata* (105 mm). *Morphol. Jahrb.* **49**:120–178.

De Burlet, H. M. 1915b. Zur Entwicklungsgeschichte des Walschädels 4, Das Primordialcranium eines Embryo von *Lagenorhynchus albirostris*. *Morphol. Jahrb.* **49**:393–406.

De Jong, W. W. 1985. Superordinal affinities of Rodentia studied by sequence analyses of eye lens proteins, in: W. P. Luckett and J.-L. Harttenberger (eds.), *Evolutionary Relationships among Rodents: A Multidisciplinary Analysis*, pp. 211–226. Plenum Press, New York.

De Jong, W. W., Leunissen, J. A. M., and Wistow, G. J. 1993. Eye crystallins and the phylogeny of the placental orders: evidence for a macroscelid–paenungulate clade, in: F. S. Szalay, M. J. Novacek, and M. C. McKenna (eds.), *Mammal Phylogeny, Placentals*, pp. 5–12. Springer-Verlag, Berlin.

De Queiroz, K., and Gauthier, J. 1990. Phylogeny as a central principle in taxonomy: phylogenetic definitions of taxon names. *Syst. Zool.* **39**(4):307–322.

Dennstedt, A. 1904. Die Sinus durae matris der Haussäugetiere. *Anat. Hefte* **25**:1–96.

Diamond, M. K. 1992. Homology and evolution of the orbitotemporal venous sinuses of humans. *Am. J. Phys. Anthropol.* **88**:211–244.

Dormer, K. J., Denn, M., and Stone, H. L. 1977. Cerebral blood flow in the sea lion (*Zalophus californianus*) during voluntary dives. *Comp. Biochem. Physiol.* **58**(A):11–18.

Du Boulay, G. H., and Verity, P. M. 1973. *The Cranial Arteries of Mammals.* Heineman Medical Books, London.

Edinger, T. 1948. Evolution of the horse brain. *Mem. Geol. Soc. Am.* **25**:47–54.

Edinger, T. 1949. Paleoneurology versus comparative brain anatomy. *Confin. Neurol.* **9**:5–24.

Edinger, T. 1955. Hearing and smell in cetacean history. *Monatsschr. Psychiatr. Neurol.* **129**:37–58.

Edinger, T. 1964. Midbrain exposure and overlap in mammals. *Am. Zool.* **4**:5–19.

Fitch, W. M., and Beintema, J. J. 1990. Correcting parsimonious trees for unseen nucleotide substitutions: the effect of dense branching as exemplified by ribonuclease. *Mol. Biol. Evol.* **7**(5):438–443.

Fordyce, R. E. 1994. *Wapatia maerewhenua*, new genus and new species (Waipatiidae, new family), an archaic Late Oligocene dolphin (Cetacea: Odontoceti: Platanistoidea) from New Zealand, in: A. Berta and T. A. Deméré (eds.), *Contributions in Marine Mammal Paleontology Honoring Frank Whitmore, Jr.*, pp. 147–176. Proceedings of the San Diego Society of Natural History **29**.

Fraas, E. 1904. Neue Zeuglodonten aus dem unteren Mitteleocän vom Mokattam bei Cairo. *Geol. Paläontol. Abh. N. F.* **6**:199–220.

Fraser, F. C., and Purves, P. E. 1960. Hearing in cetaceans, evolution of the accessory air sacs and the structure and function of the outer and middle ear in recent cetaceans. *Bull. Br. Mus. Nat. Hist. Zool.* **7**(1):1–140.

Galliano, R. E., Morgane, P. J., McFarland, W. L. Nagel, E. L., and Catherman, R. L. 1966. The anatomy of the cervicothoracic arterial system in the bottlenose dolphin (*Tursiops truncatus*) with a surgical approach suitable for guided angiography. *Anat. Rec.* **155**:325–338.

Gaskin, D. E. 1985. *The Ecology of Whales and Dolphins.* Heinemann, London.

Gatesy, J. 1997. More DNA support for a Cetacea/Hippopotamidae clade: the blood-clotting protein gamma-fibrinogen. *Mol. Biol. Evol.* **14**(5):537–543.

Gatesy, J., Hayashi, C., Cronin, A., and Arctander, P. 1996. Evidence from milk casein genes that cetaceans are close relatives of hippopotamid artiodactyls. *Mol. Biol. Evol.* **13**(7):954–963.

Gauthier, J., Kluge, A. G., and Rowe, T. 1988. Amniote phylogeny and the importance of fossils. *Cladistics* **4**:105–209.

Gazin, C. L. 1965. A study of the early Tertiary condylarthran mammal *Meniscotherium. Smithson. Misc. Collect.* **149**(2):1–98.

Gazin, C. L. 1968. A study of the Eocene condylarthran mammal *Hyopsodus. Smithson. Misc. Collect.* **153**(4):1–90.

Geisler, J. H., and Luo, Z. 1996. The petrosal and inner ear of *Herpetocetus* sp. (Mammalia: Cetacea) and their implications for the phylogeny and hearing of archaic mysticetes. *J. Paleontol.* **70**(6):1045–1066.

Geisler, J. H., Sanders, A. E., and Luo, Z. 1996. A new protocetid cetacean from the Eocene of South Carolina, U.S.A.; phylogenetic and biogeographic implications, in: J. E. Repetski (ed.), *Sixth North American Paleontological Convention Abstracts of Papers. Paleontol. Soc. Spec. Pap.* **8**:139.

Gelderen, C. van. 1924. Die Morphologie der Sinus durae matris. Zweiter Teil. Die vergleichende Ontogenie der neurokraniellen Venen der Vögel und Säugetiere. *Z. Anat. Entwicklungsgesch.* **74**:434–508.

Gentry, A. W., and Hooker, J. J. 1988. The phylogeny of the Artiodactyla, in: M. J. Benton (ed.), *The Phylogeny and Classification of the Tetrapods*, Volume 2, pp. 235–272. Clarendon Press, Oxford.

Getty, R. (ed.). 1975. *Sisson and Grossman's The Anatomy of Domesticated Animals.* Saunders, Philadelphia.

Ghoshal, N. G. 1975a. Ruminant heart and arteries, in: R. Getty (ed.), *Sisson and Grossman's The Anatomy of Domesticated Animals*, pp. 960–1023. Saunders, Philadelphia.

Ghoshal, N. G. 1975b. Porcine heart and arteries (blood supply to brain by B. S. Nanda), in: R. Getty (ed.), *Sisson and Grossman's The Anatomy of Domesticated Animals*, pp. 1216–1252. Saunders, Philadelphia.

Gingerich, P. D., and Russell, D. E. 1981. *Pakicetus inachus*, a new archaeocete (Mammalia, Cetacea) from the early-middle Eocene Kuldana Formation of Kohat (Pakistan). *Contrib. Mus. Paleontol. Univ. Michigan* **25**:235–246.

Gingerich, P. D., and Russell, D. E. 1990. Dentition of early Eocene *Pakicetus* (Mammalia, Cetacea). *Contrib. Mus. Paleontol. Univ. Michigan* **28**(1):1–20.

Gingerich, P. D., Wells, N. A., Russell, D. E., and Shah, S. M. I. 1983. Origin of whales in epicontinental remnant seas: new evidence from the early Eocene of Pakistan. *Science* **220**:403–406.

Gingerich, P. D., Smith, B. H., and Simmons, E. L. 1990. Hind limbs of Eocene *Basilosaurus isis*: evidence of feet in whales. *Science* **249**:154–157.

Gingerich, P. D., Arif, M., and Clyde, W. C. 1995. New archaeocetes (Mammalia, Cetacea) from the middle Eocene

Domanda Formation of the Sulaiman Range, Punjab (Pakistan). *Contrib. Mus. Paleontol. Univ. Michigan* **29**(11):291–330.

Gould, G. C. 1995. Hedgehog phylogeny (Mammalia, Erinaceidae)—the reciprocal illumination of the quick and the dead. *Am. Mus. Novit.* **3131**:1–45.

Graur, D., and Higgins, D. G. 1994. Molecular evidence for the inclusion of cetaceans within the order Artiodactyla. *Mol. Biol. Evol.* **11**:357–364.

Gray, H. 1974. *Anatomy, Descriptive and Surgical*, 1901 ed. Running Press, Philadelphia.

Hammond, W. S. 1937. The developmental transformations of the aortic arches in the calf (*Bos taurus*), with especial reference to the formation of the arch of the aorta. *Am. J. Anat.* **62**:149–177.

Hartman, C. G. 1916. Studies in the development of the opossum *Didelphis virginiana* L. *J. Morphol.* **27**(1):1–62.

Hervé, P., and Douzery, E. 1994. The pitfalls of molecular phylogeny based on four species, as illustrated by the Cetacea/Artiodactyla relationships. *J. Mamm. Evol.* **2**(2):133–152.

Hofmann, L. 1914. Die entwicklung der kopfarterien bei *Sus scrofa domesticus*. *Morphol. Jahrb.* **48**:645–670.

Honeycutt, R. L., Nedbal, M. A., Adkins, R. M., and Janecek, L. L. 1995. Mammalian mitochondrial DNA evolution: a comparison of the cytochrome b and cytochrome c oxidase II genes. *J. Mol. Evol.* **40**:260–272.

Hunt, R. M., Jr. 1974. The auditory bulla in Carnivora: an anatomical basis for reappraisal of carnivore evolution. *J. Morphol.* **141**(1):21–76.

Irwin, D. M., and Arnason, U. 1994. Cytochrome *b* gene of marine mammals: phylogeny and evolution. *J. Mamm. Evol.* **2**(1):37–55.

Irwin, D. M., Kocher, T. D., and Wilson, A. C. 1991. Evolution of the cytochrome *b* gene of mammals. *J. Mol. Evol.* **32**:128–144.

Janke, A., Feldmaier-Fuchs, G., Kelley Thomas, W., von Haesseler, A., and Pääbo, S. 1994. The marsupial mitochondrial genome and the evolution of placental mammals. *Genetics* **137**:243–256.

Kalandadze, N. N., and Rautian, A. S. 1992. The system of mammals and historical zoogeography, in: O. L. Rossolimo (ed.), *Filogenetika Mlekopitayushchikh. Sb. Tr. Zool. Muz.* **29**:44–152.

Kampen, P. N. van. 1905. Die Tympanalgegend des Säugetierschädels. *Morphol. Jahrb.* **34**:321–720.

Kellogg, A. R. 1936. A review of the Archaeoceti. *Carnegie Inst. Washington Publ.* **482**:1–366.

Kielan-Jaworowska, Z. 1977. Evolution of the therian mammals in the Late Cretaceous of Asia. Part III. Postcranial skeleton in *Kennalestes* and *Asioryctes*. Palaeontol. Pol. **37**:65–83.

Kielan-Jaworowska, Z. 1981. Evolution of the therian mammals in the Late Cretaceous of Asia. Part. IV. Skull structure in *Kennalestes* and *Asioryctes*. *Palaeontol. Pol.* **42**:25–78.

Kielan-Jaworowska, Z., Presley, R. and Poplin, C. 1986. The cranial vascular system in taeniolabidoid multituberculate mammals. *Philos. Trans. R. Soc. London Ser. B* **313**:525–602.

Kumar, K., and Sahni, A. 1986. *Remingtonocetus harudiensis*, new combination, a middle Eocene archaeocete (Mammalia, Cetacea) from western Kutch, India. *J. Vertebr. Paleontol.* **6**(4):326–349.

Lancaster, W. C. 1990. The middle ear of the Archaeoceti. *J. Vertebr. Paleontol.* **10**(1):117–127.

Luo, Z., and Eastman, E. R. 1995. Petrosal and inner ear of a squalodontoid whale: implications for the evolution of hearing in odontocetes. *J. Vertebr. Paleontol.* **15**(2): 431–442.

Luo, Z., and Gingerich, P. D. In press. Transition from terrestrial ungulates to aquatic whales: transformation of the basicranium and evolution of hearing. *Bull. Mus. Paleontol. Univ. Michigan.*

Luo, Z., and Marsh, K. 1996. Petrosal (periotic) and inner ear of a Pliocene kogiine whale (Kogiinae, Odontoceti): implications on relationships and hearing evolution of toothed whales. *J. Vertebr. Paleontol.* **16**(2): 328–348.

MacPhee, R. D. E. 1981. Auditory regions of primates and eutherian insectivores: morphology, ontogeny, and character analysis. *Contrib. Primatol.* **18**:1–282.

MacPhee, R. D. E. 1994. Morphology, adaptations, and relationships of *Plesiorycteropus*, and a diagnosis of a new order of eutherian mammals. *Bull. Am. Mus. Nat. Hist.* **220**:1–214.

Maddison, W. P., and Maddison, D. R. 1992. MacClade Program (3.0). Sinauer Associates, Sunderland, MA.

Maddison, W. P., Donoghue, M. J., and Maddison, D. R. 1984. Outgroup analysis and parsimony. *Syst. Zool.* **33**(1):83–103.

McCrady, E., Jr. 1940. The development and fate of the urinogenital sinus in the opossum, *Didelphis virginiana*. *J. Morphol.* **66**:131–154.

McFarland, W. L., Jacobs, M. S., and Morgane, P. J. 1979. Blood supply to the brain of the dolphin, *Tursiops truncatus*, with comparative observations on special aspects of the cerebrovascular supply of other vertebrates. *Neurosci. Biobehav. Rev.* Suppl. 1 **3**:1–93.

McKenna, M. C. 1975. Toward a phylogenetic classification of the Mammalia, in: W. P. Luckett and F. S. Szalay (eds.), *Phylogeny of the Primates*, pp. 21–46. Plenum Press, New York.

McKenna, M. C. 1992. The alpha crystallin A chain of the eye lens and mammalian phylogeny. *Ann. Zool. Fenn.* **28**:349–360.

McKenna, M. C., and Bell, S. K. 1997. *Classification of Mammals Above the Species Level*. Columbia University Press, New York.

Melnikov, V. V. 1997. The arterial system of the sperm whale (*Physeter macrocephalus*). *J. Morphol.* **234**:37–50.

Meng, J., Ting, S., and Schiebout, J. A. 1994. The cranial morphology of an early Eocene didymoconid (Mammalia, Insectivora). *J. Vertebr. Paleontol.* **14**(4):534–551.

Miller, M. E., Christensen, G. C., and Evans, H. E. 1964. *Anatomy of the Dog*. Saunders, Philadelphia.

Montgelard, C., Catzeflis, F. M., and Douzery, E. 1997. Phylogenetic relationships of artiodactyls and cetaceans as deduced from the comparison of cytochrome b and 12S RNA mitochondrial sequences. *Mol. Biol. Evol.* **14**(5):550–559.

Mossman, H. W. 1987. *Vertebrate Fetal Membranes*. Rutgers University Press, New Brunswick, NJ.

Müller, O. 1898. Untersuchungen über die Veränderungen, welche die Respirationsorgane der Säugetiere durch die anpassung an das Leben im Wasser erlitten haben. *Jen. Z. Naturwiss.* **32**:95–230.

Nagel, E. L., Morgane, P. J., McFarland, W. L., and Galliano, R. E. 1968. Rete mirabile of dolphin: its pressure-dampening effect on cerebral circulation. *Science* **161**:898–900.

Nixon, K. C., and Carpenter, J. M. 1993. On outgroups. *Cladistics* **9**:413–426.

Novacek, M. J. 1977. Aspects of the problem of variation, origin and evolution of the eutherian bulla. *Mammal Rev.* **7**:131–149.

Novacek, M. J. 1980. Cranioskeletal features in tupaiids and selected Eutheria as phylogenetic evidence, in: W. P. Luckett (ed.), *Comparative Biology and Evolutionary Relationships of Tree Shrews*, pp. 35–93. Plenum Press, New York.

Novacek, M. J. 1986. The skull of leptictid insectivorans and the higher-level classification of eutherian mammals. *Bull. Am. Mus. Nat. Hist.* **183**(1):1–112.

Novacek, M. J. 1992. Fossils, topologies, missing data, and the higher level phylogeny of eutherian mammals. *Syst. Biol.* **41**(1):58–73.

Novacek, M. J., and Wyss, A. R. 1986a. Higher-level relationships of the eutherian orders: morphological evidence. *Cladistics* **2**(3):257–287.

Novacek, M. J., and Wyss, A. R. 1986b. Origin and transformation of the mammalian stapes. *Contrib. Geol. Univ. Wyoming Spec. Pap.* **3**:35–53.

Novacek, M. J., Wyss, A. R., and McKenna, M. C. 1988. The major groups of mammals, in: M. J. Benton (ed.), *The Phylogeny and Classification of the Tetrapods, Mammals*, Volume 2, pp. 31–71. Clarendon Press, Oxford.

Oelschläger, H. A. 1986. Comparative morphology and evolution of the otic region in toothed whales (Cetacea, Mammalia). *Am. J. Anat.* **177**:353–368.

O'Leary, M. A., and Rose, K. D. 1995a. New mesonychian dentitions from the Paleocene of the Bighorn Basin, Wyoming. *Ann. Carnegie Mus.* **64**(2):147–172.

O'Leary, M. A., and Rose, K. D. 1995b. Postcranial skeleton of the early Eocene mesonychid *Pachyaena* (Mammalia: Mesonychia). *J. Vertebr. Paleontol.* **15**(2):401–430.

Padget, D. H. 1948. The development of the cranial arteries in the human embryo. *Contrib. Embryol.* **32**:205–261.

Padget, D. H. 1957. The development of the cranial venous system in man, from the viewpoint of comparative anatomy. *Contrib. Embryol.* **36**:81–151.

Patterson, C. 1982. Morphological characters and homology, in: K. A. Joysey and A. E. Friday (eds.), *Problems of Phylogenetic Reconstruction*, pp. 21–74. Academic Press, New York.

Pilleri, G. 1991. Betrachtungen über das Gehirn der Archaeoceti (Mammalia, Cetacea) aus dem Fayüm Ägyptens. *Invest. Cetacea* **23**:193–211.

Presley, R. 1979. The primitive course of the internal carotid artery in mammals. *Acta Anat.* **103**:238–244.

Prothero, D. R., Manning, E. M., and Fischer, M. 1988. The phylogeny of the ungulates, in: M. J. Benton (ed.), *The Phylogeny and Classification of the Tetrapods*, Volume 2, pp. 201–234. Clarendon Press, Oxford.

Queralt, R., Adroer, R., Oliva, R., Winkfein, R. J., Retief, J. D., and Dixon, G. H. 1995. Evolution of protamine P1 genes in mammals. *J. Mol. Evol.* **40**:601–607.

Radinsky, L. B. 1965. Evolution of the tapiroid skeleton from *Heptodon* to *Tapirus*. *Bull. Mus. Comp. Zool.* **34**(3):1–106.

Radinsky, L. B. 1976a. Oldest horse brains: more advanced than previously realized. *Science* **194**:626–627.

Radinsky, L. B. 1976b. The brain of *Mesonyx*, a middle Eocene condylarth. *Fieldiana Geol.* **33**(18):323–337.

Rose, K. D. 1985. Comparative osteology of North American dichobunid artiodactyls. *J. Paleontol.* **59**(5):1203–1226.

Rose, K. D. 1996. On the origin of the order Artiodactyla. *Proc. Natl. Acad. Sci. USA* **93**:1705–1709.

Rose, K. D., and O'Leary, M. A. 1995. The manus of *Pachyaena gigantea* (Mammalia: Mesonychia). *J. Vertebr. Paleontol.* **15**(4):855–859.

Rougier, G. W., Wible, J. R., and Hopson, J. A. 1992. Reconstruction of the cranial vessels in the early Cretaceous *Vincelestes neuquenianus:* implications for the evolution of the mammalian cranial vascular system. *J. Vertebr. Paleontol.* **12**(2):188–216.

Russell, D. E. 1964. Les mammiféres Paleocenes d'Europe. *Mem. Mus. Natl. Hist. Nat. Ser. C* **13**:1–324.

Russell, D. E., and Sigogneau, D. 1965. Etude de moulages endocraniens de mammiferes Paleocenes. *Mem. Mus. Natl. Hist. Nat. Ser. C* **16**:1–34.

Russell, D. E., Thewissen, J. G. M., and Sigogneau-Russell, D. 1983. A new dichobunid artiodactyl (Mammalia) from the Eocene of north-west Pakistan. Part II: Cranial osteology. *Proc. K. Ned. Akad. Wet. Ser. B* **86**(3):285–299.

Schaeffer, J. P. (ed.). 1953. *Morris' Human Anatomy*, 11th ed. McGraw–Hill, New York.

Schreiber, K. 1916. Zur Entwicklungsgeschichte des Walschädels. Das Primordialcranium eines Embryos von *Globiocephalus melas* (13.3 cm). *Zool. Jahrb. Abt. Ontog. Tiere* **39**:201–236.

Scott, W. B. 1888. On some new and little known creodonts. *J. Acad. Nat. Sci. Philadelphia* **9**:155–185.

Shimamura, M., Yasue, H., Oshima, K., Abe, H., Kato, H., Kishiro, T., Goto, M., Munechika, I., and Okada, N. 1997. Molecular evidence from retroposons that whales form a clade within even-toed ungulates. *Nature* **388**:66–67.

Shoshani, J. 1986. Mammalian phylogeny: comparison of morphological and molecular results. *Mol. Biol. Evol.* **3**(3):222–242.

Shoshani, J. 1993. Hyracoidea–Tethytheria affinity based on myological data, in: F. S. Szalay, M. J. Novacek, and M. C. McKenna (eds.), *Mammal Phylogeny, Placentals*, pp. 235–256. Springer-Verlag, Berlin.

Simpson, G. G. 1933. Braincasts of *Phenacodus, Notostylops*, and *Rhyphodon. Am. Mus. Novit.* **622**:1–19.

Simpson, G. G. 1945. The principles of classification and a classification of mammals. *Bull. Am. Mus. Nat. Hist.* **85**:1–339.

Sinclair, J. G. 1967. Cerebral vascular system and ocular nerves of dolphin (*Stenella*) embryos. *Tex. Rep. Biol. Med.* **25**(4):551–571.

Sisson, S. 1921. *The Anatomy of the Domesticated Animals*. Saunders, Philadelphia.

Slijper, E. J. 1936. Die Cetaceen. Vergleichend-Anatomische und Systematisch. *Capita Zool.* **7**(2):1–590.

Smith, M. R., Shivji, M. S., and Waddell, V. G. 1996. Phylogenetic evidence from the IRBP gene for the paraphyly of toothed whales, with mixed support for Cetacea as a suborder of Artiodactyla. *Mol. Biol. Evol.* **13**(3):918–922.

Smut, M. M. S., and Bezuidenhout, A. J. 1987. *Anatomy of the Dromedary*. Clarendon Press, Oxford.

Stanhope, M. J., Smith, M. A., Waddell, V. G., Porter, C. A., Shivji, M. S. and Goodman, M. 1996. Mammalian evolution and the interphotoreceptor retinoid binding protein (IRPB) gene: convincing evidence for several superordinal clades. *J. Mol. Evol.* **43**:83–92.

Stannius, H. 1841. Über den Verlauf der Arterien bei *Delphinus phocaena. Arch. Anat. Physiol.* **8**:379–402.

Swofford, D. L. 1993. PAUP: Phylogenetic Analysis Using Parsimony (Version 3.1.1). Privately distributed by Illinois Natural History Survey, Champaign.

Szalay, F. S. 1969a. Origin and evolution of function of the mesonychid condylarth feeding mechanism. *Evolution* **23**:703–720.

Szalay, F. S. 1969b. The Hapalodectinae and a phylogeny of the Mesonychidae (Mammalia, Condylarthra). *Am. Mus. Novit.* **2361**:1–26.

Szalay, F. S. 1975. Phylogeny of Primate higher taxa, in: W. P. Luckett and F. S. Szalay (eds.), *Phylogeny of the Primates—A Multidisciplinary Approach*, pp. 91–125. Plenum Press, New York.

Szalay, F. S., and Gould, S. J. 1966. Asiatic Mesonychidae (Mammalia, Condylarthra). *Bull. Am. Mus. Nat. Hist.* **132**(2):128–173.

Thewissen, J. G. M. 1985. Cephalic evidence for the affinities of Tubulidentata. *Mammalia* **49**(2):257–284.

Thewissen, J. G. M. 1990. Evolution of Paleocene and Eocene Phenacodontidae (Mammalia Condylarthra). *Univ. Michigan Pap. Paleontol.* **29**:1–107.

Thewissen, J. G. M. 1994. Phylogenetic aspects of cetacean origins: a morphological perspective. *J. Mamm. Evol.* **2**(3):157–184.

Thewissen, J. G. M., and Domning, D. P. 1992. The role of phenacodontids in the origin of the modern orders of ungulate mammals. *J. Vertebr. Paleontol.* **12**(4):494–504.

Thewissen, J. G. M., and Hussain, S. T. 1990. Postcranial osteology of the most primitive artiodactyl *Diacodexis pakistanensis* (Dichobunidae). *Anat. Histol. Embryol.* **19**:37–48.

Thewissen, J. G. M., and Hussain, S. T. 1993. Origin of underwater hearing in whales. *Nature* **361**:444–445.

Thewissen, J. G. M. Russell, D. E., Gingerich, P. D., and Hussain, S. T. 1983. A new dichobunid artiodactyl (Mammalia) from the Eocene of north-west Pakistan: dentition and classification. *Proc. K. Ned. Akad. Wet. Ser. B* **86**(2):153–180.

Thewissen, J. G. M., Madar, S. I., and Hussain, S. T. 1996. *Ambulocetus natans*, an Eocene cetacean (Mammalia) from Pakistan. *Cour. Forsch.-Inst. Senckenberg* **191**:1–86.

Ting, S., and Li, C. 1987. The skull of *Hapalodectes* (?Acreodi, Mammalia), with notes on some Chinese Paleocene mesonychids. *Vertebr. PalAsiat.* **25**(3):161–186.

Uhen, M. D. 1996. *Dorudon atrox* (Mammalia, Cetacea): form, function, and phylogenetic relationships of an archaeocete from the later middle Eocene of Egypt. Ph.D. dissertation, University of Michigan, Ann Arbor, 608 pp.

Van Valen, L. 1966. Deltatheridia, a new order of mammals. *Bull. Am. Mus. Nat. Hist.* **132**:1–126.

Vogl, A. W., and Fisher, H. D. 1981a. Arterial circulation of the spinal cord and brain in the Monodontidae (order Cetacea). *J. Morphol.* **170**:171–180.

Vogl, A. W., and Fisher, H. D. 1981b. The internal carotid artery does not directly supply the brain in the Monodontidae (order Cetacea). *J. Morphol.* **170**:207–214.

Walmsley, R. 1938. Some observations on the vascular system of a female finback. *Contrib. Embryol.* **27**:107–178.

Whitmore, F. C. 1953. Cranial morphology of some Oligocene Artiodactyla. U.S. Geol. Surv. Prof. Pap. **243H**:117–160.

Wible, J. R. 1984. The ontogeny and phylogeny of the mammalian cranial arterial pattern. Ph.D. dissertation, Duke University, Durham, 705 pp.

Wible, J. R. 1986. Transformations in the extracranial course of the internal carotid artery in mammalian phylogeny. *J. Vertebr. Paleontol.* **6**(4):313–325.

Wible, J. R. 1987. The eutherian stapedial artery: character analysis and implications for superordinal relationships. *Zool. J. Linn. Soc.* **91**:107–135.

Wible, J. R. 1990. Petrosals of late Cretaceous marsupials from North America, and a cladistic analysis of the petrosal in therian mammals. *J. Vertebr. Paleontol.* **10**(2):183–205.

Wible, J. R., and Hopson, J. A. 1995. Homologies of the prootic canal in mammals and nonmammalian cynodonts. *J. Vertebr. Paleontol.* **15**(2):331–356.

Wible, J. R., and Novacek, M. J. 1988. Cranial evidence for the monophyletic origin of bats. *Am. Mus. Novit.* **2911**:1–19.

Wible, J. R., and Zeller, U. 1994. Cranial circulation of the pen-tailed shrew *Ptilocercus lowii* and relationships of Scandentia. *J. Mamm. Evol.* **2**(4):209–230.

Williamson, T. E., and Lucas, S. G. 1992. *Meniscotherium* (Mammalia, "Condylarthra") from the Paleocene–Eocene of western North America. *N. M. Mus. Nat. Hist. Bull.* **1**:1–75.

Witmer, L. M. 1992. Ontogeny, phylogeny, and air sacs: the importance of soft-tissue inferences in the interpretation of facial evolution in Archosauria. Ph.D. dissertation, Johns Hopkins University School of Medicine, Baltimore, 461 pp.

Witmer, L. M. 1995. The extant phylogenetic bracket and the importance of reconstructing soft tissues in fossils, in: J. Thomason (ed.), *Functional Morphology in Vertebrate Paleontology*, pp. 19–33. Cambridge University Press, London.

Wortman, J. L. 1901. Studies of the Eocene Mammalia in the Marsh collection, Peabody Museum. *Am. J. Sci.* **11**:1–90.

Xu, X., Janke, A., and Arnason, U. 1996. The complete mitochondrial sequence of the greater Indian rhinoceros, *Rhinoceros unicornis*, and the phylogenetic relationship among Carnivora, Perissodactyla, and Artiodactyla (+Cetacea). *Mol. Biol. Evol.* **13**(9):1167–1173.

Zhou, X., Zhai, R., Gingerich, P. D., and Chen, L. 1995. Skull of a new mesonychid (Mammalia, Mesonychia) from the late Paleocene of China. *J. Vertebr. Paleontol.* **15**(2):387–400.

CHAPTER 7

Middle Eocene Cetaceans from the Harudi and Subathu Formations of India

SUNIL BAJPAI and J. G. M. THEWISSEN

1. Introduction

Eocene cetaceans from the Indian subcontinent were first reported over 25 years ago by Tandon (1971), who recovered specimens in the District Kachchh (formerly spelled Cutch or Kutch) in the west Indian state of Gujarat. Following this report, Mishra made extensive collections of Tertiary vertebrates, including cetaceans, from Kachchh, and these were published by Sahni and Mishra (1972, 1975). These collections changed a widely held view of cetacean origins that is exemplified by Kellogg's (1936, p. 270) statement that "the archaeocetes evidently did not reach the Indian ocean during Eocene time." Around the time of Mishra's collecting in Kachchh, geologists from the Geological Survey of India (GSI) also reported Eocene cetaceans from Kachchh (Satsangi and Mukhopadhyay, 1975). Subsequently, Kumar and Sahni (1986) described the cranial anatomy of one of the endemic cetaceans from Kachchh, *Remingtonocetus*, in detail, and erected a new family for it and *Andrewsiphius*. Since then, collections from the Harudi Formation of Kachchh have been substantially enhanced by the efforts of one of us, and these were reported in his unpublished thesis (Bajpai, 1990).

Kachchh fossils described in this chapter were found in two areas (Fig. 1). The first is the western base of Babia Hill, where most fossils come from units that contain abundant gypsum. This locality has yielded the holotype of *Babiacetus indicus* Trivedy and Satsangi, 1984. Additional mammals from Babia Hill were described by Bajpai *et al.* (1989) and Bajpai (1990).

The second area with abundant Harudi Formation fossils is a modern rivercut that extends from an area about 2 km north of the village of Harudi to several kilometers west of here. It is the type area for the Harudi Formation and its topographic name is Rato Nala (nala means riverbed, most nalas do not carry water in the dry season). In this area, the gyp-

---

SUNIL BAJPAI • Department of Earth Sciences, University of Roorkee, Roorkee 247 667, Uttar Pradesh, India. J. G. M. THEWISSEN • Department of Anatomy, Northeastern Ohio Universities College of Medicine, Rootstown, Ohio 44272.

The Emergence of Whales, edited by Thewissen. Plenum Press, New York, 1998.

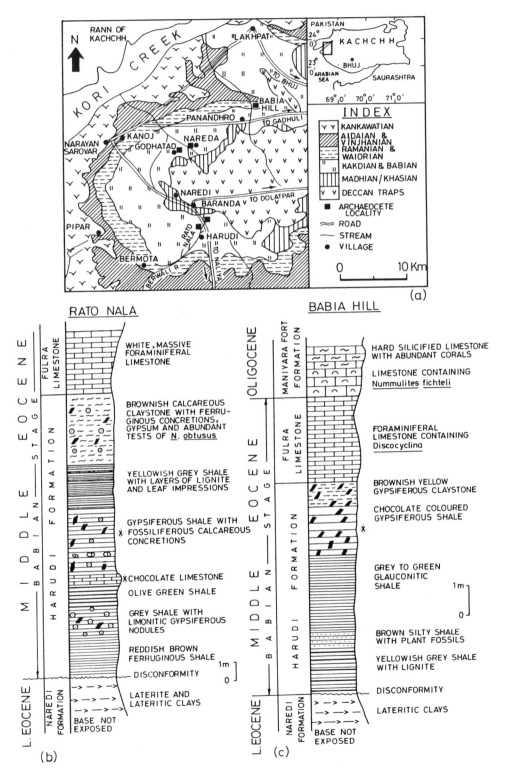

(a)

(b)

(c)

siferous shales are present, but most fossils are recovered in a limestone unit near the base of the formation. This unit, the chocolate limestone, has not been found at Babia Hill.

Kumar and Sahni (1986) identified three units yielding cetaceans in Kachchh, the Upper Gypsiferous Clay, the Grey Marl, and the Chocolate Limestone. Preliminary evaluation of these units in the field has shown that fossils and gypsum occur at several levels in the Babia Hill section and that limestone may or may not be present in different localities that have an otherwise complete section. These units are thus probably not stratigraphically meaningful.

Cetaceans are also known from the Subathu Formation of the states Jammu and Kashmir (J & K) and Himachal Pradesh (H. P.) of northwestern India. These include an unpublished mandible from the lower Eocene of the type area of the Subathu Formation (H. P.) that was recovered from an oyster limestone unit near the base of the formation as exposed along Kuthar Nala. This unit has yielded another marine mammal of disputed affinities, the type specimen of *Ishatherium subathuensis* Sahni and Kumar, 1981. Fossil fishes from Kuthar Nala area were described by Kumar and Loyal (1987).

The Subathu Formation of H. P. also yielded an isolated tooth that is described here. It was recovered from middle Eocene sediments near Dharampur, and may be contemporaneous with classical middle Eocene vertebrate faunas of Kalakot (J & K). Several teeth from Kalakot (J & K) were referred to *Ichthyolestes* by Kumar and Sahni (1985), but these were reidentified as *Pakicetus* by Thewissen and Hussain (1998).

As part of a larger, ongoing study of Eocene cetaceans from India that involve fieldwork as well as museum study, we here describe parts of the collections from Kachchh discussed by Bajpai (1990), as well as the single tooth from Dharampur. These specimens are catalogued in the collections of the Vertebrate Paleontology Laboratory (VPL) of Panjab University, Chandigarh (Panjab) and the Department of Earth Sciences of the University of Roorkee (RUSB), Roorkee, Uttar Pradesh. All specimens are housed at the latter institution and curated by the senior author.

2. Systematic Paleontology

Family Remingtonocetidae Kumar and Sahni, 1986

Remingtonocetus Kumar and Sahni, 1986

Remingtonocetus harudiensis (Sahni and Mishra, 1975)

Figures 2A–D and 3A

Referred Specimens.—RUSB 2016 (fragment of posterior maxilla with M^{2-3} and anterior orbits), RUSB 2018 (rostrum fragment with right P^2 and roots for left and right P^{1-2}).

◄——

FIGURE 1. Geological map of western Kachchh showing middle Eocene cetacean localities and relevant chronostratigraphic units, and stratigraphic sections at Rato Nala (2 km north of Harudi) and Babia Hill. Madhian/Khasian represents Paleocene, Kakdian and Babian represent Eocene, Ramanian and Waiorian represent Oligocene, and Aidaian and Vinjhanian represent Miocene. The Kankawatian represents the undifferentiated late Cenozoic. Map and units are based on Biswas and Deshpande (1970) and Biswas (1992). Asterisks in sections represent cetacean-yielding horizons.

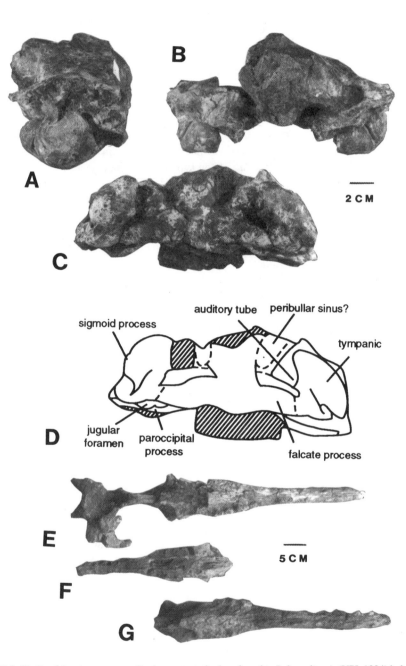

FIGURE 2. Skulls of *Remingtonocetus*. Braincase tentatively referred to *R. harudiensis* (VPL 1004) in lateral (A), anterior (B), and ventral (C) view, with explanatory diagram of the ventral view (D). (E–G) Rostra and skull of *R. sloani* in ventral view (E: VPL 1001; F: RUSB 2022; G: VPL 15003).

Tentatively referred specimens: RUSB 2017 (central fragment of braincase with left occipital condyle, but lacking ear regions), VPL 1004 (braincase with left and right ear regions, caudal aspect missing. All these specimens were found in the chocolate limestone of Rato Nala, Harudi Formation, Kachchh.

Age and Distribution.—Middle Eocene (Lutetian), Harudi Formation, Kachchh, Gujarat. Gingerich *et al.* (1995a) described *Remingtonocetus* cf. *R. harudiensis* from the Domanda Formation of the Sulaiman Range, Pakistan.

Description.—P^2 is the longest tooth in the upper dentition (54 mm in RUSB 2018). On the wall of the rostrum, dorsal to P^2, is a pronounced groove (mesorostral gutter of Kumar and Sahni, 1986) extending anterior from the infraorbital foramen. The suture between premaxilla and maxilla converges on the suture between premaxilla and nasal. The nasals taper rostrally.

A fragment with poorly preserved M$^{2-3}$ (RUSB 2016) shows that M^2 has a lingual root near the midline of the tooth whereas in M^3 the lingual root is posterior and has fused to the metacone root. The cross section of the snout immediately anterior to M^2 is 93 mm wide and 64 mm high. Paranasal sinuses make up approximately one-third of the width as well as the height of this cross section (Fig. 3A). The nasals roof this section and are flat, left and right side combined are 50 mm wide. The lacrimal forms a large triangular process on the lateral side of the face as described by Kumar and Sahni (1986). A small lacrimal foramen opens into the orbit.

Two braincases of *Remingtonocetus* (RUSB 2017 and VPL 1004, Fig. 2) cannot be confidently identified to species, but their slightly larger size than VPL 1001 (see *R. sloani*) suggests that they are most likely *R. harudiensis*. These braincases add some details to the description of Kumar and Sahni (1986), whose specimen lacked much of the lateral side of the braincase. The dominant feature of the basicranium is a laterally projecting process of the basioccipital, the falcate process of Kumar and Sahni (1986). This process tapers laterally and its extremity articulates with the posteromedial part of the petrosal. A sulcus occurs caudal to the falcate process separating it from the occipital condyle. Two foramina occur in the medial part of this sulcus (RUSB 2017), the hypoglossal and condylar foramen. The posterior of these is larger, whereas the anterior foramen is on the suture between falcate process and paroccipital process. This suture ascends the anterior wall of the sulcus and divides into two diverging sutures. The petrosal is wedged between these sutures (RUSB 2017). Laterally (VPL 1004), the entire space between paroccipital process and falcate process is taken by the jugular foramen. This foramen is divided in two parts by a small process emanating from the falcate process. Anterior to the falcate process, immediately lateral to the basioccipital, is a deep fossa indicated but not labeled by Kumar and Sahni (1986, their Fig. 3). These authors may have considered this fossa as part of a peribullar sinus. Lateral to this fossa and anterior to the falcate process is a shallow groove that houses a foramen that opens laterally into the braincase. The foramen may have carried the ventral petrosal sinus or the internal carotid artery. The tympanic (VPL 1004) is oval in ventral view, its long axis extending from mediorostral to laterocaudal. Mediocaudally, the tympanic shares a broad contact with the falcate process, and mediorostrally the tympanic is indented by the auditory tube. Caudally, the tympanic is fused extensively to the periotic. The sigmoid process is the most conspicuous feature laterally.

Discussion.—The cranial morphology of *Remingtonocetus* was described in detail by Kumar and Sahni (1986). This description is correct in most respects, but our new material indicates that their specimen (VPL/K 15001) pertained to *R. sloani* and not *R. harudiensis*. The new specimens also add some details that were not previously known, and allow for a rediagnosis of the species of *Remingtonocetus* (see *R. sloani*).

Remingtonocetus sloani (Sahni and Mishra, 1975)

Figures 2E–G, 3B, 4A–C, and 5A

Referred Material.—RUSB 2019 (premaxilla with roots for left and right (I^{1-3}; chocolate limestone of Rato Nala), RUSB 2020 (maxilla fragment with bases for left M^{2-3} and endocasts of paranasal sinuses; chocolate limestone of Rato Nala), RUSB 2022 (rostrum with anterior orbits and bases for left and right $C^1–M^1$, fragmentary crown of right M^1; gypsiferous shale of Rato Nala), RUSB 2025 (left maxilla fragment with bases for M^{1-3} and endocasts of infraorbital canal and paranasal sinuses; chocolate limestone of Rato Nala), RUSB 2026 (rostrum fragment with bases for left and right P^2; chocolate limestone of Rato Nala), VPL 1001 (poorly preserved skull with alveoli or roots of all teeth, inferior braincase missing but supraoccipital, parietal, squamosal, and bulla are preserved; gypsiferous shale of Babia Hill), VPL 15003 (rostrum with roots for all teeth, crown of left P^4, and anterior orbit; gypsiferous shale of Rato Nala). All specimens are from the Harudi Formation of Kachchh.

Age and Distribution.—Middle Eocene (Lutetian), Harudi Formation, Kachchh, Gujarat.

Description.—Left and right I^1 are juxtaposed (RUSB 2019) and are much smaller than I^2, the greatest cross section of I^1 is 12 mm, that of I^2 27 mm. I^2 is implanted obliquely and projects strongly anteriorly. I^3 is more vertical and is smaller than I^2. A similar incisor arrangement is also present in VPL 1001, whereas in RUSB 2022 the size difference

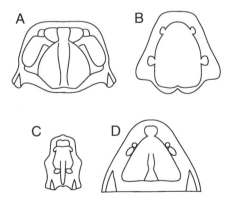

FIGURE 3. Cross sections through the rostrum of Kachchh cetaceans. (A) *Remingtonocetus harudiensis* (RUSB 2016), section immediately anterior to M^2. (B) *R. sloani* (RUSB 2020), section immediately anterior to M^2. (C) *Andrewsiphius kutchensis* (RUSB 2021), section at posterior root of P^4. (D) *Indocetus ramani* (LUVP 11034), section at anterior P^3. Internal nasal structures are extremely delicate and their variation in these specimens is probably related in part to poor preservation.

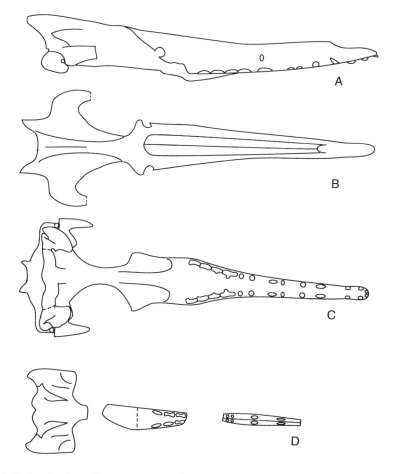

FIGURE 4. Outline drawings of Kachchh cetacean skulls. (A–C) *Remingtonocetus sloani* in lateral, dorsal, and ventral views, mainly based on VPL 1001. (D) *Andrewsiphius kutchensis* in ventral view, based on VPL 1007; dashed line represents posterior extent of preserved part of palate.

between I^1 and I^2 is less pronounced and the two teeth are separated by a diastema. The nasal opening is over the canine (VPL 1001 and 15003).

P^2 is narrow and elongate (RUSB 2026, length approximately 34 mm). It is similar in shape to *R. harudiensis*, but smaller. As in *R. harudiensis*, a pronounced groove extends anteriorly from the infraorbital foramen on the side of the maxilla.

The cross section of the skull in the molar region (Fig. 3B) is higher than wide: In RUSB 2020, its height anterior to M^1 is 58 mm, and its bilateral width 33 mm; in RUSB 2025, approximate height anterior to M^2 is 60 mm and width 33 mm. The snout is narrow in this region because of the small size of the paranasal sinuses. This is unlike the paranasal sinuses of *R. harudiensis*, which apparently extend farther anteriorly and/or are much larger, causing the snout to be wider. The small size of the paranasal

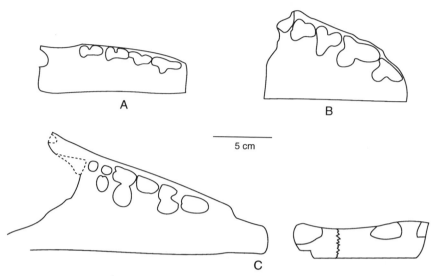

FIGURE 5. Outline drawings of maxillae of Kachchh cetaceans. (A) *Remingtonocetus sloani* with basal outline for M^{1-3} (RUSB 2025) and P^4 (VPL 15003). (B) *Indocetus ramani* (VPL 1023) with basal outline of P^3–M^2, and fragment of M^3. (C) *Gaviacetus sahnii* (VPL 1021) with basal outline of P^3–M^1 (M^2 fragment not visible). Anterior fragment includes alveoli for I^2–C^1, lower edge of drawing represents midsagittal plane.

sinuses over the anterior molars in *R. sloani* also causes the nasals to slope strongly laterally, which can be observed clearly in VPL 1001 and RUSB 2022.

The orbits are small and placed well posterior to M^3 (VPL 1001, VPL 15003). The postorbital process projects farther laterally than the anterior edge of the orbit, causing the eyes to face anterolaterally. The supraorbital region is broad and the temporal lines converge gradually on the sagittal crest by extending posteromedially from the postorbital process. The braincase resembles that of *Remingtonocetus* as described by Kumar and Sahni (1986), but the new specimens add several details not previously known. The mandibular joint is elongate anteroposteriorly, and is narrower than long (VPL 1001). The zygomatic processes of the squamosal and the postorbital process are robust. The external auditory meatus is immediately behind the postorbital process. The caudal aspect of the skull is low and wide, with the exoccipital located far laterally. In the median plane, the supraoccipitals are depressed and lateral to this, the nuchal crests project far caudally.

RUSB 2025 also preserves much of the bases of M^{1-3} (Fig. 5A), although most of the crowns are missing. The outline of M^1 is strikingly different from that of M^{2-3}. The protocone lobe of M^1 is large and is positioned in the center of the tooth. Its root is not fused to the labial roots. The protocone has a strong cingulum. M^2 and M^3 are similar in shape and differ from M^1. The protocone lobe is smaller and lacks a separate root in the posterior molars. Instead the posterolabial root has a lingual extension. It is not clear whether a protocone is present on M^{2-3} of *Remingtonocetus sloani*. As in all *Remingtonocetus*, embrasure pits are weak and the palate extends well caudal to the last molar.

Discussion.—Sahni and Mishra (1975) referred a number of cranial specimens to *R. sloani*, but Kumar and Sahni (1986) questioned the identification of several of these. We here identify several crania as *R. sloani*, including the skull described by Kumar and Sahni (1986) as *R. harudiensis* (VPL/K 15001). *R. sloani* is found in both chocolate limestone and gypsiferous shale facies of the Harudi Formation, as indicated by Sahni and Mishra (1975), and unlike the statement of Kumar and Sahni (1986). *R. sloani* differs mainly from *R. harudiensis* in having domed nasals over M^2 and a wide rostrum over M^{1-2} (Fig. 3). The nasals of *R. harudiensis* are flat and the snout is narrower near the anterior molars. *R. sloani* is also smaller than *R. harudiensis*, although the overall ranges of these species probably overlap. The size difference is more pronounced in certain teeth (e.g., P^2) and less in others (e.g., the molars).

Dentally, M^2 and M^3 of *R. sloani* are similar in the posterior position of the protocone lobe, whereas in M^1 the protocone lobe is near the center of the tooth. M^1 and M^3 of *R. harudiensis* follow the same pattern, but its M^2 resembles M^1 more closely.

Andrewsiphius kutchensis Sahni and Mishra, 1975

Figures 3C, 4D, 6, and 7

Referred Material.—VPL 1007 (cranium in three fragments: complete braincase; orbital region with posterior rostrum and posterior palate, including crown of M^3; and rostrum fragment with alveoli for left and right (C^1–P^2 and fragmentary crown for P^2; gypsiferous shale, Babia Hill), VPL 1019 (poorly preserved rostrum fragment with alveoli of fragmentary crowns for left and right P^3–M^3, and a second fragment with alveoli for three single-rooted teeth on left and right side, possibly I^3–P^1; gypsiferous

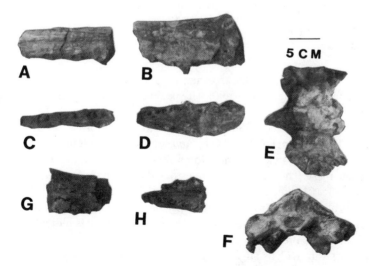

FIGURE 6. Skull and rostrum of *Andrewsiphius kutchensis*. (A, B) Rostral fragments in left lateral view. (C–E) Skull in ventral view. (F) Braincase in caudal view. (G, H) Posterior palate in right lateral and occlusal views. (A–F = VPL 1007; G–H: RUSB 2021).

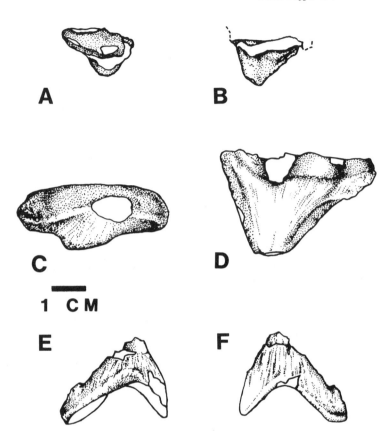

FIGURE 7. Teeth of Eocene cetaceans from India. (A, B) Left M^3 of *Andrewsiphius kutchensis* (VPL 1007) in occlusal and lateral views. (C, D) Right P^1 of *Gaviacetus sahnii* (VPL 1021) in occlusal and lingual views. (E, F) Cetacean deciduous premolar from the Subathu Formation (RUSB 2002), in lingual and labial views.

shale, Babia Hill), RUSB 2021 (rostrum fragment with base for all left and right molars; gypsiferous shale, Rato Nala). All specimens are from the Harudi Formation of Kachchh.

Age and Distribution.—Middle Eocene (Lutetian), Harudi Formation, Kachchh, Gujarat.

Description.—The rostrum (VPL 1019) of *Andrewsiphius* is extremely narrow, and higher than wide (Fig. 3C). In areas where tooth-roots are present, the maxilla widens to accommodate these. There are three single-rooted teeth on the left side of one of the fragments of VPL 1019, separated by diastemata of 30 mm. An endocast for the nasal cavity extends along the entire fragment, implying that the nasal opening is anterior to the most anterior tooth (probably I^3, as P^2 is two-rooted which is indicated by VPL 1007). Alveoli for P^2 are present on the left and right side.

The rostrum widens slightly caudal to P^1 (VPL 1007, Fig. 6). Alveoli for I^3–P^1 are similar in size and the diastemata separating them are approximately 30 mm long. P^2 is double-rooted, its length is 24 mm, its width 10 mm. A diastema of 20 mm separates P^1 from P^2. A groove extends along the lateral side of the rostrum, marking the course of the infraorbital vessels and nerves. The entire length of the palate bears a prominent midline crest (VPL 1007 and 1019).

P^3 and P^4 are double-rooted teeth that are separated by a diastema of 20 mm (VPL 1019). The alveoli for the molars are separated by a short diastema, but their crowns were probably juxtaposed. M^1 has two roots (VPL 1007) that are similar in size. M^2 also has two roots, but the posterior of these is clearly wider than the anterior. M^3 has a true protocone lobe and has two fused posterior roots. The crown is only preserved for M^3 (VPL 1007, Fig. 7A,B), and it bears a single cusp, the paracone. The palate is narrow, bilateral width is 24 mm at the anterior root of M^1, and 60 mm at the posterior root of M^3. The height of the rostrum over M^1 is 67 mm (excluding the crown height of the teeth). A large palatine foramen occurs medial to M^1. The endocast of the infraorbital canal indents the nasal cavity. Frontal sinuses also extend anterior to P^4 (RUSB 2021). The nasals (VPL 1007) form a sharp edge dorsally and slope strongly lateral. Posteriorly they widen.

The braincase of *Andrewsiphius* (VPL 1007) differs from that of *Remingtonocetus* (VPL 1003) in being narrower, with bullae that are closer to the midline. The present specimen (VPL 1007) is encrusted with gypsum, and many details, if preserved at all, cannot be observed at present. The braincase has a narrow intertemporal constriction and widens from this point posteriorly. The paroccipital process is large and directed laterally and ventrally. There is no falcate process. Pterygoid processes extend to the anterior part of the bulla. The bulla resembles that of *Remingtonocetus*, but its long axis is directed more anteroposteriorly and not obliquely.

Discussion.—The new skull of *Andrewsiphius* shows that this genus is very different from *Remingtonocetus* and that the Remingtonocetidae as a group are strongly specialized. Remingtonocetids share long and narrow rostra, small eyes, and reduced teeth. They are surprisingly different from protocetids in the anterior position of the nasal opening, the shape of the bulla, the low, broad braincase, and the morphology of the molars. *Andrewsiphius* differs from *Remingtonocetus* in that its snout is narrower, its nasal opening is more anterior, its M^3 is double-rooted (three-rooted in *Remingtonocetus*), and the falcate process is absent.

Family Protocetidae Stromer, 1908

Indocetus ramani Sahni and Mishra, 1975

Figures 3D, 5B, and 8A,B

Referred Material.—VPL 1014 (right tympanic), VPL 1017 (endocast of nasal cavity near orbits, including roots of M^{2-3}, endocast of cranial cavity, described by Bajpai *et al.*, 1996), VPL 1018 (endocast of cranial cavity), VPL 1023 (right maxilla with fragments of P^3–M^3). All of these specimens are from the chocolate limestone of the Harudi Formation, Rato Nala, Kachchh.

Age and Distribution.—Middle Eocene (Lutetian), Harudi Formation, Kachchh, Gujarat. Gingerich *et al.* (1993) described *Indocetus* from the Sulaiman Range of Pakistan, but

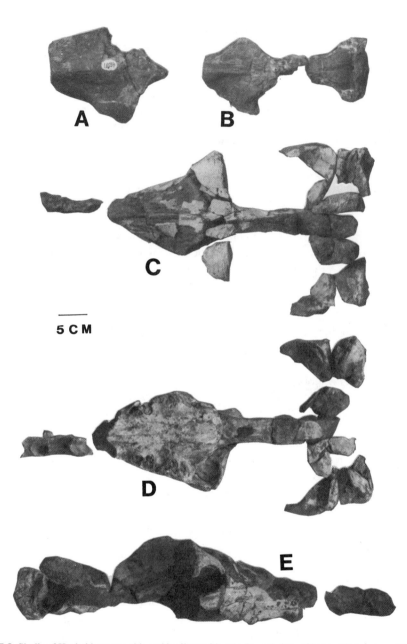

FIGURE 8. Skulls of Kachchh protocetids and basilosaurids. (A) Dorsal view of interorbital region of *Indocetus ramani* (LUVP 11034). (B) Dorsa view of nasal and cranial endocasts of *Indocetus ramani* (VPL 1017). (C–E) Skull of *Gaviacetus sahnii* (VPL 1021) in dorsal, ventral, and lateral views.

most or all of these specimens are referable to *Remingtonocetus* or *Rodhocetus* (Gingerich *et al.*, 1995a).

Description.—The new specimens cover some anatomical areas previously unknown in *Indocetus* (Sahni and Mishra, 1975; Bajpai *et al.*, 1996). VPL 1017 shows the endocast of the nasal cavity and paranasal sinuses (Fig. 8B). The cribriform plate is located between the orbits. Posterolateral to it is a small frontal or ethmoid sinus and lateral to the cribriform plate is a large maxillary sinus that extends anterior to the area of M^1. The case of the infraorbital canal is lateral to the maxillary sinus. The nasal cavity is narrow dorsally but widens inferiorly. None of the paranasal sinuses invades the supraorbital shelf of the frontal. The endocast of the cranial cavity of this specimen was published by Bajpai *et al.* (1996).

VPL 1023 adds significantly to our knowledge of the dental morphology of *Indocetus*, preserving the bases of the cheek teeth, though most of the crowns are missing (Fig. 5B). P^3 has three roots, the lingual of which is well posterior to the middle of the tooth. Given the size of this root, it probably had a protocone. Length of P^3 is approximately 36 mm. P^4 is also three rooted. Its lingual root is immediately lingual to the posterolabial root. The lingual root is larger than the posterior root and it is also larger than the posterior root of P^3. A protocone was probably present on P^4, and the preserved enamel near the base of this cusp was crenulated. There was a strong cingulum lingually. P^4 is 32 mm long, its width is 26 mm. The molars are conspicuously smaller than the premolars and decrease in size from M^1 to M^3. M^1 and M^2 molars have a large protocone lobe. In M^1, the protocone lobe is immediately posterior to the center of the tooth. In M^2, this lobe is farther posterior, at the level of the posterolabial root. M^1 preserves the enamel of the protocone, it is developed as a low crest, with a clear trigon basin labial to it. The paracone was probably large, as a shelf of enamel that formed its lingual wall crowds into the protocone. Only the anterior portion of M^3 is preserved, and it shows that the lingual root is located anteriorly, very close to the lingual root of M^2. There are deep embrasure pits between P^3 and P^4, and M^1 and M^2, but not between P^4 and M^1, or between M^2 and M^3. These pits are deeper than in the holotype (LUVP 11034, Fig. 8A). The length of M^1 is 23 mm, its width is approximately 23 mm. The length of M^2 is 22 mm, its width is 20 mm. The width of M^3 was approximately 18 mm.

Discussion.—At present the genus *Indocetus* is only known from Kachchh. Gingerich *et al.* (1993) referred material from the Sulaiman Range in Pakistan to *Indocetus*, but most or all of this material has since been withdrawn from *Indocetus* (Gingerich *et al.*, 1995a,b). On the other hand, Sulaiman *Rodhocetus* is similar to *Indocetus* in most respects, and Gingerich *et al.* (1995a) listed no characters differentiating between the two genera. Most of the characters listed by Gingerich *et al.* (1993) as distinguishing between these genera were postcranial and based on postcranial material now referred to *Remingtonocetus*.

Postcranial material for *Indocetus* cannot be identified with certainty, although it is likely that some undescribed postcranials from Kachchh pertain to this taxon. Based on size, it is virtually impossible that the vertebrae that are here referred to *Gaviacetus* (Fig. 9) pertain to *Indocetus*.

FIGURE 9. Referred caudal vertebrae of *Gaviacetus sahnii* (RUSB 2027). These are probably subsequent verte-brae, anterior toward left in B–D, toward right in A. A and C are lateral views, B and D are dorsal views. Dorsal is toward top of page in A and C.

Babiacetus Trivedy and Satsangi, 1984

Babiacetus mishrai new species

Figure 10

Holotype and Only Known Specimen.—RUSB 2512 (dentaries with left I_2, root of C, base of P_2, roots of P_3, P_4, M_1 fragment, M_{2-3}, and right I_2, C, P_{2-3} base of P_4 and M_1, and complete M_{2-3}) from Babia Hill, Kachchh.

Diagnosis.—Unlike *Babiacetus indicus, B. mishrai* has a shorter jaw, one-rooted P_1, lacks diastemata between P_{1-4}, has narrow cheek-teeth, and a small M_3.

Age and Distribution.—Middle Eocene (Lutetian), Harudi Formation, Kachchh, Gujarat.

Etymology.—Named for Dr. V. P. Mishra, who initiated comprehensive vertebrate paleon-tological work in Kachchh.

Description.—The total length of the mandible is 60 cm, the distance between the posteri-or side of M_3 and the condyle is 17 cm. It is not clear whether I_1 was present as the jaw is damaged here, but the mandible could only have housed a small tooth. Left and right mandible are fused in a solid synostosis that extends to the posterior root of P_1, and by a tapering syndesmosis from P_1 to P_3. The segment of the jaw anterior to P_1 is massive and does not taper, left and right mandible here are more than 4 cm wide as well as high. The dentaries diverge posterior to P_2, but they remain robust. The coro-noid process rises gradually above the last molar, reaching only 1 cm above the tip of M_3. The ascending ramus is strongly concave medially, and this concavity continues into the mandibular foramen. This foramen was large but does not cover the entire depth of the jaw. Its ventral part is not preserved. The mandibular condyle is rounded in outline, convex anteroposteriorly, and slightly convex mediolaterally. It faces dor-soposteriorly. The tips of all teeth are worn bluntly, and the broken surfaces of left C, P_{2-3}, and right M_1, are worn smooth, suggesting that these teeth broke well before the animal died. The I_2 (length: 27 mm, width: 15 mm) is pointed, slightly curved, and

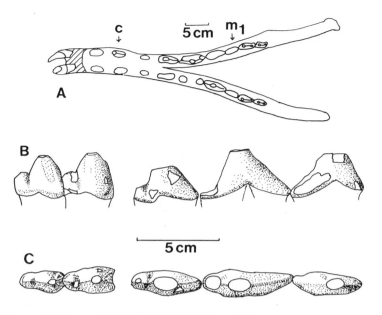

Figure 10. Holotype of *Babiacetus mishrai* (RUSB 2512). A, left and right dentary in occlusal view. B and C, labial and occlusal view of right M_3, M_2, P_4 (reversed from left), P_3, and P_2. Scale bar between B and C pertains to both of these.

oval in cross-section. It is followed by a diastema of 14 mm. The alveolus for I_3 is 25 mm, and the diastema following it is 13 mm. The C is similar to I_2 in shape; length at its base is 28 mm, and width is 19 mm. P_1 was one-rooted and smaller than the preserved incisors. Its alveolus is 19 mm long and the diastema following it is 15 mm. P_2 (length: 43 mm, width 15 mm) is double-rooted and has single, blunt cusp with crests extending anterior and posterior from it. The anterior crest descends steeply and reaches a strongly crenulated area near the base of the tooth. The posterior crest descends more gradually, giving the tooth an asymmetrical outline. There are weak anterior and posterior cingula. No diastemata occur between P_2–M_3.

P$_3$ (length: 54 mm, width: 15 mm) has a triangular crown, with anterior and posterior crests and weak cingula. The posterior crest slopes more steeply than the anterior crest. The anterior crest crosses a small elevation approximately halfway on the crown. The posterior crest has a number of crenulations and ends in an area where all enamel is missing, possibly due to wear. This area may have borne a cusp. P$_4$ (length: 44 mm, width: 16 mm) is similar to P$_3$, but smaller and stouter. The protoconid is high and robust, with a posterior crest that descends steeply to the talonid. One cusp apparently occurred on the talonid; it is worn in the RUSB 2512.

M$_2$ (length: 32 mm, width: 16 mm) and M$_3$ (length: 28, width: 14 mm) are similar in shape. The molar trigonid is high with a single, blunt cusp. A crest with fine crenulations extends anterolingually from it, ending in the cingulum. A second crest with fine crenulations is on the anterolabial side of the protoconid; it is weaker than the first crest. The anterolabial crest reaches the cingulum, but not the tip of the protoconid.

The area between the two crests is concave. A large, oval wear facet covers the area labial to the labial crest, reaching from the tip of the protoconid to just above the base of M_3, but no such facet occurs in M_2. The talonid bears a single cusp, the hypoconid. There is no talonid basin, and weak cingula are present labially and lingually.

Discussion.—RUSB 2512 is easily identified as a protocetid on the basis of the large mandibular foramen and lingually concave ascending ramus (unlike pakicetids and ambulocetids), the distinct trigonid and talonid, each with one cusp only (unlike remingtonocetids and basilosauroids). Among protocetids, a long, synostotic mandibular symphysis only occurs in *Babiacetus*.

Two mandibles for *Babiacetus indicus* have been described briefly by Gingerich et al. (1995), the holotype from Babia Hill (Geological Survey of India 19647) and a referred specimen from the Pakistani Sulaiman Range (GSP-UM 3005). Most of the teeth of *B. mishrai* are similar in length to those of *B. indicus*, but those of the former are consistently 20% narrower. Exceptions are P_4 and M_3, which are narrower as well as shorter (10% in P_4, 20% in M_3) in *B. mishrai*. A similar size difference occurs in the dentaries.

The holotype of *B. mishrai* is interesting because it is the only specimen of the genus that preserves the posterior part of the dentary. It shows that the coronoid process is lower than in many other early cetaceans. This implies that the temporal muscle had a poor lever arm and was a weak jaw closer. On the other hand, dental wear, fused symphysis, and overall robusticity suggest that powerful closing of the mouth was an important part of the feeding behavior of *Babiacetus*.

Family Basilosauridae Cope, 1868

Gaviacetus Gingerich *et al.*, 1995a

Revised Diagnosis.—Rostrum narrow, constricted strongly over P^2. *Gaviacetus* is unlike protocetids in that it lacks M^3. In this respect it is similar to other basilosaurids.

Discussion.—Gingerich et al. (1995a) proposed the genus *Gaviacetus* for a skull with an exceptionally narrow rostrum. The molar region of the teeth was incompletely preserved in their specimen, but left some doubt as to the presence of M^3. The specimens referred to *Gaviacetus* here show that the posterior extremity of the maxilla housed a very small M^2, and that the region of the maxilla housing M^3 was so narrow that this tooth must have been absent. The recovery of caudal vertebrae of a basilosaurid that are of an animal of the size of *Gaviacetus*, whereas no similar-sized cetaceans are known from Kachchh, further supports this interpretation. These vertebrae are basilosaurid because they are massive, elongate, and have short pedicles.

Gaviacetus sahnii new species

Figures 5C, 7C,D, 8C–E, and 9

Holotype.—VPL 1021 (fragmentary skull, with orbital regions, posterior palate, rostrum fragment, and fragmentary braincase, lacking all tooth-crowns except P^1; chocolate limestone of the Harudi Formation, Rato Nala, Kachchh).

Diagnosis.—*Gaviacetus sahnii* is similar in skull size to the type species *Gaviacetus razai*,
   but the former has larger teeth, between 150 and 200% larger in linear dimensions.
Referred Material.—RUSB 2024 (fragment of the interorbital region; chocolate limestone,
   Rato Nala, Kachchh), RUSB 2023 (partial braincase with supraoccipital and parietal;
   chocolate limestone of Rato Nala). RUSB 2027 (four and a half associated caudal ver-
   tebrae; gypsiferous shale of Babia Hill). All specimens are from the Harudi Formation
   of Kachchh.
Age and Distribution.—Middle Eocene (Lutetian), Harudi Formation, Kachchh, Gujarat.
Etymology.—Named for Professor Ashok Sahni, in recognition of his great contributions
   to the study of Eocene cetaceans from India.
Description.—The premaxilla–maxilla contact is preserved showing alveoli for I^{2-3} and
   C^1. The suture between premaxilla and maxilla is immediately anterior to the canine.
   The fragments preserving these elements (Fig. 5C, on left) also shows the sagittal plane
   and indicates that the palate is narrow. At I^3, the width of one side of the palate is only
   28 mm, and 15 mm of this accommodates the alveolus of I^3. This alveolus is 28 mm
   long. The dorsal side of this fragment shows that the nasal opening is caudal to the canine.
       Unilateral palatal width (including the parts housing the alveoli) at the posterior
   root of P^2 is 70 mm. The palate widens caudal to P^2, and at the anterior root of M^3 it
   is 144 mm (Fig. 5C). The greater palatine foramen opens medial to the anterior root of
   P^3 and a deep groove extends anteriorly from it. The left and right maxillary–palatal
   sutures reach in the midline as far anteriorly as the anterior root of M^1. From there,
   these sutures diverge and enter the sphenopalatine region medial to M^2. A shallow embrasure pit occurs between P^3 and P^4, and a deeper one between M^1 and M^2. There is
   no embrasure pit between P^4 and M^1. The zygomatic arch is slender and rugose for articulation with the posterior part of the maxilla. A small triangular piece of maxilla is
   missing from the specimen. It would have held most of the roots of M^2, but judging
   from the shape of the posterior palate and anterior zygoma, the maxilla caudal to M^2
   is so narrow that M^3 must be absent.
       P^1 is the only tooth for which a crown is preserved (Fig. 7C–D). There is a sin-
   gle, large cusp, worn apically and there are two roots. Blunt crests with crenulations
   extend anteriorly and posteriorly from this cusp. The anterior crest is steeper and extends toward the heavily crenulated anterior cingulum. The posterior crest reaches the
   heavily crenulated area over the posterior root. A very small tubercle is raised over the
   profile of the posterior crest. Medially, the enamel flares over a lingual expansion of
   the posterior root, but there is no protocone. The posterior part of the roots for P^2 is
   preserved. It indicates that the lingual expansion of this tooth is larger than that of P^1.
   A diastema of 20 mm separates P^2 and P^3, whereas more posterior teeth are juxtaposed.
   Only the roots are preserved for P^3–M^2. P^3 and P^4 are similar in basic outline. Both
   have three roots and their lingual roots are placed medial to the metacone. The protocone lobe is larger in P^4. The anteroposterior length of the roots of P^3 is 40 mm, its
   width 26 mm, and length of the roots of P^4 is 42 mm and width 30 mm.
       The shape of the roots of M^1 is unusual (Fig. 5C). The anterolabial root is the
   smallest, whereas the lingual and posterolabial root are similar in size. The lingual root
   is anteriorly displaced, and actually projects farther anteriorly than the anterolabial
   root. M^2 was much smaller than M^1. On the face, the infraorbital canal opens over the
   diastema between P^2 and P^3. The suture between lacrimal and maxilla is visible on the

side of the face and the lacrimal foramen opens inside the orbit. The lacrimal flares laterally immediately dorsal to this foramen.

The supraorbital process is massive and projects laterally (Fig. 8C–E). Its caudal side is flat and extends straight laterally. The temporal crest bends caudal to fuse into the sagittal crest halfway between distal extremity and sagittal plane. The interorbital region is narrow and its lateral side is dominated by a deep groove leading to the orbit. This groove probably extended anteriorly from the sphenorbital fissure, which is preserved caudally on the ventral aspect of the endocast of the skull.

Fragments of the braincase are preserved. Its greatest width is at the zygoma where it is approximately 160 mm wide (unilaterally). In dorsal aspect, the endocast of the cerebrum is visible. It shows a pronounced impression for the dorsal sagittal sinus and the poorly preserved cast for the cerebellum reaches farther dorsal than that for the cerebrum. Only fragments of the lateral wall of the skull remain. The zygoma flares laterally leaving a broad temporal fossa. The postglenoid process projects ventrally and medial to it is a small (approximately 3 mm) venous foramen that could be interpreted as the postglenoid foramen. The caudal aspect of the postglenoid process is concave and forms the anterior wall of the external auditory meatus. Caudal to this meatus is a large irregular process made up by three separate elements. The mastoid forms the anterior part of the process. Its medial contacts are not preserved, but it does form the superior and caudal walls of the meatus. A deep vertical depression occurs in the meatus laterally. The exoccipital forms the caudal part of the process. Its caudal aspect is flat, but rugose for the origin of the neck muscles. Petrosal and tympanic are not preserved.

Few details are preserved on the ventral aspect of the braincase. The lateral part of the basioccipital slopes gradually ventral toward the petrosal. It is somewhat rugose here and its caudal part is grooved, probably for one of the structures leaving the jugular foramen. The basioccipital did not articulate with the petrosal. Anterior to the ear region is a convex process. Its anterolateral part is smooth and provided the origin of the lateral pterygoid muscle. Caudally, this process is rounded and did not articulate with the petrosal. A large canal (approximately 8 mm midline) leaves the braincase at the posteromedial side of this process and extends laterorostrally. It represents the canal for the mandibular branch of the trigeminal nerve.

RUSB 2027 consists of four and a half caudal vertebrae that are referred to *Gaviacetus* here (Fig. 9). The vertebral bodies are elongate, with short processes, and neural arches that are restricted to the anterior half of the body. The two most anterior vertebrae have a single transverse process, protruding from the middle of the vertebra and at their base approximately half as long (anteroposteriorly) as the length of the body. The remaining specimens have a transverse process that is divided into an anterior and a posterior process. It is not clear if anterior and posterior processes meet distally on the more anterior vertebrae, as the extremity is only preserved on one vertebra. In this vertebra the anterior and posterior transverse processes do not meet distally. Fragments of neural arches are preserved on two vertebrae. The laminae are short anteroposteriorly, only bridging part of the neural canal. The vertebral bodies are much narrower in the center than at their extremities. These constrictions become more prominent in the more posterior vertebrae. Hemal processes are clearly visible on the caudal side of each vertebra. The cortical surface of the vertebrae is poorly preserved, precluding exact measurements. Vertebral length for all specimens varies between 145 and 155 mm, width of the anterior articular surface between 95 and 100 mm, and height (anteriorly) between 85 and 95 mm.

Discussion.—The holotype of *G. sahnii* has a high braincase and a well-developed lingual root of the upper molars, indicating that it is not a remingtonocetid. M^3 is probably absent in this specimen, indicating that it is not an ambulocetid or protocetid. *Gaviacetus* is by far the largest cetacean known from Kachchh and the referred vertebrae match only *Gaviacetus* in size, and the simplest interpretation at present is that they pertain to this taxon. Two protocetids are known from Kachchh, *Indocetus* and *Babiacetus*. The orbital region of *Indocetus* is low and the overall size of both is much smaller than that of the new skull. Both of these protocetids retain a large M^3. *Gaviacetus* was named by Gingerich *et al.* (1995a) for a very narrow-snouted cetacean from central Pakistan. Their specimen lacked the caudal part of the maxilla, and it was not clear whether M^3 was present. The Kachchh skull is similar to their specimen, but allows reinterpretation of the familial attribution based on the realization that it is similar to basilosaurids and different from protocetids in the absence of M^3. The referred vertebrae confirm this diagnosis, as they resemble basilosaurids and no other Eocene cetacean family in the proportions and profile of their body and the short pedicles.

Cetacea indet.

Figure 7E,F

Referred Specimen.—RUSB 2002 (premolar, possibly dP; Subathu Formation, Dharampur area, Simla Hills, Himachal Pradesh). The specimen was found at a roadcut on the Kalka-Shimla road, just south of the village of Dharampur. The tooth was recovered in a dark gray bed of mudstone that is several meters thick.

Age.—Lutetian (middle Eocene), as discussed by Sahni (1983).

Description.—Two-rooted tooth, with a single high cusp. The anterior part of the tooth is not preserved, but the preserved parts of the tooth suggest that anterior and posterior slopes were more or less symmetrical. There is no talonid, but a small cusp occurs on the crest leading from the protoconid to the caudal base of the tooth. A cingulum is present along the entire preserved perimeter of the tooth, and is weaker on one side (presumed labial). Overall, the tooth is mostly symmetrical around its anteroposterior axis. The enamel–dentin junction is far more occlusal near the center of the tooth than near its anterior and posterior extremity. The deep incision that this arrangement causes suggests that this tooth is a deciduous premolar.

Discussion.—This specimen is the only cetacean known from the Dharampur area of the Simla Hills. Sahni (1979, 1983), Sahni *et al.* (1980), and Kumar *et al.* (1994) discussed some outcrops of the Subathu Formation in the Dharampur area. Kumar and Loyal (1987) presented a map of this area that included the outcrop where the tooth was found.

This tooth is not diagnostic of any family of cetaceans, but shows that cetaceans were present in this part of the Subathu Formation. It is the first cetacean from the upper part of this formation found in clearly marine sediments (as evidenced by the recovery of the pelvis of a marine turtle from the same locality). It shows that further collecting effort in the Subathu Area will almost certainly yield a rich cetacean fauna.

3. Summary and Conclusions

The Harudi Formation fauna is the classical cetacean fauna from South Asia, and shows a remarkable diversity of early marine cetaceans. Described from the Harudi Formation are the protocetids *Indocetus ramani, Babiacetus indicus* and *B. mishrai*, the remingtonocetids *Remingtonocetus sloani, R. harudiensis, Andrewsiphius kutchensis*, and *A. minor*, and the basilosaurid *Gaviacetus sahnii*. This faunal composition is different from the poorly known cetacean faunas of the Subathu Formation of northwestern India, which at this point includes a single named taxon, *Pakicetus attocki*. The Subathu fauna is older than the Harudi Fauna. No other regions of India have yielded Eocene cetaceans.

The cetacean faunas from the Domanda and Drazinda Formations of the Sulaiman Range are similar in age to those of the Harudi Formation of Kachchh in the presence of *Remingtonocetus, Babiacetus*, and *Gaviacetus*. Kachchh *Andrewsiphius* has not been found in Pakistan and Sulaiman *Takracetus* is not known from India. Sulaiman *Dalanistes* appears to be similar to *Remingtonocetus*, and Gingerich *et al.* (1995a) could not identify differences between *Indocetus* and *Rodhocetus*.

The present chapter does document the recent advances in our understanding of Eocene cetaceans from India. It is largely based on the collecting effort of the senior author (Bajpai, 1990), but shows that further collecting effort in western India will almost certainly lead to the recovery of new taxa as well as more complete specimens of already known cetaceans. It is not unlikely that the stratigraphic ranges of Indian cetaceans can also be expanded: A 1-hour survey of the Ypresian Naredi Formation in Kachchh by the authors in 1996 led to the recovery of several marine vertebrates.

Acknowledgments

We thank Ashok Sahni and Philip Gingerich, the dissertation advisers of the first and second author, respectively, for their help and encouragement in the study of fossil mammals and for access to collections under their care. We also thank the curators of the fossil whales at the U.S. National Museum and the University of Michigan for access to collections.

References

Bajpai, S. 1990. Geology and paleontology of some late Cretaceous and middle Eocene sequences of Kachchh, Gujarat, western India. Ph.D. dissertation, Punjab University, Chandigarh, India, 287 pp.
Bajpai, S., Srivastava, S., and Jolly A. 1989. Sirenian–moerithere dichotomy: some evidence from the middle Eocene of Kachchh, western India. *Curr. Sci.* **58**:304–306.
Bajpai, S., Thewissen, J.G.M., and Sahni, A. 1996. *Indocetus* (Cetacea, Mammalia) endocasts from Kachchh (India). *J. Vertebr. Paleontol.* **16**:582–584.
Biswas, S. K. 1992. Tertiary stratigraphy of Kutch. *J. Palaeontol. Soc. India* **37**:1–29.
Biswas, S. K., and Deshpande, S. V. 1970. Geological and tectonic maps of Kutch. *Oil Nat. Gas Comm. (India), Bull.* **7**:115–120.
Gingerich, P. D., Raza, S. M., Arif, M., Anwar, M., and Zhou, X. 1993. Partial skeletons of *Indocetus ramani* (Mammalia, Cetacea) from the lower middle Eocene Domanda Shale in the Sulaiman Range of Punjab (Pakistan). *Contrib. Mus. Paleontol. Univ. Michigan* **28**(16):393–416.

Gingerich, P. D., Arif, M., and Clyde, W. C. 1995a. New archaeocetes (Mammalia, Cetacea) from the middle Eocene Domanda Formation of the Sulaiman Range, Punjab (Pakistan). *Contrib. Mus. Paleontol. Univ. Michigan* **29**(11):291–330.

Gingerich, P. D., Arif, M., Bhatti, M. A., Raza, H. A., and Raza, S. M. 1995b. *Protosiren* and *Babiacetus* (Mammalia, Sirenia and Cetacea) from the middle Eocene Drazinda Formation, Sulaiman Range, Punjab (Pakistan). *Contrib. Mus. Paleontol. Univ. Michigan* **29**(12):331–357.

Kellogg, R. 1936. A review of the Archaeoceti. *Carnegie Inst. Washington Publ.* **482**:1–366.

Kumar, K., and Loyal, R. S. 1987. Eocene ichthyofauna from the Subathu Formation, northwestern Himalaya, India. *J. Palaeontol. Soc. India* **32**:60–84.

Kumar, K., and Sahni, A. 1985. Eocene mammals from the upper Subathu Group, Kashmir Himalaya, India. *J. Vertebr. Paleontol.* **5**:153–168.

Kumar, K., and Sahni, A. 1986. *Remingtonocetus harudiensis*, new combination, a middle Eocene archaeocete (Mammalia, Cetacea) from western Kutch, India. *J. Vertebr. Paleontol.* **6**:326–349.

Kumar, K., Loyal, R. S., and Srivastava, R. 1994. Middle Eocene rodents from the Subathu Formation, Solan District, Himachal Pradesh (northwest Himalaya), in: *14th Indian Colloquium on Micropalaeontology and Stratigraphy, Madras*, Abstract.

Sahni, A. 1979. An Eocene mammal from the Subathu–Dagshai transition zone, Dharampur, Simla Hills. *Bull. Indian Geol. Assoc.* **12**:259–262.

Sahni, A. 1983. Eocene vertebrate fauna from the Salt Range (Pakistan), Kashmir and Simla Himalayas, and China. *Himalayan Geol.* **11**:1–17.

Sahni, A., and Kumar, K. 1981. Lower Eocene Sirenia, *Ishatherium subathuensis* gen. et sp. nov. from the type area, Subathu Formation, Subathu, Simla Himalayas, H.P. *J. Palaeontol. Soc. India* **23**:132–135.

Sahni, A., and Mishra, V. P. 1972. A new species of *Protocetus* (Cetacea) from the middle Eocene of Kutch, western India. *Palaeontology* **15**:490–495.

Sahni, A., and Mishra, V. P. 1975. Lower Tertiary vertebrates from western India. *Palaeontol. Soc. India Monogr.* **3**:1–48.

Sahni, A., Kumar, K., and Tiwari, B. N. 1980. Lower Eocene marine mammal (Sirenia) from Dharampur, Simla Himalayas, H.P. *Curr. Sci.* **49**:270–271.

Satsangi, P. P., and Mukhopadhyay, P. K. 1975. New marine Eocene vertebrates from Kutch. *J. Geol. Soc. India* **16**:84–86.

Stromer, E. 1908. Die Archaeoceti des Ägyptischen Eozäns. *Beitr. Paläontol. Geol. Österreich-Ungarns Orients* **21**:106–177.

Tandon, K. K. 1971. On the discovery of mammalian and reptilian remains from the middle Eocene rocks of S.W. Kutch, India. *Curr. Sci.* **40**:436–437.

Thewissen, J.G.M., and Hussain, S. T. 1998. Systematic review of the Pakicetidae, early and middle Eocene Cetacea (Mammalia) from Pakistan and India. *Bull. Carnegie Mus. Nat. His.* **34**:220–238.

Trivedy, A. N., and Satsangi, P. P. 1984. A new archaeocete (whale) from the Eocene of India. *27th Int. Geol. Congr. Abstr.* **1**:322–323.

CHAPTER 8

Postcranial Osteology of the North American Middle Eocene Protocetid *Georgiacetus*

RICHARD C. HULBERT, JR.

1. Introduction

Archaeocete whales are a paraphyletic assemblage of species that represent the evolutionary intermediaries between modern, fully marine cetaceans (Odontoceti and Mysticeti) and their terrestrial ancestors, the Mesonychia (Van Valen, 1966; Fordyce and Barnes, 1994; Thewissen, 1994). For over 80 years the holotype specimen of *Protocetus atavus* from the middle Eocene of Egypt was the oldest known postcranial skeleton of a cetacean. Necessarily it figured prominently in studies on the early evolution of the postcranial skeleton of cetaceans (e.g., Kellogg, 1936). Several critical specimens collected over the past two decades have greatly improved this record. Notable among these are partial postcranial skeletons of *Ambulocetus natans* (Thewissen *et al.*, 1994, 1996); *Rodhocetus kasrani* (Gingerich *et al.*, 1994), *Remingtonocetus harudiensis* (Gingerich *et al.*, 1993; listed therein as *Indocetus ramani* but partly reidentified by Gingerich *et al.*, 1995), and *Georgiacetus vogtlensis* (Hulbert *et al.*, 1998). A fifth specimen, the Habib Rahi whale, remains undescribed (Gingerich, 1991). All of these are more complete than the holotype of *P. atavus* and include some appendicular elements. The postcrania of *Ambulocetus* and *Remingtonocetus* have been thoroughly described and illustrated (Gingerich *et al.*, 1993, 1995; Thewissen *et al.*, 1996). The purpose of this report is to describe the postcranial remains of *G. vogtlensis* and compare them with those of other archaeocetes.

    Georgiacetus vogtlensis was recently named by Hulbert *et al.* (1998) based on three specimens collected from the Upper Coastal Plain of Georgia. Its postcrania were only perfunctorily described by Hulbert *et al.* (1998). *G. vogtlensis* is the first formally described protocetid cetacean from North America; all other species originated from either Africa or the Indo-Pakistani region. Several other North American protocetids are under study by oth-

RICHARD C. HULBERT, JR. • Department of Geology and Geography, Georgia Southern University, Statesboro, Georgia 30460-8149.

The Emergence of Whales, edited by Thewissen. Plenum Press, New York, 1998.

er workers, however (Geisler *et al.*, 1996; McLeod and Barnes, 1996; Uhen, 1996b). Stratigraphic and biochronologic evidence provides a late middle Eocene age for *G. vogtlensis*, approximately 40 to 41 Ma (Hulbert *et al.*, 1998), making it one of the youngest known protocetids. Phylogenetic analysis of *Georgiacetus* has determined that it and the African protocetids *Pappocetus* and *Eocetus* are more closely related to basilosaurid archaeocetes than they are to *Protocetus* or older whales (Hulbert, 1994, 1996). As both *Pappocetus* and *Eocetus* are very incompletely known (Kellogg, 1936; Barnes and Mitchell, 1978, Uhen, in press), *Georgiacetus* is critical in understanding the transition from protocetid- to basilosaurid-grade cetaceans.

The most complete of the three specimens of *G. vogtlensis* is Georgia Southern Museum (GSM) 350, the holotype. It is an associated partial skeleton that includes a nearly complete skull, much of the left dentary, 23 vertebra, 12 ribs, and both innominates (Fig. 1). Among described protocetids it is surpassed in completeness only by the holotype of *Rodhocetus kasrani* (Gingerich *et al.*, 1994). Two much less complete specimens of *G. vogtlensis* are known, GSM 351 and GSM 352. GSM 351 includes three thoracic vertebrae and portions of at least four ribs. GSM 352 consists of a single lumbar vertebra. The majority of the elements of GSM 350 and 351 are relatively well preserved with little postmortem destruction other than cracking and loss of surficial bone at the ends of spines and processes. A few were crushed by adjacent elements as the sediments compacted. Several vertebra of GSM 350 were damaged as the specimen was initially exposed by construction workers and others were probably destroyed completely.

2. Materials and Methods

This study is based on direct observations of specimens housed in the Georgia Southern Museum, Statesboro (*Georgiacetus*) and the Staatliches Museum für Naturkunde, Stuttgart (*Protocetus*). These are augmented by descriptions in the literature, most notably the studies of Fraas (1904), Kellogg (1936), Gingerich *et al.* (1993, 1994, 1995), and Thewissen *et al.* (1996). Terminology and measurements for vertebrae follow Zhou *et al.* (1992). The abbreviations C, T, L, S, and Ca are used for cervical, thoracic, lumbar, sacral, and caudal vertebrae, respectively, and are followed by a numeral representing the position in the vertebral series. For example, T4 refers to a fourth thoracic vertebra. As discussed by Hulbert *et al.* (1998), the term *sacral* is used for protocetid vertebrae that are homologous with those forming the sacrum in mesonychids and remingtonocetids, even if they no longer directly articulate with the ilium. Sacral vertebrae retain a distinctive morphology different from the lumbars and caudals, unlike the condition in basilosaurid, mysticete, and odontocete cetaceans.

Because the vertebrae of *Georgiacetus* were not preserved in articulation (Fig. 1), their original sequencing was reconstructed based on "best fit" between zygapophyses and centra, quantitative trends, and comparisons with identified vertebrae in other archaeocetes (Fraas, 1904; Kellogg, 1936; Gingerich, 1991; Gingerich *et al.*, 1993). Even after combining the vertebrae from GSM 350, 351, and 352, it is evident that some are missing, i.e., they do not form a complete, continuous series. The exact locations of some of the breaks in the sequence are obvious, but others are less clear because of poor preservation and/or crushing. There is also the question of the number of vertebrae missing in each break. This is es-

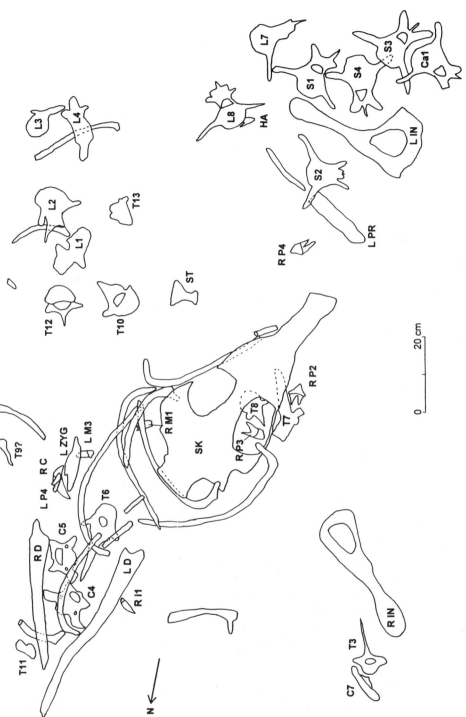

FIGURE 1. Sketch showing in plan view the relative positions of the recovered skeletal elements of GSM 350, holotype of *Georgiacetus vogtlensis*, as they were preserved. Unlabeled elements are ribs or rib fragments. Abbreviations: R, right; L, left; SK, skull; D, dentary; PR, premaxilla; ZYG, zygomatic arch (jugal); IN, innominate; HA, hemal arch; ST, sternebra; i, lower incisor; C, upper canine; P, upper premolar; M, upper molar. Vertebrae are identified as to their most likely position along the column using the abbreviations C for cervical, T for thoracic, L for lumbar, S for sacral, and Ca for caudal plus an Arabic numeral (e.g., T6 is the sixth thoracic vertebra). Construction work destroyed any bones to the north or east of those shown here, so the specimen was originally even more complete.

pecially critical in the thoracic, lumbar, and sacral series where the number of vertebrae can vary between taxa. Although the relative order of the vertebrae has been reconstructed with a very high degree of confidence, their exact identifications are subjective and may change as additional associated vertebral columns of protocetids are described.

For vertebrae, the term *length* is used to describe measurements taken in the antero-posterior direction, *width* the transverse, mediolateral direction, and *height* the dorsoventral direction. For ribs, *breadth* refers to the anteroposterior dimension and *width* the mediolateral dimension.

3. Postcrania of *Georgiacetus vogtlensis*

3.1. Vertebrae

3.1.1. Cervical Vertebrae

Three cervical vertebrae are present in GSM 350: a nearly complete, well-preserved C4; a badly crushed C5; and a well-preserved C7 whose transverse processes and centrum were damaged during excavation. The identification of these vertebrae is considered relatively secure based on their close similarity with cervicals of *Protocetus atavus*.

The C4 centrum is relatively short and has obliquely set (anterodorsal–posteroventral) articular surfaces relative to the craniocaudal axis (Figs. 2A,B and 3B). The anterior articular surface of the centrum is slightly taller than wide (Table I), producing a relatively circular outline truncated dorsally by the relatively flat border of the neural canal. A strong, central transverse groove divides the anterior surface of the centrum into distinct dorsal and ventral halves. The dorsal half has two, low, convex projections separated by a slight medial groove, while the ventral half has a transverse convex ridge that lacks a medial groove and does not project as far as the dorsal convexities. The posterior articular surface of the centrum is flatter, with a concave center and raised margins, and is hexagonal in outline (Fig. 2B). The ventral surface of the centrum consists of a raised, very weakly keeled medial region flanked laterally by large concavities. The surface of the medial region is rugose, bears several small nutrient foramina, and there is a moderately large hypapophysis at the posterior end. The dorsal surface of the centrum, which forms the ventral wall of the neural canal, is flat except for a pair of large, deep pits separated by a 6-mm-wide medial bridge. Each of the pits has a small foramen at its base. The neural canal is oval, wider than it is tall (Table I), and is bounded laterally by robust pedicles that are approximately as thick transversely (19.7 mm) as they are long anteroposteriorly (20.0 mm). The large, oval zygapophyses overlap each other, as the posterior third of the prezygapophysis lies dorsal to the anterior third of the postzygapophysis. The prezygapophysis faces dorsally, medially, and anteriorly; its articular surface is slightly concave and measures ca. 33 by 24 mm. The postzygapophysis faces ventrally, laterally, and posteriorly; its articular surface is slightly convex and measures ca. 30 by 22 mm. The laminae meet medially to the zygapophyses to form the anteroposteriorly narrow roof of the neural arch. Its dorsal surface has a small, central, posterior depression located behind a very small neural spine. The neural spine is not bilaterally symmetrical as is the rest of the vertebra, but is located primarily on the left side of the neural arch. The transverse process extends out straight laterally ca. 47.5 mm from

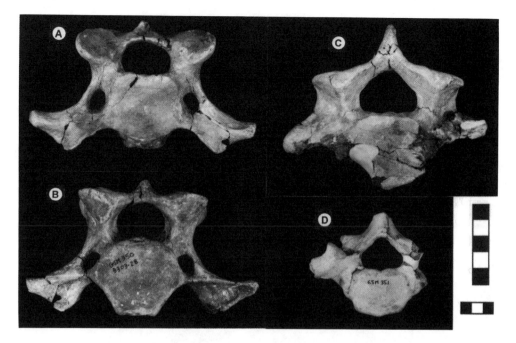

FIGURE 2. Vertebrae of *Georgiacetus vogtlensis*. (A–C) GSM 350. (A) Anterior and (B) posterior views of fourth cervical vertebra. (C) Posterior view of seventh cervical vertebra. (D) Anterior view of first thoracic vertebra of GSM 351. Scale bars in cm; upper bar for A–C, lower bar for D.

the middle of the centrum (in posterior view, Fig. 2B), with a blunt, knoblike terminus and an anteroposterior length of ca. 12 mm. Total width of C4, across both transverse processes, is 146 mm, while the total height from the top of the neural spine to the base of the centrum is 84.5 mm. The anteroventral edge of the transverse process has a thin, curving flange whose edge is continuous with the anteroventral rim of the centrum. In anterior view, the edge of this flange curves first up, then down, and finally up again to meet the terminus of the transverse process. In lateral or ventral view it curves first anteriorly, then posteriorly, and finally straight laterally. This produces a complex, curving surface that is strongly concave on its anteroventral and posterodorsal sides. The deepest portion of the posterodorsal concavity contains a series of pits or grooves separated by dorsoventrally oriented ridges, while the anteroventral concavity is smooth. A thin, 8- to 9-mm-wide bar extends ventrolaterally from the pedicle just anterior to the postzygapophysis to join with the dorsal surface of the transverse process, forming an oval vertebrarterial canal with dimensions 12.3 by 11.4 mm posteriorly. There is a deep pit on the wall of the centrum that forms the medial side of the vertebrarterial canal. Its anterior margin is steeper than the posterior, and it has a small foramen at its bottom. Another small (pin-sized) foramen is located ca. 3 mm dorsomedially to the anterior opening of the vertebrarterial canal, immediately adjacent to the articular surface of the centrum.

The morphology of C5 is generally similar to that of C4, with the following differences: The dual conical projections on the dorsal half of the anterior articular surface of the cen-

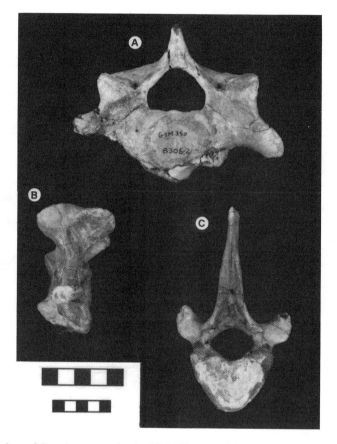

FIGURE 3. Vertebrae of *Georgiacetus vogtlensis*, GSM 350. (A) Anterior view of seventh cervical vertebra. (B) Left lateral view of fourth cervical vertebra. (C) Posterior view of seventh thoracic vertebra. Scale bars in cm; upper bar for A and B, lower bar for C.

trum are more convex and stronger; there is neither a keel nor a hypapophysis on the ventral surface of the centrum; the posterior articular surface of the centrum is oval and not hexagonal; the transverse process is longer, ca. 50 mm; the anteroventral flange on the transverse process is dorsoventrally thicker and bears a distinct anterior-projecting process; the bar forming the dorsolateral border of the vertebrarterial canal is wider (ca. 10 mm); and the small neural spine is located more posteriorly on the neural arch, not in front of a depression.

C7 differs considerably from C4 and C5 (Figs. 2C and 3A). The body of its centrum is short, with obliquely set articular surfaces. The anterior articular surface is mostly flat with a small central depression and two very low, widely separated dorsal convexities. There are two small nutrient foramina just lateral to the anterior articular surface. The posterior articular surface is slightly concave with a medially located dorsal protuberance. Most of the ventral surface of the centrum is missing, but there is a broad, low keel and no hypapophysis. The dorsal surface of the centrum bears two pits, as does C4, but they are shal-

Table I. Measurements (in mm) of Vertebrae of the Middle Eocene Archaeocete Whale *Georgiacetus vogtlensis* from Burke County, Georgia[a]

	Centrum					Neural canal				Pedicle L		Neural spine height
	L	AW	AH	PW	PH	AW	AH	PW	PH	Left	Right	
GSM 350												
C4	28.1	46.3	42.8	53.2	a46	30.9	21.2	37.5	21.3	20.3	20.0	15.1
C5	28.6	a42	41.6	—	—	—	—	—	—	21.5	—	—
C7	28.5	45.6	—	—	45.8	36.4	26.6	38.1	34.0	—	—	23.4
T3	37.2	54.2	40.5	65.3	a42	a30	26.4	a35	26.4	23.0	—	121+
T6	40.5	54.1	44.4	62.6	a47	30.0	29.3	38.8	30.6	24.4	24.4	a138
T7	40.4	54.0	46.2	64.8	46.6	34.0	32.4	38.6	29.4	25.3	23.8	124+
T8	40.7	57.1	43.3	65.9	42.8	35.6	33.0	37.9	28.6	25.8	24.7	85+
T10	—	—	—	a69	45.1	a36	31.1	43.3	29.7	—	—	—
T12	47.8	68.2	43.9	73.9	47.7	38.4	27.0	38.5	29.5	35.3	33.1	68+
T13	48.3	a76	47.1	72.9	52.5	a38	29.2	—	—	—	38.5	—
L1	50.3	78.4	52.4	78.8	a52.5	38.1	27.2	41.0	27.7	40.5	41.4	69+
L2	54.8	—	—	a77	—	40.5	26.5	42.8	26.3	39.9	39.4	a80
L3	—	a76	53.3	—	—	—	—	—	—	—	—	—
L4	59.1	78.0	—	79.6	a55	41.7	27.3	—	a27	41.7	—	a89
L7	71.2	81.5	—	77.9	—	a41	21.8	44.1	26.9	39.2	39.7	45+
L8	72.3	a78	58.2	a83	65.0	43.2	23.1	41.3	29.0	41.1	—	—
S1	—	74.8	58.8	a72	a67	45.3	21.4	37.5	—	43.2	—	49+
S2	73.5	77.2	67.2	a76	—	39.8	25.7	37.0	a29	46.8	46.1	—
S3	77.3	a74	69.9	78.8	—	36.4	21.3	33.5	28.4	46.7	47.0	38+
S4	78.6	72.4	a70	73.1	a77	33.4	21.7	30.6	25.4	46.1	46.2	37+
Ca1	80.0	72.4	77.5	79.0	a80	33.8	21.0	29.2	27.4	47.8	46.9	—
GSM 351												
T1	34.2	50.7	40.3	a53	—	37.1	31.9	a38	30.8	18.1	—	—
T3	38.0	57.5	41.8	64.3	45.1	32.1	25.1	35.8	26.0	25.1	25.4	120+
T9	37.9	56.8	42.7	68.5	42.1	36.4	—	38.8	31.0	30.7	28.5	—
GSM 352												
L4?	a62	a80	a55	a82	a58	a43	26.8	—	—	a43	—	a90

[a]An "a" preceding a value indicates an approximation or estimate; a "+" after a value indicates that there was breakage and that the actual value is larger than that listed. AW, anterior width; AH, anterior height; L, length; PW, posterior width; PH, posterior height. Neural spine height measured on posterior side.

lower, separated by a 6.7-mm-wide bridge, and each contains two foramina. The neural canal is large, more triangular than oval, as the laminae extend dorsomedially from the pedicles. The ventral surface of the laminae that faces posteriorly is rugose. The flat to slightly convex, oval prezygapophysis faces dorsomedially and slightly anteriorly, with a concave region immediately to its anteromedial surface. The flat, oval postzygapophysis faces ventrolaterally and very slightly posteriorly (it is almost not visible in posterior view, Fig. 2C). There is a small amount of dorsoventral overlap between the higher prezygapophysis and the lower postzygapophysis. The short neural spine has a broad base located medially to the postzygapophyses and in the middle of the neural arch. Although small, it is significantly larger than the neural spines of C4 and C5. The pedicles are posteriorly located on the centrum, while those of C4 and C5 are more centrally positioned. Both transverse processes of C7 are broken, but are apparently shorter than those of C4 and C5. The process from the pedicle is not a thin bar, but instead is an oblique, anteroposteriorly flat plate whose lateral surface is continuous with the anterior margin of the prezygapophysis. This plate thickens laterally, and the surface of its posterodorsal face is irregular with gentle concave and convex regions. Its anteroventral face is smooth. Breakage prohibits determination of the presence or absence of the vertebrarterial canal. The posterodorsal regions of the centrum that would have housed the capitular foveae for the first ribs are both broken.

3.1.2. Thoracic Vertebrae

Fairly complete representatives are known for 9 of the most likely 13 thoracic vertebrae of *Georgiacetus*. GSM 351 includes T1, T3, and T9, whereas the more complete GSM 350 contains T3, T6, T7, T8, T10, T12, and T13. GSM 350 also includes portions of the left pedicle with the postzygapophysis of what are most likely T9 and T11. Centrum length increases posteriorly through the thoracic series (Table I); the length of T13 is about 40% longer than T1. However, the increase in centrum length is very slight between T3 and T9, with most of the change occurring before or after this group. The centra of the thoracics are acoelous to very slightly opisthocoelous, with their articular surfaces parallel to the transverse plane of the body, not oblique. Pre- and postzygapophyses are relatively large and well developed. Their orientation changes from that of primarily dorsoventrally flat in the anterior thoracics to much more vertical in the posterior thoracics, with T11 being the diaphragmatic vertebra. The neural spines on T1 through T10 project posteriorly as well as dorsally, are relatively short anteroposteriorly, and are very long (> 100 mm). The neural spines on T11 through T13 are shorter in height, longer in anteroposterior length, and extend straight dorsally.

In some features the vertebra identified as T1 differs from the other anterior thoracics and more resembles a posterior cervical (Fig. 2D). The centrum is short, shortest of the known thoracics, but is 20% longer than that of the preceding C7 (Table I). The anterior articular surface of its centrum is generally flat with gentle undulations including a slight central depression and a pair of dorsal convexities. The latter are less pronounced than those of the cervicals. The shape of the anterior articular surface is more rectangular with rounded corners than oval. The ventral portion and the lateral rims of the posterior articular surface of the centrum are broken, but its shape is a transversely wide oval, and its surface is centrally concave with a raised rim. The ventral surface of the T1 centrum is flatter and less curved than in the other thoracics, with a distinct, narrow, low keel, and a pair of nutrient foramina on both the left and right sides. The dorsal surface of its centrum is flat with paired nutrient foramina of which the one on the right side is much smaller and located more posterior than the one on the left. The very shallow capitular foveae are located on the lateral rims of the anterior and posterior articular surfaces of the centrum and are not well preserved. The anterior fovea for the first rib is small and located at about the midline of the centrum, slightly more ventral than the larger posterior capitular fovea. Short, broad pedicles bear the large, flat, subcircular zygapophyses. Their orientations are similar to those of C7, but there is no dorsoventral overlap between them. The prezygapophysis faces dorsomedially, is about 21 mm long and 20 mm wide, and is separated anteriorly from the dorsal surface of the neural arch by a distinct pit. The postzygapophysis faces ventrolaterally, and is about 17 mm long and 19 mm wide. A second, much smaller (7.8 by 6.5 mm) articular surface projects from the posteroventral margin of the pedicle, facing ventromedially and posteriorly, and projecting farther than the centrum. The neural canal is large, slightly wider than tall, triangular in shape with rounded corners, and roofed dorsally by narrow laminae projecting dorsomedially from the pedicles. The neural spine is broken close to its base, but appears to be slightly smaller than that of T3 but much larger than that of C7. The transverse process extends laterally and slightly posteriorly from the pedicle, and ending in a large (28.5 by 19 mm), concave, laterally facing fovea for the tuberculum of the first rib.

The orientation and size of the prezygapophyses of the remaining thoracic vertebrae from GSM 351 and 350 preclude their articulation with the postzygapophyses of the vertebra identified as T1. Therefore, T2 is not represented among the available specimens. The anteriormost of the remaining thoracics is assumed to be T3. Both GSM 350 and 351 have a vertebra identified as a T3. That of GSM 351 is more complete and better preserved, so the following description is primarily based on this specimen. The anterior articular surface of the centrum is wider than tall (Table I, Fig. 4A), with rounded lateral and ventral margins and a fairly straight dorsal margin. Its surface is generally flat with a slight central depression and dorsal and ventral convexities. The posterior articular surface of the centrum has a similar quasioval shape and is even flatter, but with a distinct central concavity. The capitular foveae are located at the dorsolateral margins of both articular surfaces; the posterior fovea is the larger of the two. Their slightly concave articular surfaces are not well separated from those of the articular ends of the centrum. The ventral surface of the centrum is smoothly rounded with no concave regions, a slight keel that is strongest at the anterior end, and four nutrient foramina. Three of these are arranged along a line extending away from the midline and in the anterior half of the centrum; the fourth is on the posterior side and about halfway up the centrum. The dorsal surface of the centrum is slightly concave with an irregular surface and several small foramina. The squat pedicles are located at the anterior end of the centrum. The prezygapophysis juts anteriorly from the pedicle past the anterior articular surface of the centrum, and faces primarily dorsally, but also slightly anteriorly and laterally. Its articular surface is about as long (19–20 mm) as it is wide (ca. 21 mm), and very slightly arched or convex anteroposteriorly. The postzyg-apophysis is located on the ventrolateral margin of the neural arch, faces primarily ventrally but also slightly posteriorly and medially, and its surface is very slightly concave anteroposteriorly. It projects beyond the centrum posteriorly. The neural canal is oval, wider than tall (Table I). The tall neural spine extends posterodorsally from the neural arch (Fig. 4B). It is triangular in cross section at its base with paired anterolateral sides made of smooth bone

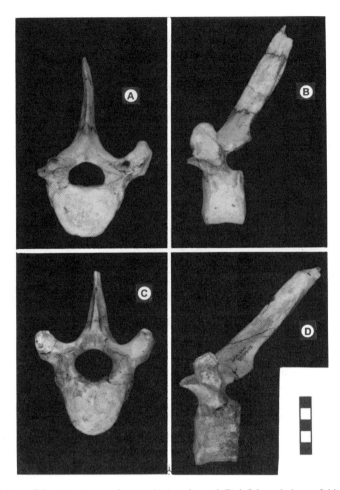

FIGURE 4. Vertebrae of *Georgiacetus vogtlensis.* (A) Anterior and (B) left lateral views of third thoracic verte-bra, GSM 351. (C) Anterior and (D) left lateral views of seventh thoracic vertebra, GSM 350. Scale bar in cm.

and a posterior side that is markedly concave at the base with a highly rugose surface that begins medial to the postzygapophyses. The surfaces of the anterolateral sides are flat to slightly concave, and continuous with the dorsal surfaces of the laminae. They meet ante-riorly to form a sharp ridge. The posterior side of the spine flattens out after ca. 30 mm, but remains rugose and flat for an additional 70 mm, after which the spine becomes laterally compressed, elongate-oval in cross section with two pointed ends. There is a sharp, narrow, 43-mm-long medial ridge present on the posterior side of the neural spine that begins about 50 mm from its base. Two concave regions are located anterior to the prezygapophy-sis, with the deeper, more lateral one forming a pocket into the anterior side of the transverse process. The surface of the neural arch between the two medial concavities is very rugose. The short transverse process projects dorsally, laterally, and slightly anteriorly from the pedicle, posterior to the prezygapophysis. It bears a large, slightly concave articular surface for the tuberculum of the third rib that faces laterally and ventrally. The two vertebrae identified as T3 differ significantly in only two respects. The neural spine of the T3 of GM 350 projects slightly more posteriorly and less dorsally than that of GSM 351. The epiphyseal lines between the articular epiphyses and the body of the centrum are still visible on GSM 351, but not GSM 350, indicating that GSM 350 represents an older adult than GSM 351.

T3 of GSM 350 does not articulate well with the anteriormost of the remaining thoracic vertebrae. Its postzygapophyses are set too narrow to match the latter's wider prezygapophyses, and its centrum is too short dorsoventrally to match the latter's. Therefore, a second break is assumed to exist in the anterior thoracic series, consisting of at least one, and more probably two (or possibly even more) vertebrae. The vertebra identified as T6 represents the beginning of a successive series of five thoracics, all relatively complete except for the last one.

T6 generally resembles T3, but its centrum is longer and higher (Table I) and its transverse process is narrower anteroposteriorly. The articular surfaces of the centrum are "heart shaped" in outline, with rounded ventral margins. The most significant differences with T3 are on the neural arch and spine. The concave depression anterior and lateral to the prezygapophysis is much shallower, and there is no anteromedial depression. However, the anterior surface of the neural arch between the prezygapophyses around the base of the neural spine is very rugose and hollowed out. The anterolateral walls of the neural spine are concave, and form a very thick anterior medial margin for the spine. The posterior face of the neural spine is like that of T3, but is transversely wider and the raised medial ridge is located about 45 mm from the base and runs for about 50 mm.

T7 greatly resembles T6, but is better preserved allowing the observation of more details (Fig. 4C–D). There is no keel on the ventral surface of the centrum. The neural canal in this region of the vertebral column is about as tall as it is wide (Table I). The capitular foveae are larger than those of T1 or T3, about equal in size, but the posterior fovea is more concave. The transverse process projects directly dorsolaterally, without the slight anterior component found in T6 or T3. The facet for the rib tuberculum faces laterally and slightly ventrally, is only slightly concave, and is separated from the dorsal knob of the transverse process by a shallow groove. There is no concavity or pit on the anterior surface of the transverse process anterior to the prezygapophysis. The rugose, concave surfaces at the posterior and anterior bases of the neural spine are more highly developed than on T6, especially on the posterior surface where a thick, medial ridge runs from the base of the spine for about 47 mm, finally forming the posterior margin of the distal end of the spine as it becomes laterally narrow (Fig. 3C).

T8 is a near duplicate of T7, except that the transverse process is slightly smaller, the thicker anterior ridge of the neural spine is bordered laterally by shallower depressions, and there is a slight concavity posterior to the prezygapophysis at the origin of the transverse process. Also, the centrum is a little wider but of lesser height (Table I). T9 (from GSM 351) continues and accentuates these trends. The concavity at the anterior base of the transverse process on T9 is larger and deeper, and the anterior edge of the neural spine is thicker than that of T8 and bordered laterally by slight depressions. On T9, the posterior face of the neural spine is still concave and rugose, but not to the degree of the two preceding vertebrae, and the posterior median ridge is weaker. This is not the result of ontogenetic difference between GSM 351 and 350, as T10 from GSM 350 continues these trends. A fragment of an anterior thoracic vertebra consisting of the left transverse process, the postzygapophysis, and portions of the neural arch from GSM 350 may also represent T9. Its size and morphology are more consistent with that position than T4 or T5, while its posteriorly projecting neural spine precludes a position posterior to T10.

T10 is represented by a broken specimen; the anterior half of its centrum and pedicles (including both prezygapophyses) are missing as is the right transverse process. It is generally similar to T9, except that the neural spine projects much less posteriorly and more dorsally, although the spine still extends past the centrum posteriorly. The posterior capitular fovea is either absent or vestigial. A sharp ridge runs posteriorly from the prezygapophysis that then curves medially, forming the lateral border of a concave, rugose region around the anterior base of the neural spine. A blunt, anteriorly projecting process is located at the midpoint of this region at the very base of the spine just dorsal to the neural canal. This process is much larger than on any of the preceding thoracics, but smaller than those of the posterior thoracics or anterior lumbars. Greater centrum width is another feature of T10 and T9 that is intermediate between the anterior and posterior thoracics (Table I). However, in most respects T10 resembles the preceding rather than subsequent thoracics, most importantly in the orientation of its zygapophyses and neural spine.

A minimum of three vertebrae form the next morphologically distinctive group, collectively referred to as the *posterior thoracics*. The anteriormost, T11, is represented only by a fragment, but it is complete enough to be distinctive. It consists of the lower portion of the neural spine with parts of the left lamina and postzygapophysis. The postzygapophysis faces ventrally and laterally, about 30° from vertical. The estimated width across the postzygapophyses is ca. 65 mm, a greater value than in all other posterior thoracics and anterior lumbars. As this parameter decreases posteriorly in this series, this fragment must be the anteriormost of this group. Furthermore, when its postzygapophysis is articulated with the prezygapophysis of T12, the midlines of the two neural spines correspond exactly. Therefore, the fragment must represent the vertebra that directly preceded the one identified as T12. The posterior face of the neural spine of T11 is concave at its base with a rugose surface. About 15 mm dorsal to the level of the postzygapophysis, the posterior face of the neural spine becomes flat to slightly convex, with a faint medial ridge. The neural spine is teardrop shaped in cross section, with a sharp narrow anterior edge and an anteroposterior length of about 29 mm. In lateral view, the posterior edge of the spine is oriented vertically while the anterior edge is obliquely oriented with a slight posterior tilt. If this interpretation of vertebral sequence is correct, then the fragment identified as T11 is both the diaphragmatic and anticlinal vertebra.

T12 (and the succeeding T13) are morphologically distinct from T3–T10, and in many respects are more like the anterior lumbars (Fig. 5). The centrum of T12 is 20% longer than that of T9 (Table I). Its anterior and posterior articular surfaces are kidney shaped, mostly flat with slight central depressions, and wider than tall. In lateral view, the ventral surface is smooth and notably arched or concave, more so than in T3–T10, which are either flat or slightly concave. The dorsal surface of the centrum bears a longitudinal groove. Pedicles are long and low, so that the oval neural canal is both lower and wider than that of T10. The zygapophyses are obliquely oriented, about 35° from vertical, and their surfaces are flat, subcircular, and extend past the body of the centrum. The prezygapophysis faces dorsomedially and very slightly anteriorly, while the postzygapophysis faces ventrolaterally. The prezygapophysis is borne on short, dorsolateral projections from the pedicle. The neural

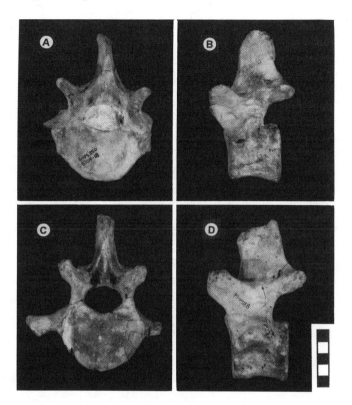

FIGURE 5. Vertebrae of *Georgiacetus vogtlensis*, GSM 350. (A) Anterior and (B) right lateral (reversed) views of 12th thoracic vertebra. (C) Anterior and (D) right lateral (reversed) views of first lumbar vertebra. Scale bar in cm.

spine projects dorsally and does not extend beyond the centrum (in lateral view, Fig. 5B). Although the distal end is broken, it is relatively short dorsoventrally. The lateral sides of the neural spine are convex posteriorly and concave anteriorly, making it teardrop shaped in cross section. The anterior base of the spine features a large, medial, anterior projection that is separated laterally from the prezygapophyses by oval depressions. Ridges running dorsomedially from the prezygapophyses converge to form the anterior edge of the neural spine; between these ridges is a triangular region with an exceedingly rugose surface. The region between the postzygapophyses is deeply recessed and concave. The capitular foveae are not located at the dorsolateral corners of the articular surfaces of the centrum as in T1–T9, but are instead on a very short transverse process that extends from the dorsal half of the lateral side of the centrum and the basal portion of the pedicle. This process bears two large concavities. The anteroventral of the two is larger, faces anterolaterally, and articulated with the single head of the 12th rib. A deep groove with a nutrient foramen transversely splits the fovea. The posterodorsal concavity on the transverse process is shallower, with a rugose surface, and a strongly rimmed anterodorsal margin that also overlies the fovea for the rib dorsally. The second hollow is probably not for articulation with a rib but rather for attachment of muscle or connective tissue.

A pathologic asymmetry in the location of the zygapophyses appears in T12. The postzygapophysis on the right side is located slightly higher (more dorsal) than the left postzygapophysis. This asymmetry is continued in the pre- and postzygapophyses of the succeeding vertebrae until S1.

The centrum of T13 is larger in all dimensions than that of T12 (Table I). Both articular surfaces of the centrum are similar, except the ventral boundary of the anterior surface is located at a more dorsal level than the posterior surface (evident in lateral view). The ventral surface of the centrum is anteroposteriorly concave but somewhat flattened transversely. The pedicle is long and narrow, with the prezygapophysis located on the medial surface of an anterodorsally projecting process. The surface of the prezygapophysis is flat and faces dorsomedially. The surface of the postzygapophysis is transversely convex, elongate-oval shaped, and faces ventrolaterally. It is distinctly more vertical than that of T12, and extends well beyond the centrum. Maximum width across the postgypapophyses is 50.1 mm, less than on T12 (56.5 mm). The neural spine is broken off near its base. The medial anterior projection at the base of the spine is strong and appears larger than that of T12 because the pits to each side of the projection are much deeper and pocketed by dorsal rims. The transverse process is not well preserved but similar to that of T12 except smaller, with less of a sharp lateral projection, and much shallower, ungrooved fovea for articulation with the 13th rib.

3.1.3. Lumbar Vertebrae

There are fairly complete examples of five of the probable total of eight lumbar vertebrae from GSM 350: the L1, L2, L4, L7, and L8. The more fragmentary L3 has a nearly complete but badly crushed centrum along with the bases of the pedicles and most of the right transverse process. A minimum gap of two vertebrae (i.e., L5 and L6) is indicated by the differences in size and shape of L4 and L7. A lumbar vertebra from GSM 352, probably L4, gives some indication of intraspecific variation.

In general, the centra of the lumbar vertebrae increase in length posteriorly, while maintaining near-constant widths and heights (Table I). Articular surfaces of the centra are near vertical, heart or kidney shaped on L1–L4 and elongate-oval on L7 and L8, and flat with central concavities. Ventral surfaces are anteroposteriorly concave and transversely flat. The flattened medial region progressively increases in width posteriorly, and bears numerous nutrient foramina. There is no keel. The transverse processes project from the lateral sides of the centra, are dorsoventrally flattened, and increase in size posteriorly. They also project or curve anteriorly on L1–L4, but not on L7 and L8. The large, interlocking zygapophyses are elongate-oval in shape and project beyond the centra. The prezygapophyses face dorsomedially, while the postzygapophyses face ventrolaterally. The neural spines are shorter than those of the thoracics, but remain tall back to the sacral series and project dorsally.

The centrum of L1 is about 5% larger than that of T13 in all dimensions (Fig. 5C–D; Table I), but its neural canal is of about the same width and of less height, and retains its oval shape. The long, thin pedicle begins at the anterior margin of the centrum. A short, blunt metapophysis projects anteriorly, dorsally, and laterally from the prezygapophysis. The latter faces dorsomedially, has a trapezoid-shaped articular surface with a maximum width of 25 mm at the anterior end and a length of 22 mm, and is distinctly concave transversely. The postzygapophysis faces ventrolaterally, with an oval articular surface (ca. 23 by 16.5 mm), and is transversely convex. Maximum width across both postzygapophyses is 55.7 mm. The neural spine is rectangular in lateral view (Fig. 5D), teardrop shaped in cross section, with nearly vertical anterior and posterior edges, and an anteroposterior length of ca. 41 mm. The anterior and posterior bases of the neural spine have rugose surfaces, similar to T13. The transverse process has a longer base (ca. 35.5 mm anteroposteriorly) than the distance it extends laterally from the centrum, ca. 29 mm. The dorsal surface of the transverse process is strongly convex, and the ventral surface concave, but to a lesser degree so that the process is thickest in its middle. The transverse process also projects anteriorly, but with its short length does not extend past the anterior margin of the centrum. The anterior margin of the transverse process extends more ventrally than does the posterior margin.

L2 is like L1 in most features, although it has a longer centrum, larger transverse process, and wider neural canal (Table I). The neural spine is thinner transversely, but of similar length (42 mm). The maximum width across the postzygapophyses is 55.5 mm. The transverse process has a basal length of about 36 mm, but it extends for more than 35 mm from the centrum (as the distal end is broken, it was actually considerably longer than this). It projects anterolaterally and slightly ventrally, past the anterior end of the centrum, with a slightly convex dorsal surface and a flat to very slightly convex ventral surface.

L3 is represented by an incomplete specimen. Its centrum is about 57 mm long, intermediate of the vertebrae identified as L2 and L4 (Table I). The transverse process of L3 projects laterally, anteriorly, and slightly ventrally from the lateral wall of the centrum, and extends beyond the anterior end of the centrum. The distal end is broken, but the transverse process extends at least 60 mm from the centrum. Its dorsal and ventral surfaces are mostly flat to slightly convex, and it is thicker posteriorly than anteriorly. The transverse process is about 43 mm long anteroposteriorly at its base, but narrows to 28.3 mm at a point 25 mm lateral to the centrum, and then broadens distally.

L4 of GSM 350 resembles L2 in most features, with the following notable differences: the more vertically oriented zygapophyses (Fig. 6), the taller neural spine, and the walls of the centrum ventral to the transverse processes contain large, shallow depressions. The transverse process resembles but is less complete than that of L3; its basal anteroposterior length is ca. 42 mm, and its minimum length is 26.5 mm at a point 23 mm lateral to the centrum. GSM 352 is a lumbar vertebra from a slightly larger individual than GSM 350 (Table I). The morphology of the centrum, neural spine, and transverse processes is most like that of L4 of GSM 350. The prezygapophysis is strongly concave, and the postzygapophysis strongly convex, both more so than their counterparts in GSM 350. The large, dorsally projecting neural spine has a narrow anterior edge and a much broader posterior margin with a strong medial ridge.

L7 differs considerably from the more anterior lumbars in both size (Table I) and morphology (Figs. 7 and 8A). Its centrum is much longer, but still wider than long. The articular surfaces are oval, relatively flat with central depressions, and the posterior surface is more concave than the anterior. The height of the neural canal is relatively low (Table I), and it is triangular in outline. The length of the pedicle is similar to those of L1–L4, thus relatively short compared with the centrum. The prezygapophyses are not well preserved but appear to be small and face dorsomedially, less vertical than in L4, more similar to those of L2 in orientation. The postzygapophysis faces ventrolaterally, about 40° from vertical,

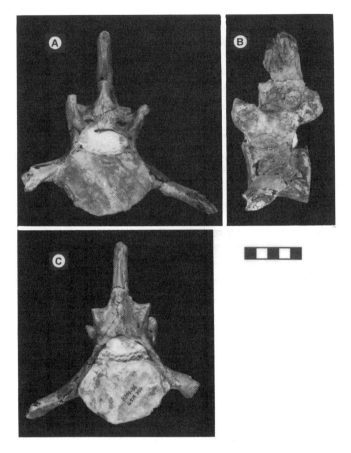

FIGURE 6. Fourth lumbar vertebra of *Georgiacetus vogtlensis*, GSM 350. (A) Anterior, (B) left lateral, and (C) posterior views. Scale bar in cm.

with an oval, flat articular surface. The maximum width across the postzygapophyses is 53.1 mm. The prezygapophysis is located more posteriorly than in L4, lateral to the base of the neural spine, and does not project past the anterior face of the centrum. The postzygapophysis does project past the posterior face of the centrum, but to a much lesser degree than in L4. The concave, rugosely textured, triangular-shaped region at the anterior base of the neural spine with a medial projection and deep lateral pits that is present between the prezygapophyses of L1 through L4 is not present on L7, although the surface of the bone in this region is still rugose. Similarly, there is no deep recess between the postzygapophyses as on L1–L4. The neural spine of L7 is very thin (much thinner than on the preceding lumbars), laterally compressed, and has an oblique anterior edge that runs posterodorsally and a vertical posterior margin. The height of the spine appears to have been less than on L4, but cannot be measured because of breakage. The transverse process projects almost directly laterally from the middle of the centrum (Fig. 7A), with only slight anterior and ventral components to its direction. It is about 45 mm long anteroposteriorly at its base, and

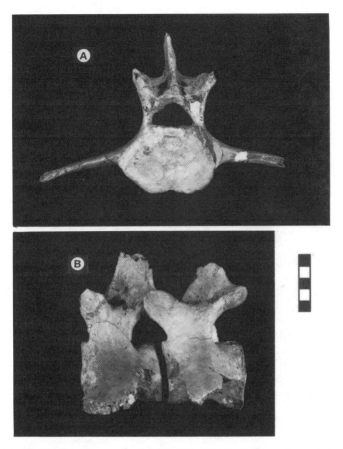

FIGURE 7. Vertebrae of *Georgiacetus vogtlensis*, GSM 350. (A) Anterior view of seventh lumbar vertebra. (B) Right lateral (reversed) view of seventh and eighth lumbar vertebrae. Scale bar in cm.

extends at least 80 mm (its distal end is broken so this is a minimum value) from the centrum. Near its base, the ventral and dorsal surfaces are slightly convex; these become flat moving away from the centrum, and the dorsal surface finally becomes slightly concave near the anteroposteriorly expanded distal end. The transverse process is dorsoventrally thin, thickest at its base, and thinning distally. In dorsal view (Fig. 8A), the anterior and posterior margins of the transverse process are both initially concave, so the anteroposterior length narrows to a minimum of ca. 35 mm about 26 to 28 mm lateral to the centrum; then the margins become convex distally as the process expands.

   L8 is similar to L7, especially in the size and shape of the centrum and the neural canal, but differs in some respects. The prezygapophysis is more anteriorly located, extending past the centrum, and faces dorsomedially at a less vertical angle than either L7 or L4. Its articular surface is flat, about as long as wide, and is located ventromedial to a moderate metapophysis. The articular surface of the postzygapophysis is also flat, but elongate-oval in shape, and faces ventrolaterally (more vertically than that of L7). The maximum width

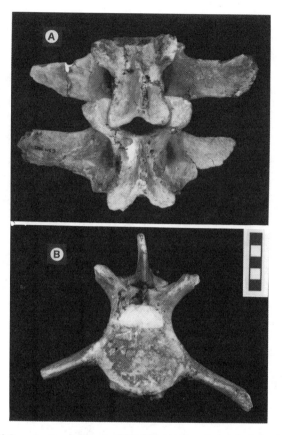

FIGURE 8. Vertebrae of *Georgiacetus vogtlensis*, GSM 350. (A) Dorsal view (anterior to top of figure) of seventh and eighth lumbar vertebrae. (B) Anterior view of first sacral vertebra. Scale bar in cm.

across the postzygapophyses is 57.4 mm. As in L7, the neural spine of L8 is transversely thin, but is shorter in height and its anterior margin more oblique, projecting at a 60° angle from vertical (Fig. 7B). The transverse process is large and long, extending at least 75 mm from the lateral side of the centrum and probably much more as the distal end is incomplete (Fig. 8A). This process differs from that of L7 in not projecting anteriorly (only laterally and ventrally), is more arched transversely (that of L7 is relatively straight), its dorsal surface is more anteroposteriorly convex, and it does not thin as greatly, at least on the preserved portion of the process. The ventrolateral wall of the centrum below the transverse process is markedly concave.

3.1.4. Sacral Vertebrae

The sacral vertebrae of GSM 350 are generally characterized by having centra longer than L8 (Table I), but centrum width similar to or less than that of the lumbars. In S3 and more posterior vertebrae the length of the centrum is greater than its width. Centrum height also increases posteriorly, so that the articular surfaces are round rather than oval through the sacral series. The neural canal retains the low height of the posterior lumbars, and narrows transversely along the sacral series. The obliquely oriented zygapophyses are similar to those of L8. Very large, dorsolaterally projecting metapophyses are a peculiar feature of the sacrals. The transverse processes are large, dorsoventrally thick relative to those of the lumbars, transversely arched, do not project anteriorly, and, where preserved, thicken distally. Neural spines are very thin transversely, but are at least 30 mm tall (and probably more but each is broken).

S1 is crushed and poorly preserved (Fig. 8B, Table I), but the following are evident: The articular surfaces of the centrum are wider than tall, perfectly flat without central depressions, and slightly inclined dorsoventrally. The articular surface of the prezygapophysis is large, oval (30.5 by 26.5 mm), flat, and faces dorsomedially at an angle of ca. 50° relative to vertical, the metapophysis is large, and the articular surface of the postzygapophysis is elongate-oval (ca. 27 by 21 mm), flat, and faces ventromedially. The maximum width across the postzygapophyses is 50.6 mm. The thin neural spine has slightly concave, dorsally erect anterior and posterior margins, and a short anteroposterior length of ca. 36 mm. The base of the transverse process extends almost along the entire lateral side of the centrum (for at least 53 mm; because of breakage it cannot be measured accurately). The process projects ventrolaterally with slight transverse arching (less than in L8), and both its dorsal and ventral surfaces are convex anteroposteriorly, the dorsal surface more so than the more nearly flat ventral surface. The transverse process thins distally from the base to a point about 27 mm lateral from the centrum, and then begins to gradually thicken. At the broken distal end, the process is 75 mm lateral to the centrum wall. In dorsal view, the anteroposterior length of the transverse process rapidly decreases distally, but whether or not this trend was reversed and it expanded in this dimension at the distal end as well as dorsoventrally cannot be determined.

S2 is more intact and better preserved than S1, although not entirely complete (Fig. 9A,B). Its centrum is only slightly wider than long (Table I), with round articular surfaces that are as tall as wide. The prezygapophysis and metapophysis are like those of S1, but the postzygapophysis is smaller (ca. 25 by 18 mm) and located more medially. The maximum width across the postzygapophyses is only 43.7 mm. The asymmetry in the dorsoventral location of the zygapophyses observed in the posterior thoracics and lumbars returns in S2 (and continues through S4), but in the opposite direction: In the sacrals it is the left zygopophyses, not those on the right side, that are located more dorsally. The neural arch of S2 is notably taller than in the other sacrals (Table I). The transverse process is similar to that of S1, but the left process is more complete than in S1. The process is downturned slightly more ventrally than in S1, extending past the ventral base of the centrum (Fig. 9A). As in S1, it thickens dorsoventrally at the distal end, to about 17 mm, but continually narrows anteroposteriorly to the distal end, ending bluntly about 79 mm lateral to the centrum wall.

S3 is in many respects the best preserved of the sacral vertebrae, especially with regard to the processes of the neural arch (Fig. 9C). It is very similar to S2 in most features although the centrum is about as long as wide, the anterior articular surface of the centrum is less flat with a slight dorsomedial concavity overlying a shallow ventromedial convexity, and the neural canal is much smaller. The articular surface of the prezygapophysis is large, oval (ca. 33 by 23.5 mm), flat, and faces dorsomedially. The surface of the postzygapophysis is small, elongate-oval (ca. 22 by 15 mm), flat, and faces ventrolaterally. The

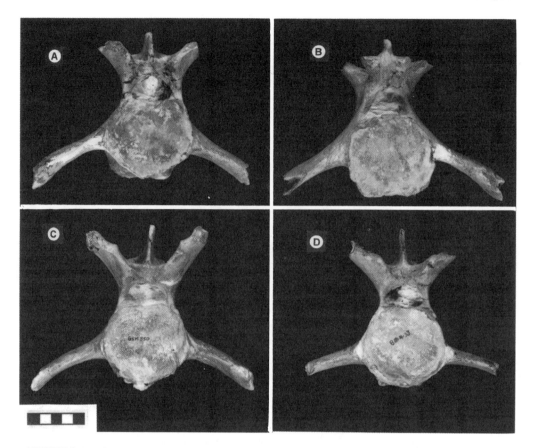

FIGURE 9. Vertebrae of *Georgiacetus vogtlensis*, GSM 3500. (A) Anterior and (B) posterior views of second sacral vertebra. (C) Anterior view of third sacral vertebra. (D) Anterior view of fourth sacral vertebra. Scale bar in cm.

maximum width across the postzygapophyses is 39.6 mm. A large metapophysis extends dorsolaterally 25 mm beyond the articular surface of the prezygapophysis, with a blunt, rounded end. The neural spine is broken distally but is like the others of the sacral series in its narrowness and dorsal projection. The transverse process is similar to that of S2, but is less complete distally, so the degree of expansion cannot be determined.

The centrum of S4 is longer than wide (Table I), with flat, circular articular surfaces. The pre- and postzygapophyses are smaller than those of S3 (Fig. 9D). The articular surface of the prezygapophysis is elongate-oval (ca. 29.5 by 15 mm), flat, and faces dorsomedially. That of the postzygapophysis is very narrow (ca. 25 by 10 mm) and faces ventrolaterally. The maximum width across the postzygapophyses is only 29.5 mm. The metapophyses are large, like those of S3. The transverse process is straighter than those of S2 or S3, projecting only slightly ventrally (Fig. 9). It is relatively thick and both its dorsal and ventral surfaces are convex anteroposteriorly. While it narrows distally in the anteroposterior dimension, it retains constant thickness, at least until where each side is broken.

3.1.5. Caudal Vertebra

GSM 350 includes a vertebra that articulates posteriorly with S4, and that is therefore identified as Ca1. No other caudal vertebrae are known for *Georgiacetus*. The centrum of Ca1 is longer and taller than wide (Table I, Fig. 10), with circular articular surfaces. In many respects it is similar to S4. However, the large metapophysis projects mostly laterally, as opposed to those of the sacrals that project more dorsally. Corresponding to the narrow articular surface of the postzygapophysis of S4, the prezygapophysis of Ca1 is long and narrow, with dimensions of its articular surface about 26.5 by 13.5 mm. It faces dorsomedially, but more dorsally than in S4. The postzygapophysis is much wider than that of S4, about 27 by 17.5 mm, faces ventromedially, and is more vertical than those of the sacrals. The maximum width across the postzygapophyses is 41.9 mm. The neural spine is like those of the sacrals. The transverse process projects almost straight laterally with a slight downturn at its distal end. Its dorsal and ventral surfaces are convex. The dorsoventral thickness of the transverse process remains constant until near the distal end, where there is a slight increase, but not as much as in S2. There is a broad, shallow, unrimmed concavity on the sur-

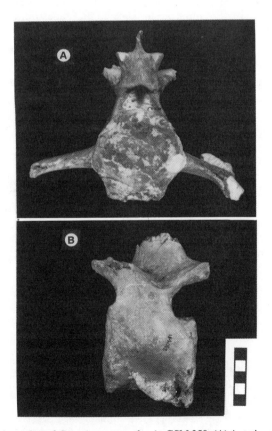

FIGURE 10. First caudal vertebra of *Georgiacetus vogtlensis*, GSM 350. (A) Anterior and (B) left lateral views. Scale bar in cm.

face of the centrum dorsal to the transverse process and ventral to the pedicle. There are two large, bulbous convexities on the posterior end of the ventral surface of the centrum that may have supported a hemal arch. However, the compact bone is very thin in this region and has been destroyed along with some of the underlying spongy bone, so the presence or absence of an articular surface for a hemal arch cannot be determined. Similar but much smaller swellings are present on the posteroventral surfaces of the centra beginning with the vertebra identified as L7, and continue through the sacral series, becoming larger posteriorly. These regions are consistently poorly preserved, with loss of their surficial bone, so their original morphology and function are difficult to ascertain.

3.2. Hemal Arch

A single hemal arch or chevron bone is present in GSM 350. The specimen is incomplete, crushed, and poorly preserved, with much of the hemal spine lost (terminology follows Kellogg, 1936). In general, it resembles the anterior hemal arch of *Zygorhiza* figured by Kellogg (1936, Fig. 69), although it is much larger. The arm of the hemal arch is transversely very thin, about 16 mm long anteroposteriorly, and slightly expanded at its proximal (dorsal) end with a small, oval articular surface that faces dorsomedially. The posterior edge of the arm is sharper than the rounder anterior edge, and the transverse diameter of the arm doubles just dorsal to the hemal spine. The hemal canal is a deep (ca. 54 mm), narrow, V-shaped opening between the hemal arms, and bounded ventrally by a large hemal spine. The posterior surface of the hemal spine is concave, while its anterior side is convex.

3.3. Ribs

GSM 350 includes seven left and five right ribs with their proximal ends, as well as several fragments of rib shafts. At least two from the right side represent positions not found among those on the left, so a minimum of nine morphologically distinct ribs occur with this specimen. However, unlike the vertebrae, the exact relative sequence of these ribs has not been determined, although the description of the ribs of *Zygorhiza* in Kellogg (1936, pp. 166–173) does allow their provisional separation into anterior versus posterior groups. The very short and stout first rib is distinctive and easily identified. As all of the ribs preserved with GSM 350 are "two-headed' (i.e., possess both a capitulum and a tuberculum), none represent the ribs that articulated with the last two thoracic vertebrae (T12 and T13), which both lack a tubercular facet and would have had "single-headed" ribs. GSM 351 includes the shaft of the first rib and the proximal ends of two others, adding little beyond that represented by GSM 350.

The tuberculum, capitulum, and distal end of the shaft of each rib are composed of the usual very thin layer of compact bone underlain by an irregular lattice of thin trabeculae forming the spongy bone. The remainder of the shaft on each rib, as well as the angle and neck, consist of a very thick layer of compact bone surrounding a small core of spongy bone (Fig. 11F). The trabeculae here are at least twice as thick as those found near the articular

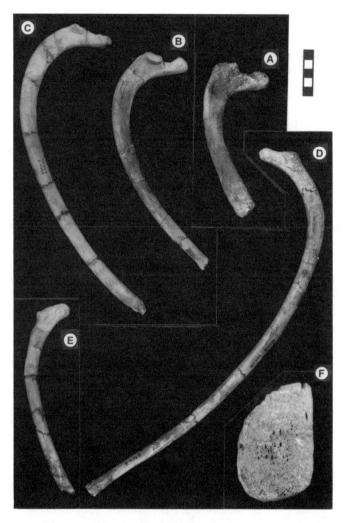

FIGURE 11. Ribs of *Georgiacetus vogtlensis*. (A–E) GSM 350. (A) Nearly complete first rib. (B) Incomplete anterior rib. (C) Nearly complete anterior rib. (D) Complete midthorax rib. (E) Incomplete posterior rib, possibly the 10th or 11th. (F) Cross section through the osteosclerotic shaft of a rib of GSM 351. Scale bar (for A–E only) in cm. Actual dimensions of F are 14.0 by 21.7 mm.

surfaces. The thickness of the layer of compact bone in these osteosclerotic regions ranges from 2 to 6 mm.

The first rib is very short (Fig. 11A), with a large tuberculum (27.4 by 20.2 mm) separated from the small, subspherical capitulum by a short (ca. 15 mm), oval neck. The breadth of the neck (ca. 13.5 mm) is less than its dorsoventral height (17 mm). The angle has a relatively sharp curve, and is anteroposteriorly flattened. Both the anterior and posterior surfaces on the angle are concave, but the posterior concavity has a deeper, more pronounced groove that continues distally almost to the midpoint of the shaft. It becomes shallower and narrower distally. The shaft is slightly curved or bowed outward and downward, its width is greater than its breadth, and it is not twisted. The shaft gradually widens all the way to its distal end, at which the rib has a breadth of 21.8 mm and a width of ca. 23.5 mm (on GSM 350). The shaft of the first rib of GSM 351 is more compressed anteroposteriorly than that of GSM 350, then widens suddenly at its distal end. It has a distal breadth of 26.3 mm and a distal width of 21.2 mm. On both specimens, the distal end is flattened with a pockmarked surface of spongy bone.

Two examples of what are considered to represent anterior ribs are shown in Fig. 11B,C. The angle and proximal end are compressed anteroposteriorly. Distal to the sharply downturned angle, the shaft twists such that the posterior surface at the proximal end is continuous with the medial surface at the distal end; likewise the anterior surface at the proximal end is continuous with the lateral surface at the distal end. Along the distal third of the shaft, the medial surface is strongly convex with a distinct crest or ridge on the middle of the shaft. This crest fades at the swollen, very distal end where the cross-sectional shape of the rib becomes more oval and less elongate. The rib shown in Fig. 11B has a large tuberculum with a saddle-shaped articular surface that faces posterodorsally and medially, and that is separated from the subspherical capitulum by a short (ca. 15 mm long) neck. The neck is not anteroposteriorly compressed, and is much thicker ventrally than dorsally because of a ridge on the posterior side that extends out from the capitulum. On the more complete specimen (Fig. 11C), the proximal end is less well preserved, but apparently has a smaller tuberculum and capitulum, and a slightly longer and more slender neck.

Figure 11D shows a nearly complete rib judged to originate from near the middle of the series. The well-separated tuberculum and capitulum are smaller than on the more anterior ribs, and the smaller shaft flares only very slightly distally, lacks the strong medial ridge, and is not twisted. Finally, one of the ribs, shown in Fig. 11E, differs from all of the others in lacking a true neck, although a distinct capitulum and tuberculum are both present but separated by only a few millimeters. Compared with the other ribs, the angle has a much lesser downturned arch, the shaft is neither twisted nor compressed (breadth and width about equal) with a flat ventral and medial margin and a strongly convex dorsal–lateral margin. There are distinct grooves on both the posterior and anterior sides ventral to the tuberculum, and a pit lateral to the tuberculum on the dorsal surface. This specimen is provisionally identified as the 10th or 11th rib.

3.4. Pelvis

Both the right and left innominates of the pelvis are represented in GSM 350. The left is more complete, but the right better preserves details of the acetabulum. On both, the iliac crest, the posterior border of the ischium (including most of the ischial tuberosity), and portions of the pubic symphyses were destroyed prior to burial, with loss of surficial compact bone and some inner spongy bone. The remaining portions of the pelvis are generally well preserved, with only minor cracking and loss of surficial bone in a few limited areas. The innominate is elongate (Fig. 12), with a maximum greatest length of more than 360 mm. The ilium is long, ca. 50% of the total length of the innominate, with slight anterior flare. The relatively flat iliac blade lacks the lateral flexion reported in some aquatic and semiaquatic mammals (Tarasoff *et al.*, 1972). The external surface of the iliac blade faces

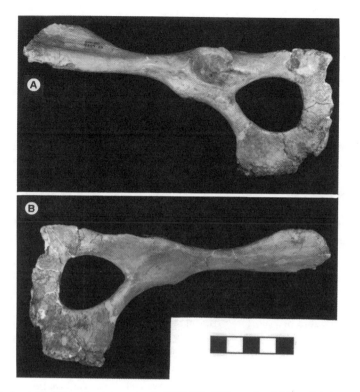

FIGURE 12. Left pelvis of *Georgiacetus vogtlensis*, GSM 350. (A) Lateral and (B) medial views. Length of scale bar 10 cm.

dorsomedially and its surface is slightly concave; the internal surface of the iliac blade faces ventromedially and is slightly convex. The dorsal and ventral borders of the ilium are rounded and thick where preserved. The inner surface of the iliac blade is relatively smooth, lacking an auricular surface for articulation with the sacral vertebrae. However, there is a slightly rugose region in the middle of the blade that probably attached indirectly to one or more transverse processes of the sacrals via connective tissue. There is no sharp external angle on the dorsal edge of the ilium to demarcate the iliac body from the iliac blade; instead the ilium expands gradually anterior to the acetabulum. The ilium is constricted and narrow between the iliac blade and acetabulum, with a flat medial and strongly convex lateral surfaces. The dorsal and ventral edges of the ilium are here rounded and not sharp. There is a raised, rugose region on the lateral surface just anterior to the acetabulum for the origin of the rectus femoralis muscle. The acetabulum is large (Fig. 12A), about 46 mm in diameter, with a deep, circular articular surface that opens widely posteroventrally. The deep cotyloid (or acetabular) notch is directed posteriorly, and roofed by the dorsal lunate surface and a short ridge on the ischium posterior to the acetabulum. The development of this notch indicates the presence of a large, well-formed ligamentum teres. The dorsal and anterior edges of the acetabulum are not sharp, but combine to form a thick, flat, rugose ridge that gradually thins to a sharp crest along the acetabulum's ventral margin. Posterior to the acetabulum, the ischium is transversely very thin, its lateral surface convex, medial surface flat, and lacking any trace of a distinct lesser sciatic notch on its dorsal border. The large, oval obturator foramen (ca. 81.5 mm long, 69 mm tall) is formed by the pubic and ischium. Its pubic border is rounder and less sharp than the ischial margin. The ischial tuberosity is relatively small and not prominent. The ischial spine is also reduced, forming a thin ridge. The pubic body extends posteroventrally and medially from the acetabulum, is oval in cross section, and has a moderately sharp ventral edge. Ventromedial to the obturator foramen, the pubic ramus is greatly widened with the minimum distance between the pubic symphysis and the obturator foramen about 63 mm. The length of the pubic symphysis is relatively short, ca. 77 mm. A large, rugose iliopectineal eminence is located on the ventral margin below and slightly anterior to the acetabulum; it was the site of origin of the pectineus muscle.

4. Comparisons between *Georgiacetus* and Other Archaeocetes

4.1. Introduction

During the Eocene, cetaceans evolved from semiaquatic, fluvial, relatively small-bodied mammals capable of both terrestrial and aquatic locomotion into much larger, highly specialized, fully marine animals incapable of walking on land. Adaptations to the postcranial skeleton were an essential component of this transformation. The fossil record of archaeocetes, although still incomplete, demonstrates the stepwise nature and rate of this transformation. *Georgiacetus* represents a grade of cetaceans far along in this evolutionary series, one in which adults probably spent nearly all of their lives in the ocean, returning briefly to land only to give birth. While on land they were incapable of standing erect or walking on their limbs, but instead must have relied on undulations of the body.

In this section postcranial skeletal elements of *Georgiacetus* are compared with those of other archaeocetes. Discussion is limited to elements known from *Georgiacetus*, namely, vertebrae, ribs, and pelvis. The systematic framework used here generally follows Thewissen *et al.* (1996) and Fordyce and Barnes (1994) in recognizing six grades of archaeocetes: pakicetids, ambulocetids, remingtonocetids, protocetids, dorudontines, and basilosaurines. Although postcranials of pakicetids have been recovered (Thewissen, 1995), they remain undescribed. According to Thewissen *et al.* (1996), pakicetids were primarily terrestrial with few skeletal adaptations for a specialized aquatic lifestyle. If so, the relatively well-known postcranial skeleton of mesonychids (Wortman, 1901–1902; Zhou *et al.*, 1992; O'Leary and Rose, 1995) can serve as a proxy for those of pakicetids as a starting point for cetaceans.

Postcranials corresponding (at least partially) to those known in *Georgiacetus* have been described in the other five archaeocete grades. These include the ambulocetid *Ambulocetus* (Thewissen *et al.*, 1996), the remingtonocetids *Remingtonocetus* and *Dalanistes* (Gingerich *et al.*, 1993, 1995), the protocetids *Rodhocetus, Gaviacetus,* and *Protocetus* (Fraas, 1904; Kellogg, 1936; Gingerich *et al.*, 1994, 1995), the dorudontines *Zygorhiza, Dorudon,* and *Ancalecetus* (Kellogg, 1936; Gingerich and Uhen, 1996; Uhen 1996a, this volume), and the basilosaurine *Basilosaurus* (Kellogg, 1936; Buffrénil *et al.*, 1990; Gingerich *et al.*, 1990; Uhen, this volume). Unsurprisingly, the postcranials of *Georgiacetus* are most similar to those of members of its family, the Protocetidae. Elongated, *Basilosaurus*-like vertebrae were referred to the protocetid *Eocetus* by Fraas (1904) and Kellogg (1936). As there are serious doubts about their association with the type cranium of *Eocetus schweinfurthi* (Hulbert *et al.*, 1998; but see Uhen, in press), and thus their systematic allocation, these vertebrae are not included in the following discussion. The vertebrae assigned to the protocetid *Pappocetus* by Halstead and Middleton (1974) also lack direct association with diagnostic cranial or dental remains and are similarly excluded from this study.

4.2. Vertebrae

Mesonychids have 12 thoracic, 7 lumbar, and 3 sacral vertebrae (Wortman, 1901–1902; Zhou *et al.*, 1992), the latter fused to form a typical mammalian sacrum. Fossils complete enough to provide similar vertebral counts are lacking for pakicetids, ambulocetids, and remingtonocetids, although it is known that the latter had sacra consisting of 4 fused vertebrae (Gingerich *et al.*, 1995). *Rodhocetus* has the same number of thoracolumbar vertebrae as mesonychids, 19, but they were apportioned differently, with 13 thoracics and only 6 lumbars (Gingerich *et al.*, 1994). As the diaphragmatic thoracic was T10 in mesonychids and T13 in *Rodhocetus*, the two differ significantly in the number of postdiaphragmatic, presacral vertebrae (PDPSV), 9 in mesonychids and 6 in *Rodhocetus*. How widespread this reduction in PDPSV was among early archaeocetes must await discovery of additional complete specimens. The Habib Rahi whale (Gingerich, 1991), which is older than *Rodhocetus*, also has 13 thoracic vertebrae of which the last is diaphragmatic. Thus, the reduction in PDPSV is not entirely limited to *Rodhocetus*. Later protocetids reversed this trend and increased the number of PDPSV both by increasing the number of lumbars and by moving the diaphragmatic vertebra cranially within the thoracic series. Conservatively, *Georgiacetus* has 13 thoracic, 8 lumbar, and 4 sacral vertebrae, and there are at least 2 postdiaphragmatic thoracics, producing a likely total of 10 PDPSV. *Protocetus* similarly has a minimum of 2 postdiaphragmatic thoracics, and at least 7 and possibly more lumbars. This moderate increase in numbers of lumbars and PDPSV observed in *Protocetus* and *Georgiacetus* is taken to more extreme measures in both dorudontines and basilosaurines (Kellogg, 1936; Gingerich *et al.*, 1990; Gingerich and Uhen, 1996).

In addition to simply increasing the numbers of lumbar vertebrae and PDPSV, later archaeocetes also increased the total functional length of the lumbar region by increasing the lengths of the individual centra and by freeing the sacral vertebrae (allowing them and the anterior caudals to function with the lumbars as a unit). Length-of-vertebrae profiles for *Rodhocetus*, *Protocetus*, and *Georgiacetus* (Fig. 13) generally parallel one another back through about the 23rd position (to the middle of the lumbus), but *Georgiacetus* differs from the other two by more greatly increasing the lengths of its posterior lumbar and sacral vertebrae. Length-of-vertebrae profiles for basilosaurines and dorudontines are quite different from those of protocetids. In their profiles, centrum length increases (much more so in basilosaurines than dorudontines) sharply within the thorax, back to position 20 (Gingerich *et al.*, 1990; Gingerich and Uhen, 1996; Uhen, this volume). Centrum length remains relatively constant from there through the lumbus until the beginning of the caudals, after which it declines. Vertebral columns of remingtonocetids are less well known than either proto-

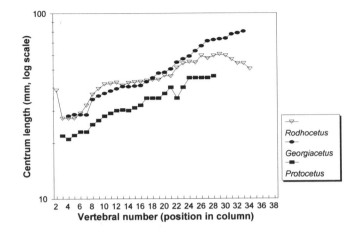

FIGURE 13. Length-of-vertebrae profiles of three middle Eocene protocetid archaeocetes, *Protocetus atavus* (SMNS 11084; profile begins with third cervical), *Rodhocetus kasrani* (after Gingerich *et al.*, 1994; profile begins with axis), and *Georgiacetus vogtlensis* (GSM 350 and 351; profile begins with fourth cervical). All three protocetids have similarly shortened cervical vertebrae. Their profiles parallel each other through the thorax, but beginning in the middle of the lumbar series (about vertebral number 23), the vertebrae of *Georgiacetus* are much longer both absolutely and relatively compared with *Protocetus* and *Rodhocetus*.

cetids or basilosaurids, but appear to differ from protocetids (including *Georgiacetus*) in that there is much less increase in centrum length from the anterior thorax back through to the sacrals (Gingerich *et al.*, 1993).

As demonstrated by Gingerich *et al.* (1993, 1995) and Thewissen *et al.* (1996), ambulocetids and remingtonocetids have relatively long centra in the cervical vertebrae, proportionally like those of mesonychids. Protocetids, in contrast, have relatively short cervicals. This was previously known in *Rodhocetus* and *Protocetus* (Gingerich *et al.*, 1993, 1994), and the cervicals of *Georgiacetus* are most similar to those two in terms of relative length (Fig. 13). Dorudontines and basilosaurines have even shorter cervicals than protocetids, relative to the length of the anterior thoracics. Shortening of the neck in whales is usually interpreted as a mechanism to stabilize the body when propulsion is generated by the tail. The morphology of the cervical vertebrae in *Georgiacetus* is also like that of *Protocetus*.

The thoracic vertebrae of *Georgiacetus* are very similar to those of *Protocetus* (and other protocetids), except for their larger size and greater robustness. The neural spines of the anterior thoracics are broader and project more posteriorly than those of dorudontines or basilosaurines. Protocetid thoracics also differ from the latter by having larger zygopophyses and lacking the broad, rib-bearing transverse processes on the posterior thoracics. However, the thoracics, particularly the more anterior ones, show less overall morphologic change than any other portion of the vertebral column within the Archaeoceti.

Considerable attention has been focused in recent years on the sacral vertebrae of archaeocetes. Sacrals are known (at least partially) in *Remingtonocetus, Dalanistes, Rodhocetus, Gaviacetus, Protocetus, Georgiacetus*, and various dorudontines and basilosaurines.

They are as yet undescribed for both pakicetids and ambulocetids. The robustness of the hind-limb elements of *Ambulocetus* and its inferred ability to walk on land (Thewissen *et al.*, 1996) suggest that, like remingtonocetids, it would have retained the fused sacrum of its mesonychid ancestors. Interestingly, the four known examples of protocetid sacrals are all slightly different, forming an exemplary morphocline between the fused, plesiomorphic state of remingtonocetids and the condition observed in basilosaurids. In the latter, there is so little morphologic distinction between the posterior lumbars, sacrals, and caudals that use of the term *sacral* to refer to a morphologically distinct vertebra or series of vertebrae between the lumbars and caudals is no longer appropriate. Indeed, the lumbar/caudal boundary can only be fixed by the presence/absence of facets for articulation with the hemal arches and a slight thickening in the transverse processes (Gingerich and Uhen, 1996; Uhen, this volume).

The morphology of the sacral vertebrae in protocetids forms a continuum that spans the differences observed in remingtonocetids and basilosaurids. The four sacrals of *Rodhocetus* differ from those of remingtonocetids (and mesonychids) by lacking fusion between their centra, neural spines, and transverse processes (Gingerich *et al.*, 1994). However, the transverse processes are otherwise plesiomorphically very narrow, retain pleuropophyses that surround sacral foramina, and, on S1, have a large auricular surface for articulation with the ilium. The laminae of the S1 and S2 neural arches articulate, as do the pleuropophyses of adjoining sacrals (Gingerich *et al.*, 1995). Only S1 of *Gaviacetus* is known. It is derived relative to that of *Rodhocetus* in lacking articulating laminae between S1 and S2, pleuropophyses are absent, and the transverse processes are slightly wider, extending about 33 mm from the centrum (Gingerich *et al.*, 1995). S1 of *Protocetus* has a much wider transverse process than that of *Gaviacetus* (extending ca. 47 mm from the centrum), and it is also considerably thinner close to its origin on the centrum. Distally, it expanded anteroposteriorly and dorsoventrally. The transverse process of S1 of *Georgiacetus* differs from that of *Protocetus* in being even wider (extending more than 75 mm laterally from the centrum), having less distal expansion, and in lacking an auricular surface (inferred from the morphology of the ilium). S2 through S4 of *Georgiacetus* (which are unknown in *Protocetus* and *Gaviacetus*) similarly bear large, arched transverse processes with slightly thickened distal ends. They can be distinguished from the posterior lumbars only by their larger metapophyses and dorsoventrally thicker transverse processes.

4.3. Ribs

The ribs of archaeocetes are much more poorly known than the vertebrae, except for those of basilosaurines and dorudontines that are well studied (Kellogg, 1936; Buffrénil *et al.*, 1990; Gingerich and Uhen, 1996). Ribs of *Ambulocetus* have been described (Thewissen *et al.*, 1996), but not those of pakicetids, remingtonocetids, or most protocetids. Among the latter, most of the ribs belonging to the type specimen of *Protocetus atavus* were destroyed during its preparation and only a few proximal ends survived (Fraas, 1904). The rib cage of the type of *Rodhocetus kasrani* is evidently relatively complete (Gingerich *et al.*, 1994, Fig. 1a), but is as yet virtually undescribed. The referred proximal rib of *Pappocetus* (Halstead and Middleton, 1974, Fig. 5) is completely unlike that of any archaeocete and is most likely misidentified. Gingerich *et al.* (1995) figured and described a single, incomplete midthoracic rib of *Gaviacetus*.

The ribs of *Ambulocetus* differ from those of *Georgiacetus* (and other protocetids) in their greater robustness and curvature (Thewissen *et al.*, 1996). Unlike all other known archaeocete ribs, those of *Ambulocetus* are in combination pachyostotic but not osteosclerotic. The distal ends of the anterior ribs of basilosaurines and *Zygorhiza* are swollen or clublike (less so in other dorudontines), and this also appears to be the case in *Georgiacetus* but to a much lesser degree. In *Basilosaurus* and *Zygorhiza* this pachyostotic region of the rib shaft is also osteosclerotic (Buffrénil *et al.*, 1990), but in *Georgiacetus* this region of the ribs has a very thin cortex and is not osteosclerotic. However, the angle and proximal portion of the rib body are osteosclerotic (but not pachyostotic) in *Georgiacetus* (Fig. 11F).

4.4. Pelvis

No pelves have as yet been described for either pakicetids or ambulocetids. Partial pelves of remingtonocetids have been described by Gingerich *et al.* (1993, 1995); all consist primarily of the acetabular region and lack most of the pubis, ilium, and iliac blade. There was broad articulation between the ilium and sacrum. Other than that of *Georgiacetus*, the pelvis of protocetids is known only from *Rodhocetus* (Gingerich *et al.*, 1994). Among basilosaurids, the best known pelvis is that of *Basilosaurus* (Lucas, 1900; Kellogg, 1936; Gingerich *et al.*, 1990; Uhen, this volume). Although there are differences among the pelves of the two protocetids (to be discussed below), in general they both retain most of the overall morphology and size of pelves of typical terrestrial mammals when compared with basilosaurid pelves. In addition to smaller size, related to the vestigial nature of the hind limb, basilosaurid pelves have greatly reduced ilia and ischia, very small obturator foramina, poorly developed acetabula, and very wide pubes (Gingerich *et al.*, 1990). Despite the fact that the sacral vertebrae of *Georgiacetus* closely approach the condition observed in those of basilosaurids, and that its iliosacral articulation is lost, the pelvis of *Georgiacetus* is surprisingly plesiomorphic in appearance (Fig. 12). The only derived features it shares with basilosaurid pelves are a reduced ischium and a widened pubis, and both of these are at far less derived states than those observed in basilosaurids.

The pelvis of *Georgiacetus* differs from that of *Rodhocetus* (Gingerich *et al.*, 1994, 1995) in having a relatively longer ilium, a shorter, more reduced ischium, a well-developed acetabular notch, and a much wider pubis. Also, the medial articular surface on the ilium for attachment with the first sacral vertebra is absent in *Georgiacetus*. The widened pubes are needed to enclose the much larger transverse processes of the sacral vertebrae. Overall, the younger *Georgiacetus* possesses more derived character states throughout its skeleton related to aquatic adaptations relative to *Rodhocetus*. However, the longer ilium and possession of an acetabular notch of *Georgiacetus* are plesiomorphic states. This suggests that disparate protocetid lineages existed in the middle Eocene (perhaps separated geographically), each independently adapting to a marine way of life. Together with their differences in number and length of lumbar vertebrae, and the different location of the diaphragmatic vertebra, the dissimilar pelves of *Rodhocetus* and *Georgiacetus* imply different styles of aquatic locomotion in the two taxa.

5. Conclusions

   G. vogtlensis is the youngest, most aquatically adapted protocetid for which postcranial skeletal elements are associated with a skull. The thoracic, lumbar, and sacral vertebrae are robust and bear large zygapophyses and processes. The general trend in archaeocete evolution to increase the functional length of the postdiaphragmatic vertebral series has in *Georgiacetus* progressed well beyond the grade observed in *Rodhocetus* by four methods: increased number of lumbar vertebrae;increased length of posterior lumbar, sacral, and anterior caudal vertebrae; an anterior shift in the position of the diaphragmatic vertebra in the thorax; and by morphologic changes to the sacral and anterior caudal vertebrae (most notably enlarging their transverse processes) that allowed them to function more like lumbars. A necessary consequence of the latter was loss of direct articulation between the pelvis and sacrum. This in turn, along with increases in body size, made movement on land even more limited. Concomitant with this was a decrease in use of the hind limb in aquatic locomotion. Gingerich *et al.*, (1994) and Thewissen *et al.* (1996) documented how cetacean aquatic locomotion evolved through four major stages during the Eocene: (1) quadrupedal paddling (pakicetids); (2) dorsoventral undulations of the lower back using the hind feet as the main hydrofoil (ambulocetids); (3) shifting the main hydrofoil from the hind feet to the (fluked?) tail (protocetids); and (4) complete loss of locomotor function in the hind limb with total dependence on caudally propelled swimming (basilosaurids, odontocetes, mysticetes). *Georgiacetus* neatly (for the most part) fills the morphologic gap in this sequence between early protocetids (such as *Rodhocetus*) and basilosaurids. Another postcranial link between *Georgiacetus* and basilosaurids is common possession of osteosclerotic ribs, although this character needs to be studied in other protocetids. This presumably implies a common method to solve problems related to buoyancy. The pelvis of *Georgiacetus*, on the other hand, is overall decidedly plesiomorphic relative to that of known basilosaurids, with the exception of the crucial loss of iliosacral articulation. This likely relates to retention of some, probably limited, locomotor function for the hind limb in *Georgiacetus*. The bizarre morphology of the basilosaurid pelvis could not be attained until this was completely lost.

Acknowledgments

   I thank Hans Thewissen for inviting me to contribute to this volume and for his editorial assistance. Collection and preparation of the specimens of *Georgiacetus* were facilitated by the Georgia Power Company, and supervised by the late R. M. Petkewich. Support for travel and supplies were provided by the Georgia Southern University Office of Research Services and Sponsored Programs and the Department of Geology and Geography. I especially wish to thank my many colleagues with whom I have had useful discussions during the last 7 years concerning archaeocete anatomy, ecology, and systematics.

References

Barnes, L. G., and Mitchell, E. 1978. Cetacea, in: V. J. Maglio and H.B.S. Cooke (eds.), *Evolution of African Mammals*, pp. 582–602. Harvard University Press, Cambridge, MA.

Buffrénil, V. de, Ricqlés, A. de, Ray, C. E., and Domning, D. P. 1990. Bone histology of the ribs of the archaeo-cetes (Mammalia: Cetacea). *J. Vertebr. Paleontol.* **10**:455–466.

Fordyce, R. E., and Barnes, L. G. 1994. The evolutionary history of whales and dolphins. *Annu. Rev. Earth Planet. Sci.* **22**:419–455.

Fraas, E. 1904. Neue Zeuglodonten aus dem unteren Mitteleocän vom Mokattam bei Cairo. *Geol. Palaeontol. Abh. N.F.* **6**:199–220.

Geisler, J., Sanders, A. E., and Luo, Z. 1996. A new protocetid cetacean from the Eocene of South Carolina, U.S.A.; phylogenetic and biogeographic implications in: J. E. Repetski (ed.), *Sixth North American Paleontological Convention Abstracts of Papers. Paleontol. Soc. Spec. Pap.* **8**:139.

Gingerich, P. D. 1991. Partial skeleton of a new archaeocete from the earliest middle Eocene Habib Rahi Limestone, Pakistan. *J. Vertebr. Paleontol.* **11**:31A.

Gingerich, P. D., and Uhen, M. D. 1996. *Ancalecetus simonsi*, a new dorudontine archaeocete (Mammalia, Cetacea) from the early late Eocene of Wadi Hitan, Egypt. *Contrib. Mus. Paleontol. Univ. Michigan* **29**(13):359–401.

Gingerich, P. D., Smith, B. H., and Simons, E. L. 1990. Hind limbs of Eocene *Basilosaurus*: evidence of feet in whales. *Science* **249**:154–157.

Gingerich, P. D., Raza, S. M., Arif, M., Anwar, M., and Zhou, X. 1993. Partial skeletons of *Indocetus ramani* (Mammalia, Cetacea) from the lower middle Eocene Domanda Shale in the Sulaiman Range of Punjab (Pakistan). *Contrib. Mus. Paleontol. Univ. Michigan* **28**(16):393–416.

Gingerich, P. D., Raza, S. M., Arif, M., Anwar, M., and Zhou, X. 1994. New whale from the Eocene of Pakistan and the origin of cetacean swimming. *Nature* **368**:844–847.

Gingerich, P. D., Arif, M., and Clyde, W. C. 1995. New archaeocetes (Mammalia, Cetacea) from the middle Eocene Domanda Formation of the Sulaiman Range, Punjab (Pakistan). *Contrib. Mus. Paleontol. Univ. Michigan* **29**:(11)291–330.

Halstead, L. B., and Middleton, J. A. 1974. New material of the archaeocete whale, *Pappocetus lugardi* Andrews, from the middle Eocene of Nigeria. *J. Min. Geol.* **8**:81–85.

Hulbert, R. C. 1994. Phylogenetic analysis of Eocene whales ("Archaeoceti") with a diagnosis of a new North American protocetid genus. *J. Vertebr. Paleontol.* **14**:30A.

Hulbert, R. C. 1996. Dental and basicranial anatomy of a late middle Eocene protocetid cetacean from the southeastern United States. *Sixth North American Paleontological Convention Abstracts of Papers. Paleontol. Soc. Spec. Pap.* **8**:186.

Hulbert, R. C., Petkewich, R. M., Bishop, G. A., Bukry, D., and Aleshire, D. P. 1998. A new middle Eocene protocetid whale (Mammalia: Cetacea: Archaeoceti) and associated biota from Georgia. *J. Paleontol.* **72**:905–925.

Kellogg, R. 1936. A review of the Archaeoceti. *Carnegie Inst. Washington Publ.* **482**:1–366.

Lucas, F. A. 1900. The pelvic girdle of zeuglodon *Basilosaurus cetoides* (Owen), with notes on other portions of the skeleton. *Proc. U.S. Natl. Mus.* **23**:327–331.

McLeod, S. A., and Barnes, L. G. 1996. The systematic position of *Pappocetus lugardi* and a new taxon from North America (Archaeoceti: Protocetidae). *Sixth North American Paleontological Convention Abstracts of Papers. Paleontol. Soc. Spec. Pap.* **8**:270.

O'Leary, M. A., and Rose, K. D. 1995. Postcranial skeleton of the early Eocene mesonychid *Pachyaena* (Mammalia: Mesonychia). *J. Vertebr. Paleontol.* **15**:401–430.

Tarasoff, F. J., Bisaillon, A., Pierard, J., and Whit, A. P. 1972. Locomotory patterns and external morphology of the river otter, sea otter, and harp seal (Mammalia). *Can. J. Zool.* **50**:915–929.

Thewissen, J.G.M. 1994. Phylogenetic aspects of cetacean origins: a morphological perspective. *J. Mamm. Evol.* **2**:157–184.

Thewissen, J.G.M. 1995. Cetacean diversity in fluvial Eocene sediments of Punjab, Pakistan. *J. Vertebr. Paleontol.* **15**:56A.

Thewissen, J.G.M., Hussain, S. T., and Arif, M. 1994. Fossil evidence for the origin of aquatic locomotion in archaeocete whales. *Science* **263**:210–212.

Thewissen, J.G.M., Madar, S. I., and Hussain, S. T. 1996. *Ambulocetus natans*, an Eocene cetacean (Mammalia) from Pakistan. *Cour. Forsch.-Inst. Senckenberg* **191**:1–86.

Uhen, M. D. 1996a. Composition and characteristics of the subfamily Dorudontinae (Archaeoceti, Cetacea). *Sixth North American Paleontological Convention Abstracts of Papers. Paleontol. Soc. Spec. Pap.* **8**:403.

Uhen, M. D. 1996b. New protocetid archaeocete (Mammalia, Cetacea) from the late middle Eocene Cook Mountain Formation of Louisiana. *J. Vertebr. Paleontol.* **16**:70A.

Uhen, M. D. In press. New species of protocetid archaeocyte(Mammalia, Cetacea) from the middle Eocene of North Carolina. *J. Paleontol.*

Van Valen, L. 1966. Deltatheridia, a new order of mammals. *Bull. Am. Mus. Nat. Hist.* **132**:1–126.

Wortman, J. L. 1901–1902. Studies of Eocene Mammalia in the Marsh Collection. Part I. Carnivora, suborder Creodontia Cope. *Am. J. Sci.* **12**:281–296, 377–382, 421–432, **13**:39–46.

Zhou, X., Sanders. W. J., and Gingerich, P. D. 1992. Funtional and behavioral implications of vertebral structure in *Pachyaena ossifraga* (Mammalia, Mesonychia). *Contrib. Mus. Paleontol. Univ. Michigan* **28**:289–319.

CHAPTER 9

Homology and Transformation of Cetacean Ectotympanic Structures

ZHEXI LUO

1. Introduction

The ectotympanic bullae of whales and land mammals have many differences that resulted from the morphological divergence of whales from their ungulate ancestors. These differences in bullar structure are the product of functional adaptation of cetaceans to the aquatic environment, and ontogenetic alteration of the ancestral precursor condition of the terrestrial ungulates.

The bulla is crucial for hearing in many therian mammals. Its structure varies, at least in part, as a result of functional adaptation to the acoustic environment (Fleischer, 1978). Water is very different from air as a medium to transmit sound, namely, it has greater density and much faster speed of sound. The physical difference of water from air results in a greater sound pressure, longer wavelength, and decreased particle velocity underwater (Reysenbach de Haan, 1960; Hawkins and Myrberg, 1983; Lancaster, 1990). Several specialized bullar structures of cetaceans clearly have acoustic significance for underwater hearing. These bullar characters of cetaceans may thus be attributable to evolutionary adaptation.

Evolutionary adaptation to underwater hearing, however, is only one of several factors that could have determined the anatomical structure of the cetacean bullae. Ontogenetic alterations (for no obvious adaptive value) and phylogenetic constraints by the preexisting precursor structures in land mammal ancestors can also strongly influence the anatomical structure of the cetacean bullae.

Several prominent bullar characters of cetaceans, such as the sigmoid process, develop from embryonic precursors that are also present in the development of terrestrial ungulates. Their apomorphic conditions in cetaceans are attributable to seemingly small ontogenetic alterations from these embryonic precursors (Ridewood, 1922). For example, extant odontocetes have only one apomorphy that is universally absent in nonodontocete mammals—the presence of the bullar accessory ossicle (see the section on odontocetes). The ac-

ZHEXI LUO • Section of Vertebrate Paleontology, Carnegie Museum of Natural History, Pittsburgh, Pennsylvania 15213-4080.
The Emergence of Whales, edited by Thewissen. Plenum Press, New York, 1998.

cessory ossicle develops from the same precursor as, thus homologous to, the embryonic accessory ossicle of ungulates (*sensu* van Kampen, 1905). The conspicuous difference between an apomorphic accessory ossicle of adult odontocetes and a plesiomorphic processus tubarius (derived from the accessory ossicle) in ungulates is formed by a minor ontogenetic alteration. The embryonic accessory ossicle retains a neotenic condition of independent ossification in adult odontocetes, whereas it is incorporated into the bullar body in adult ungulates. The evolutionary origin of this odontocete apomorphy may have nothing to do with the sophisticated hearing adaptation of odontocetes. If considered within the broader framework of cetacean–ungulate phylogeny, only a few bullar characters in the earliest cetaceans are truly derived characters. Most characters in the bullae of the earliest cetaceans are primitive features shared by ungulate mammals. Some other highly modified cetacean bullar structures have little or no demonstrable function for underwater hearing. These are best considered to be secondary modifications correlated with the transformation of other skull structures.

In the literature of basicranial anatomy of modern cetaceans, several bullar features were commonly regarded as "unique cetacean characters." These include: (1) a posterior process enlarged from the posterior crus of the ectotympanic ring, forming the primary bullar articulation to the basicranium; (2) a sigmoid process enlarged from the anterior crus of the ectotympanic ring; (3) a conical apophysis modified from the ectotympanic ring; (4) a conical tympanic ligament derived from tympanic membrane in correlation with a modification of the ectotympanic ring; (5) an involucrum formed from the pachyosteosclerosis of the bulla. To understand the evolution of these cetacean bullar structures, it is useful to distinguish two categories of structures: (1) those bullar structures developed from the precursor condition in ancestral ungulates by changes in the timing and pace of ontogeny, versus (2) the structures attributable to the evolutionary adaptation for underwater hearing.

The goals of this chapter are: (1) to review the literature on the embryogenesis of major bullar structures in extant mysticetes and odontocetes, in an effort to establish the homology (or the lack thereof) between the different bullar structures of adult cetaceans and their ungulate relatives (see Table I); (2) to describe the bullar structures of pakicetids, the transitional group between terrestrial ungulates and the more derived whales, and the protocetids that were amphibious; (3) to survey the systematic diversity of the bullar structure in the main cetacean groups, such as basilosaurids, mysticetes, and odontocetes (Table II); and (4) to elucidate the pattern of evolutionary transformation by distinguishing the cetacean apomorphies that could have derived from adaptive evolution from the apomorphies with no apparent adaptive value.

Institutional abbreviations are listed in Table III.

2. Development

Ridewood (1922) provides a detailed description of the development of the ectotympanic bulla in balaenopterid mysticetes. He traces the development of various bullar characters in an ontogenetic series of four fetal skulls of balaenopterids on the basis of two different species (*Megaptera novaeangliae* and *Balaenoptera musculus*). Ridewood's observations on the embryogenesis of the bulla and tympanic membrane ("tympanum") are represented below, supplemented from other studies on *Monodon* (Eales, 1951) and other cetaceans (Kernan, 1916; Fraser and Purves, 1960). These embryological observations can shed considerable light on the development of the different bullar structures in the adults of various cetacean groups (Figs. 1 and 2).

The ectotympanic, an intramembranous ossification, forms a simple ring at the earliest available embryonic stage of *Megaptera* at a body length of 152 mm (Ridewood, 1922). The anterior crus ("anterior process" of some authors) of the ectotympanic ring is connected to the meckelian cartilage (Fig. 1A). The posterior crus is anchored to the crista parotica, a crest on the cartilaginous otic capsule (MacPhee, 1981) and near the base of Reichert's cartilage, a precursor to the tympanohyal of the adult. At this stage the ectotympanic ring is similar to the ectotympanic of all other placental mammals (de Beer, 1937; MacPhee, 1981).

At this early stage, the tympanic membrane is attached to the ectotympanic ring on the anterior, ventral, and posterior sides (Fig. 1A). Its dorsal edge is attached to the squamosal. The medial side of the membrane is attached to the manubrium of the malleus via a fibrous tissue tract that later develops into the conical tympanic ligament. The tympanic membrane at this stage is slightly concave toward the outside and the whole membrane is oriented nearly horizontal. In all essential aspects, the tympanic membrane at this embryonic stage of cetaceans is similar to the tympanic membrane of both embryonic and adult terrestrial mammals (Fraser and Purves, 1960), except for slightly different orientation of the tympanic membrane in the later development of terrestrial mammals.

At the next embryonic stage of *Megaptera* (Fig. 1B; body length of 406 mm), the middle segment of the ectotympanic ring grows to form a crescentic plate. The concave dorsal side of the plate forms the floor of the tympanic cavity (Ridewood, 1922). The tympanic membrane is flat to slightly concave.

In the embryo of *Megaptera* with a body length of 686 mm (Fig. 1C), the ectotympanic starts to grow anteriorly. This new growth forms the Eustachian part of the bulla, and is anterior to the base of the anterior crus. As a result, the anterior crus becomes a dorsolaterally projecting process in the midportion of the bulla, and starts to bear a strong resemblance to the sigmoid process in the adult bulla (Fig. 1C).

The soft tissues around the malleus and ectotympanic ring start to develop into two interrelated structures. The first is an external glove finger. This glove finger is an external membrane strongly convex toward the outside ("thimble-shaped") and forms the proximal (or inner) end of the fibrous plug that occludes the external auditory meatus. The glove finger at this embryonic stage starts to assume the morphology or the condition of the adult mysticetes (Lillie, 1910). Much of the inner edge of the glove finger is attached to the ectotympanic ring, while its dorsal edge is attached to the squamosal.

The second structure to appear simultaneously at this embryonic stage is an internal conical tympanic ligament between the glove finger membrane and the manubrium of malleus. The distal end of internal conical tympanic ligament is attached to the ectotympanic ring and to the inner surface of the glove finger. The proximal end of the conical tympanic ligament is attached to the manubrium of the malleus. Fraser and Purves (1960) consider that both the glove finger membrane and the conical tympanic ligament are homologous to the tympanic membranes of other mammals. However, Lancaster (1990) argues that only the conical tympanic ligament is homologous to the tympanic membrane. This study assumes that the conical tympanic ligament of mysticetes is a modified form of a more or less planar tympanic membrane of terrestrial mammals, as suggested by Lancaster (1990). The

Table I. Homology of Bullar Structures of Ungulate Mammals and Cetaceans[a]

Bullar character	Cetacean apomorphies	Ungulate mammal plesiomorphies
1	Sigmoid process	Both the hypertrophied anterior crus of ectotympanic ring and the anterior wall of meatus
	1-a. Platelike (= anterior wall of meatus) (pakicetids only)	*
	1-b. Swelling and involuted margins (post-pakicetid cetaceans)	*
2	Articulation of sigmoid process to basicranium at:	Articulation of anterior crus to the squamosal
	2-a. Squamosal (archaeocetes only)	Same as above
	2-b. Tegmen tympani of petrosal (most extant cetaceans)	*
	2-c. With hypertrophied ventrolateral tuberosity (some odontocetes)	*
3	Mallear process	Anterior crus of ectotympanic contacting the long process of malleus, also incorporating the goniale (mallear accessory ossicle)
4	Tubular portion of external auditory meatus of ectotympanic	Present
	4-1. Present but reduced (pakicetids)	*
	4-2. Lost (basilosaurids, mysticetes, odontocetes)	*
5	Conical apophysis (medial conical process)	A folded middle portion of ectotympanic ring
	5-a. Absent or weakly developed (pakicetids)	
	5-b. Conical apophysis is strongly developed (basilosaurids)	*
	5-c. Conical apophysis is appressed to sigmoid process (mysticetes and odontocetes)	*
6	Homologues of processus tubarius / accessory ossicle / anterior pedicle	Processus tubarius of bulla (*sensu* van Kampen, 1905)
	6-a. Accessory ossicle as a new ossification split from processus tubarius (odontocetes)	Neomorphic feature derived from processus tubarius
	6-b. Accessory ossicle develops into the anterior pedicle and fused to the petrosal (mysticetes)	*
7	Processus tubarius contacting basicranium at:	Processus tubarius contacting basicranium
	7-a. Squamosal (archaeocetes only)	Present
	7-b. Petrosal (basilosaurids, mysticetes, odontocetes)	*
8	Accessory ossicle contacting basicranium at:	*
	8-a. Fovea epitubaria of the petrosal (delphinoids)	*
	8-b. Anterior bullar facet [separate from fovea epitubaria (*Zarhachis*, eurhinodelphids)]	*
	8-c. Fusion with anterior process of petrosal (physeteroids, ziphiids, some delphinoids)	*

No.	Character	Apomorphy / comment
9	Pedicle for anterior process of tympanic (mysticetes)	Derived from processus tubarius via an intermediate condition of accessory ossicle
10	Base of posterior process—posterior crus	Posterior crus of the ectotympanic ring
10-a.	Broad (pakicetids)	Also present in mesonychids and other ungulates
10-b.	Constricted neck (pedicle) of posterior process	*
10-c.	Double posterior pedicles with a pedicle foramen (protocetids and basilosaurids)	*
11	Morphology of posterior process—posterior crus	Posterior crus tapers posteriorly
11-a.	Thin and vertical plate (posterior wall of external auditory meatus of ectotympanic) (pakicetids)	Expanded from the contact of posterior crus of ectotympanic and crista parotica of petrosal
11-b.	Thick and horizontal plate covering the entire posterior process of petrosal (all postpakicetid cetaceans)	*
11-c.	Suture contact to posterior process of petrosal (archaeocetes and odontocetes)	*
11-d.	Smooth surface contact to posterior process of petrosal (kogiids)	*
11-e.	Fusion to posterior process of petrosal (mysticetes)	*
12	Involucrum	Pachyosteosclerotic medial rim of bulla
13	Eustachian tube	Eustachian tube
13-a.	With recoiled rim (nonodontocete cetaceans)	Adult condition in artiodactyls
13-b.	With sharp edge (odontocetes only)	Juvenile condition in artiodactyls
14	Median furrow/ridge	*
14-a.	Furrow (protocetids, basilosaurids, odontocetes)	*
14-b.	Ridge (mysticetes)	*
15	Interprominental notch	Sulcus for tympanohyal (Reichert's cartilage)
16	Petrotympanic fissure	Gap formed by the separation of the medial rim of bulla from the basicranium

[a]The anatomical terminology for the basicranium of noncetacean eutherians is adopted from Ridewood (1922), Van der Klaauw (1931), and MacPhee (1981). The synonymous terms for the cetaceans are adopted from Pompecki (1922), Ridewood (1922), Kellogg (1928, 1936), and Kasuya (1973). For phylogenetic distribution of the apomorphies of bullar charecters among cetaceans see Table II. Some bullar characters are labeled in Fig. 8.

*Cetacean neomorphies (or "apomorphies") for which there may be no morphological counterpart (or homologous character) in ungulate mammals.

Table II. Ectotympanic Apomorphies at Different Hierarchical Levels of Cetacean Phylogeny[a]

1. Cetacea (including Pakicetidae)
   A. Presence of the involucrum
   B. Presence of the lateral furrow on the bulla
   C. Sigmoid process is a simple plate (equivocal, present in the artiodactyl *Diacodexis*)
   D. Absence of the ventral floor to external meatal tube (equivocal)
2. Monophyletic group of "post-Pakicetidae" Cetacea
   (Protocetidae (Basilosauridae + Dorudontidae (Mysticeti + Odontoceti)))
   A. Incipient conical apophysis
   B. Reduced tympanic opening for the external meatus
   C. Sigmoid process is twisted and has involuted margins
   D. Elongate posterior process of the ectotympanic to cover the entire length of the mastoid (posterior) process of the petrosal. The distal (posterior) portion of the posterior process is a horizontal plate (not vertical)
   E. Presence of median furrow on the ventral surface of the bulla
   F. Presence of double pedicles for the posterior process of tympanic (equivocal: secondarily lost in crown group of cetaceans)
3. Monophyletic group of (Basilosauridae + Dorudontidae (Mysticeti + Odontoceti))
   A. Annulus for tympanic membrane vestigal or absent
   B. Conical apophysis is fully developed
   C. Broader contact of the processus tubarius of the bulla with the anterior process of the petrosal than in the Protocetidae
4. Crown group of Cetacea (Mysticeti + Odontoceti)
   A. Processus tubarius lost contact with the squamosal
   B. Sigmoid process detached from the squamosal
   C. loss of double posterior pedicles (equivocal—reversal to Pakicetidae)
5. Odontoceti
   A. Presence of the accessory ossicle in adults [equivocal—reversal to the embryonic (but not adult) condition in modern mysticetes and the artiodactyl *Ovis*]
   B. Sharp margin around the anterior Eustachian aperture [equivocal—reversal to the embryonic (but not adult) condition in the artiodactyl *Ovis; Xenorophus* is an exception to odontocetes in this character]
6. Mysticeti
   A. Long pedicle for the anterior process of the ectotympanic
   B. Dorsal end of the anterior process of ectotympanic fused to the tegmen tympani of petrosal
   C. Presence of the median keel (in the place of the median furrow)

[a]Some apomorphies are labeled in Fig. 8.

Table III. Abbreviations of Institutions

AMNH	American Museum of Natural History (New York)
CMNH	Carnegie Museum of Natural History (Pittsburgh)
H-GSP	Joint collection of Howard University (Washington, DC) and Pakistani Geological Survey (Islamabad, Pakistan)
IVPP	Institute of Vertebrate Paleontology and Paleoanthropology, Chinese Academy of Sciences (Beijing, China)
LUVP	Lucknow University, Vertebrate Palaeontology Laboratory (India)
UM	Museum of Paleontology, University of Michigan (Ann Arbor)
UM-GSP	Joint collection of University of Michigan (Ann Arbor) and Geological Survey of Pakistan (Islamabad, Pakistan)
USNM	United States National Museum of Natural History, Smithsonian Institution (Washington, DC)
VPL	Vertebrate Palaeontology Laboratory, Punjab University (Chanigarh, India)

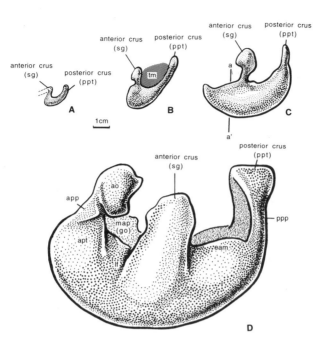

FIGURE 1. Development of the ectotympanic bullar structures in balaenopterids (composite series of bullae in four fetal skulls of balaenopterids, all after Ridewood, 1922). (A) *Megaptera* (ventral view, 152 mm in body length); (B) *Megaptera* (ventral view, 406 mm in body length); (C) *Megaptera* (lateral view, 686 mm in body length); (D) *Balaenoptera* (lateral view, 1930 mm in body length). Abbreviations: ao, accessory ossicle (also present in adult odontocetes); app, pedicle of the anterior process of the bulla (derived from an embryonic accessory ossicle); apt, anterior process of the tympanic (= processus tubarius); eam, external auditory opening; go, goniale; lf, lateral furrow of the ectotympanic bulla; map, mallear process of the tympanic and the site of attachment for malleus (homologous to the goniale element associated with the anterior process of malleus); ppp, pedicle of the posterior process of ectotympanic; ppt, posterior process of the ectotympanic bulla; sg, sigmoid process (= anterior crus of the ectotympanic ring of ungulate mammals); tm, tympanic membrane. Line a'a in panel C indicates the approximate position of transverse sections of Fig. 2.

glove finger is absent in extant odontocetes, and is therefore considered to be a neomorph of mysticetes. The possible evolutionary transformation of the glove finger will be discussed later in this chapter.

In the embryo of *Balaenoptera musculus* at a body length of 1194 mm, two new bony structures have emerged on the anterior part of the bulla. The first structure is the mallear process of the bulla near the base of the sigmoid process [Fig. 1D: map(go)]. This structure is fused in a later stage to the anterior process of the malleus, and to the membranous ossification that is homologous to the goniale.

The second new structure to appear in the cetacean bulla at this stage is the accessory ossicle (Fig. 1D: ao), a structure that is distinctive from the adjacent region at this stage. The accessory ossicle is present not only in the development of mysticetes and odontocetes, but also in the bulla of the fetus of *Ovis* (van Kampen, 1905). The accessory ossicle has slightly different development fates in mysticetes, odontocetes, and *Ovis* (in later stages beyond that of Fig. 1D).

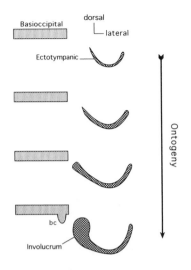

FIGURE 2. Development of the involucrum from the medial edge of the ectotympanic bulla of cetaceans (modified from Hanke, 1914). Schematic transverse sections across the anterior part of the embryonic bulla (indicated by line a′a in Fig. 1C). bc, basioccipital crest (= falcate process of the basioccipital).

In adult mysticetes, the accessory ossicle develops into a gracile pedicle, and the distal (dorsal) end of the pedicle will be eventually fused with the anterior process (= anterior extension of the tegmen tympani) of the petrosal. By contrast, the accessory ossicle in odontocetes, once formed in the embryo, remains more or less the same in the adults. The ossicle of an adult odontocete retains the embryonic condition of a mysticete fetus. Relative to the anterior pedicle of adult mysticetes, the accessory ossicle of adult odontocetes is a neotenic feature.

The accessory ossicle (ao) is distinctive from the adjacent region of the bulla in the fetus of *Ovis* (van Kampen, 1905). The ossicle fuses later to the adjacent region of the bulla in adult *Ovis*. It is then known as the processus tubarius, or the anterior process of the bulla (van Kampen, 1905). Thus, the embryonic accessory ossicle in the newborn of *Ovis* is incorporated into the processus tubarius, which is indistinguishable from the rest of the bullar body in the adult.

The main differences among ungulates (such as *Ovis*), mysticetes, and odontocetes are as follows. In *Ovis*, the embryonic accessory ossicle is incorporated into the processus tubarius and the latter becomes indistinguishable from the bullar body. In mysticetes, however, the accessory ossicle is separated from (but still connected to) the processus tubarius by a slender anterior pedicle. The dorsal part of the accessory ossicle is fused to the anterior process (a part of the tegmen tympani) of the petrosal in adults. In adult odontocetes, the accessory ossicle is similar to the anterior pedicle of mysticetes in being a distinctive structure but still attached to the processus tubarius (= anterior process of the ectotympanic). But the accessory ossicle differs from the pedicle of mysticetes in that the former is not incorporated into the tegmen tympani (there are some exceptions, see the section on odontocetes). The incorporation of the accessory ossicle into the tegmen tympani in extant mysticetes, and the independent accessory ossicle of odontocetes are both considered to be derived conditions by comparison with the processus tubarius that is fully incorporated in the bullar body in *Ovis* (see Section 4).

The medial margin of the bulla starts as a thin edge during ontogeny (Fig. 2). Then the margin becomes increasingly thickened and involuted. This thick and involuted medial margin of the bulla becomes extremely dense (pachyostotic) and forms the involucrum in adults (Hanke, 1914; Ridewood, 1922).

The external auditory openings in the bullae of basilosaurids, odontocetes, and mysticetes differ from those of terrestrial ungulates in possessing a conical apophysis, a triangular projection protruding into the external auditory meatal opening. This apophysis is often tightly pressed to the base of the sigmoid process, and blocks much of the auditory meatal opening. As a result, the external auditory opening is extremely small in proportion to the whole bulla (Lillie, 1910; Kasuya, 1973; Pilleri *et al.*, 1987). The conical apophysis (median conical process) of adults develops from a W-shaped folding in the middle section of the embryonic ectotympanic (Hanke, 1914; see Fig. 3). The conical apophysis serves as the lateral point of attachment of the conical tympanic ligament (homologous to the tympanic membrane of ungulates). The annulus for suspending the planar tympanic membrane in terrestrial mammals disappears, or becomes a vestigial trace on the internal surface of the bullar wall in some cetaceans (e.g., a juvenile *Megaptera*, USNM 268230; a juvenile *Balaenoptera physalus*; and a juvenile *Eschrichtius*, an uncataloged specimen of USNM). Even if the vestigial annulus is present, it does not suspend the conical tympanic ligament.

In contrast to the cetacean condition, the bony external auditory meatus in terrestrial ungulates opens up like a funnel toward the interior of the tympanic cavity. The annulus for suspending the tympanic membrane at the inner edge of the meatal tube is larger in diameter than the size of the external tube, allowing a tympanic membrane with a larger surface area than the external auditory meatal opening. The modification of the adult conical tympanic ligament from a planar embryonic tympanic membrane is directly correlated with the folding of the embryonic ectotympanic into a conical apophysis (Fig. 3), and degeneration of the annulus during the development of cetaceans.

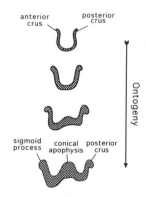

FIGURE 3. Ontogeny of the conical apophysis (middle conical process) from the ectotympanic ring in cetaceans (lateral view, modified from Hanke, 1914). The conical apophysis develops from a W-shaped fold in the midsection of the ring. The conical apophysis serves as the main point of the lateral attachment of the conical tympanic ligament.

3. Systematic Diversity (Table II)

3.1. Pakicetids

Ichthyolestes (Fig. 4D,E, H-GSP 18391). The most prominent feature in the medial aspect is the involucrum, which is the greatly inflated (pachyostosed) medial rim of the bulla (Figs. 4–6: iv). The involucrum has a slight depression on its dorsal aspect, although the size of this depression is not clear because of the breakage of the specimen. The dorsoposterior part of the involucrum bears a protuberance for articulation with the exoccipital (Figs. 4, 5: eoc). Posteriorly the external bulge of the involucrum is demarcated by a broad sulcus for the tympanohyal (Figs. 4–6: ths). The ventral side of the bulla does not form a medial prominence as in protocetids or basilosaurids. The anterior part of the involucrum is rounded and it borders anterolaterally on the broad trough for the Eustachian tube (Figs. 4–6: et).

The processus tubarius (anterior process of the tympanic, Fig. 4: apt) is a bulging part on the lateral side of the bulla. The processus is formed by a thin bullar wall, and its interior is completely hollow, as evidenced by the broken wall of the processus tubarius (Figs. 4–6). The processus tubarius of *Ichthyolestes* is very similar to that of *Ovis* (CMNH G995, G997). In *Pakicetus inachus* in which the bulla is preserved *in situ* (UM-GSP 081, Gingerich *et al.*, 1983; Luo and Gingerich, in press), the processus tubarius articulates with the entoglenoid process of the squamosal, as is the case for mesonychids and extant artiodactyls (Zhou *et al.*, 1995; Luo and Gingerich, in press).

The external auditory meatus of pakicetids is formed by a deep trough in the squamosal and posterior to the glenoid (Gingerich *et al.*, 1983). The distal portion of the meatus is an open groove. The proximal (medial) portion of the meatus is ventrally enclosed by the ectotympanic. In the detached ectotympanic bulla of *Ichthyolestes* (Figs. 4 and 5: eam), the external auditory meatus is a tubular structure and is trough shaped in dorsal view.

The free inner edge of the meatal tube protrudes internally into the bullar cavity. The under side of the free edge of the meatal tube is buttressed by three trabeculae (or septa) in the bulla (Figs. 5, 6: sep). The trabeculae are partitions that separate the lateral part of the bullar cavity into several compartments. The freestanding inner edge of the meatus in *Ichthyolestes* is identical to those of terrestrial placental mammals, such as *Ovis* (Fig. 7:an). Trabeculae in the bullar cavity are also present in certain terrestrial mammals, such as many rodents, some carnivores, and some primates (Fleischer, 1978).

The annulus for the suspension of the tympanic membrane is formed by a free inner edge and a corresponding concentric groove parallel to the inner edge (Figs. 5–7: an). The internal portion of the meatal tube is like a funnel, and the annulus is much larger in diameter than the tube of the auditory meatus. The annulus is similar to that of a terrestrial ungulate mammal in morphology and size. This suggests that *Ichthyolestes* had a planar tympanic membrane similar to that of a terrestrial mammal, but that it lacked the conical tympanic ligament of the extant cetaceans.

In most terrestrial mammals, the tympanic membrane is slightly concave toward the outside. This is especially true for the pars tensa of the membrane that attaches to the manubrium of the malleus. The degree of concavity of the membrane depends on how much the manubrium of the malleus is rotated inwards and away from the plane of the annulus. Although the malleus is still unknown in pakicetids, their incudo-mallear joint as preserved on the incus (Thewissen and Hussain, 1993) is more rotated inwards than in terrestrial mam-

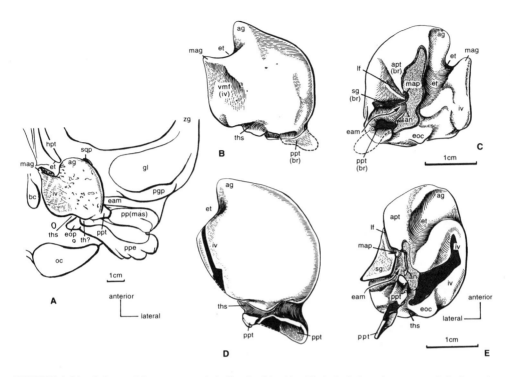

FIGURE 4. Morphology of the ectotympanic bulla of pakicetids. (A) Articulation of ectotympanic bulla to the basicranium of *Pakicetus inachus* (UM-GSP 081; left side, ventral view; modified from Luo and Gingerich, in press). (B, C) Ventral and dorsal views of the bulla of *Pakicetus attocki* (H-GSP 96334, the specimen has broken sigmoid process and processus tubarius, the missing posterior process is reconstructed after UM-GSP 081). (D, E) Ventral and dorsal views of the bulla of *Ichthyolestes* (H-GSP 18391; Thewissen and Hussain, 1998). Abbreviations: ag, anterior angle of the bulla; an, annulus of the ectotympanic ring (for suspending the tympanic membrane); apt, processus tubarius or anterior process of the tympanic; bc, basioccipital crest; (br), broken or incomplete structure; eam, external auditory meatus; eoc, exoccipital contact (on the ventral side of involucrum); eop, exoccipital bullar process; et, Eustachian tube; gl, glenoid fossa; hpt, hamulus of pterygoid; iv, involucrum; lf, lateral furrow; mag, medio-anterior angle of involucrum; map, mallear process of ectotympanic bulla; oc, occipital condyle; pgp, postglenoid process; pp(mas), posterior (mastoid) process of petrosal; ppe, paroccipital process of exoccipital; ppt, posterior process of tympanic; sg, sigmoid process (a structure equivalent to the anterior wall of the meatus and derived from the anterior crus of the embryonic ectotympanic); sqp, squamosal process for bulla; th, attachment site of tympanohyal; ths, tympanohyal sulcus (= interprominential notch of the bulla); vmf(iv), ventromedial facet of the involucrum; zg, zygoma (of squamosal).

mals, but less so than in basilosaurids (Lancaster, 1990) and modern cetaceans (Fleischer, 1978). This indicates, albeit indirectly, that the tympanic membrane may be slightly more concave than in terrestrial mammals. Yet it had not achieved the acute conical shape of the conical tympanic ligament of the more derived basilosaurids (Lancaster, 1990) and extant cetaceans (Fleischer, 1978; Thewissen, 1994).

The anterior and posterior walls of the meatus are formed by the sigmoid process and the posterior process of the bulla, respectively. The sigmoid process is a simple and transverse plate. Its anterior face is slightly convex and its posterior face is flat to slightly con-

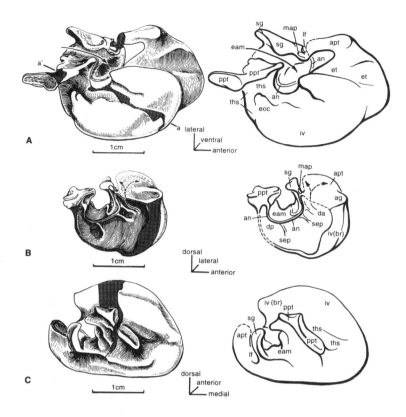

FIGURE 5. Ectotympanic bulla of *Ichthyolestes* sp. (H-GSP 183931; Thewissen and Hussain, 1998). (A) Dorsal view (line a′a indicates approximately the plane of breakage). (B) Internal (and anterior) view of the lateral half of the broken bulla (broken along plane a′a). (C) Posterior view. Hatch pattern indicates the breakage of the specimen. Abbreviations: ag, anterior angle; an, annulus (for suspension of tympanic membrane); apt, anterior tympanic process (processus tubarius); (br), broken structure; da, diverticulum anterior (antrum, or anterior chamber of bullar cavity); dp, diverticulum posterior (posterior chamber of bullar cavity); eam, external auditory meatus; eoc [exoccipital contact (on the side of involucrum)]; et, Eustachian tube; iv(br), involucrum (breakage); lf, lateral furrow; map, mallear process (of bulla); ppt, posterior process of tympanic; sep, septum or trabecula within the bullar cavity; sg, sigmoid process (a structure equivalent to the anterior wall of the meatus and derived from the anterior crus of the embryonic ectotympanic); ths, sulcus for tympanohyal on the bulla.

cave. The margin of the sigmoid process forms a sharp edge, very much unlike the sigmoid process in the more derived whales that have a thick and involuted rim. The base of the sigmoid process is separated anteriorly from the processus tubarius by the Glaserian fissure, the gap through which the chorda tympani exits the bulla. The Glaserian fissure leads to the lateral furrow (Figs. 4, 5: lf) on the lateral surface of the bulla. At the dorsal end of the furrow and near the base of the sigmoid process is the mallear process of the bulla—the site of attachment for the anterior process of the malleus.

The posterior process of the ectotympanic bulla is a thin and vertical plate between the sulcus of the tympanohyal and the external auditory meatus (Fig. 4E: ppt). The proportion

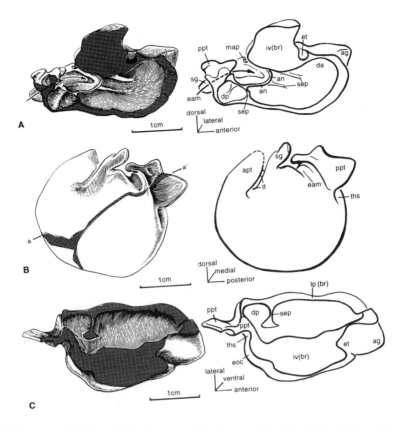

FIGURE 6. Ectotympanic bulla of *Ichthyolestes* sp. (Thewissen and Hussain, 1998). (A) Internal (and medial) view of the lateral half of the broken bulla (broken along line a′a as indicated in panel B). (B) Ventrolateral view of the bulla (broken along plane a′a). (C) Internal (and lateral) view of the medial half of the broken bulla. Hatch pattern indicates the breakage of the specimen. Abbreviations: ag, anterior angle, an, annulus (for suspension of tympanic membrane); apt, anterior process of tympanic (processus tubarius); (br), broken structure; da, diverticulum anterior (antrum or anterior chamber of bullar cavity); dp, diverticulum posterior (posterior chamber of bullar cavity); eam, external auditory meatus; eoc, exoccipital contact on bulla; et, Eustachian tube; iv(br), involucrum (broken); lf, lateral furrow; lp (br), lateral lip of bulla (broken); map, mallear process (of bulla); ppt, posterior process of tympanic; sep, septum or trabecula within the bullar cavity; sg, sigmoid process (a structure derived from the anterior crus of the embryonic ectotympanic); ths, sulcus for tympanohyal on the bulla.

of the posterior process relative to the entire bulla of *Ichthyolestes* is about the same as in *Pakicetus inachus* (Fig. 4). The dorsoposterior face of the posterior process of the ectotympanic is slightly concave and it contacts the mastoid (= posterior) process of the petrosal (Fig. 4D). Because it is much shorter than the mastoid process of the petrosal, the posterior process only covers the proximal portion of the mastoid process of the petrosal (Fig. 4A). Near the base of the posterior process of the ectotympanic is an internal projection that arches over the external auditory meatus, reaching toward (but not contacting) the sigmoid process (Fig. 5). This projection is probably homologous to the posterior crus of the ectotympanic ring.

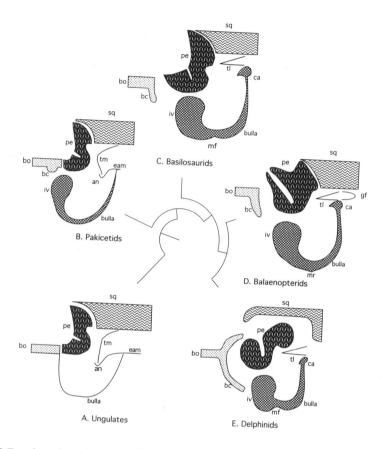

FIGURE 7. Transformations of external auditory meatus, tympanic annulus, and tympanic membrane (schematic transverse section). The transformation of a functional planar tympanic membrane is correlated to the reduction of the tympanic annulus and the tubular portion of external auditory meatus. Transformation of a planar tympanic membrane into the conical tympanic ligament occurred as the annulus for membrane suspension was reduced and the ectotympanic ring is folded into the conical apophysis (medial process). (A) Generalized structure of ungulates (based on *Mesonyx* and the artiodactyl *Ovis*. (B) Pakicetids (apomorphies: presence of the involucrum, reduction of the lateral portion of the external auditory meatus; by inference, the tympanic membrane was planar but more concave than in ungulates). (C) Basilosaurids (apomorphy: presence of the conical apophysis; the conical tympanic membrane is inferred to be present). (D) Balaenopterids (as representative of mysticetes). (E) Delphinids (as representative of extant odontocetes; apomorphy: absence of the bony external auditory meatus in the squamosal). The spiral cladogram suggests the phylogenetic relationships of ungulates and cetaceans. Abbreviations: an, annulus for suspension of tympanic membrane; bc, basioccipital crest (falcate process); bo, basioccipital; ca, conical apophysis (medial process); eam, external auditory meatus; gf, glove finger; iv, involucrum; mf, median furrow of bulla; mr, median ridge of bulla; pe, petrosal; sq, squamosal; tl, conical tympanic ligament; tm, tympanic membrane.

Pakicetus attacki (H-GSP 96344; Fig. 4B,C). The bullae of *Pakicetus* are similar to those of *Ichthyolestes*, although some features related to the involucrum differ between the two taxa. As in *Ichthyolestes*, the bulla of *Pakicetus* has an annulus for the suspension of the tympanic membrane. It also possesses a platelike sigmoid process (present but broken in H-GSP 96344, complete in UM-GSP 081) and a short and platelike posterior process

(UM-GSP 081). This posterior process is much shorter than the mastoid (posterior) process of the petrosal such that it covers the proximal portion but not the distal portion of the latter. In both taxa, the processus tubarius is formed by a thin crust of the bullar wall. The anterior part of the processus forms a lateral angle (Fig. 4: ag). On the dorsal aspect of the bulla, a broad trough for the Eustachian tube is present between the processus tubarius and the involucrum (Fig. 4E).

The isolated bulla of *Pakicetus* (H-GSP 96344) is less complete than the bulla of *Ichthyolestes* (H-GSP 18391). The bulla of *Pakicetus inachus* type specimen (UM-GSP 081) remains attached to the skull and its interior structures are not visible. However, as far as can be ascertained, the involucrum differs between *Pakicetus* and *Ichthyolestes*. The anterior end of the involucrum forms a pointed medial angle in *Pakicetus* (Fig. 4: mag). The comparable part is fully rounded in *Ichthyolestes*. Related to this feature is the difference in the anterior aperture of the Eustachian tube between the two taxa. A prominent anterior notch for the Eustachian tube is present in *Pakicetus inachus* (Gingerich *et al.*, 1983; Luo and Gingerich, in press) but absent in *Ichthyolestes*. The involucrum forms a prominent posteromedial corner in *Pakicetus* but is fully rounded posteromedially in *Ichthyolestes*. The ventromedial surface of the involucrum is more flat in *Pakicetus* but the comparable area is more convex in *Ichthyolestes* (Fig. 4).

Cetaceans including pakicetids have only one unambiguous bullar synapomorphy that is absent in all noncetacean mammals—the involucrum, or the pachyosteosclerosis of the medial margin of the bulla. Other diagnostic characters, such as the sigmoid process, as discussed below, are now open to question in the wake of the new fossil evidence from *Pakicetus* and *Ichthyolestes*.

The sigmoid process is a prominent process in front of the external auditory meatal opening. It is so named because its distal (dorsal) end is slightly twisted from its base in modern cetaceans (see Kellogg, 1928, 1936). The presence of the sigmoid process has been widely accepted as a major diagnostic character for cetaceans (Kellogg, 1936; Gingerich *et al.*, 1983; Barnes, 1984; Oelschläger, 1990; Lancaster, 1990; Thewissen *et al.*, 1994; Berta, 1994; Fordyce and Barnes, 1994).

This character is equivocal. In mammals with a fully developed ectotympanic floor for the meatal tube, such as *Mesonyx* (Geisler and Luo, this volume) and *Sinonyx* (Zhou *et al.*, 1995; Luo and Gingerich, in press), the anterior crus becomes incorporated into the anterior wall of the external auditory meatus. Thus, it does not appear as an independent process. By contrast, the sigmoid process is prominent in cetaceans in which the meatal floor adjacent to the process was poorly developed relative to terrestrial mammals. The anterior wall of the meatal tube becomes more isolated and prominent, forming the sigmoid process as an independent projection. In view of the new evidence from *Ichthyolestes* and *Pakicetus* (that the sigmoid is a simple plate, not S-shaped, and lacking the involuted margins) and its development from the anterior crus of the ectotympanic ring, the sigmoid process should be redefined as a systematic character, and its value as a cetacean synapomorphy should be reconsidered.

The cetacean sigmoid process is an enlarged version of the anterior crus of the ectotympanic ring of most therian mammals. As shown by Ridewood (1922), the sigmoid process can be clearly traced to the anterior crus (process) of the ectotympanic ring in early ontogeny. In terrestrial mammals with an ectotympanic meatal tube, such as rodents (Webster, 1975), lagomorphs (Novacek, 1977), and ungulates, the anterior crus of the ring is annexed by the meatal tube and the crus is incorporated into anterior wall of the meatal tube. In extant cetaceans, the lateral growth of the ventral meatal floor (but not anterior crus and the anterior wall of the meatus) of the ectotympanic is arrested early in embryogenesis. As a result, the sigmoid process remains as an independent process as did its embryonic precursor the anterior crus. As such, the sigmoid process of the adult bulla is neotenic and represents the retention of the embryonic condition of the ectotympanic in the noncetaceans. The most crucial development that contributes to the formation of an independent anterior crus (= sigmoid process) is the early arrest of embryonic growth of the ectotympanic meatal floor.

The sigmoid process in pakicetids is a simple plate. Its edge lacks the involuted and thick margin of basilosaurids and extant cetaceans. It does not have a "sigmoid" outline as in extant odontocetes and mysticetes. Its prominence as an independent process is correlated with the absence of the ventral floor of the meatal tube. Since the cetacean sigmoid process is only an enlarged version of the anterior crus of ectotympanic ring of other mammals, the absence of the ventral floor to the meatal tube would be a better apomorphy for cetaceans in sharp contrast to the meatal tube floor in mesonychids (Zhou *et al.*, 1995; Luo and Gingerich, in press; Geisler and Luo, this volume). Therefore, it may be better to use "the absence of the ventral meatal floor" as a cetacean apomorphy than "the presence of a sigmoid process." More properly, the sigmoid process should be defined as "an enlargement of the anterior crus in the adult bulla" or "anterior wall of the external auditory meatus."

The distribution of the sigmoid process among ungulates and cetaceans suggests that this character may be homoplastic. A sigmoid process is present in the early artiodactyl *Diacodexis* (Russell *et al.*, 1983). In mesonychids (*Sinonyx*, Zhou *et al.*, 1995; *Mesonyx*, Luo and Gingerich, in press; Geisler and Luo, this volume; *Harpagolestes*, Geisler and Luo, this volume), the ectotympanic bulla has a long and tubular external meatus. The anterior crus and the anterior wall of the meatus are incorporated into this external meatal tube. Because there is a large body of evidence to support the sister-group relationship of mesonychids and cetaceans to the exclusion of artiodactyls (Geisler and Luo, this volume; Luo and Gingerich, in press), the sigmoid process in *Diacodexis* could be considered to be a "convergence" to cetaceans.

Alternatively, the presence of an independent sigmoid process can be interpreted as the synapomorphy of the Paraxonia (a clade that includes artiodactyls, mesonychids, and cetaceans, *sensu* Geisler and Luo, this volume). The lack of an independent sigmoid process in mesonychids can be attributed to the incorporation of this process into a robust meatal tube in mesonychids. The absence of the sigmoid process in mesonychids could be considered to be a secondary loss within the Paraxonia. No matter how the distribution of the sigmoid process is optimized among the artiodactyl *Diacodexis*, mesonychids, and cetaceans, it is clear that this character (i.e., an independent and enlarged projection derived from the embryonic anterior crus) has some degree of homoplasy, compromising its utility as a "dead ringer" apomorphy for cetaceans.

Other bullar features of pakicetids are primitive characters (plesiomorphies) with wide distribution among noncetaceans. These characters are: presence of a tympanic annulus for a fully developed tympanic membrane, protrusion of the external meatus into the bullar cavity, a large aperture for the Eustachian tube, a short posterior process (crus), and a hollow processus tubarius that contacts the entoglenoid process of the squamosal.

3.2. Protocetids

The bullae of protocetids are represented by several taxa: *Indocetus* (LUVP 11034, Kumar and Sahni, 1986), *Gaviacetus* (Gingerich *et al.*, 1995; Luo and Gingerich, in press), and *Protocetus* (Fraas, 1904). All postpakicetid cetaceans appear to form a monophyletic group (Hulbert, 1993; Thewissen *et al.*, 1996; Geisler and Luo, this volume; Luo and Gingerich, in press). However, the monophyly of all taxa that were formerly assigned to protocetids has not been established by a parsimony analysis. The interrelationships of the taxa formerly placed in the "protocetid group" have been established to some extent (Hulbert, 1993; Thewissen *et al.*, 1996; Geisler and Luo, this volume; Hulbert, this volume). It appears that some protocetids are more closely related to the more derived whales than other protocetids (Thewissen *et al.*, 1996; Luo and Gingerich, in press; Geisler and Luo, this volume; Hulbert, this volume). As a simple matter of expedience, protocetids are treated as a group in this paper. Yet it is clear that the "protocetids" (*sensu* Kellogg, 1936) are a grade.

Protocetids (including the more derived basilosaurids) are characterized by several apomorphies of the bulla. The first is the enlargement of the posterior process of the tympanic. The posterior process extends to cover the entire ventral aspect of the mastoid process of the petrosal. By contrast, the posterior process of the tympanic is short and the mastoid process is broadly exposed ventrally in pakicetids and mesonychids (Luo and Gingerich, in press).

The second apomorphy of protocetids (including basilosaurids) is that the posterior process is joined to the bullar body by two pedicles (Kellogg, 1936; Hulbert, 1993; Geisler *et al.*, 1996; Luo and Gingerich, in press). The two pedicles between the distal portion of the posterior process and the bullar body are separated by a posterior cleft, the pedicular foramen (Fig. 8C). Of the two pedicles, the external pedicle appears to correspond in position to the posterior process of the pakicetid bulla; thus, it is hypothesized here that the external pedicle is homologous to the posterior process of pakicetids, whereas the internal pedicle is considered to be a neomorph. It is also possible that both pedicles of protocetids are homologous to the single posterior pedicle of pakicetids (Geisler *et al.*, 1996; Geisler, personal communication). The double posterior pedicles of protocetids could have resulted from the excavation by the better-developed posterior pterygoid sinus in protocetids than in pakicetids.

Two other apomorphies of protocetids and more derived cetaceans are related to a greater expansion of the involucrum. The larger involucrum forms a medial prominence in the posteroventral corner of the bulla in protocetids. By contrast, the same area in pakicetids is almost flat. Related to the medial prominence is a longitudinal median furrow on the ventral surface of the bulla. In correlation with the expansion of the involucrum, the Eustachian notch, which is prominent in pakicetids, has been modified in protocetids. The Eustachian tube is smaller in proportion to the entire bulla, than in pakicetids. The external opening of the Eustachian tube is on the medial side of the bulla in protocetids, not on the anteromedial corner of the bulla as in pakicetids (Luo and Gingerich, in press).

Among the protocetids available in this study, none has preserved or exposed the internal structures inside the bulla. The condition of the tympanic annulus and the internal morphology of the sigmoid process are not known. However, from a cast of *Indocetus* basicranium (LUVP 11034), it appears that the processus tubarius is formed by a hollow crust of bone, similar to the condition in pakicetids. Another apomorphy is equivocal because it is only present in some "protocetids." The medial prominence of the involucrum has a con-

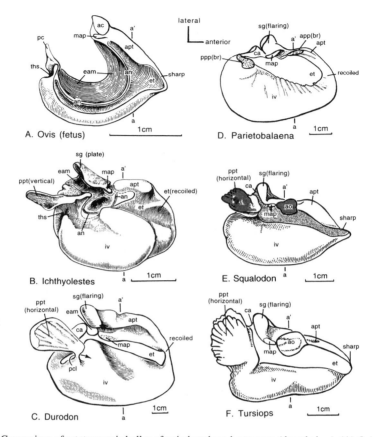

FIGURE 8. Comparison of ectotympanic bullae of artiodactyls and cetaceans (dorsal view). (A) *Ovis aries* or domestic sheep (fetus, modified from van Kampen, 1905, on the basis of newborn specimens of CMNH G995, G997). (B) *Ichthyolestes* (Pakicetidae; H-GSP 18391). (C) *Dorudon* (UM 94812). (D) *Parietobalaena* (modified from Kellogg, 1965; the long and gracile anterior and posterior pedicles are only shown in cross section). (E) *Squalodon*. (F) *Tursiops*. Line a′a indicates the approximate position of transverse sections of Fig. 9. Abbreviations: ac, anterior crus of ectotympanic ring; an, annulus for suspending tympanic membrane; ao, accessory ossicle; app(br), broken stump of anterior pedicle; apt, anterior process of the tympanic (= processus tubarius); ca, conical apophysis (middle process) of tympanic; eam, wall of external auditory meatus; et, Eustachian tube; iv, involucrum; map, mallear process of the tympanic and the site of attachment for malleus (homologous to the goniale element associated with the anterior process of malleus); pc, posterior crus of the ectotympanic ring; pcl, posterior cleft (foramen between the double pedicles for the posterior process of ectotympanic); ppt, posterior process of the ectotympanic bulla; ppp(br), broken posterior pedicle; sg, sigmoid process (= anterior crus of the ectotympanic ring of ungulate mammals); ths, tympanohyal sulcus on the ectotympanic bulla.

tact with the basioccipital crest (the falcate process of basioccipital) in a majority of protocetids (*Protocetus*, Fraas, 1904; Kellogg, 1936; *Indocetus*, LUVP 11034; *Gaviacetus*, Luo and Gingerich, in press). As an exception, this bullar contact is not present in a protocetid from the Eocene of Georgia, *Georgiacetus* (Hulbert, 1993), because of the better developed sinus between the basioccipital and the bulla in *Georgiacetus*. Protocetids primitively have an articulation between the involucrum and the basioccipital falcate process, but this articulation is lost as a consequence of posterior expansion of the pterygoid sinus (Luo and Gingerich, in press). This contact is also present in a remingtonocetid (VPL 1004).

3.3. Basilosaurids and Dorudontids

The most significant apomorphies of the bullae of basilosaurids and dorudontids are associated with the modification of the external auditory opening of the ectotympanic. The annulus of the ectotympanic, which is present in pakicetids and ungulates, is not only greatly reduced in size, but also modified in morphology in basilosaurids (Pompeckj, 1922; Kellogg, 1936; Lancaster, 1990; Uhen, 1996; Luo and Gingerich, in press). The edge and groove associated with the annulus in pakicetids and ungulate mammals are absent in basilosaurids. The rim of the external auditory opening of the bulla has a thickened and involuted edge, and a conical apophysis (medial process of bulla, Kellogg, 1936; see character 5b in Fig. 8C). As discussed earlier, this conical apophysis is developed as a W-shaped fold of the embryonic ectotympanic ring in modern cetaceans (Fig. 3).

The sigmoid process of the tympanic bulla in basilosaurids and dorudontids differs from the condition in pakicetids (the condition in protocetids is not sufficiently known for comparison). Unlike pakicetids, in which the sigmoid process is a simple plate (Fig. 8B; Table II: character state 1a), the sigmoid process of basilosaurids is an inflated structure (Fig. 8C; Table II: state 1b). Its anterior face is convex, and its posterior surface is concave and with an involuted and thickened edge. The distal (dorsal) end of the process is slightly twisted from its base, giving it an S-shaped outline in lateral view (Kellogg, 1928, 1936; Lancaster, 1990; Luo and Gingerich, in press). The malleus is attached to the bulla medial of the base of the sigmoid process. The larger size of the sigmoid process and the twist of its body have displaced the malleus more medially and away from the rim of the external auditory opening (Lancaster, 1990; Luo and Gingerich, in press).

The internal rim of the external auditory opening of the basilosaurids is identical to those of extant cetaceans in possessing (1) an involuted (or recoiled) rim, (2) a conical apophysis, (3) a much smaller aperture, and (4) a wider separation of the malleus from the rim of the meatal opening. Based on these similarities, Lancaster (1990) hypothesized that basilosaurids had developed a conical tympanic ligament as in extant cetaceans. The lateral end of the conical tympanic ligament would be attached to the conical apophysis and its periphery. The ligament would stretch to reach the manubrium of the malleus that is separated far from the rim of the external meatal opening. The inward rotation of the incus and malleus add to the distance between the manubrium of the malleus and the conical apophysis, thereby lengthening the conical tympanic ligament (Lancaster, 1990; Thewissen, 1994). Lancaster's (1990) observation is supported by the additional comparative evidence from this study.

Two features that used to be considered as the "diagnostic characters" of basilosaurids (e.g., Kellogg, 1936; Lancaster, 1990) are now considered to be plesiomorphies. The double pedicles for the posterior process and the posterior cleft of the bulla are also present in protocetids (Geisler *et al.*, 1996).

3.4. Cetacean Crown Group

The extant cetaceans are diagnosed by two bullar apomorphies. First, the processus tubarius (or anterior process) has lost its contact with the entoglenoid process of the squamosal. In all extant cetaceans, the processus tubarius of the bulla directly articulates with the anterior process of the tegmen tympani but does not contact the squamosal. By contrast, the processus tubarius of basilosaurids (Fig. 8C: character 6) partially contacts the anterior process of the petrosal's tegmen tympani, and partially overlaps a crest on the entoglenoid process of the squamosal (see Luo and Gingerich, in press). In pakicetids, the processus tubarius articulates entirely with the bullar process on the entoglenoid squamosal, and has little contact with the anterior process of the petrosal. In protocetids, the processus tubarius (anterior process) has no contact with the anterior process of the petrosal in *Gaviacetus* (Luo and Gingerich, in press), and a small contact in other protocetid taxa (Hulbert, 1993; Geisler *et al.*, 1996).

The posterior process of the ectotympanic in the crown group of cetaceans has a mosaic of primitive and derived characters. Among the primitive characters is the broad base of the posterior process connected to the bullar body, a character present in pakicetids. A derived character for odontocetes and mysticetes is the absence of the double pedicles for the posterior process in protocetids, basilosaurids, and dorudontids.

Another important apomorphy for the crown group of cetaceans is that the sigmoid process is detached from the entoglenoid process of the squamosal in the crown group of cetaceans. Because of the increased size of the pedicle of the posterior process (more so in mysticetes than in odontocetes), the sigmoid process is removed and detached from the squamosal (except for kogiids, see Luo and Marsh, 1996). The articulation of the sigmoid process of the ectotympanic with the squamosal, as seen in pakicetids and basilosaurids (Luo and Gingerich, in press), is a primitive character. The anterior crus of the ectotympanic ring—the homologue of the sigmoid process—is consistently articulated with the squamosal in a majority of placental mammal groups (Van der Klaauw, 1931; MacPhee, 1981; Novacek, 1986).

3.5. Odontocetes

Despite their great morphological diversity, all odontocete families share two prominent bullar apomorphies: an accessory ossicle and a thin edge of the Eustachian tube opening.

The accessory ossicle is an independent ossification derived from the processus tubarius (Figs. 8E,F, 9E). This structure is no longer a part of the bullar body in odontocetes, in contrast to the situation in basilosaurids, protocetids, pakicetids, and terrestrial mammals. The accessory ossicle is clearly demarcated from the bullar body and usually connected to the bulla only by a very fragile sheet of bone (Kasuya, 1973; Fordyce, 1981, 1983, 1994; Pilleri *et al.*, 1987; Muizon, 1987, 1994; Oelschläger, 1990; Luo and Marsh, 1996). As a consequence of this thin connection, the accessory ossicle is frequently broken off from the thin anterior edge of the bulla in isolated bullae of both extant and fossil odontocetes.

In several groups of odontocetes, including all physeterids, kogiids, ziphiids, and some delphinids, the accessory ossicle is secondarily attached or even fused to the anterior process of the petrosal in the adult, even though it is embryonically developed as a part of the ectotympanic bulla. If the bulla becomes separated from the petrosal during fossil preservation, the accessory ossicle is invariably attached to the petrosal, but not the bulla (Luo and Marsh, 1996). The secondary attachment of the accessory ossicle may occasionally occur on the individual specimens of eurhinodelphids and other primitive extinct river dolphin groups (Muizon, 1987, 1988; Fordyce, 1994), but this does not occur in all individuals of the same taxon, or in all genera within a family, except for physeterids, kogiids, and ziphiids (Kasuya, 1973; Luo and Marsh, 1996).

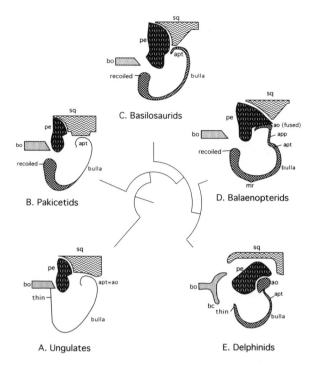

FIGURE 9. Transformation of processus tubarius, anterior process, accessory ossicle, and anterior pedicle. Schematic transverse section across line a′a in the anterior part of the bulla (shown in Fig. 8). (A) Ungulates: anterior process (= processus tubarius) is a hollowed crust, contacting the squamosal. (B) Pakicetids (retaining the primitive condition of ungulates). (C) Basilosaurids (apomorphy: anterior process of the tympanic is shifted to contact the petrosal). (D) Balaenopterids (as the representative of mysticetes; apomorphies: dorsal part of the accessory ossicle fused to the petrosal and has lost its contact with the squamosal, ventral part of the accessory ossicle develops into a gracile pedicle). (E) Delphinids (as the representative of odontocetes; apomorphy: accessory ossicle remains distinctive as an adult structure). The spiral cladogram suggests the phylogenetic relationships of ungulates and cetaceans. Abbreviations: ao, accessory ossicle; app, pedicle of anterior process of tympanic bulla; apt, anterior process of tympanic, or processus tubarius; bc, basioccipital crest; bo, basioccipital; mr, median ridge; pe, petrosal; sq, squamosal.

The anterior bullar lip forms a sharp edge around the aperture for the Eustachian tube in all odontocetes (Figs. 8E,F, 9E). This is a derived character by contrast to the recoiled edge of the bullar lip of the Eustachian opening in nonodontocete cetaceans. In ungulates, such as *Ovis*, the Eustachian tube has a thin margin in early developmental stages (Fig. 8A), which later becomes recoiled or involuted in adults (not illustrated).

3.6. Mysticetes

The most striking apomorphy of mysticetes is the complete fusion of the anterior process of the bulla to the anterior process of the petrosal (Fraser and Purves, 1960; Kasuya, 1973; Geisler and Luo, 1996). The anterior process has a very gracile pedicle that separates the lateral lip of the bulla from the fused anterior process of the petrosal and the anterior process of the tympanic (Fig. 1D).

As documented by Ridewood (1922), the embryonic accessory ossicle is an independent ossification center differentiated from but still attached to the main body of the ectotympanic bulla. Although the ossicle appears to be the same in the early development of both odontocetes and mysticetes, the later development of this embryonic ossicle differs in the two groups. Once the accessory ossicle forms in the embryos of odontocetes, it remains more or less unchanged in adult odontocetes, except becoming slightly more robust and spherical. By contrast, the accessory ossicle in embryos of mysticetes will further develop into a ventral pedicle connected to the bullar body, and a distal (dorsal) part formed by a crust of thin bone in the juvenile specimens. The dorsal part of the accessory ossicle is broadly overlapped and sutured to the anterior process (an extension of the tegmen tympani) of the petrosal (e.g., an uncatalogued specimen in the Kellogg Collection of USNM). In adult mysticetes, the suture between the anterior bullar pedicle and the tegmen tympani is completely fused (Fraser and Purves, 1960; Geisler and Luo, 1996). The accessory ossicle is also present in the newborn of *Ovis* but is incorporated into the bullar body in adults. The adult structure derived from the accessory ossicle is termed the *processus tubarius* by van Kampen (1905). It is hypothesized that the anterior bullar pedicle in mysticetes is either a peramorphic or a hypermorphic development beyond the terminal adult condition of a distinctive accessory ossicle of odontocetes.

The involucrum in mysticetes differs from all nonmysticete cetaceans in having a ventral keel (Geisler and Luo, 1996), instead of a median furrow as in protocetids, basilosaurids, and odontocetes. The median furrow is obviously a primitive condition because of its broad distribution in several major clades (e.g., odontocetes) and grades (e.g., protocetids).

Extant mysticetes have developed an extremely long posterior process of the tympanic (Geisler and Luo, 1996), in correlation with the gigantism of their skulls. As the skulls of mysticetes become exceedingly large relative to body length and in absolute size, the posterior process of the petrosal and the posterior process of the tympanic become disproportionally long, especially in extant balaenopterids (Geisler and Luo, 1996). The posterior process of the tympanic is tightly fused with the posterior process of the petrosal, a condition unique to mysticetes but absent in archaeocetes (including pakicetids) and fairly uncommon in odontocetes.

4. Character Evolution

4.1. Tympanic Membrane and Its Homologues

As an adaptation to the aquatic environment, extant cetaceans have extensive modifications of their external ears. As a result, the tympanic membrane is so highly modified that it becomes nonfunctional for hearing, as demonstrated in the physiological studies of McCormick *et al.* (1970, 1972). Most anatomical studies accept that the highly modified external ears and conical tympanic ligaments in extant cetaceans have no function for hearing (Huber, 1934; Norris, 1966, 1980; Fleischer, 1978; Ketten, 1992). Alternatively, two workers (Purves, 1966; Pilleri, 1983; Purves and Pilleri, 1983) argue that the external auditory meatus and the tympanic ligament have some capacity for hearing despite their extreme modification in response to the high pressure of the underwater environment. However, there is no disagreement that the modification of the tympanic membrane into the conical tympanic ligament is correlated to the changes in the mode of hearing as a result of the aquatic adaptation by cetaceans.

The tympanic membrane ("tympanum") of a terrestrial mammal has two parts, the pars flaccida and the pars tensa. The pars tensa has a slight concavity toward the outside because the mallear menubrium attaching to the center (umbo) of the pars is slightly offset from the plane of the annulus that suspends the periphery of the pars tensa (Fig. 7A). This slight concavity is far less developed than in the highly conical tympanic ligament of the extant cetaceans. The tympanic membrane of a terrestrial mammal as a whole may correspond to two structures in mysticetes (Fraser and Purves, 1960): the glove finger and the tympanic ligament (Fig. 7D). (1) The glove finger is a tube-shaped membrane that opens internally to the tympanic cavity, with much of its inner circumference attached to the margin of the external meatal opening of the tympanic bulla. The external side of the membrane is convex and the apex of the tube-shaped membrane points into the external meatus of the squamosal (Lillie, 1910; Ridewood, 1922). Ridewood (1922) shows that the flat tympanic membrane is the embryonic precursor to the glove finger. The flat membrane in early embryogenesis becomes a cone-shaped tympanic membrane at an intermediate stage before it forms the long and tubular membrane of the glove finger in adults. (2) The conical tympanic ligament is the connection from the malleus to the internal side of the glove finger in balaenopterids (Lillie, 1910; Ridewood, 1922). In balaenids, a substantial part of the tympanic ligament also attaches to the margin of the external meatal opening. Fraser and Purves (1960) propose that the tympanic membrane of terrestrial mammals is homologous to both the glove finger and the conical tympanic ligament in balaenopterids (but see Ridewood, 1922). Lancaster (1990) suggests that the tympanic membrane of the terrestrial mammal is homologous only to the conical tympanic ligament, although he did not address the issue of the glove finger in mysticetes.

Extant odontocetes (Fig. 7E) have the conical tympanic ligament but lack the glove finger (Huber, 1934; Fraser and Purves, 1960, Fig. 27). The ligament is shaped like a cone concave toward the outside. Its inner apex attaches on the manubrium of the malleus. The external edge of the ligament attaches to the conical apophysis and the periphery of the external meatal opening of the tympanic, although the extent of this attachment may vary widely among the odontocete taxa (Fraser and Purves, 1960). In both mysticetes and odontocetes, the incus–malleus complex is rotated inward from the conical apophysis, further elongating the conical tympanic ligament between the mallear manubrium and the conical apophysis (Fleischer, 1978; Lancaster, 1990).

Several characters of the bulla and other parts of the mammalian basicranium are closely related to the tympanic membrane and the external auditory meatus. Because the bony characters are preserved in transitional taxa, such as mesonychids and archaeocetes, these provide fossil evidence for the evolution of the tympanic membrane and the external auditory meatus, both of which are crucial for hearing in terrestrial mammals and fundamentally modified in cetaceans.

In adult terrestrial mammals, the tympanic membrane ("tympanum") is suspended by an annulus at the inner (proximal) edge of the external auditory meatus. The annulus for the tympanic membrane develops from the embryonic ectotympanic ring during embryogenesis (de Beer, 1937; MacPhee, 1981). In some mammalian groups, such as rodents, lagomorphs, and many ungulates (see the review by Novacek, 1977), the ectotympanic ring develops into the entire bulla. The ectotympanic also makes up a major part of the bulla in several other groups of eutherians (Van der Klaauw, 1931; Novacek, 1977; MacPhee, 1981). Regardless of the variation in the bullar wall, the tympanic membrane retains its attachment to the tympanic annulus (Van der Klaauw, 1931; MacPhee, 1981). The suspension of the tympanic membrane by the annulus in the adult mammalian skull remains the same as was established in early embryogenesis.

Based on this very consistent relationship of the ectotympanic annulus and the tympanic membrane in all extant mammals, the condition of the tympanic membrane in fossil mammals can be reliably inferred from the annulus as preserved in the fossils. The presence versus absence of the annulus would indicate the presence or absence of the tympanic membrane in extinct cetaceans. If present, the size of the annulus ring provides an accurate estimate of the size (surface area) of the tympanic membrane.

Mesonyx (AMNH 12643) has an annulus typical of a terrestrial mammal, indicating that it had a fully developed planar tympanic membrane for hearing airborne sound. Pakicetids show an intermediate condition between the terrestrial ungulate mammals and the more derived basilosaurids. Pakicetids have an annulus about the same proportion as in *Mesonyx* (AMNH 12643) and extant artiodactyls (*Ovis*, CMNH G995, G997), suggesting that pakicetids had a large surface area of tympanic membrane as ungulates do. However, unlike ungulates but reminiscent of basilosaurids, the incus of pakicetids already developed some degree of rotation in the incudomallear joint (as preserved on the incus, see Thewissen and Hussain, 1993), although the malleus of pakicetids is still not known. This suggests that the large tympanic membrane in pakicetids may be slightly more concave than those of terrestrial mammals, but far different from the acute conical tympanic ligament of extant cetaceans (Fig. 7B), and the inferred condition in basilosaurids as discussed below.

Bullae in basilosaurids have a very reduced aperture for the external auditory opening. The rim around the aperture is slightly involuted and has a conical apophysis. There is no annulus for the attachment of a fully developed tympanic membrane (Pompeckj, 1922; Kellogg, 1936; Lancaster, 1990; Luo and Gingerich, in press). The rotation of the incudomalleus complex from the plane of the annulus is very well developed in basilosaurids (Lancaster, 1990). Basilosaurids resemble extant mysticetes and odontocetes that have the conical tympanic ligament, but differ from such terrestrial mammals as extant artiodactyls that have the tympanic membrane in all of these characters (the involuted rim of the meatal opening, the presence of the conical apophysis, and the rotation of the incudomalleus from the plane of the annulus). This strongly suggests that basilosaurids (Fig. 7C) modified their tympanic membrane and achieved the same specialization in the tympanic structures as in extant mysticetes and odontocetes (see also Lancaster, 1990).

Although some portion of the cartilaginous external auditory meatus is still present in extant cetaceans, the squamosal canal for the external auditory meatus is entirely absent in extant odontocetes (Lillie, 1910; Hanke, 1914; Huber, 1934; Fraser and Purves, 1960; Purves, 1966; Pilleri, 1983). Absence of the glove finger in extant odontocetes is correlated with the loss of the bony external auditory meatus in the squamosal, as the latter houses the glove finger in extant mysticetes. As such, presence or absence of the glove finger, a soft tissue membrane, can be inferred from the presence or absence of the bony canal in the squamosal for the external auditory meatus, an osteological structure that is usually preserved in fossils.

In extant mysticetes, the bony canal of the external auditory meatus in the squamosal is much more reduced than in archaeocetes. The external face of the glove finger is permanently blocked by a thick plug of wax (Lillie, 1910; Fraser and Purves, 1960). All protocetids and basilosaurids have broad canals in the squamosal for the external auditory meatus. The squamosal canal is comparable in size to that in early mysticetes but slightly better developed than that in mesonychids and pakicetids. It cannot be ruled out that these basilosaurids had the glove finger as well.

Several stem groups of odontocetes still have a well-developed canal for the external auditory meatus in the squamosal (Whitmore and Sanders, 1976; Fordyce, 1981, 1994; Muizon, 1987, 1988, 1994). However, this meatal canal is lost in the extant families of odontocetes. The loss of the squamosal meatal canal most probably occurred independently in several clades of extant odontocetes, and after the divergence of the extant odontocete families from the stem groups. The presence or absence of the glove finger membrane is correlated with the squamosal canal in extant cetaceans (Huber, 1934; Fraser and Purves, 1960; Purves, 1966; Pilleri, 1983). It is possible that the glove finger membrane was present in early odontocetes but independently lost in extant families.

4.2. Sigmoid Process

There are two prominent differences in the bulla between mesonychids and cetaceans. In mesonychids the external auditory meatus is ventrally covered by a tubular floor of the ectotympanic. By contrast, this tubular floor is absent in all cetaceans, and as a result, the bony squamosal canal for the external meatus is exposed ventrally in nonodontocete cetaceans. The cetacean bulla has a sigmoid process anterior to the opening of the meatus into the bullar cavity, whereas this process is absent in most (but not all) ungulates including mesonychids. A hypothesis on the homology of the sigmoid process to a structure of other mammals and its evolutionary transformation must account for both morphological differences between mesonychid ungulates and cetaceans. Luo and Gingerich (in press) note that the sigmoid process of the cetacean bulla could correspond to three different structures in noncetacean ungulate mammals. Given the distribution of characters among fossil ungulates and cetaceans, all three hypotheses of homology of the cetacean sigmoid process are possible.

4.2.1. Homology to the Anterior Meatal Wall and the Anterior Crus

The sigmoid process in cetaceans and the anterior wall of the meatal tube in ungulates occupy the same position and have the same topographic relationships to the surrounding structures. Both structures are derived embryonically from the anterior crus. Doran (1878) first suggested the possible homology of the sigmoid process with the anterior crus of the ectotympanic ring (Ridewood, 1922).

The homology of the anterior crus (and anterior meatal wall) to the sigmoid process is most plausible on three criteria: (1) ontogenetic sequence, (2) positional relationships, and (3) phylogenetic transformation series. The sigmoid process develops from the anterior crus of the embryonic tympanic ring (Fig. 1). The only modification of the sigmoid process of adult cetaceans from its embryonic precursor, the anterior crus, is a slightly larger size of the process in adults (Ridewood, 1922). The anterior crus of the ectotympanic ring has remained as an independent process throughout later development. The topographical relationships of the malleus and the sigmoid process in cetaceans are almost identical to the relationships of the malleus and the anterior crus of the ectotympanic in noncetacean mammals (Fig. 1). The sigmoid process in pakicetids is platelike (Figs. 4, 5), and represents an intermediate condition between the simple anterior wall of the meatal tube of an ungulate, such as *Mesonyx*, and the inflated and twisted sigmoid process as in basilosaurids and extant cetaceans. This gives further credence to the homology of the sigmoid process to the anterior wall of the meatal tube, and by extension, to the anterior crus that grows into the anterior wall of the meatal tube. This phylogenetic transformation is corroborated by the strong sister-group relationships of mesonychids and cetaceans (Van Valen, 1966; Barnes and Mitchell, 1978; Thewissen, 1994; Geisler and Luo, this volume).

The preferred hypothesis is that the sigmoid process is homologous to the anterior crus of the ectotympanic, as the sigmoid process is directly traceable to this structure during embryogenesis in baleen whales (Ridewood, 1922; see Fig. 1). In the light of the embryonic evidence from baleen whales, the following two alternative hypotheses are less likely.

4.2.2. Homology to the Folian Process plus the Anterior Crus

This hypothesis suggests that the sigmoid process of cetaceans is homologous to a compound structure fused from the Folian process (developed from the goniale), the anterior process of the malleus, plus the anterior crus in noncetacean mammals (Van der Klaauw, 1931). Under this hypothesis, the sigmoid process is derived from several independent embryonic precursors in noncetacean ungulates. The main difficulty with this hypothesis is that the goniale ossification and the Folian process are very small in mammals, and could not account for the large size of the sigmoid process of cetaceans.

If the sigmoid process is considered to be homologous to the anterior crus of the ectotympanic or homologous to a compound structure of the goniale and the anterior process of the malleus, then the absence of the ectotympanic floor to the meatus in cetaceans must be interpreted as a secondary loss of this floor in cetaceans from the ancestral condition (the presence of the floor) in shared ancestor to mesonychids and cetaceans. The lack of the tubular floor to the meatus in cetaceans can be interpreted as the consequence of an early arrest of the lateral growth of the ectotympanic ring, which would form the ventral floor of the external auditory meatus in the later embryogenesis of ungulates.

4.2.3. Homology to the Entire Meatal Tube

The third alternative hypothesis is that the sigmoid process is homologous to the entire meatal tube of the ectotympanic, not just the anterior wall of the tube derived from the anterior crus of the ectotympanic ring in noncetacean ungulates. This is supported by the similarities of the inner surface of the sigmoid process of cetaceans and the meatal tube in ungulates. The internal (posteromedial) surface of the cetacean sigmoid process is a trough with rounded rim, very similar to the inner surface of the tubelike ectotympanic floor to the external auditory meatus in terrestrial mammals.

This hypothesis requires some putative morphological changes for the tubular floor. If the meatal tube decreases in size, and rotates anterodorsally in early embryogenesis, the tube could be transformed to the sigmoid process. The rotation of the meatal part of the ectotympanic away from the external auditory meatus and a simultaneous reduction in size would have modified a tubelike external auditory meatus in ungulates into a ventrally open groove in early cetaceans. This hypothesis does not have to postulate any loss of any major bony element through the ungulate–cetacean transition. There is no embryonic evidence for an anterior rotation of the tubular floor as predicted by this hypothesis. Instead, it shows that the tubular floor fails to develop in cetaceans.

4.3. Posterior Process

The posterior process of the ectotympanic bulla in adult cetaceans develops from the posterior crus of the ectotympanic ring at an earliler embryonic stage (Fig. 1: ppt; Ridewood, 1922). Based on this ontogenetic evidence, it is hypothesized here that the enlarged posterior process in extant cetaceans is a neomorphic character state of the posterior crus of the ectotympanic ring of other mammals.

In terrestrial mammals, the posterior crus articulates with the crista parotica of the petrosal and in juxtaposition with the base of the tympanohyal (de Beer, 1937; MacPhee, 1981). The contact between the posterior crus and the petrosal is small. In ungulate mammals, the ectotympanic ring grows laterally to form a tubular floor to the external auditory meatus. The posterior crus is modified into a plate that forms a part of the posterior wall of the external auditory meatus tube. The posterior crus, and the posterior meatal wall derived from the crus are interlocked with the tympanohyal. The posterior meatal tube has a more extensive contact with the squamosal than with the petrosal (*Ovis*, van Kampen, 1905; *Sinonyx* and *Mesonyx*, Zhou et al., 1995; Luo and Gingerich, in press).

The posterior process in pakicetids is comparable to the posterior process (= posterior wall of the meatus) in adult *Ovis* and mesonychids in its platelike morphology, and in that it flanks the posterior side of the external auditory meatus (Fig. 8). However, it differs from ungulates in its articulation to the petrosal, rather than the squamosal.

The posterior process of the tympanic in pakicetids represents an intermediate condition between the primitive condition of the posterior crus of the ectotympanic of terrestrial eutherians (Novacek, 1977, 1986; MacPhee, 1981) and the more derived condition in protocetids, basilosaurids, and extant cetaceans (e.g., Fig. 8). The process of pakicetids is larger and has more extensive contact to the petrosal than in ungulates, but it lacks the distal extension and enlargement, which occur in protocetids and more derived cetaceans. As pointed out earlier, the posterior process in pakicetids is not large enough to cover the entire ventral surface of the posterior (mastoid) process of the petrosal, in contrast to the conditions of protocetids and more derived cetaceans.

4.4. Involucrum

As shown in Figs. 7–9, the presence of the involucrum in all cetaceans including pakicetids suggests that this is a major apomorphy of cetaceans. Because the sigmoid process is morphologically variable in pakicetids and more derived cetaceans, its utility as a key character for cetaceans is now a bit equivocal (see the previous section). The involucrum becomes the only unequivocal diagnostic character for the cetaceans including *Pakicetus*.

The involucrum marks the beginning of the development of pachyostosis (massiveness and hypertrophy) and pachyosclerosis (high density and heavy mineralization) of the tympanic complex in cetaceans. The involucrum is essentially a pachyosteosclerotic medial part of the bulla. The massive and dense bone of the involucrum, coupled with the vascular and pneumatic sinuses medial to the bulla, could serve as a crude barrier to acoustic interference via bone conduction between the left and right ears.

It is noteworthy that pachyosteosclerosis of the tegmen tympani of the petrosal occurred before pachyostosis of the tympanic bulla in the ungulate–cetacean transition. The bulla of mesonychids has a thin crust and lacks the pachyostosis of the cetacean bullae

(*Mesonyx obtusidens*, AMNH 12643; *Sinonyx*, IVPP V10760; see Fig. 9A). Both mesony-chids and cetaceans share a pachyostotic tegmen tympani of the petrosal. This indicates that pachyostosis of the tegmen tympani is a primitive condition shared by mesonychids and cetaceans, and it phylogenetically preceded the pachyostosis in the bullae of cetaceans.

In pakicetids and protocetids, the promontorium has not yet developed pachyostosis, whereas the bullar involucrum has already become heavy and massive. In contrast, in basilosaurids and more derived cetaceans, the promontorium of the petrosal is also pachyostotic. By inference, the pachyosteosclerosis of the involucrum would have occurred prior to the pachyostosis of the promontorium (Luo and Gingerich, in press).

Development of skeletal pachyosteosclerosis is a convergence in many unrelated marine vertebrates that descended from different terrestrial ancestors. Dense and massive middle ear bones, such as the hypertrophied quadrate, evolved convergently in marine ichthyosaurs and mosasaurs (Pompeckj, 1922). Pachyosteosclerosis is common in marine mammals (Domning and Buffrénil, 1991). The petrosal and tympanic in sirenians have developed pachyostosis. Although to a lesser extent than in cetaceans, the tympanic ring is massive and the mastoid process of the petrosal is hypertrophied in sirenians (Ketten *et al.*, 1992).

The petrosal is usually pachyosclerotic (dense) in extant odontocetes but not always pachyostotic (inflated). It is always pachyosteosclerotic (dense, more mineralized, in addition to being massive and hypertrophied) in extant mysticetes. The ectotympanic bullae of both odontocetes and mysticetes are not only pachyostotic but also pachyosclerotic.

The increased mineralization of the tympanic bulla in cetaceans gives the bulla a greater density, and a better stiffness that results in a high modulus of elasticity (Currey, 1979, 1984). The microstructure of the bulla is more randomly arranged than in other types of bones, such as long bones (Currey, 1979), and the bone of the bulla is more isotropic. Isotropic bones transmit sound vibration without distortion. The hardness of the bulla borders on the lower range of enamel (Vincent, 1990; Rensberger, personal communication). The bone structure of the bulla is second only to dental enamel in density and stiffness among all vertebrate skeletal tissues (Vincent, 1990; Rensberger, personal communication). This has resulted in their better preservation in fossils.

The pachyosteosclerosis of the petrotympanic complex in cetaceans increases the density differential between the petrotympanic complex and the surrounding soft tissues and the vascular sinuses. The greater the differential, the more effective is the sound deflection at the interface between the petrotympanic complex and its surrounding soft tissues. The greater the contrast in density between the petrotympanic complex and the surrounding soft tissues, the less acoustic interference via conduction would be possible from the rest of the cranium to the petrotympanic complex. In addition to their utility for balance and buoyancy (Domning and Buffrénil, 1991), pachyosclerosis of the ear bones may be correlated with the development of underwater hearing in cetaceans and sirenians. The pachyosteosclerosis of the petrosal and tympanic (or other bony elements of the ear) is probably a prerequisite to specializations for underwater hearing of both low and high frequencies (Luo and Gingerich, in press).

4.5. Hearing of Pakicetids

Pakicetids have a crucial position in the phylogenetic transition from the terrestrial ungulate ancestry to fully aquatic whales. Some features of their ear region, such as the complete annulus for the tympanic membrane and the diverticula within the tympanic cavity, are the primitive characteristics of terrestrial mammals. The tympanic annulus strongly indicates that pakicetids were fully capable of hearing airborne sound, but did not achieve the highly specialized hearing for the waterborne sound of more derived cetaceans. Although the malleus of *Pakicetus* remains unknown, the incus of *Pakicetus* clearly indicates that the animal could hear airborne sound, as in land mammals (Thewissen and Hussain, 1993).

Diverticula partitioned by trabeculae or septa within the bullar cavity are a characteristic of many land mammals. It is noteworthy that the volume of the tympanic bullar cavity of pakicetids is very large relative to the size of the skull. The anterior diverticulum, a hollow chamber, is enclosed by the crust of the processus tubarius (Fig. 5B: da). This diverticulum considerably increases the space of the bullar cavity (more so in *Ichthyolestes* than in *Pakicetus*). The diverticulum is morphologically comparable to (although not homologous to) the enlarged antrum of some land mammals with large bullar cavities.

A larger bullar cavity partitioned by internal septa is an important attribute of low-frequency-hearing mammals, most of which have ground-dwelling or burrowing adaptations (Webster and Webster, 1975; Fleischer, 1978; Webster and Plassmann, 1992). Based on these analogous comparisons, it is hypothesized here that the pakicetid bullae were adapted to low-frequency hearing. Fleischer (1978) also suggested that the involucrum, first to appear in pakicetids in the cetacean evolution, was probably for the reception of low-frequency sounds.

Thewissen (1994, p. 177) suggested that "*Pakicetus* and other archaeocetes may have been specialized for high frequency sound reception, and this was probably part of the cetacean morphotype." Later, Thewissen *et al.* (1996) interpreted the hearing functions of primitive archaeocetes differently. They suggested that *Ambulocetus*, a protocetid more advanced than pakicetids, was adapted to hearing substrate-borne sounds of low frequencies; and that the fat pad in the mandibular canal of protocetids, such as *Ambulocetus* and *Rodhocetus*, was initially developed for receiving low-frequency sounds from the ground. Their homologues in odontocetes became coopted for hearing high-frequency sounds of echolocation later in phylogenetic evolution.

The new evidence from the bullae (especially the septa and the diverticula) of pakicetids suggests unequivocally that these earliest and the most primitive cetaceans lacked the specialized hearing for underwater sounds. Their middle ear structures were fully capable of hearing airborne sound, not unlike a generalized terrestrial mammal. The best hearing sensitivity for pakicetids was probably in the low-frequency ranges.

5. Conclusions

The embryonic precursors for the ectotympanic bullae are similar for both cetaceans and terrestrial ungulate mammals, allowing the recognition of the homology of the highly specialized cetacean bullar structures to their counterparts in other mammals. The prominent differences between the posterior process, and the involucrum of cetaceans and their homologues in ungulates are attributable to peramorphic or hypermorphic growth of these features in cetaceans beyond the adult condition of ungulates. The development of the conical apophysis and the modification of the tympanic membrane into the cetacean conical tympanic ligament result from the fold of the ectotympanic ring in early development of cetaceans. The sigmoid process is homologous to the anterior crus of the ectotympanic and the anterior wall that grows from the anterior crus. Its prominence in cetaceans is caused by the developmental arrest of the ventral floor of the external auditory meatus at an early stage.

Bullar characters are useful for cetacean systematics. Cetaceans (including pakicetids) are diagnosed by the presence of the involucrum. Postpakicetid cetaceans (including protocetids, basilosaurids, and dorudontids) are diagnosed by the distal enlargement of the posterior process and double posterior pedicles (probably also the loss of the annulus and the presence of a conical apophysis). Basilosaurids, dorudontids, and more derived whales are diagnosed by a complete detachment of bullae from the basioccipital crest and the exoccipital. Modern odontocetes are diagnosed by the presence of the accessory ossicle, a neotenic structure by comparison with the development of nonodontocetes. Mysticetes are diagnosed by the presence of gracile anterior and posterior pedicles of the ectotympanic.

The earliest cetaceans—the pakicetids—have several bullar structures for hearing airborne sound. Based on the annulus for suspending the tympanic membrane, it is inferred that pakicetids had a large tympanic membrane as do terrestrial mammals (but the membrane is more concave in pakicetids) and lacked the conical tympanic ligament of modern cetaceans. The large bullar cavity with its internal septa and diverticula suggests low-frequency hearing in pakicetids.

Acknowledgments

First and foremost, I thank Dr. J. G. M. Thewissen for his generosity in allowing me to study the pakicetid bullae, which were collected under NSF Grant EAR 9526686 (to Dr. Thewissen). I am very grateful to Professor P. D. Gingerich for graciously allowing me access to the magnificent archaeocete collection at the University of Michigan. I must also thank J. H. Geisler for his interest in the basicranial anatomy of mesonychids and archaeocetes and his valuable help with this work. Drs. John R. Wible, J. G. M. Thewissen, and Mr. J. H. Geisler have reviewed and helped to improve this manuscript. Many thanks to Amy Henrici and Ellen Williams for their preparation of the pakicetid bullae.

For access to the comparative collections during this study, I acknowledge the following institutions and individuals: the Pakistani Geological Survey, the United States National Museum (F. C. Whitmore, B. Purdy, D. Bohaska, J. Mead, C. Potter), the American Museum of Natural History (R. D. E. MacPhee, N. B. Simmons, J. H. Geisler), the Calvert Marine Museum (M. Gottfried), the Charleston Museum (A. E. Sanders), and the University of Michigan Museum of Paleontology (P. D. Gingerich, M. D. Uhen, W. Sanders). This research is supported by grants DEB 941898 and DEB 9527892 to the author from the National Science Foundation.

References

Barnes, L. G. 1984. Whales, dolphins and porpoises: origin and evolution of the Cetacea, in: P. D. Gingerich and C. E. Badgley (eds.), *Mammals—Notes for a Short Course*, pp. 139–153. University of Tennessee, Department of Geological Sciences, Studies in Geology 8.
Barnes, L. G., and Mitchell, E. 1978. Cetacea, in: V. J. Maglio and H. B. S. Cooke (eds.), *Evolution of African Mammals*, pp. 582–602. Harvard University Press, Cambridge, MA.
Berta, A. 1994. What is a whale? *Science* **263**:180–181.

Currey, J. D. 1979. Mechanical properties of bone with greatly differing functions. *J. Biomech.* **12**:313–319.

Currey, J. D. 1984. *Mechanical Adaptations of Bones.* Princeton University Press, Princeton, NJ.

De Beer, G. R. 1937. *The Development of the Vertebrate Skull.* Clarendon Press, Oxford. (Reprinted by the University of Chicago Press, 1985)

Domning, D. P., and Buffrénil, V. de. 1991. Hydrostasis in Sirenia: quantitative data and functional interpretations. *Mar. Mammal Sci.* **7**:331–368.

Doran, A. 1878. Morphology of mammalian ossicula auditus. *Trans. Linn. Soc. (London)* Series 2 (Zoology) **1**:371–497.

Eales, N. B. 1951. The skull of the foetal narwhal, *Monodon monoceros* L. *Philos. Trans. R. Soc. London* **235B**:1–33.

Fleischer, G. 1978. Evolutionary principles of the mammalian middle ear. *Adv. Anat. Embryol. Cell Biol.* **55**:1–70.

Fordyce, R. E. 1981. Systematics of the odontocetes *Agorophius pygmaeus* and the family Agorophiidae (Mammalia, Cetacea). *J. Paleontol.* **55**:1028–1045.

Fordyce, R. E. 1983. Rhabdosteid dolphins (Mammalia: Cetacea) from the middle Miocene, Lake Frome area, South Australia. *Alcheringa* **7**:27–40.

Fordyce, R. E. 1994. *Waipatia maerewhenua*, new genus and new species (Waipatiidae, New Family), an archaic late Oligocene dolphin (Cetacea: Odontoceti: Platanistoidea) from New Zealand, in: A. Berta and T. Deméré (eds.), *Contributions in Marine Mammal Paleontology Honoring Frank C. Whitmore, Jr.*, pp. 147–176. Proceedings of the San Diego Society of Natural History, No. 29.

Fordyce, R. E., and Barnes, L. G. 1994. The evolutionary history of whales and dolphins. *Annu. Rev. Earth Planet. Sci.* **22**:419–455.

Fraas, E. 1904. Neue Zeuglodonten aus dem unteren Mitteleocän vom Mokattam bei Cairo. *Geol. Paläontol. Abh. N. F.* **6**:199–220.

Fraser, F. C., and Purves, P. E. 1960. Hearing in cetaceans. Evolution of the accessory air sacs and the structures of the outer and middle ear in recent cetaceans. *Bull. Br. Mus. Nat. Hist. Zool.* **7**:1–140.

Geisler, J., and Luo, Z. 1996. The ear structure of an archaic mysticete whale, and its bearings on relationships and hearing evolution of mysticetes. *J. Paleontol.* **70**:1045–1066.

Geisler, J., Sanders, A. E., and Luo, Z. 1996. A new protocetid cetacean from the Eocene of South Carolina, U.S.A.; phylogenetic and biogeographic implications, in: J. E. Repetski (ed.), *Sixth North American Paleontological Convention Abstracts of Papers. Paleontol. Soc. Spec. Pap.* **8**:139.

Gingerich, P. D., Wells, N. A., Russell, D. E., and Shah, S. M. I. 1983. Origin of whales in epicontinental remnant seas: new evidence from the early Eocene of Pakistan. *Science* **222**:403–406.

Gingerich, P. D., Arif, M., and Clyde, W. C. 1995. New archaeocetes (Mammalia, Cetacea) from the middle Eocene Domanda Formation of the Sulaiman Range, Punjab (Pakistan). *Contrib. Mus. Paleontol. Univ. Michigan* **29**(11):291–330.

Hanke, H. von. 1914. Ein Beitrag zur Kenntnis der Anatomie des äusseren und mittlere Ohres der Bartenwale. *Jena. Z. Naturwiss.* **3**:487–524.

Hawkins, A. D., and Myrberg, A. A. 1983. Hearing and sound communication under water, in: B. Lewis (ed.), *Bioacoustics, a Comparative Approach*, pp. 347–429. Academic Press, New York.

Huber, E. 1934. Anatomical notes on Pinnipedia and Cetacea. *Carnegie Inst. Washington Publ.* **447**(IV):105–136.

Hulbert, R. C., Jr. 1993. Craniodental anatomy and systematics of a middle Eocene protocetid whale from Georgia. *J. Vertebr. Paleontol.* **13**(Suppl. to No. 3):42A.

Kasuya, T. 1973. Systematic consideration of recent toothed whales based on morphology of tympano-periotic bone. *Sci. Rep. Whale Res. Inst. (Tokyo)* **25**:1–103.

Kellogg, A. R. 1928. The history of whales—their adaptation to life in water. *Q. Rev. Biol.* **3**:29–76, 174–208.

Kellogg, A. R. 1936. A review of the Archaeoceti. *Carnegie Inst. Washington Publ.* **482**:1–366.

Kellogg, A. R. 1965. Fossil marine mammals from the Miocene Calvert Formation of Maryland and Virginia. *U.S. Natl. Mus. Bull.* **247**:1–63.

Kernan, J. D. 1916. The ear, in: H. W. von Schulte (ed.), Monographs of the Pacific Cetacea. II.—The Sei Whale (*Balaenoptera borealis* Lesson): 2. Anatomy of a foetus of *Balaenoptera borealis. Mem. Am. Mus. Nat. Hist. N. S.* **1, 6**:389–502.

Ketten, D. R. 1992. The marine mammal ear: specializations for aquatic audition and echolocation, in: D. B. Webster, R. R. Fay, and A. N. Popper (eds.), *The Evolutionary Biology of Hearing*, pp. 717–750. Springer-Verlag, Berlin.

Ketten, D. R., Domning, D. P., and Odell, D. K. 1992. Structure, function, and adaptation of the manatee ear, in: J. Thomas *et al.* (eds.), *Marine Mammal Sensory Systems*, pp. 77–95. Plenum Press, New York.

Kumar, K., and Sahni, A. 1986. *Remingtonocetus harudiensis*, new combination, a middle Eocene archaeocete (Mammalia, Cetacea) from western Kutch, India. *J. Vertebr. Paleontol.* **6**:326–349.

Lancaster, W. C. 1990. The middle ear of the Archaeoceti. *J. Vertebr. Paleontol.* **10**:117–127.

Lillie, D. G. L. 1910. Observations on the anatomy and general biology of some members of the larger Cetacea. *Proc. Zool. Soc. (London)* **2**:769–791.

Luo, Z., and Gingerich, P. D. In press. Transition from terrestrial ungulates to aquatic whales: transformation of the basicranium and evolution of hearing. *Bull. Mus. Paleontol. Univ. Michigan*

Luo, Z., and Marsh, K. 1996. The petrosal and inner ear structure of a fossil kogiine whale (Odontoceti, Mammalia). *J. Vertebr. Paleontol.* **16**:328–348.

MacPhee, R. D. E. 1981. Auditory regions of primates and eutherian insectivores: morphology, ontogeny and character analysis. *Contrib. Primatol.* **18**:1–282.

McCormick, J. G., Wever, E. G., Palin, J., and Ridgway, S. H. 1970. Sound conduction in the dolphin ear. *J. Acoust. Soc. Am.* **48**:1418–1428.

McCormick, J. G., Wever, E. G., Ridgway, S. H., and Palin, J. 1972. Sound reception in the porpoise as it relates to echolocation, in: R.-G. Busnel and J. F. Fish (eds.), *Animal Sonar Systems*, pp. 449–467. Plenum Press, New York.

Muizon, C. de. 1987. The affinities of *Notocetus vanbenedeni*, an early Miocene platanistoid (Cetacea, Mammalia) from Patagonia, southern Argentina. *Am. Mus. Novit.* **2904**:1–27.

Muizon, C. de. 1988. Le polyphyléstisme des Acrodelphidae Odontocete longirostre du Miocene européen. *Bull. Mus. Natl. Hist. Nat. 4e Ser. 10 Sect. C* **1**:31–88.

Muizon, C. de. 1994. Are the squalodonts related to the platanistoids? in: A. Berta and T. Deméré (eds.), *Contributions in Marine Mammal Paleontology Honoring Frank C. Whitmore, Jr.*, pp. 135–146. Proceedings of the San Diego Society of Natural History 29.

Norris, K. S. (ed.) 1966. *Whales, Dolphins, and Porpoises*. University of California Press, Berkeley.

Norris, K. S. 1980. Peripheral sound processing in odontocetes, in: R.-G. Busnel and J. F. Fish (eds.), *Animal Sonar Systems*, pp. 495–509. Plenum Press, New York.

Novacek, M. J. 1977. Aspects of the problem of variation, origin, and evolution of the eutherian auditory bulla. *Mammal Rev.* **7**:13–149.

Novacek, M. J. 1986. The skull of leptictid insectivorans and the higher-level classification of eutherian mammals. *Bull. Am. Mus. Nat. Hist.* **183**:1–112.

Oelschläger, H. A. 1990. Evolutionary morphology and acoustics in the dolphin skull, in: J. A. Thomas, and R. A. Kastelein (eds.), *Sensory Abilities of Cetaceans*, pp. 137–162. Plenum Press, New York.

Pilleri, G. 1983. Concerning the ear of the narwhal, *Monodon monoceros*. *Invest. Cetacea* **15**:175–179.

Pilleri, G., Gihr, M., and Kraus, C. 1987. The organ of Corti in cetaceans—1. Recent species. *Invest. Cetacea* **20**:43–125.

Pompeckj, J. F. von. 1922. Das Ohrskelett von *Zeuglodon*. *Senckenbergiana* **3**(3/4):44–100.

Purves, P. E. 1966. Anatomy and physiology of the outer and middle ear in cetaceans, in: K. S. Norris (ed.), *Whales, Dolphins, and Porpoises,* pp. 320–376. University of California Press, Berkeley.

Purves, P. E., and Pilleri, G. 1983. *Echolocation in Whales and Dolphins*. Academic Press, New York.

Reysenbach de Haan, F. W. 1960. Some aspects of mammalian hearing underwater. *Proc. R. Soc. London* **152B**:54–62.

Ridewood, W. G. 1922. Observations on the skull in foetal specimens of whales of the genera *Megaptera* and *Balaenoptera*. *Philos. Trans. R. Soc. London Ser. B* **211**:209–272.

Russell, D. E., Thewissen, J. G. M., and Sigogneau-Russell, D. 1983. A new dichobunid artiodactyl (Mammalia) from the Eocene of north-west Pakistan. *Proc. K. Ned. Akad. Wet. Ser. B* **86**:285–300.

Thewissen, J. G. M. 1994. Phylogenetic aspect of cetacean origins: a morphological perspective. *J. Mamm. Evol.* **2**:157–184.

Thewissen, J. G. M., and Hussain, S. T. 1993. Origin of underwater hearing in whales. *Nature* **361**:444–445.

Thewissen, J. G. M., and Hussain, S. T. 1998. Systematic review of the Pakicetidae, Early and Middle Eocene Cetacea from Pakistan and India, in: K. C. Beard and M. R. Dawson (eds.), *Carnegie Mus. Nat. Hist. Bull.* **34**:220–238.

Thewissen, J. G. M., Hussain, S. T., and Arif, M. 1994. Fossil evidence for the origin of aquatic locomotion in archaeocete whales. *Science* **263**:210–212.

Thewissen, J. G. M., Madar, S. I., and Hussain, S. T. 1996. *Ambulocetus natans*, an Eocene cetacean (Mammalia) from Pakistan. *Cour. Forsch.-Inst. Senckenberg* **191**;1–86.

Uhen, M. D. 1996. *Dorudon atrox* (Mammalia, Cetacea): form, function, and phylogenetic relationships of archaeocetes from the late middle Eocene of Egypt. Ph.D. dissertation, University of Michigan, Ann Arbor, 608 pp.

Van der Klaauw, C. J. 1931. The auditory bulla in some fossil mammals, with a general introduction to this region of the skull. *Bull. Am. Mus. Nat. Hist.* **62**:1–352.

van Kampen, P. N. 1905. Die Tympanalgegend des Säugetierschädel. *Morphol. Jahrb.* **34**:321–722.

Van Valen, L. M. 1966. Deltatheridia, a new order of mammals. *Bull. Am. Mus. Nat. Hist.* **132**:1–126.

Vincent, J. F. V. 1990. *Structural Biomaterials*. Princeton University Press, Princeton, NJ.

Webster, D. B. 1975. Auditory systems of the Heteromyidae: postnatal development of the ear of *Dipodomys merriami. J. Morphol.* **146**:377–394.

Webster, D. B., and Plassmann, W. 1992. Parallel evolution of low-frequency sensitivity in Old World and New World desert rodents, in: D. B. Webster, R. R. Fay, and A. N. Popper (eds.), *The Evolutionary Biology of Hearing*, pp. 633–636. Springer-Verlag, Berlin.

Webster, D. B., and Webster, M. 1975. Auditory systems of Heteromyidae: functional morphology and evolution of the middle ear. *J. Morphol.* **146**:343–376.

Whitmore, F. C., and Sanders, A. E. 1976. Review of the Oligocene Cetacea. *Syst. Zool.* **25**:304–320.

Zhou, X., Zhai, R., Gingerich, P. D., and Chen, L. 1995. Skull of a new mesonychid (Mammalia, Mesonychia) from the late Paleocene of China. *J. Vertebr. Paleontol.* **16**:387–400.

CHAPTER 10

Biomechanical Perspective on the Origin of Cetacean Flukes

FRANK E. FISH

1. Introduction

The evolution of aquatic forms from terrestrial ancestors has been a reoccurring event in the history of the vertebrates. As these animals adapted to the aquatic environment, the most derived representatives developed structures and mechanisms for high-performance propulsion in water. These organisms converged on propulsive modes that utilized oscillating hydrofoils for rapid and sustained swimming (Howell, 1930; Webb, 1975; Webb and Buffrénil, 1990; Fish, 1993a).

Swimming is the only mode of locomotion for cetaceans, which moved into the water and abandoned the terrestrial environment in the Eocene (Gingerich *et al.*, 1983, 1994; Thewissen *et al.*, 1994). For modern whales, the horizontally oriented tail flukes represent a hydrofoil. Whales use their axial muscles to propel themselves by vertical oscillations of flukes (Parry, 1949a; Slijper, 1961; Strickler, 1980; Bello *et al.*, 1985). These oscillations produce hydrodynamic thrust that opposes the drag experienced by the body as a result of the viscosity and the flow pattern of the medium. The thrust generated is a resultant force from lift created by the hydrofoil (Webb, 1975; Fish, 1993a; Vogel, 1994). An effective hydrofoil produces a large lift (force acting perpendicular to the flow) while minimizing drag (force acting parallel to flow; Weihs, 1989; Vogel, 1994). Although the sirenians also developed tail flukes, it is the cetaceans for which flukes have been most associated with rapid and relatively high-powered swimming (Fish and Hui, 1991). The force generated by the flukes is sufficient to drive a dolphin through the water at speeds exceeding 10 m/s or launch a 30-ton whale out of the water.

This chapter will focus on how thrust is generated by the action of the flukes in accordance with the varied morphological designs exhibited by whales. A biomechanical analysis of extant species is required to evaluate mechanisms by which the flukes of cetaceans would have evolved. Such an understanding permits a functional interpretation of the limited fossil remains of primitive cetaceans. Although we understand the evolution

FRANK E. FISH • Department of Biology, West Chester University, West Chester, Pennsylvania 19383.
The Emergence of Whales, edited by Thewissen. Plenum Press, New York, 1998.

of terrestrial locomotion because of the available physical evidence such as skeletal remains and footprints, no such record exists for swimming by cetaceans as no fossilized imprints of the flukes have been unearthed and the sea leaves no tracks.

2. Morphology Design and Construction of Flukes

2.1. Structure

The flukes are lateral extensions of the distal tail. Structurally, the flukes are composed of a cutaneous layer, a subcutaneous blubber layer, a ligamentous layer, and a core of dense fibrous tissue (Felts, 1966). Both the cutaneous and subcutaneous layers are continuations of their respective layers covering the rest of the body.

The bulk of the fluke is composed of the core of fibrous tissue. The core of collagen fibers forms a thick solid attachment on the numerous, short caudal vertebrae and intervertebral disks (Felts, 1966). This attachment unites the caudal vertebrae associated with the flukes into a single resilient element. Within the fluke, the collagen fibers are arranged in horizontal, vertical, and oblique bundles (Felts, 1966; Purves, 1969). Horizontal fibers radiate out through the fluke. The pattern of fiber bundles indicates an orientation appropriate for incurring high tensile stresses.

Similarly, the ligamentous layer is arranged to resist tension of the flukes particularly at the trailing edge and tips. In these regions, the core of fibrous tissue is thin, whereas the ligamentous layer is relatively thick (Felts, 1966). In addition, the ligamentous layer is thickest at the tips and inserts perpendicularly at the trailing edge. This architecture during the stroke cycle would limit bending, which is variable between species. The harbor porpoise (*Phocoena phocoena*) displays almost no bending at either the fluke tips or the trailing edge, whereas the white-sided dolphin (*Lagenorhynchus acutus*) with larger flukes shows 35 and 13% deflections across the chord (i.e., distance from leading to trailing edges) and tip-to-tip span, respectively (Curren *et al.*, 1994). Such differences in flexibility may reflect modification of the fibrous layers, which could affect swimming performance. Flexibility across the chord can increase propulsive efficiency by 20%, compared with a rigid propulsor executing similar oscillations (Katz and Weihs, 1978).

The posteriormost caudal vertebrae continue into the flukes and end immediately anterior to the fluke notch (Rommel, 1990). Vertebrae anterior to the flukes are laterally compressed, whereas vertebrae within the flukes are dorsoventrally compressed. The peduncle–fluke junction is characterized by relatively large intervertebral spacings (Rommel, 1990; Long *et al.*, 1997). The intervertebral joint at the base of the flukes mechanically functions as a low-resistance hinge acting as a center of rotation about the sagittal plane (Parry, 1949b; Long *et al.*, 1997). In addition to the low stiffness of the joint, rotation is aided by the "ball" vertebrae (Watson and Fordyce, 1993). Located at the peduncle–fluke junction, the ball vertebrae have convex cranial and caudal surfaces. Fluke rotation is controlled by the epaxial m. extensor caudae lateralis and the hypaxial m. hypaxialis lumborum (Pabst, 1990).

Resistance to fossilization of the soft anatomy of the flukes poses difficulties for reconstruction of the ancestral form. Modern cetaceans, however, possess various skeletal features associated with the flukes that may be compared with fossils to indicate the course of fluke evolution. This evolution has led to designs that provide high thrust production and efficiency in modern cetaceans.

2.2. Design and Physics of Flukes

Biomechanically, the flukes act like a pair of wings (Vogel, 1994). However, unlike the static wings of airplanes and jets, the flukes oscillate to generate a lift-derived thrust. As winglike structures, the flukes can be analyzed as engineered air- and hydrofoils to determine their effectiveness in lift generation. The shape of the flukes influences the energy requirements for swimming.

Fluke planforms have a sweptback tapered design that varies between different species with respect to hydrodynamically relevant parameters (Figs. 1, 2). For the fluke planform, measurements can be made on the span (*S*: tip-to-tip distance), root chord (*RC*: distance from base of fluke to trailing edge), sweep (Λ: angle between a perpendicular to *RC* and one-quarter chord position), and planform area (*A*).

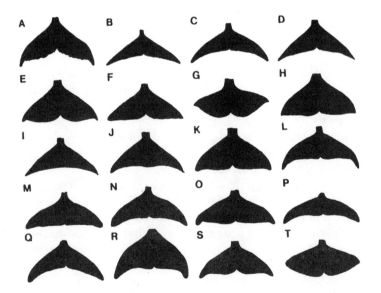

FIGURE 1. Planforms of flukes from representative cetacean species: A, humpback whale (*Megaptera novaeangliae*); B, blue whale (*Balaenoptera musculus*); C, minke whale (*Balaenoptera acutorostrata*); D, right whale (Eubalaena glacialis); E, gray whale (*Eschrichtius robustus*); F, sperm whale (*Physeter macrocephalus*); G, narwhal (*Monodon monoceros*); H, beluga (*Delphinapterus leucas*); I, Sowerby's beaked whale (*Mesoplodon bidens*); J, northern bottlenose whale (*Hyperoodon ampullatus*); K, Amazon river dolphin (*Inia geoffrensis*); L, long-finned pilot whale (*Globicephala melaena*); M, bottle-nose dolphin (*Tursiops truncatus*); N, Pacific white-sided dolphin (*Lagenorhynchus obliquidens*); O, killer whale *(Orcinus orca)*; P, false killer whale (*Pseudorca crassidens*); Q, Heaviside's dolphin (*Cephalorhynchus heavisidii*); R, northern right whale dolphin (*Lissodelphis borealis*); S, harbor porpoise (*Phocoena phocoena*); T, Dall's porpoise (*Phocoenoides dalli*).

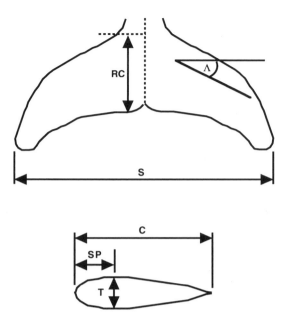

FIGURE 2. Fluke dimensions of planform (above) and cross-sectional profile (below). Explanation of dimensions is given in text.

Data from 34 cetacean species show that S and A are directly related to increasing body length (Fig. 3). This trend is expected because the fluke span and area determine the mass of water that is affected for thrust generation. Larger A would generate more thrust. Because thrust developed by the flukes is necessary to counter the drag incurred by the body as determined by its surface area (Bose *et al.*, 1990), S and A are associated with body length (BL) where S is proportional to BL and A is proportional to BL^2. Fluke span displayed a slight positive allometry according to the relationship $S = 0.111\ BL^{1.128}$, whereas fluke projected area displayed a slight negative allometry according to the relationship $A = 0.017\ BL^{1.946}$. Large whales would have a relatively larger S with smaller A than smaller dolphins.

When the relationships between S, A, and BL are compared between life history stages within a species, differences between juveniles and adults are evident that would affect performance (Amano and Miyazaki, 1993; Curren *et al.*, 1993). In neonates and prepubescent dolphins, the increase in S with respect to BL is not as rapid as observed for adults (Perrin, 1975; Amano and Miyazaki, 1993). Therefore, young animals may be at a disadvantage when swimming, thus requiring the use of free-riding behaviors to maintain speed with the parent (Lang, 1966).

The interaction between S and A as related to effectiveness of the hydrofoil design is expressed as the aspect ratio (AR). AR is calculated as S^2/A (Webb, 1975; Vogel, 1994). High AR indicates long narrow flukes, whereas low AR indicates broad flukes with a short S. High-AR hydrofoils are characteristic of relatively fast swimmers.

AR varies from 2.0 for the Amazon river dolphin (*Inia geoffrensis*) to high values of 6.1 and 6.2 for the fin whale (*Balaenoptera physalus*) and false killer whale (*Pseudorca*

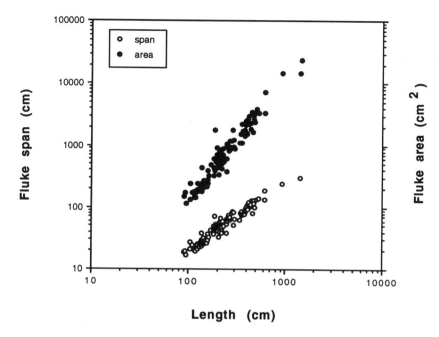

FIGURE 3. Relationship of planar surface area and fluke span versus body length. Data from Videler and Kamermans (1985), Bose and Lien (1989), Bose *et al.* (1990), Curren (1992), and Fish (1993b, unpublished).

crassidens), respectively (Bose and Lien, 1989; Fish, unpublished). These values correspond to the swimming performance in these species. *Pseudorca* moves at maximum speeds of 7.5 m/s (Fish, unpublished) with speeds in schools of 3 m/s (Norris and Prescott, 1961) and *Balaenoptera* attains maximum speeds of 10 m/s with sustained speeds of 2.5 m/s (Bose and Lien, 1989). *Inia* is relatively slower with a maximum speed of 3.9 m/s and routine speed of about 0.4–0.9 m/s (Best and da Silva, 1989).

Well-performing flukes maximize the ratio of life (L) to drag (D) generated by their action (Webb, 1975). An increase in the maximum L/D with increasing size is achieved by increasing S more rapidly than the square root of A, thereby increasing AR (von Mises, 1945; Lighthill, 1977; van Dam, 1987). The lift for flukes of a given area and motion would be greatest when AR is highest (Bose *et al.*, 1990; Daniel *et al.*, 1992). The longer trailing edge of a high-AR fluke increases the mass of water deflected posteriorly augmenting the thrust component. However, AR above 8–10 provides little further advantage and may be structurally limited (Webb, 1975).

Drag incurred by the flukes is inversely dependent on AR related primarily to the induced drag component (Webb, 1975). Induced drag is produced as a consequence of the lift generated by the flukes. As the flukes are canted at an angle to the water flow, lift is produced by deflection of the water and pressure difference between the dorsal and ventral surfaces of the flukes (Webb, 1975; Blake, 1983). The pressure difference produces spanwise cross flows that go around the fluke tips resulting in the formation of spiraling vortical flow. The flow is shed from the fluke tips as longitudinal tip vortices. The energy dissipated by the vortices represents the induced drag. High *AR* and tapering of the flukes reduce tip vorticity and induced drag (Webb, 1975; Rayner, 1985; Webb and Burrfénil, 1990; Daniel *et al.*, 1992).

Induced drag also is limited by the sweep (Λ) of the flukes. It was shown by van Dam (1987) that a tapered wing with sweptback or crescent design could reduce the induced drag by 8.8% compared with a wing with an elliptical planform. Minimal induced drag is fostered by a sweptwing planform with a root chord greater than the chord at the tips giving a triangular shape (Küchermann, 1953; Ashenberg and Weihs, 1984). This optimal shape approximates the planform of cetacean flukes. Flukes have sweep angles ranging from lows in killer whale (*Orcinus orca*) and Dall's porpoise (*Phocoenoides dalli*) of 4.4 and 5.4°, respectively, to a maximum value of 47.4° for white-sided dolphin (*Lagenorhynchus acutus*) (Bose *et al.*, 1990; Fish, unpublished).

Sweep of the fluke together with taper has the effect of concentrating the surface area toward the trailing edge. This would effectively shift the lift distribution posterior of the center of gravity affecting pitching equilibrium (von Mises, 1945; Webb, 1975). Lighthill (1970) and Wu (1971b) suggested that a minimum in wasted energy would be realized when the pitching axis was moved to the 0.75 chord position. Proximity of the pitch axis close to the trailing edge was supported by Chopra (1975).

The sweep angle is inversely related to the aspect ratio of the flukes (Bose *et al.*, 1990; Curren *et al.*, 1993; Fig. 4). The combination of low sweep with high *AR* allows for high-efficiency rapid swimming (Azuma, 1983). High sweep may compensate for the reduced lift production of low-*AR* flukes. Highly sweptback, low-*AR* wings produce maximum lift

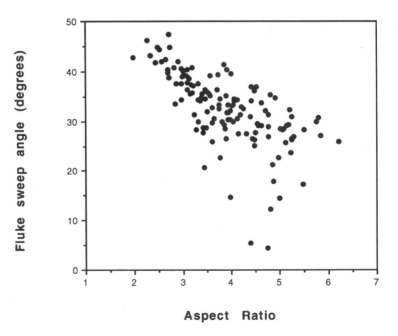

FIGURE 4. Relationship between sweep angle (A) and aspect ratio (*AR*). Data from Bose and Lien (1989), Bose *et al.* (1990), Curren (1992), and Fish (1993b, unpublished).

when operating at large angles of attack (see below), when low-sweep, high-*AR* designs would fail (Hurt, 1965). However, the maximum lift is reduced with increasing sweep angle for a given *AR*, whereas efficiency increases (Liu and Bose, 1993). Mathematical analysis by Chopra and Kambe (1977), however, found that a sweep angle exceeding about 30° leads to a reduction in efficiency. The relationship between sweep and *AR* also indicates a structural limitation to the strength and stiffness of the flukes (van Dam, 1987; Bose *et al.*, 1990). The ability to sustain certain loads without breaking is considered a major constraint on increasing span and *AR* (Daniel, 1988). Because the fibrous composition not only strengthens the flukes but also increases flexibility, extreme increase in span with increased *AR*, although potentially generating higher lift, would exaggerate the bending of the appendage in an oscillatory mode and reduce performance.

Flukes, however, do show some degree of both spanwise and chordwise flexibility. The center of the flukes is more rigid than the tips. During the upstrokes fluke tips are bent down slightly from the plane of the fluke and lag behind the center, whereas bending in the opposite direction occurs during the downstroke. Bose *et al.* (1990) suggested that the phase difference related to this spanwise flexibility would prevent the total loss of thrust at the end of the stroke. While the tips would be ending the stroke and effectively generating no thrust, the center would have started the next stroke and begun thrust generation. On the other hand, chordwise flexibility at the trailing edge of the flukes potentially can increase the efficiency of the flukes by up to 20% with only a moderate decrease in the overall thrust (Katz and Weihs, 1978).

Sections of the fluke along the longitudinal axis display a conventional streamlined foil profile with a rounded leading edge and long tapering trailing edge (Fig. 2). The sharp trailing edge and rounded leading edge are crucial for generating lift and minimizing drag (Lighthill, 1970; Vogel, 1994). The streamlined cross section is maintained by the core of fibrous material (Felts, 1966). The flukes are symmetrical about the chord (Lang, 1966; Bose *et al.*, 1990). In examining the flukes of the common dolphin (*Delphinus bairdi*), Lang (1966) reported that some warpage was evident. This may explain the contradictory results of Purves (1969) who noted an asymmetry in the fluke cross sections. Whereas an asymmetry would have supported thrust production through only half of the stroke cycle of the dolphin, the symmetrical design of the flukes indicates that thrust is generated on both up- and downstrokes.

For any section through the fluke in the parasagittal plane, measurements can be made on the chord (*C*), maximum thickness (*T*), and shoulder position (*SP*: distance of *T* from leading edge expressed as a percentage of *C*) (Fig. 2). *SP* and the thickness ratio ($TR = T/C$) indicate hydrodynamic performance relating to the generation of lift and drag for foil sections (von Mises, 1945; Hoerner, 1965). Flukes range from 25 to 40% for *SP* and between 0.16 and 0.25 for *TR* (Lang, 1966; Bose *et al.*, 1990). These sections are similar to engineered foils, which are classified by the National Advisory Committee for Aeronautics (NACA). The NACA 63_4-021 foil (Abbott and von Doenhoff, 1949) provides a reasonable facsimile of the fluke sections. This resemblance suggests that the flukes would be able to produce high lift with low drag at angles of attack up to 20°. This would be possible because the shape of the fluke section does not promote extremes in the chordwise pressure distribution which cause separation of flow from the foil surface and increase drag (Lang, 1966). Thus, cavitation that could damage the animal is not expected to occur even at routine swimming speeds of dolphins.

The flukes of modern cetaceans have a design to act as a hydrofoil for lift production. The combination of moderate aspect ratio, sweep, cross-sectional design, and flexibility of the flukes furnishes a morphology capable of the generation of high lift with low drag performance. To realize the potential of a lift-based propulsion, the flukes must be moved in a fashion that optimally orients the flukes into a flow.

3. Kinematics and Hydrodynamic Performance

3.1. Propulsive Movement

Whales generate thrust exclusively with caudal flukes (Fish and Hui, 1991). The tail flukes, which act as the hydrofoil, oscillate dorsoventrally. Although different in the orientation of the hydrofoil, the propulsive motions of cetaceans are similar to those of some of the fastest marine vertebrates, including scombrid fishes and laminid sharks. These motions are characteristic of the so-called carangiform with lunate tail or thunniform mode (Lighthill, 1969).

The posterior one-third of the body is bent to effect dorsoventral movement of the flukes (Fig. 5; Parry, 1949b; Slijper, 1961; Lang and Daybell, 1963; Videler and Kamermans, 1985; Yanov, 1991; Fish, 1993b). Although these heaving motions vertically displace the flukes through an arc, the flukes do not move as a simple pendulum. Rather, superimposed on the motion, the flukes are pitched at a joint at their base. Pitching at the base of the flukes occurs because of the double hinge mechanism of the caudal vertebrae including the "ball" vertebrae (Watson and Fordyce, 1993).

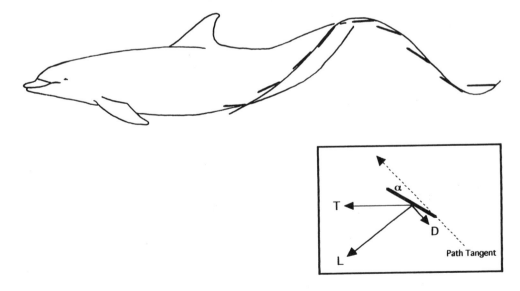

FIGURE 5. Path of oscillating dolphin flukes through a stroke cycle. The tips of the flukes move along a sinusoidal path. Fluke position along the path is illustrated as a straight line. The box shows the relationship between the tangent to the path of the flukes and the angle of attack, α. From Fish (1993b).

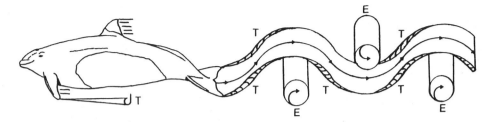

FIGURE 6. Pattern of vorticity shed in wake of dolphin. Tip vortices (T) and trailing edge vortices (E) generated from the flukes are shown. From Fish (1993a).

The heaving and pitching motions of the flukes result in a varying angle (pitch angle) between the flukes and the horizontal plane (Figs. 5, 6). At its maximum vertical displacement, the flukes have a pitch angle of zero so that the axis of the fluke chord is parallel to the axis of progression (i.e., horizontal when the animal is swimming at constant depth). This orientation effectively minimizes the drag on the flukes but generates no thrust. As the flukes are downswept, the pitch angle increases by flexion at the peduncle–fluke junction. Through the middle of the downward excursion of the stroke, the pitch angle is maintained at a maximum (Fig. 5). The end of the stroke is accompanied by a decrease in the pitch angle with the flukes again oriented parallel to the direction of forward progression. The average pitch angles for the harbor porpoise (*Phocoena phocoena*) and white-sided dolphin (*Lagenorhynchus acutus*) were 34 and 33°, respectively (Curren *et al.*, 1994).

The ability to rotate the flukes about a pitching axis allows for control of angle of attack. Angle of attack is defined as the angle between the tangent of the fluke's path and the axis of the fluke's chord (Fig. 5) (Fierstine and Walters, 1968; Fish, 1993b). Maintenance of a positive angle of attack ensures thrust generation throughout the majority of the stroke cycle (Lang and Daybell, 1963; Lighthill, 1969, 1970; Videler and Kamermans, 1985; Goforth, 1990).

The magnitude of the attack angle will affect the propulsive efficiency and the thrust generated in lift-based swimming (Webb, 1975). As angle of attack is increased for a hydrofoil, there is increase in both lift and drag. Lift will increase faster than the drag with increasing angle of attack up to a critical level. Further increase of angle of attack leads to an increase in drag and precipitous loss of lift in a condition called *stall*. Stall is caused by separation of the flow from the foil surface, which is unavoidable at a critical angle of attack.

Angle of attack of oscillating dolphin flukes increases rapidly at the initiation of up- and downstrokes reaching a maximum within the first third of the stroke (Fish, 1993b). Maximum angle of attack varies indirectly with swimming speed in bottlenose dolphins (*Tursiops truncatus*). Maximum values ranging from 12 to 21° for *Tursiops* and 22.5 to 24° for *Lagenorhynchus* were reported by Fish (1993b) and Lang and Daybell (1963), respectively. Such ranges are below the 30° angle for an oscillating foil at which stall occurs (Triantafyllou and Triantafyllou, 1995).

Maximum heave amplitude is confined to the tips of the flukes. At low swimming speeds (<2.2 m/s for *Lagenorhynchus*), heave amplitude appears to increase with speed (Curren *et al.*, 1994). Heave amplitude, however, is independent of swimming speed at routine and sprint speeds (Fish, 1993b). Maximum heave amplitude remains a constant proportion of body length at approximately 20%.

Alternatively, the stroke frequency varies directly with swimming speed. Maximum frequency corresponds to tailbeat frequency at maximum voluntary muscular effort for dolphins (Goforth, 1990). The positive linear relationship of frequency with swimming speed for *Tursiops* is consistent with observations on other marine mammals and fish that use hydrofoil propulsion (Webb, 1975; Feldkamp, 1987; Fish *et al.*, 1988). Modulation of frequency with constant amplitude would prevent excessive distortion of the body, which would increase overall drag and decrease locomotor efficiency. This trend differs from semiaquatic paddlers, which modulate amplitude and maintain a constant frequency to achieve higher swimming speeds (Williams, 1983; Fish, 1984).

Flukes follow a sinusoidal pathway (Fig. 5) that is symmetrical about the longitudinal axis of the body and in time (Videler and Kamermans, 1985; Goforth, 1990; Fish, 1993b). Previously it was assumed that cetaceans swam with an asymmetrical propulsive stroke. This assumption was predicated on differences in the epaxial and hypaxial muscle masses (Purves, 1963). Parry (1949a) was able to confirm differences in stroke duration between upstroke and downstroke from counts of film frames of a dolphin swimming away from a camera. Unfortunately, the film records of the swimming dolphin showed that the animal was giving birth at the time. However, control of the flukes does permit variable movements during up- and downstrokes so that a stroke cycle can be divided into power and recovery phases (Purves, 1963).

As already discussed, the propulsive movements of cetaceans are confined to the posterior one-third of the body, with the greatest amplitude at the flukes (Fish and Hui, 1991). This action restricts thrust production to the flukes. By restricting bending to the peduncle and base of the flukes, this permits rotational motion to maintain a positive angle of attack to the oncoming flow (Webb, 1975). Thus, the flukes are able to produce a high lift and nearly continuous thrust over the entire stroke cycle (Fish, 1992). Such movements, classified as thunniform swimming, will engender high performance in thrust production and efficiency.

3.2. Thrust Production and Efficiency

Although measurements and calculations of lift and drag performance by static hydrofoils and wings permit a rudimentary understanding of the development of thrust by flukes, such estimates are not directly translated to thrust production by the flukes because of their oscillatory motions. To comprehend the dynamic production of lift-based thrust, hydromechanical models were employed (Lighthill, 1969; Wu, 1971b; Chopra and Kambe, 1977). Because the flukes are connected to the body by a narrow attachment, the caudal peduncle, that oscillates in the direction of its minimum resistance, the flukes are essentially separated from the body (Lighthill, 1969, 1970; Fish and Hui, 1991). This allows analysis of thrust production by the flukes to be made separate of the body and its actions.

Estimates of thrust based on the motion of the flukes alone have been used to independently assess the drag related to body form and swimming motions. A number of studies have used kinematic data (Norris and Prescott, 1961; Lang and Daybell, 1963; Fish, 1993b) to help develop hydromechanical models based on oscillating plates or hydrofoils

(Parry, 1949a; Lighthill, 1969, 1970; Wu, 1971b; Chopra and Kambe, 1977; Yates, 1983; Romanenko, 1995).

As the flukes oscillate ventrally, they are pitched at a positive angle of attack to the oncoming flow. The angle of attack and fluke velocity are determined by the vertical velocity of the flukes and horizontal velocity of the body. The streamlines of fluid are deflected above and below the flukes imparting a higher velocity to the upper flow. By the Bernoulli theorem, a pressure difference results with a lower pressure on the dorsal aspect of the flukes. The net pressure produces a pressure force that is resolved into drag tangent to the axis of motion of the flukes and a lift perpendicular to the axis of motion (Webb, 1975). The center of lift is relatively near the leading edge at or anterior to the maximum thickness (Vogel, 1994). The pressure force is reversed on the upstroke.

The orientation of the flukes throughout the stroke produces lift directed forward and upward during the downstroke and forward and downward during the upstroke. The anteriorly directed component from the mean forward tilt of lift represents the thrust (Daniel *et al.*, 1992). Thrust from lift increases directly with increases in angle of attack. However, low angles of attack increase hydromechanical efficiency while reducing the probability of stalling and decreased thrust production (Chopra, 1976).

Lift also depends on the frequency of oscillation of the flukes. Thrust increases with frequency whereas efficiency decreases (Daniel, 1991). The reciprocating action of the flukes means that the flow pattern is reversed through the stroke, and because the water has inertia, the flow pattern will take time to redevelop potentially affecting performance (Wu, 1971b; Daniel, 1991; Daniel *et al.*, 1992). The importance of the oscillatory motion to thrust generation and efficiency is determined by the reduced frequency parameter, which is the ratio of oscillatory to forward motion (Daniel *et al.*, 1992). A reduced frequency less than 0.1 indicates nearly steady motion (Yates, 1983; Daniel *et al.*, 1992). High values of reduced frequency indicate the dominance of unsteady effects, which incur lower lift than steady motion (Lighthill, 1970). Reduced frequencies of 0.51–1.15 for *Tursiops* and 0.4 for *Lagenorhynchus* were reported by Fish (1993b) and Webb (1975), respectively, indicating the dominance of unsteady effects.

Unsteady effects may contribute to thrust production by increasing the relative velocity and thus the lift (Daniel *et al.*, 1992). In addition, accelerational flows fostered by the unsteady effects may generate thrust.

As thrust from lift is produced, momentum is transferred from the flukes to the water. The momentum is proportional to the mass of the affected water and velocity of the flukes. The water is pushed back in a direction opposite to the swimming direction with a net rate of change of momentum that according to Newton's third law is equal and opposite to the thrust (Wu, 1971a; Chopra, 1975; Videler, 1993). The thrust produced balances the viscous and pressure drag of the body and flukes.

The momentum imparted to the fluid is concentrated in a jet of fluid directed on average opposite to the swimming direction (Wu, 1971a; Rayner, 1985; Videler, 1993). The jet induces the resting water around it to generate a vortex wake. A wake is necessary to produce thrust. The wake is visualized as a trail of connected alternating clockwise and anticlockwise vortex rings with the jet directed through the center of the rings (Fig. 6). This vortex pattern is generated at the bottom and top of the stroke as vortices shed from the fluke with opposite circulation (Vogel, 1994). Tip vortices that roll off the fluke tips connect the shed vortices to form the ring.

In addition to the lift, leading edge suction contributes to thrust (Lighthill, 1970; Chopra and Kambe, 1977; Ahmadi and Widnall, 1986). A suction is created as the flow becomes highly accelerated as it moves around a sharp corner (Yates, 1983). The high acceleration locally decreases the pressure and produces the suction. At the trailing edge where vorticity is shed, flow around the edge is forbidden (Videler, 1993; Vogel, 1994) The rounded leading edge promotes the suction force (Lighthill, 1970; Wu, 1971b). The effect of leading edge suction is to tilt the pressure force forward by an angle equal to the angle of attack (Weihs, personal communication). The total lift force, which is typically normal to the fluke axis, is tilted perpendicular to the direction of fluke motion and thus increases the thrust component. However, excessive leading edge suction could induce stalling via boundary layer separation and reduce thrust. The lunate configuration of the leading edge of the flukes reduces leading edge suction without a decrease in total thrust (Chopra and Kambe, 1977; van Dam, 1987).

Efficiency is defined as the ratio of the mean thrust power required to overcome the drag on the animal divided by the mean rate at which the animal is doing work against the surrounding water (Lighthill, 1970). Efficient thrust production requires high lift production while minimizing energy loss into the wake (Blake, 1983).

Wu (1971b) estimated that the propulsive efficiency of a dolphin could be as high as 0.99. This efficiency was assumed to be an overestimate because Wu used a two-dimensional analysis that underestimated trailing vorticity and wake energy loss (Ahmadi and Widnall, 1986; Karpouzian et al., 1990). Competing three-dimensional models of lunate tail swimming predict efficiencies lower than 0.99 but above 0.7 (Chopra and Kambe, 1977; Yates, 1983; Bose and Lien, 1989; Karpouzian et al., 1990; Fish, 1993b, 1996). Efficiencies in this range are considered good, because few engineered propellers achieve efficiencies higher than 0.7 (Liu and Bose, 1993). The high efficiencies associated with swimming by dolphins are dependent on a fluke design that enhances high thrust with reduced drag and on fluke oscillation that maintains continuous thrust production (Fish, 1992).

The design and mechanics of flukes of modern cetaceans have produced a highly efficient propulsor for large thrust production. However, the ancestors of cetaceans were terrestrial (Gingerich et al., 1983; Thewissen et al., 1994) and would not have developed flukes prior to entering the water. How then did flukes evolve, and what were the transitional stages and their levels of performance?

4. Evolution

The evolution of cetacean flukes has been a matter of speculation for over 100 years (Flower, 1883; Howell, 1930). The soft tissue composition of the flukes has left no record of how or why they arose and how they may have evolved to the high-efficiency and large-thrust-generating propulsive structures. This is contradictory to the situation for ichthyosaurs, which are another highly derived secondarily aquatic vertebrate with a high-aspect-ratio caudal propulsor (Motani et al., 1996). These extinct marine reptiles left not only a rich store of fossilized skeletons, but in a number of cases, particularly during the Jurassic, they fortuitously left imprints of lunate tail fins. The recent discovery of a Lower Triassic ichthyosaur, *Chensaurus chaoxianensis*, with an elongate body and caudal fin shape allowed investigators to construct a possible pathway examining the transition from anguilliform, through subcarangiform to more derived thunniform swimmers (Motani *et al.*, 1996).

Despite recent discoveries of early cetaceans, such as *Pakicetus, Ambulocetus*, and *Rodhocetus* (Gingerich *et al.*, 1983, 1994; Thewissen *et al.*, 1994, 1996a), there still remains a paucity of tangible physical evidence on the evolution of the flukes. Tail vertebrae in these fossils are lacking or incomplete, especially for the most terminal portions. To this add that (1) modern cetacean species exhibit the highly derived thunniform swimming mode and design, (2) no series of intermediate fluke designs exist, and (3) they are phylogenetically disjunct from their closest living relative (i.e., ungulates), which have specialized for terrestrial locomotion; thus, little direct information is available to answer the evolutionary questions regarding the transition of the flukes.

Substitutes to direct observations of morphological transition in fossil lineages are the use of ontogenetic information and the use of model specimens. By examining embryonic development within an individual, the ontogeny can be used as an additional character for understanding the phylogenetic framework. Similarities between ontogenetic and phylogenetic sequences reflect a possible pathway related to shared developmental patterns. Such inferences have been used previously to lend support to the evolution of complex anatomical changes (Rowe, 1996) and have been suggested as a means of reconstructing design modifications in the flukes of whales (Folkens and Barnes, 1984). Alternatively, a model can be constructed that draws on the swimming performance of modern species as analogues of the primitive intermediate forms (Fish, 1996). By testing living organisms of similar mechanical design to extinct species, performance characteristics can be examined to judge their importance in order to construct a mechanically plausible evolutionary scenario. Because the mechanics of swimming is used to generate the model, the modern species chosen need not be closely related to the group of interest or to each other. Lauder (1995) justified this technique of using extant surrogates to determine performance in extinct forms from a different clade.

4.1. Ontogeny

Despite their importance to the evolution and locomotion of whales, development of the flukes has been largely overlooked. The ontogeny of the flukes was the main focus of study over 100 years ago (Flower, 1883; Ryder, 1885; Howell, 1930). The argument at that time was whether the flukes represented a secondarily acquired structure or were the vestiges of the hind limbs. Morphologists of the time including Huxley, Flower, and Owen supported the former view, while Ryder defended the idea that the flukes were integumentary limb-folds that had migrated caudally (Flower, 1883; Ryder, 1885; Howell, 1930).

Much of the evidence quoted by Ryder (1885) was based on the assumption that whales had a seal-like ancestor that used its hind flippers for propulsion similar to modern seals. With the hind limbs extended posteriorly, these limbs were believed to have fused to the body rendering them immobile and eventually causing the muscles and bones of the pelvis and extremities to atrophy. However, the integument of the feet would be in position on the lateral aspect of the tail to form the rudiments of flukes. In addition, the flukes would develop as folds distally rather than as lateral ridges extending the entire length of the tail.

Ryder's scenario has been discounted and the flukes are considered to represent merely outgrowths of the skin and connective tissue of the tail (Slijper, 1979). The occurrence of whales exhibiting external hind limbs (Andrews, 1921), the presence of embryonic limb buds anterior of the tail (Rice and Wolman, 1971; Slijper, 1979), and fossil cetaceans with mobile but reduced hind limbs (Gingerich *et al.*, 1990, 1994) indicate separate ontogenies of the flukes and the hind limbs.

Initially, the tail in cetacean embryos is drawn out to a point with no trace of flukes. For example, in *Stenella attenuata*, the absence of flukes occurs in embryos approximately 45 mm in total length (Meyer *et al.*, 1995; Perrin, 1997). By 63 mm, flukes appear as a diamond or spadelike design with rounded edges (Perrin, 1997). This stage is reached in the gray whale (*Eschrichtius robustus*) at a size of 120 mm, which is estimated to be an age of 87 days (Rice and Wolman, 1971). When the *Stenella* fetus has reached a size of 165 mm, the flukes have expanded laterally reminiscent of the fluke shape of slow-swimming species. *Stenella* exhibits a deeply notched, lunate fluke design similar to adults by 225 mm (Meyer *et al.*, 1995).

Examination of other whale embryos showed similar developmental patterns (Ryder, 1885; Rice and Wolman, 1971). Such patterns possibly reflect the evolutionary transition of cetacean flukes. This developmental-transitional sequence is displayed in Fig. 7 as a reproduction of the figure original provided by Ryder (1885). Such a generalized sequence, although theoretically possible, is conjectural as a reconstruction of the evolution of flukes (Folkens and Barnes, 1984).

4.2. Functional Model

Prior to the evolution of flukes, the ancestors of modern cetaceans would have needed extensive changes not only to their morphology but also to the mechanism used to pro-

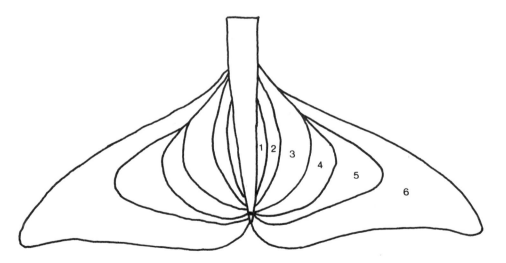

FIGURE 7. Successive contours of six stages of fluke development redrawn from Ryder (1885). Fluke outlines were compiled from different sources and scaled to equivalent size.

duce thrust for swimming (Fish, 1996; Thewissen *et al.*, 1996a). This is no trivial matter in that the swimming modes of terrestrial and semiaquatic ancestors would have used not only different appendages from the flukes of cetaceans, but also the force necessary to generate thrust would have had different derivations (Howell, 1930; Fish, 1992, 1996).

Paddling using alternate motions of the legs is the primitive mode of aquatic propulsion in mammals (Williams, 1983; Rayner, 1985; Fish, 1992, 1993a,c). Combinations of forefeet, hind feet, or all four feet are used to generate thrust (Tarasoff *et al.*, 1972; Williams, 1983, 1989; Fish, 1984, 1993c). The stroke cycle is composed of power and recovery phases.

During the power phase the foot is stroked posteriorly while the digits are abducted (spread) to maximize the planar area by the inclusion of interdigital webbing or lateral fringe hairs. The increased area and movement of the foot produces a drag with an anteriorly directed component. Thus, thrust for the animal is generated from the integration of the drag component from all paddling feet. This drag-based mechanism is analogous to paddling a canoe or rowing a boat.

During recovery, the foot is repositioned without generating thrust. Despite adduction of the digits and other conformational changes that reduce the area of the limb during recovery, some posteriorly oriented drag will be generated that reduces the net thrust (Fish, 1984). As a result, drag-based paddling has a low mechanical efficiency of < 40% (Fish, 1993a). Therefore, more energy is required to produce the same thrust by paddling than by lift-based propulsion (Weihs and Webb, 1983). Propulsors using lift can be twice as effective in generating thrust and may be up to five times more effective (Weihs, 1989).

How then could lift-based swimming with flukes have evolved from an inefficient drag-based limbed propulsion? To address this question, as well as the evolution of the derived swimming modes for other aquatic clades, Fish (1993a, 1996) developed a functional model based on kinematics, energetics, and hydrodynamics of swimming in modern mammals. The underlying assumption of the model is that neuromotor patterns for locomotion are conservative (Jenkins and Goslow, 1983; Smith, 1994). The idea of conservatism permits large-scale changes in swimming kinematics and performance with only slight modification of the neuromotor pattern originating from terrestrial locomotion. Therefore, despite the dissimilarities between the motions of the body and appendages for drag-based and lift-based aquatic propulsion, the modes can be related as modifications for swimming from the neuromotor patterns associated with terrestrial gaits (Fish, 1996). Symmetrical terrestrial gaits, such as walk and trot, are the basis of paddling modes; whereas lift-based swimming arises from asymmetrical gaits, such as gallops and bound, which utilize spinal flexion in concert with simultaneous limb motion for rapid progression (Rayner, 1985; Williams, 1989; Fish, 1993a, 1994, 1996; Gutmann, 1994).

A subset of the model by Fish (1996) pertinent to the present discussion is shown in Fig. 8. The model assumes that the terrestrial ancestor to cetaceans (i.e., mesonychian condylarths) would first swim using a modified quadrupedal gait by paddling with all four limbs as has been overly characterized as a "dog paddle." Mesonychians were terrestrial and displayed a morphology for cursorial locomotion (Gingerich *et al.*, 1994; Thewissen, 1994; O'Leary and Rose, 1995). In addition, lift provided by the paddling motions of the forelimbs would aid in maintaining trim to keep the nares above the water surface for continuous respiration (Fish, 1993c).

With adaptation to a semiaquatic existence, a shift to exclusive use of the hind limbs for paddling would occur. Pelvic paddling prevents mechanical and hydrodynamic inter-

Locomotor Mode	Performance Developments	Morphological Developments
Caudal Oscillation ↑	Lift-based propulsion; increased efficiency, power and speed; possible elastic energy storage; reduced pitching movements	High aspect ratio flukes; control of joint at peduncle; loss of external hindlimbs
Undulation ↑	Reduced drag; increased efficiency; maintain constant speed	Reduction of hindlimbs; expansion of tail at tip
Simultaneous Pelvic Paddling ↑	Submerged swimming; increased power and speed	Elongate, dorsoventrally flattened tail
Alternate Pelvic Paddling ↑	Increased paddling efficiency; buoyancy control	Enlargement of hind feet; interdigital webbing; non-wettable fur
Quadrupedal Paddling ↑	Move to water; drag-based propulsion	
Terrestrial Quadruped		

FIGURE 8. Functional model showing changes in swimming mode, performance, and morphology associated with increasing aquatic habits within the cetacean lineage.

ference between ipsolateral limbs and frees the forelimbs for tactile reception, prey capture, food manipulation, maneuverability, and locomotor stabilization (Fish, 1993c, 1996). Abandonment of forelimb use for propulsion and trim control would be compensated by the acquisition of enlarged hind feet and nonwettable fur. Increased thrust would have been fostered by elongation of the digits and addition of interdigital webbing permitting an increased foot area to affect a larger mass of water (Fish, 1984). Air entrapped in the fur in conjunction with the air in the lungs would maintain a positive buoyancy to prevent submergence of respiratory openings and provide a nearly horizontal trim for improved streamlining (Fish, 1993c). In addition, the nonwettable fur increases the insulation over the body to reduce heat loss to the highly thermally conductive aquatic environment.

Increased buoyancy, however, would eventually be a disadvantage as the cetacean ancestors commenced underwater foraging. The greater up-force would have made it difficult and energetically demanding to submerge. To generate sufficient force for diving to counteract the positive buoyancy, simultaneous strokes of the hind limbs could have been used to submerge to a sufficient depth where compression of the air in the fur and lungs would reduce the up-force. In addition, the simultaneous strokes increase speed when submerged, particularly in pursuit of prey. Such a swimming pattern is used when diving by otters (Tarasoff *et al.*, 1972; Williams, 1989; Fish, 1994). Because this swimming mode represents a modified bounding gait, the motion of the limbs is coordinated with flexion and extension of the spine (Fish, 1994). As observed for the river otter (*Lutra canadensis*), the tail displays an undulatory pattern that has the capability of thrust generation.

The addition of the tail in thrust generation becomes important to maintain speed underwater, because the simultaneous strokes of the limbs will incur periods of no thrust and increase drag when both feet are in the recovery phase of the stroke. Undulatory movements of the tail can generate nearly continuous thrust over the stroke cycle reducing accelerations and increasing efficiency (Fish, 1993a). For *Lutra*, undulation will be limited because the tail tapers posteriorly. Increased thrust production from tail undulation is facilitated by expansion of the tail tip perpendicular to the plane of motion. This morphology is displayed by the giant river otter (*Pteronura brasiliensis*), which possesses webbed hind feet and a broad, flattened tail (Howell, 1930; Chanin, 1985). From a film by Cousteau and Cousteau, Fish (1994) made observations of swimming *Pteronura* that showed the otter to rapidly swim submerged by simultaneous strokes of the hind limbs and undulation of the tail in the vertical plane. The morphology and swimming mode of *Pteronura* suggest an intermediate design between drag-based paddling and lift-based undulation (Howell, 1930; Sanderson, 1956) and a modern analogue of the primitive cetacean (Flower, 1883).

As suggested from the swimming patterns of otters, the transition from paddling to axial undulation required the presence of a long tail. This has been a long-held view (Flower, 1883; Ryder, 1885), although there has never been a mechanism to explain the tail's occurrence. One possible explanation for the possession of a long tail may be its use for thermoregulation in warm climates. The high surface-to-volume ratio and elaborate vascularization of mammalian tails act as effective thermoregulatory devices for the control of body insulation by regional heterothermia (Fish, 1979; Hickman, 1979). In warm climates where low insulation is warranted, a long tail would be an advantage to prevent overheating by elimination of excess metabolic heat to the environment. Indeed, cetaceans originated in tropical and subtropical climates where a long tail for temperature regulation would be advantageous (Barnes *et al.*, 1985; Fordyce, 1992; Gingerich *et al.*, 1994; Thewissen *et al.*, 1994).

In the final stages of the sequence, undulation of the tail would replace limb propulsion (Fish, 1993a, 1996). By removing the appendages from propulsion, drag is reduced, thrust can be continuously generated, and efficiency is increased. Further increase in performance using caudal undulation is achieved by expanding the tail distally, which would culminate in the rapid evolution of high-efficiency caudal hydrofoils (i.e., flukes) (Webb and Buffrénil, 1990; Fish, 1993a, 1994, 1996; Gutmann, 1994). The inclusion of a joint at the base of the flukes to control pitch would complete the transition to lift-based propulsion. The implication of the model is that the orientation and propulsive movements of the flukes are the evolutionary result of axial motions associated with terrestrial gaits.

When early archeocetes are examined with respect to the functional model, the morphology of these early whales suggests changes corresponding to the proposed sequence. *Pakicetus inachus* from the early Eocene is recognized as the oldest and most primitive cetacean known (Gingerich *et al.*, 1983; Thewissen, 1994). Although only the cranial and jaw structures have been described, *Pakicetus* does exhibit some features associated with movement into water and submerged foraging, including its ear structure and dentition (Gingerich *et al.*, 1983; Thewissen and Hussain, 1993). The degree of aquatic adaptation in *Pakicetus*, however, indicates that this species was amphibious and as such most likely would have propelled itself by paddling.

Further evolution toward increased aquatic habits is displayed by *Ambulocetus natans*, which had well-developed limbs and a prominent tail (Thewissen *et al.*, 1994). This early semiaquatic cetacean appears to occupy a place on the functional model where propulsion is achieved by a combination of simultaneous pelvic paddling and caudal undulation (Fig. 8). The hind foot of *Ambulocetus* had elongate metapodials and phalanges, indicating increased surface area for paddling (Thewissen *et al.*, 1994). The spine was considered primarily restricted to movement in the sagittal plane (Thewissen *et al.*, 1996a). The tail of *Ambulocetus* was long and powerful and was composed of few (possibly 15) long caudal vertebrae similar to those of mesonychians (Thewissen *et al.*, 1996a). The length of the tail suggested to Thewissen *et al.* (1996a) that a well-developed hydrofoil was lacking. The bending moments for such a tail would be effectively too low to move fully developed flukes. Compared with *Ambulocetus*, modern cetaceans have short caudal vertebrae providing better leverage for the longitudinal muscles.

Thewissen and Fish (1997) considered *Ambulocetus* to be analogous in swimming performance to *Pteronura*. As such, the tail of *Ambulocetus* may have supported at least a rudimentary lateral flange to increase thrust production by caudal undulation. Flower (1883) suggested that an ancestral cetacean would show similarities with *Pteronura*. However, Ryder (1885) argued against any analogy with *Pteronura* on ontogenetic grounds. He cited that the lateral ridges extended the entire length of the tail of *Pteronura*, whereas embryonic cetaceans showed fluke folds developed only at the tail tip. In addition, Thewissen *et al.* (1996a) suggested that *Ambulocetus* was an ambush hunter, relying on burst swimming to capture prey, rather than an agile pursuit predator like otters. Despite such differences, otters represent a suitable model for swimming performance in early cetaceans based on their similar morphology.

Flukes apparently evolved by the middle of the Eocene as observed indirectly from the fossils in the various cetacean families. In the Protocetidae, *Rodhocetus kasrani* displayed a number of spinal characteristics associated with generating large forces from dorsoventral movements of the tail, including robust vertebrae with high neural spines and unfused sacral vertebrae (Gingerich *et al.*, 1994). In addition, the femur was markedly reduced indicating a reduction or abandonment of paddling and perhaps terrestrial movements. Based on examination of oxygen isotopes, the protocetid *Indocetus* ingested sea water and lived in a neritic habitat similar to modern cetaceans (Thewissen *et al.*, 1996b). Because the Protocetidae were marine pursuit hunters (Thewissen *et al.*, 1996a), fluke swimming would be required for sustained locomotion in the open ocean, particularly where prey is dispersed widely (Webb, 1984).

By the end of the Eocene and throughout the Oligocene, swimming was exclusively with flukes with the hind limbs being inconsequential or absent (Gingerich *et al.*, 1990). The rapid radiation of cetaceans with increasing body size favored mechanisms that maximized swimming performance (Webb and Buffrénil, 1990; Fordyce, 1992).

The inference from the ontogenetic data and functional model is that in early cetaceans the flukes would have been represented by small rounded, lateral outgrowths toward the tip of the tail. Such a structural modification would have been advantageous for the small ancestral cetaceans. As cetaceans became larger and more proficient in the water, the outgrowths would have expanded farther, increasing the propulsive surface area to compensate for the increased drag. The flukes would have evolved into their present form as speed and efficiency increased.

5. Summary and Conclusions

The flukes of modern cetaceans are high-performance oscillatory propulsors with a design to produce high lift for thrust generation with low drag and high efficiency. The collagenous internal structure provides the framework for a flexible hydrofoil with relatively high aspect ratio and moderate sweepback. The evolutionary sequence by which flukes arose is still speculative, because there is no direct evidence that reveals origin. The ontogeny of the flukes suggests a derivation from lateral integumentary folds at the end of the tail. Using modern analogues of transitional stages with increasingly aquatic habits, a functional model proposes that ancestral cetaceans had long tails that were used for swimming in conjunction with modified terrestrial gaits of the limbs. Flukes would have developed to increase power output and efficiency by lift-based propulsion as the inefficient drag-based paddling was abandoned. Recent fossil discoveries of early cetaceans support this scenario. However, until impressions of flukes early in their evolution are unearthed, this inquiry will never be fully resolved.

Acknowledgments

I am indebted to J. Beneski, R. Z. German, J. G. Mead, A. J. Nicastro, D. A. Pabst, J. G. M. Thewissen, D. Weihs, and an anonymous reviewer for their advice and comments on sections of the manuscript. I also wish to thank J. G. Mead, C. W. Potter, the Smithsonian Institution, Sea World, and the National Aquarium for permitting me to examine their collections for fluke morphometrics and V. Zelenetskaya for Russian translations. Support for the preparation of this paper was funded by ONR grant N00014-95-1-1045.

References

Abbott, I. H., and von Doenhoff, A. E. 1959. *Theory of Wing Sections*. Dover, New York.

Ahmadi, A. R., and Widnall, S. E. 1986. Energetics and optimum motion of oscillating lifting surfaces of finite span. *J. Fluid Mech.* **162**:261–282.

Amano, M., and Miyazaki, N. 1993. External morphology of Dall's porpoise (*Phocoenoides dalli*): growth and sexual dimorphism. *Can. J. Zool.* **71**:1124–1130.

Andrews, R. C. 1921. A remarkable case of external hind limbs in a humpback whale. *Am. Mus. Novit.* **9**:1–6.

Ashenberg, J., and Weihs, D. 1984. Minimum induced drag of wings with curved planform. *J. Aircr.* **21**:89–91.

Azuma, A. 1983. Biomechanical aspects of animal flying and swimming, in: H. Matsui and K. Kobayashi (eds.), *Biomechanics VIII-A: International Series on Biomechanics*, Volume 4A, pp. 35–53. Human Kinetics Publishers, Champaign, IL.

Barnes, L. G., Domning, D. P., and Ray, C. E. 1985. Status of studies on fossil marine mammals. *Mar. Mamm. Sci.* **1**:15–53.

Bello, M. A., Roy, R. R., Martin, T. P., Goforth, H. W., Jr., and Edgerton, V. R. 1985. Axial musculature in the dolphin (*Tursiops truncatus*): some architectural and histochemical characteristics. *Mar. Mamm. Sci.* **1**:324–336.

Best, R. C., and da Silva, V. M. F. 1989. Amazon river dolphin, boto *Inia geoffrensis* (de Blainville, 1817), in: S. H. Ridgeway and R. Harrison (eds.), *Handbook of Marine Mammals*, Volume 4, pp. 1–23. Academic Press, London.

Blake, R. W. 1983. *Fish Locomotion*. Cambridge University Press, London.

Bose, N., and Lien, J. 1989. Propulsion of a fin whale (*Balaenoptera physalus*): why the fin whale is a fast swimmer. *Proc. R. Soc. London Ser. B* **237**:175–200.

Bose, N., Lien, J., and Ahia, J. 1990. Measurements of the bodies and flukes of several cetacean species. *Proc. R. Soc. London Ser. B* **242**:163–173.

Chanin, P. 1985. *The Natural History of Otters.* Facts on File, New York.

Chopra, M. G. 1975. Lunate-tail swimming propulsion, in: T. Y. Wu, C. J. Brokaw, and C. Brennen (eds.), *Swimming and Flying in Nature,* Volume 2, pp. 635–650. Plenum Press, New York.

Chopra, M. G. 1976. Large amplitude lunate-tail theory of fish locomotion. *J. Fluid Mech.* **74**:161–182.

Chopra, M. G., and Kambe, T. 1977. Hydrodynamics of lunate-tail swimming propulsion. Part 2. *J. Fluid Mech.* **79**:49–69.

Curren, K. C. 1992. Designs for swimming: morphometrics and swimming dynamics of several cetacean species. M.S. thesis, Memorial University of Newfoundland.

Curren, K. C., Bose, N., and Lien, J. 1993. Morphological variation in the harbour porpoise (*Phocoena phocoena*). *Can. J. Zool.* **71**:1067–1070.

Curren, K. C., Bose, N., and Lien, J. 1994. Swimming kinematics of a harbor porpoise (*Phocoena phocoena*) and an Atlantic white-sided dolphin (*Lagenorhynchus acutus*). *Mar. Mamm. Sci.* **10**:485–492.

Daniel, T. 1988. Forward flapping flight from flexible fins. *Can. J. Zool.* **66**:630–638.

Daniel, T. 1991. Efficiency in aquatic locomotion: limitations from single cells to animals, in: R. W. Blake (ed.), *Efficiency and Economy in Animal Physiology,* pp. 83–95. Cambridge University Press, London.

Daniel, T., Jordan, C., and Grunbaum, D. 1992. Hydromechanics of swimming, in: R. M. Alexander (ed.), *Advances in Comparative and Environmental Physiology,* Volume 11, pp. 17–49. Springer-Verlag, Berlin.

Feldkamp, S. D. 1987. Foreflipper propulsion in the California sea lion, *Zalophus californianus. J. Zool.* **212**:43–57.

Felts, W. J. L. 1966. Some functional and structural characteristics of cetaceans' flippers and flukes, in: K. S. Norris (ed.), *Whales, Dolphins and Porpoises,* pp. 255–276. University of California Press, Berkeley.

Fierstine, H. L., and Walters, V. 1968. Studies of locomotion and anatomy of scombrid fishes. *Mem. South. Calif. Acad. Sci.* **6**:1–31.

Fish, F. E. 1979. Thermoregulation in the muskrat (*Ondatra zibethicus*): the use of regional heterothermia. *Comp. Biochem. Physiol.* **64**:391–397.

Fish, F. E. 1984. Mechanics, power output and efficiency of the swimming muskrat (*Ondatra zibethicus*). *J. Exp. Biol.* **110**:183–201.

Fish, F. E. 1992. Aquatic locomotion, in: T. E. Tomasi and T. H. Horton (eds.), *Mammalian Energetics: Interdisciplinary Views of Metabolism and Reproduction,* pp. 34–63. Cornell University Press, Ithaca, NY.

Fish, F. E. 1993a. Influence of hydrodynamic design and propulsive mode on mammalian swimming energetics. *Aust. J. Zool.* **42**:79–101.

Fish, F. E. 1993b. Power output and propulsive efficiency of swimming bottlenose dolphins (*Tursiops truncatus*). *J. Exp. Biol.* **185**:179–193.

Fish, F. E. 1993c. Comparison of swimming kinematics between terrestrial and semiaquatic opossums. *J. Mammal.* **74**:275–284.

Fish, F. E. 1994. Association of propulsive swimming mode with behavior in river otters (*Lutra canadensis*). *J. Mammal.* **75**:989–997.

Fish, F. E. 1996. Transitions from drag-based to lift-based propulsion in mammalian swimming. *Am. Zool.* **36**:628–641.

Fish, F. E., and Hui, C. A. 1991. Dolphin swimming—a review. *Mammal Rev.* **21**:181–195.

Fish, F. E., Innes, S., and Ronald, K. 1988. Kinematics and estimated thrust production of swimming harp and ringed seals. *J. Exp. Biol.* **137**:157–173.

Flower, W. H. 1883. On whales, past and present, and their probable origin. *Nature* **28**:226–230.

Folkens, P. A., and Barnes, L. G. 1984. Reconstruction of an archaeocete. *Oceans* **17**:22–23.

Fordyce, R. E. 1992. Cetacean evolution and Eocene/Oligocene environments, in: D. R. Prothero and W. A. Berggren (eds.), *Eocene–Oligocene Climatic and Biotic Evolution,* pp. 368–381. Princeton University Press, Princeton, NJ.

Gingerich, P. D., Wells, N. A., Russell, D. E., and Shah, S. M. I. 1983. Origin of whales in epicontinental remnant seas: new evidence from the early Eocene of Pakistan. *Science* **220**:403–406.

Gingerich, P. D., Smith, B. H., and Simons, E. L. 1990. Hind limbs of Eocene *Basilosaurus isis:* evidence of feet in whales. *Science* **249**:154–157.

Gingerich, P. D., Raza, S. M., Arif, M., Anwar, M., and Zhou, X. 1994. New whale from the Eocene of Pakistan and the origin of cetacean swimming. *Science* **368**:844–847.

Goforth, H. W. 1990. Ergometry (exercise testing) of the bottlenose dolphin, in: S. Leatherwood (ed.), *The Bottlenose Dolphin*, pp. 559–574. Academic Press, San Diego.

Gutmann, W. F. 1994. Konstruktionszwänge in der Evolution: schwimmende Vierfüsser. *Nat. Mus.* **124**:165–188.

Hickman, G. C. 1979. The mammalian tail: a review of functions. *Mammal Rev.* **9**:143–157.

Hoerner, S. F. 1965. *Fluid-Dynamic Drag*. Published by author, Brick Town, NJ.

Howell, A. B. 1930. *Aquatic Mammals*. Thomas, Springfield, IL.

Hurt, H. H., Jr. 1965. *Aerodynamics for Naval Aviators*. U.S. Navy, NAVWEPS 00-80T-80.

Jenkins, F. A., Jr., and Goslow, G. E., Jr. 1983. The functional anatomy of the shoulder of the savannah monitor lizard (*Varanus exanthematicus*). *J. Morphol.* **175**:195–216.

Karpouzian, G., Spedding, G., and Cheng, H. K. 1990. Lunate-tail swimming propulsion. Part 2. Performance analysis. *J. Fluid Mech.* **210**:329–351.

Katz, J., and Weihs, D. 1978. Hydrodynamic propulsion by large amplitude oscillation of an airfoil with chordwise flexibility. *J. Fluid Mech.* **88**:485–497.

Küchermann, D. 1953. The distribution of lift over the surface of swept wings. *Aeronaut. Q.* **4**:261–278.

Lang, T. G. 1966. Hydrodynamic analysis of cetacean performance, in: K. S. Norris (ed.), *Whales, Dolphins and Porpoises*, pp. 410–432. University of California Press, Berkeley.

Lang, T. G., and Daybell, D. A. 1963. Porpoise performance tests in a seawater tank. NOTS Technical Publication 3063. Naval Ordnance Test Station, China Lake, CA. NAVWEPS Report 8060.

Lauder, G. V. 1995. On the inference of function from structure, in: J. J. Thomason (ed.), *Functional Morphology in Vertebrate Paleontology*, pp. 1–18. Cambridge University Press, London.

Lighthill, J. 1969. Hydrodynamics of aquatic animal propulsion—a survey. *Annu. Rev. Fluid Mech.* **1**:413–446.

Lighthill, J. 1970. Aquatic animal propulsion of high hydromechanical efficiency. *J. Fluid Mech.* **44**:265–301.

Lighthill, J. 1977. Introduction to scaling of aerial locomotion, in: T. J. Pedley (ed.), *Scale Effects in Animal Locomotion*, pp. 365–404. Academic Press, New York.

Liu, P., and Bose, N. 1993. Propulsive performance of three naturally occurring oscillating propeller planforms. *Ocean Eng.* **20**:57–75.

Long, J. H., Jr., Pabst, D. A., Shepherd, W. R., and McLellan, W. A. 1997. Locomotor design of dolphin vertebral columns: bending mechanics and morphology of *Delphinus delphis*. *J. Exp. Biol.* **200**:65–81.

Meyer, W., Neurand, K., and Klima, M. 1995. Prenatal development of the integument in Delphinidae (Cetacea: Odontoceti). *J. Morphol.* **223**:269–287.

Motani, R., You, H., and McGowan, C. 1996. Eel-like swimming in the earliest ichthyosaurs. *Nature* **382**:347–348.

Norris, K. S., and Prescott, J. H. 1961. Observations on Pacific cetaceans of California and Mexican waters. *Univ. Calif. Publ. Zool.* **63**:291–402.

O'Leary, M., and Rose, K. D. 1995. Postcranial skeleton of the early Eocene mesonychid *Pachyaena*. *J. Vertebr. Paleontol.* **15**:401–430.

Pabst, D. A. 1990. Axial muscles and connective tissues of the bottlenose dolphin, in: S. Leatherwood and R. R. Reeves (eds.), *The Bottlenose Dolphin*, pp. 51–67. Academic Press, San Diego.

Parry, D. A. 1949a. The swimming of whales and a discussion of Gray's paradox. *J. Exp. Biol.* **26**:24–34.

Parry, D. A. 1949b. Anatomical basis of swimming in whales. *Proc. Zool. Soc. London* **119**:49–60.

Perrin, W. F. 1975. Variation of spotted and spinner porpoise (genus *Stenella*) in eastern tropical Pacific and Hawaii. *Bull. Scripps Inst. Oceanogr.* No. 21.

Perrin, W. F. 1997. Development and homologies of head stripes in the delphinoid cetaceans. *Mar. Mamm. Sci.* **13**:1–43.

Purves, P. E. 1963. Locomotion in whales. *Nature* **197**:334–337.

Purves, P. E. 1969. The structure of the flukes in relation to laminar flow in cetaceans. *Z. Saeugetierkd.* **34**:1–8.

Rayner, J. M. V. 1985. Vorticity and propulsion mechanics in swimming and flying animals, in: J. Riess and E. Frey (eds.), *Konstruktionsprinzipen lebender und ausgestorbener Reptilien*, pp. 89–118. University of Tubingen, Tubingen, Germany.

Rice, D. W., and Wolman, A. A. 1971. *The Life History and Ecology of the Gray Whale (Eschrichtius robustus)*. Am. Soc. Mamm. Spec. Publ. No. 3.

Romanenko, E. V. 1995. Swimming of dolphins: experiments and modelling, in: C. P. Ellington and T. J. Pedley (eds.), *Biological Fluid Dynamics*, pp. 21–33. The Company of Biologists, Cambridge.

Rommel, S. 1990. Osteology of the bottlenose dolphin, in: S. Leatherwood and R. R. Reeves (eds.), *The Bottlenose Dolphin*, pp. 29–49. Academic Press, San Diego.

Rowe, T. 1996. Coevolution of the mammalian middle ear and neocortex. *Science* **273**:651–654.

Ryder, J. A. 1885. On the development of the Cetacea, together with consideration of the probable homologies of the flukes of cetaceans and sirenians. *Bull. U.S. Fish Comm.* **5**:427–485.

Sanderson, I. T. 1956. *Follow the Whale*. Little, Brown, Boston.

Slijper, E. J. 1961. Locomotion and locomotory organs in whales and dolphins (Cetacea). *Symp. Zool. Soc. London* **5**:77–94.

Slijper, E. J. 1979. *Whales*. Cornell University Press, Ithaca, NY.

Smith, K. K. 1994. Are neuromotor systems conserved in evolution? *Brain Behav. Evol.* **43**:293–305.

Strickler, T. L. 1980. The axial musculature of *Pontoporia blainvillei*, with comments on the organization of this system and its effect on fluke-stroke dynamics in the Cetacea. *Am. J. Anat.* **157**:49–59.

Tarasoff, F. J., Bisaillon, A., Pierard, J., and Whitt, A. P. 1972. Locomotory patterns and external morphology of the river otter, sea otter, and harp seal (Mammalia). *Can. J. Zool.* **50**:915–929.

Thewissen, J. G. M. 1994. Phylogenetic aspects of cetacean origins: a morphological perspective. *J. Mamm. Evol.* **2**:157–184.

Thewissen, J. G. M., and Fish, F. E. 1997. Locomotor evolution the earliest cetaceans: functional model, modern analogues, and paleontological evidence. *Paleobiology* **23**:482–490.

Thewissen, J G. M., and Hussain, S. T. 1993. Origin of underwater hearing in whales. *Nature* **361**:444–445.

Thewissen, J. G. M., Hussain, S. T., and Arif, M. 1994. Fossil evidence for the origin of aquatic locomotion in archaeocete whales. *Science* **263**:210–212.

Thewissen, J. G. M., Madar, S. I., and Hussain, S. T. 1996a. *Ambulocetus natans*, an Eocene cetacean (Mammalia) from Pakistan. *Cour. Forsch.-Inst. Senckenberg* **191**:1–86.

Thewissen, J. G. M., Roe, L. J., O'Neil, J. R., Hussain, S. T., Sahni, A., and Bajpal, S. 1996b. Evolution of cetacean osmoregulation. *Nature* **381**:379–380.

Triantafyllou, M. S., and Triantafyllou, G. S. 1995. An efficient swimming machine. *Sci. Am.* **272**:64–69.

van Dam, C. P. 1987. Efficiency characteristics of crescent-shaped wings and caudal fins. *Nature* **325**:435–437.

Videler, J. 1993. *Fish Swimming*. Chapman & Hall, London.

Videler, J., and Kamermans, P. 1985. Differences between upstroke and downstroke in swimming dolphins. *J. Exp. Biol.* **119**:265–274.

Vogel, S. 1994. *Life in Moving Fluids*. Princeton University Press, Princeton, NJ.

von Mises, R. 1945. *Theory of Flight*. Dover, New York.

Watson, A. G., and Fordyce, R. E. 1993. Skeleton of two minke whales, *Balaenoptera acutorostrata*, stranded on the south-east coast of New Zealand. *N. Z. Nat. Sci.* **20**:1–14.

Webb, P. W. 1975. Hydrodynamics and energetics of fish propulsion. *Bull. Fish. Res. Bd. Can.* **190**:1–158.

Webb, P. W. 1984. Body form, locomotion and foraging in aquatic vertebrates. *Am. Zool.* **24**:107–120.

Webb, P. W., and Buffrénil, V. de. 1990. Locomotion in the biology of large aquatic vertebrates. *Trans. Am. Fish. Soc.* **119**:629–641.

Weihs, D. 1989. Design features and mechanics of axial locomotion in fish. *Am. Zool.* **29**:151–160.

Weihs, D., and Webb, P. W. 1983. Optimization of locomotion, in: P. W. Webb and D. Weihs (eds.), *Fish Biomechanics*, pp. 339–371. Praeger, New York.

Williams, T. M. 1983. Locomotion in the North American mink, a semi-aquatic mammal. I. Swimming energetics and body drag. *J. Exp. Biol.* **103**:155–168.

Williams, T. M. 1989. Swimming by sea otters: adaptations for low energetic cost locomotion. *J. Comp. Physiol. A* **164**:815–824.

Wu, T. Y. 1971a. Hydrodynamics of swimming propulsion. Part 1. Swimming of a two-dimensional flexible plate at variable forward speeds in an inviscid fluid. *J. Fluid Mech.* **46**:337–355.

Wu, T. Y. 1971b. Hydrodynamics of swimming propulsion. Part 2. Some optimum shape problems. *J. Fluid Mech.* **46**:521–544.

Yanov, V. G. 1991. The systematic-functional organization of the kinematics of dolphin swimming. *Rep. Acad. Sci.* **317**:1089–1093 (in Russian).

Yates, G. T. 1983. Hydromechanics of body and caudal fin propulsion, in: P. W. Webb and D. Weihs (eds.), *Fish Biomechanics*, pp. 177–213. Praeger, New York.

## CHAPTER 11

Implications of Vertebral Morphology for Locomotor Evolution in Early Cetacea

EMILY A. BUCHHOLTZ

1. Introduction

Living whales are among the small number of swimming tetrapods that locomote axially. It is clear that members of their terrestrial sister group, mesonychian condylarths (Prothero *et al.*, 1988; Thewissen, 1994), were quadrupedal. The transition from a terrestrial habitat and paraxial, oscillatory limb locomotion to an aquatic habitat and axial undulatory locomotion was accompanied by marked changes in postcranial anatomy. Aspects of this locomotor transition can be reconstructed from a study of the vertebral columns of mesonychids and fossil and living whales. Examination of regional variation within the axial skeleton is a powerful and underutilized tool for the analysis of locomotor style (Slijper, 1946, 1961; Worthington and Wake, 1972; Crovetto, 1991; Long *et al.*, 1997). Although almost invariably fragmentary, early cetacean postcranial skeletons are surprisingly informative, and can complement morphological indicators of diet, sensory specializations, and skull reorganization to fill in some of the gaps in our understanding of the dramatic transition of early cetaceans as they moved from land to water.

Cetaceans possess a large suite of derived traits associated with a marine lifestyle. Nevertheless, analysis of the terrestrial mesonychids provides a useful starting point in a discussion of whale history (Gingerich *et al.*, 1983). Previously known largely on the basis of cranial material, detailed postcranial descriptions (Zhou *et al.*, 1992; O'Leary and Rose, 1995) are now available for the species *Pachyaena ossifraga* and indicate that this animal was a hoofed, omnivorous cursor adapted for endurance instead of speed. The vertebral column (Zhou *et al.*, 1992) has a count of 7 cervical, 12 thoracic, 7 lumbar, 3 sacral, and 15+ caudal vertebrae. Posterior cervicals are longer than high or wide (Fig. 1A), and long neural spines on the anterior thoracics indicate the presence of robust musculature to support a large head. The tenth thoracic vertebra is both anticlinal and diaphragmatic, and is interpreted by Zhou *et al.* (1992) as the site of greatest precaudal mobility. Incorporation of pos-

EMILY A. BUCHHOLTZ • Department of Biological Sciences, Wellesley College, Wellesley, Massachusetts 02181.

The Emergence of Whales, edited by Thewissen. Plenum Press, New York, 1998.

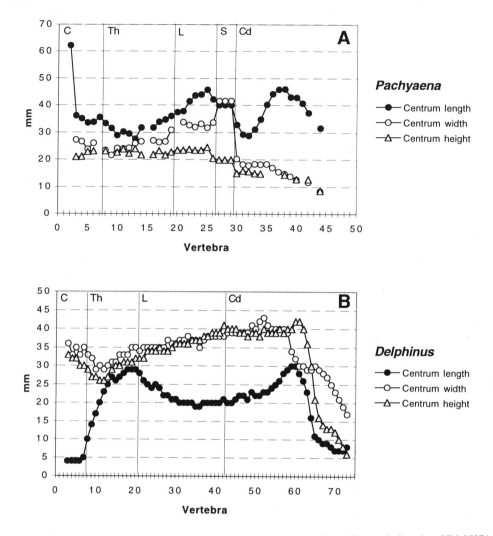

FIGURE 1. (A) Centrum length, width, and height of *Pachyaena ossifraga*. Presacrals based on UM 95074, sacrals and caudals based on YPM-PU 14708. Data from Zhou *et al.* (1992). (B) Centrum length, width, and height of *Delphinus delphis*, USNM 550868. Vertical lines identify boundaries between vertebral series. Abbreviations for figures: C, cervical; Cd, caudal; D, dorsal; L, lumbar; S, sacral; Th, thoracic.

terior thoracics into the functional lumbos is indicated by similarity of centrum dimensions. Relative to the anterior thoracics, the posterior thoracics and lumbars are not only elongate but also laterally reinforced by increased centrum width. Zygopophyseal structure suggests that the lumbos was relatively stable. Sacral vertebrae are dimensionally discontinuous with both lumbars and caudals in length, width, and height, signaling functional discontinuity. Centrum height and width, and by inference body cross section, decrease markedly in the anterior caudal vertebrae. Elongate posterior caudals suggest a considerable range of tail movement.

Vertebral anatomy and locomotion in living whales present stark contrasts to those of mesonychids. Most descriptions of whale locomotion are based on living dolphins or porpoises (Pabst, 1990, 1993; Fish and Hui, 1991; Curren *et al.*, 1994), widely recognized as exceedingly specialized odontocetes. Nevertheless, they provide important models for comparison, as do other marine mammals. The vertebral count of the *Delphinus delphis* specimen shown in Fig. 1B is 7 cervical, 13 thoracic, 22 lumbar, and 31 caudal. The exceedingly short cervical vertebrae, found in many living cetaceans, restrict the range of neck movement as the body is propelled from the posterior end. The trunk is composed of numerous, nearly identical vertebrae with short centrum lengths and large articular faces with limited intervertebral mobility. Most motion occurs at anterior (anal) and posterior (fluke) caudal nodes. The laterally compressed caudal peduncle ($\sim$ vertebrae 56–63) cuts dorsoventrally through the water with minimal resistance (Slijper, 1961; Fish and Hui, 1991). Its presence and location are identifiable in the midtail by vertebral centra that mimic the high, narrow dimensions of the body region they support. A narrow "neck" ($\sim$ vertebra 64) with a nearly spherical centrum connects the caudal peduncle and the broad, dorsoventrally compressed fluke ($\sim$ vertebrae 65–73), which is the propulsive organ (Slijper, 1961). Fluke centra are broad and low, again reflecting the dimensions of the body region they support. Although sometimes described as oscillatory in nature, the motion at both nodes is considered undulatory by Webb and Blake (1985), because both the peduncle and the fluke are deformed as waves travel down them. This terminology is accepted here. The sinusoidal deformation in *Phoecoena* was diagrammed by Slijper (1936) and first documented in films of swimming whales by Parry (1949).

The present study uses the descriptions of fossil cetacean vertebrae, published by many workers over the past century, as its primary data base. The fragmentary nature of fossil material dictates that characters chosen for emphasis are those most frequently recoverable from extinct taxa: vertebral series boundaries, vertebral counts, and centrum dimensions. These simple parameters contain a surprising amount of information about column function.

Although no archaeocetes survive, there are numerous living aquatic tetrapods, including members of Cetacea, Sirenia, Carnivora, Crocodilia, and Squamata, for which both postcranial anatomy and locomotor style can be documented. Patterns of vertebral anatomy that signal the existence of differentiated body regions (foreshortened neck, elongate torso, peduncle, fluke) or that enhance particular modes of axial movement are documented below. Although no claim is made that early cetaceans had a lifestyle identical to that of any particular living group, recognition of similar anatomical patterns in archaeocetes can be used to predict similar body regionalization and/or modes of movement in these extinct animals. Comparisons with archaeocete vertebrae indicate a range of vertebral variation that partially overlaps but is not coincident with that of the comparison taxa.

Specimen abbreviations used in this study are: AMNH, American museum of Natural History, New York; MCZ, Harvard Museum of Comparative Zoology, Cambridge; UM, University of Michigan Museum of Paleontology, Ann Arbor; H-GSP, Howard University–Geological Survey of Pakistan, Islamabad; GSM, Georgia Southern Museum, Statesboro; USNM, United States National Museum, Washington, D.C.; GSP-UM, Geological Survey of Pakistan–University of Michigan Museum of Paleontology, Islamabad; YPM-PU, Princeton University Collection at Yale Peabody Museum, New Haven.

2. Vertebral Variation and Locomotion in Aquatic Tetrapods

2.1. Functional Units of the Column

Vertebral series boundaries were originally defined in quadrupedal, terrestrial mammals, grouping vertebrae similar in both structure and function. For example, centra of sacral vertebrae in quadrupeds are typically set off by structure (shorter, broader, and lower) as well as by function (limb attachment and body support) from both preceding (lumbar) and following (caudal) vertebrae. The changes in structure along the column signal changes in function. Extent of coincidence between the location of traditional series boundaries and areas of maximal dimensional change (and by inference, functional change) are examined below for a quadrupedal swimmer and a whale.

McShea (1992) has provided a set of complexity metrics for evaluation of variation within a single vertebral column. His irregularity metric, the log of the average difference in the chosen dimensional parameter between adjacent vertebrae, does not identify the location of variation within the column, and is unsuitable for comparison of animals of different body size. As an alternative, percent change in a dimension (in this case centrum length, width, or height) between adjacent vertebrae along the column is used here. Average percent change of these three linear dimensions is used as a summary parameter.

In the column of the semiaquatic, quadrupedal otter *Lutra canadensis* (Fig. 2A,B), the largest relative changes in vertebral dimensions occur at or immediately adjacent to traditional series boundaries (vertebrae 7/8, 21/22, 26/27, and 31/32). Functional units (neck, thorax, lumbos, sacrum, tail) are therefore nearly coincident with traditional series. An additional area of marked dimensional change occurs at the posterior end of the tail, where all three linear dimensions of the centra decrease rapidly.

The radical reorganization of the cetacean vertebral column for axial locomotion has resulted in noncoincidence of typical series boundaries and functional units of the column. The column of the beluga whale *Delphinapterus leucas* (Fig. 3A,B) is used as an example. Anatomical changes producing noncoincidence are independent of, and in addition to, changes in vertebral counts. Maximum variation in average linear dimension occurs at the cervical/thoracic boundary, midway in the thoracic series, and at two points within the caudal series, dividing the column into neck, chest, torso, peduncle, and fluke. Near uniformity of centrum dimensions occurs across the thoracic/lumbar and lumbar/caudal boundaries. Note that noncoincidence of vertebral series and functional units in whales requires introduction of alternative terms for functional units. Dimensional changes that allow recognition of the functional torso, peduncle, and fluke are discussed below.

2.1.1. Recognition of the Functional Torso

In living whales, centra of posterior thoracic, lumbar, sacral, and anterior caudal vertebrae are dimensionally similar and comprise a single functional unit, the torso. In mammals the end of the thorax is signaled by the last articulated rib. Because posterior thoracic vertebrae typically mimic the lumbars in structure, the functional change from the fixed and protective chest to the mobile torso may occur as far anterior as midway in the thoracic series (at approximately vertebra 15 in *Delphinapterus*). More subtle indicators of functional chest/torso transitions may include the posterior discontinuation of ribs with costal/ster-

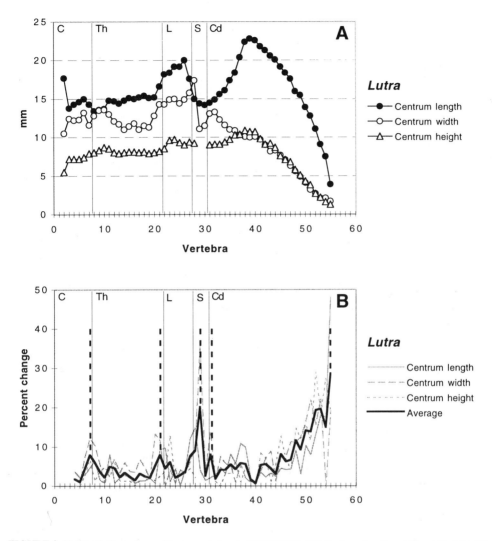

FIGURE 2. Vertebral dimensions of *Lutra canadensis*, MCZ 61308. (A) Centrum length, width, and height; (B) percent variation in centrum length, width, and height between adjacent vertebrae (grayed lines), and average of these variations (heavy line). The functional neck, thorax, lumbos, sacrum, and tail (boundaries at broken lines) are nearly identical to units defined by traditional series borders (boundaries at continuous lines). Abbreviations as in Fig. 1.

nal articulations or of ribs with medial articulations on two adjacent vertebrae. The term *thoracic lumbars* is used here for thoracic vertebrae that are functionally part of the torso.

In terrestrial vertebrates, sacral vertebrae are those that have transverse processes that articulate directly or indirectly with the pelvis and anchor the hind limb. They are often fused to each other. Lack of classically defined sacrals in modern whales typically results in classification of vertebrae serially homologous with the sacrals of quadrupeds as lumbars. The identity of these vertebrae is difficult to determine, although Slijper (1936) did so

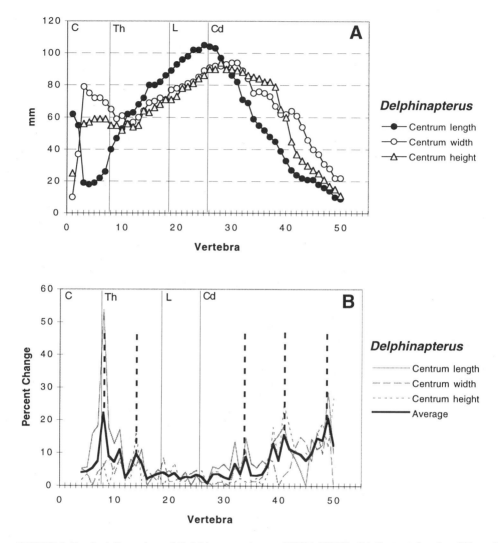

FIGURE 3. Vertebral dimensions of *Delphinapterus leucas*, USNM 571021. (A) Centrum length, width, and height; (B) percent variation in centrum length, width, and height between adjacent vertebrae (grayed lines), and average of these variations (heavy line). The functional chest, torso, peduncle, and fluke (boundaries at broken lines) are not coincident with units defined by traditional series borders (boundaries at continuous lines). Abbreviations as in Fig. 1.

for a number of species on the basis of the location of the pudendal nerve. The term *sacral lumbars* is introduced for these vertebrae ($\sim$ vertebrae 22–25 in *Delphinapterus*). Lack of easily recognizable sacrals in whales also makes identification of the first true caudal somewhat ambiguous, because all postsacral vertebrae of quadrupeds are by definition caudal. The first vertebra with a hemal arch on its anterior (Slijper, 1936) or posterior (DeSmet, 1977; this chapter) boundary is typically considered the first caudal. However, a variable number (Figs. 1B, 3A) of caudal vertebrae so defined are very similar to lumbars in centrum dimensions, and are part of the functional torso. Here they are called *caudal lumbars*.

The existence of an extensive sequence of similar vertebrae that acts as a functional torso is characteristic of living whales, and can be recognized in many fossil taxa (see below), even when columns are incomplete.

2.1.2. Recognition of the Peduncle and Fluke

Introduction of dimensional discontinuities within an anatomical series indicates its partitioning for more than one functional role. Two structural discontinuities (at approximately vertebrae 33 and 41, Fig. 3A) subdivide the anatomical caudal vertebrae of *Delphinapterus* into three functional groups: an anterior lumbarized region that is part of the functional torso, a middle laterally compressed peduncle designed for maximum excursion through the water with little resistance, and a terminal propulsive fluke.

A demonstration of the ability of dimensional variation to demarcate caudal subregions is provided by a comparison of the width and height of the centra and of the tail of the dolphin (*Delphinus delphis*), the manatee (*Trichechus manatus*), and the dugong (*Dugong dugon*). The nearly circular cross section of the long, uniform torso of the dolphin is reflected in subequal centrum width and height from the anterior thorax to the anterior tail (Fig. 4). A reduction in centrum width below centrum height in the midcaudals marks the anterior margin of the laterally compressed, nonpropulsive peduncle which carries the propulsive fluke through a wide arc. Reversal of these dimensional relationships marks the transition to the laterally broad but dorsoventrally compressed fluke. This abrupt reversal is the signature of the presence of a well-differentiated fluke, and is easily recognizable in fossil specimens (see below).

Transitions in patterns of centrum width and height are less dramatic in more massively built marine mammals with less well-defined peduncles and flukes, illustrated by the living manatee and dugong (Figs. 4, 5A,B). The manatee is a very broad-bodied estuarine and nearshore marine mammal with a fluke only subtly set off from the thick torso and anterior tail (Kaiser, 1974; Caldwell and Caldwell, 1985). This body build is reflected in the pattern of centrum dimensions. Centrum width is markedly greater than centrum height in the lumbos (vertebrae 21–23, CW/CH = 1.4–1.6). Extent of this difference actually increases somewhat in the anterior tail (vertebrae 25–40, CW/CH = 1.5–2.0), and then more dramatically in the fluke (vertebrae 41–48, CW/CH = 1.7–2.2). This pattern argues for recognition of the anterior tail itself as part of the propulsive surface, as opposed to a peduncle with reduced resistance. In contrast, the anterior tail of the dugong (Figs. 4, 5B) is more laterally compressed, and the fluke more dorsoventrally compressed and distinctively set off, than in the manatee. Correspondingly, relative centrum width is smaller in the peduncle (vertebrae 37–51, CW/CH = 1.2–1.6), and larger in the fluke (vertebrae 52–60, CW/CH = 1.7–2.6), than for *Trichechus*.

Cetacean evolution is sometimes viewed as encompassing a move toward postcervical dimensional uniformity of vertebrae, but this analysis suggests that it is more accurately interpreted as a relocation of areas of dimensional discontinuity from traditional series borders to locations within series. When analyzing archaeocete columns, uniformity in the structure of sequential vertebrae, even when located across series boundaries as conventionally defined, is interpreted here as signaling uniformity of function. Similarly, structur-

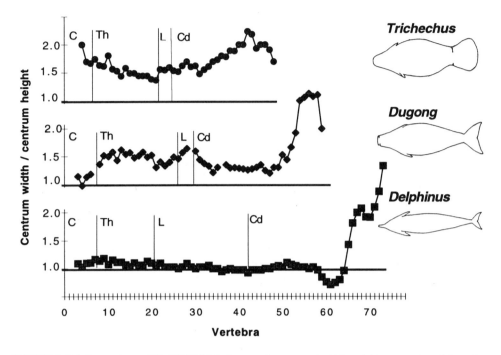

FIGURE 4. Relative centrum width (CH/CW) in three marine mammals with variable expression of the peduncle and fluke. Abbreviations as in Fig. 1.

al discontinuity in the dimensions of sequential vertebrae is interpreted as an indicator of
functional discontinuity, even when it arises within a series.

2.2. Changes in Cetacean Vertebral Counts

Cetaceans have high vertebral counts relative to their presumed terrestrial ancestors,
although the increase is unequally distributed among different regions of the column. As a
general rule, a series with a high vertebral count has a long length, a high number of intervertebral joints, and high flexibility. However, as discussed below, relative centrum dimensions dramatically affect relative movement between adjacent vertebrae, and some of
the highest vertebral counts occur in the torsos of living whales with reduced torso flexibility.

The number of vertebrae primitive to each cetacean series was discussed extensively
by Slijper (1936). He based his predictions on the vertebral counts of columns known from
terrestrial mammals he considered primitive, on the archaeocete *Protocetus atavus*, and on
living cetaceans with nonvertebral traits he considered primitive. He predicted that the "ancestral whale" had a count of 7 cervicals, 17–19 noncervical presacrals (12–13 thoracics,

FIGURE 5. Centrum length, width, and height of marine tetrapods. (A) *Trichechus manatus*, MCZ 7295; (B)
Dugong dugon, MCZ 6955; (C) *Amblyrhynchus cristalus*, MCZ 2006; (D) *Gavialis gangeticus*, AMNH 110145;
(E) *Ophisaurus apodus*, MCZ 2094; (F) *Eumetopias jubatus*, MCZ 129; (G) *Monachus tropicalis*, MCZ 7264; (H)
Enhydra lutris, MCZ 9345. Abbreviations as in Fig. 1.

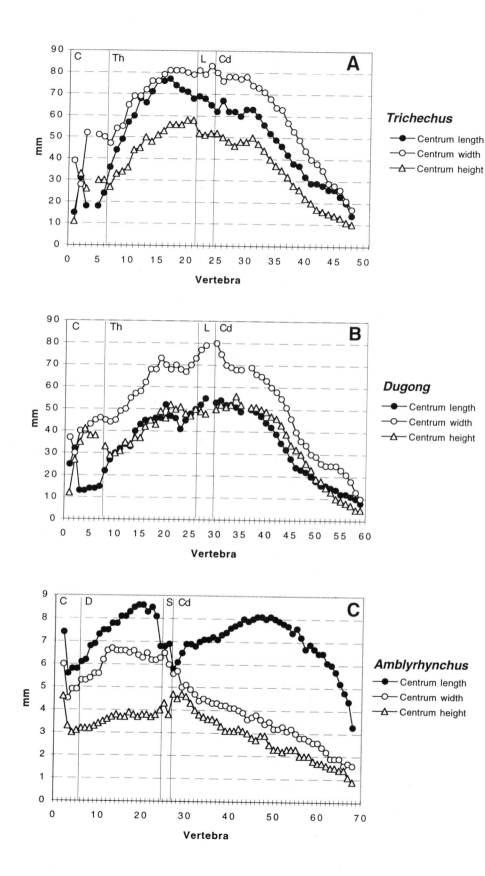

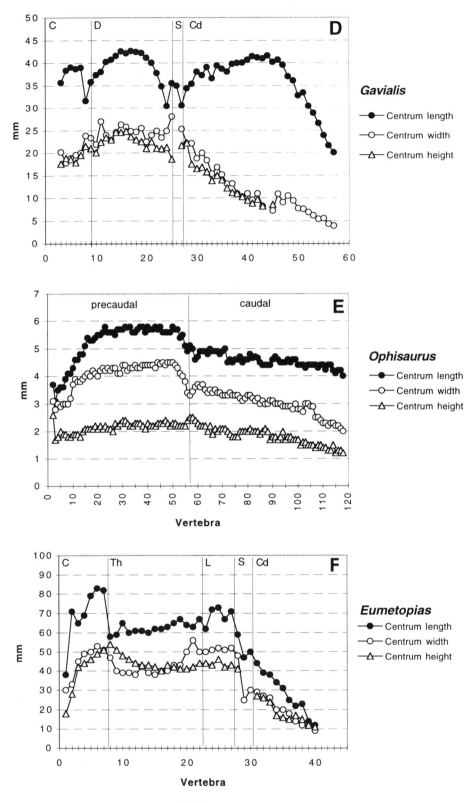

FIGURE 5. (*Continued*).

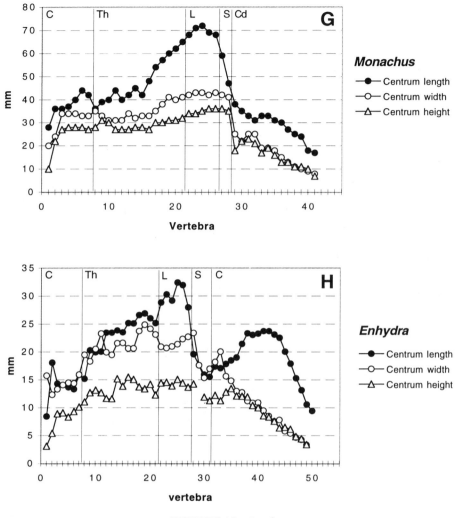

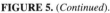

FIGURE 5. (*Continued*).

5–6 lumbars), and 2 sacrals. Sixty years and many fossil discoveries later, these predictions have been largely upheld. Possible exceptions are the allocation of thoracics and lumbars in the noncervical presacral count of 19 and the number of sacrals. The recently described early archaeocetes *Rodhocetus* (Gingerich *et al.*, 1994), *Remingtonocetus* (Kumar and Sahni, 1986; Gingerich *et al.*, 1995), and *Dalanistes* (Gingerich *et al.*, 1995) have slightly higher thoracic (13,14) and sacral (4) counts. The primitive number of caudals is unknown, but was probably in the range of 15–20. Although a small ($\sim$ 30 precaudal, 15+ caudal) total vertebral count is certainly primitive, some series have increased, and others decreased, in number during cetacean evolution.

Cervical number is stable at seven in all known mesonychids and whales, as well as in all aquatic mammals except the manatee. Although many living whales display fusion of some or all of the cervical vertebrae, patterns of fusion vary and almost certainly occurred in parallel in several lines as establishment of a posteriorly located propulsive organ increased selection pressure for a short and stable neck.

Among living cetaceans, the most common thoracic count is 13 (data from Slijper, 1936), essentially identical to the primitive count. However, its relationship to the lumbar count has reversed. Thoracic vertebrae equal or exceed the lumbar plus sacral count in all of the archaeocetes examined here. The reverse is true for many (*Delphinus*, Fig. 1B) but not all (*Delphinapterus*, Fig. 3A) living whales. In addition, the functional chest is shortened by lumbarization of posterior thoracics.

An elongate torso occurs repeatedly in tetrapods (snakes, limbless lizards) using axial undulation (Landes, 1976). It is a key anatomical feature of modern cetaceans, providing an extensive area for attachment of the muscles generating axial movements (Pabst, 1990, 1993). Elongation is variably the result of some combination of increase in centrum length, increase in the number of "true" lumbars, and/or transformation of adjacent vertebrae into functional lumbars. Among the latter are vertebrae identified as sacrals in archaeocetes, which when added to the 5–6 "true" lumbars produce a primitive torso count of 9–10. In *Delphinus*, the torso count is approximately 35 (Fig. 2, vertebrae 20–55). Increase in the vertebral count of the torso may be used as a rough measure of propulsive power, and thus of swimming ability, in archaeocetes (see below).

Reduction in the number of sacrals, loss of fusion between individual sacral vertebrae, and loss of articulation of sacral vertebrae with the pelvis undoubtedly all occurred during early whale evolution. The sacral vertebrae of living whales are part of the functional torso, and are typically included in the lumbar count.

Unfortunately, the caudal series is incomplete or absent in most archaeocete specimens. Nevertheless, it is clear that a very early transition was made from a short, nonpropulsive tail to an elongate, robust organ with propulsive function. Most of this increase in length was accomplished by the addition of vertebrae, as caudal counts of up to 49 are found in modern whales (Slijper, 1936).

2.3. Relative Centrum Length Is an Indicator of Column Flexibility

Increase in centrum length without alteration of other centrum dimensions or number results in both longer total column length and greater possible displacement of the column, either transversely or vertically, from the midline. The second point follows from the smaller relative area of the intervertebral articular surfaces and the greater excursion of the posterior end of each centrum as the length of the latter increases. The reverse trends hold with decrease in centrum length. Relative length of a given vertebra can be evaluated with reference to the height or width measurements of the same vertebra or to centrum length of a given, "standard" vertebra. Incomplete skeletons make the first method preferable when dealing with fossil material.

2.3.1. Shortened Cervical Vertebrae Reduce Neck Flexibility

A short neck enhances hydrodynamic shape and stabilizes the front end of the body in vertebrates with a posteriorly generated propulsive force. Because of the stable number of cervical vertebrae in all aquatic mammals except the manatee, shortening is the result of re-

Table I. Relative Cervical Centrum Length (CL/CH) Measurements of Living
and Fossil Marine Mammals (C1–C7 = Cervical Vertebrae 1–7)

	C1	C2	C3	C4	C5	C6	C7
Monachus tropicalis MCZ 7264	2.8	1.6	1.3	1.3	1.4	1.5	1.6
Eumetopias jubatus MCZ 129	2.1	2.5	1.5	1.6	1.7	1.7	1.6
Enhydra lutris MCZ 9345	2.7	3.3	1.6	1.6	1.6	1.4	1.6
Dugong dugon MCZ 6955		2.1	1.2	0.4	0.3	0.4	0.4
[a]*Pachyaena ossifraga* UM 95074			1.7	1.7	1.5	1.5	
[b]*Remingtonocetus* cf. *R. harudiensis* GSP-UM 3057		1.7	1.0	1.0	1.1	1.0	1.2
[c]*Rodhocetus kasrani* GSP-UM 3012		1.1	0.8	0.7	0.7	0.8	0.9
[d]*Protocetus atavus*		1.7	0.7			0.7	0.6
[e]*Georgiacetus vogtlensis* GSM 350				0.7	0.7		
[f]*Zygorhiza kochi* USNM 4748			0.5	0.4	0.5		
[f]*Basilosaurus cetoides* USNM 4675	1.2	0.7	0.5	0.4	0.5	0.6	0.7
Megaptera novaeangliae MCZ 6177	1.5	0.5	0.2	0.2	0.3	0.3	0.2
Delphinus delphis USNM 550868			0.1	0.1	0.1	0.1	0.2

Data sources: [a]Zhou *et al.* (1992); [b]Gingerich *et al.* (1995); [c]Gingerich *et al.* (1994) and Gingerich (personal communication); [d]Fraas (1904); [e]Hulbert (personal communication); [f]Kellogg (1936).

duction in centrum length, not count. Reduction in relative centrum length (and corresponding relative increase in articular surface area) to produce discoidal centra also decreases angle of possible intervertebral flexion, reducing lateral displacement of the head.

Comparative data for cervical centrum length versus centrum height measurements are presented in Table I. In mammals with terrestrial (*Pachyaena*) or transitional aquatic habits (*Monachus, Enhydra, Eumetopias*), cervical centrum length is typically greater than centrum height. Of living genera surveyed here, only those mammals with obligate aquatic habits (*Dugong, Megaptera, Delphinus*) have a cervical centrum length smaller than centrum height, and among these, the most powerful and durable swimmers (*Megaptera, Delphinus*) have the shortest relative centrum lengths. It follows that aquatic locomotion historically precedes extensive shortening of the cervical vertebrae, and that the most extreme shortening is associated with the most extensive aquatic adaptations. In almost all cases, vertebrae in the center of the cervical series (C3–C6) are shorter than those at the ends of the series.

2.3.2. Location of the Undulatory Portion of the Torso

The same principles of relative motion between adjacent vertebrae hold for the postcervical column. Locomotor style of sprawling quadrupeds is used to demonstrate that variation in centrum length allows recognition of both that portion of the column that undulates and whether the generated wave has constant amplitude. During terrestrial locomotion, the trunk of sprawling quadrupeds bends into a lateral "wave" between the pectoral and pelvic girdles (Hildebrand, 1985; Ritter, 1992). The distinctive peak in centrum length between the limb girdles corresponds to the portion of the column undergoing the undulatory wave (*Amblyrhynchus*, Fig. 5C; *Gavialis*, Fig. 5D). Vertebrae at different locations within the wave of limbed lizards can be structurally and functionally differentiated. Vertebrae located in the center of the wave are relatively long (CL/CH ~ 2.3). The relatively small articular surfaces of their "spool" shapes enhance angular displacement at intervertebral joints, and their long lengths enhance lateral excursion. Vertebrae located near the limb girdles are absolutely and relatively shorter (CL/CH ~ 2.0). The relatively large articular surfaces of their "disk" shapes limit angular displacement between vertebrae, and their short lengths limit lateral excursion. The variation in lateral excursion produces a wave with varying amplitude at different locations along the column.

A somewhat different pattern is seen in the tail of these swimming reptiles (Dawson *et al.*, 1977; Whitaker and Basu, 1982), and in both the torso and the tail of lizards with reduced or absent limbs (*Ophisaurus*, Fig. 5E), in which a sequence of lateral waves passes down the column (Gans, 1975). Here "plateaus" of relatively long (CL/CH ~ 2.5–4.0), nearly uniform centra indicate portions of the column where each vertebra sequentially performs the same range of movements as those anterior and posterior to it. The waves generated along these columns are of uniform or near-uniform amplitude.

Patterns of centrum length may be used similarly as indicators of sites of undulatory waves in the locomotion of marine mammalian carnivores. Three carnivores of distinctly different swimming mode may be used as examples. Swimming in the otariids (sea lions and fur seals, including *Eumetopias jubatus*, Fig. 5F) is accomplished via propulsive forces generated by the forelimbs (English, 1976; Feldkamp, 1987). Hind limbs and tail are not used during swimming, but terrestrial locomotion is limb-based and quadrupedal. The vertical or horizontal displacement of the column that occurs in walk, gallop, or bounding gaits (Beentjes, 1990) is reflected in the modest peak in lumbar centrum length. Swimming in phocids (earless seals including *Monachus tropicalis*, Fig. 5G) is hind limb dominated (Ray, 1963; Tarasoff *et al.*, 1972). The posterior torso swings laterally as the hind flippers execute powerful medially directed strokes. Terrestrial locomotion also involves dorsoventral lumbar undulation, with weight alternately on the pelvis and sternum. A large peak in centrum length from the midthorax to the posterior lumbos indicates the location of greatest column displacement. Swimming in the semiaquatic sea otter *Enhydra lutris* (Fig. 5H) involves both the hind limbs, which are moved by powerful dorsoventral oscillations of the torso, as well as dorsoventral sweeps of the tail (Kenyon, 1981). Forelimbs are used for manipulating food and for quadrupedal terrestrial locomotion. Centrum length in *Enhydra* is bimodal, with a peak in the thoracolumbar region and a short plateau of long centrum length in the caudals corresponding to the two sites of undulation.

In living sirenians (Fig. 5A,B), the single broad peak in centrum length encompasses almost the entire postthoracic column. In the manatee the lumbar and anterior caudal vertebrae are relatively long (CL/CH = 1.2–1.3), not structurally isolated from the fluke, and must be capable of significant intervertebral movement. In contrast, centra at the same location in the dugong column are relatively much shorter (CL/CH = 0.9–1.1), providing a zone of limited intervertebral movement against which the peduncle acts. Unfortunately, literature on sirenian locomotion is very limited. However, it is clear that the dugong is a faster and more efficient swimmer than the manatee (Hartman, 1979; Caldwell and Caldwell, 1985; Nishiwaki and Marsh, 1985).

Most archaeocete and living whales also show a single peak in pre-fluke centrum length (Figs. 1B, 3A), indicating the presence of an undulatory wave with variable amplitude. However, there is marked variation in the location and anteroposterior extent of the peak along the column. Greatest centrum length may occur in the true lumbars, sacrals (sacral lumbars), or anterior caudals, implying that maximal dorsoventral excursion occurred in different parts of the axial skeleton. A trend toward posterior translocation of the peak is apparent in archaeocetes (see below). In *Delphinapterus*, greatest centrum length of the broad peak occurs at the lumbar/caudal boundary, whereas in *Delphinus*, the peak is restricted to midcaudal vertebrae. Correspondingly, movement of the torso is largely restricted to this small area of the column, as documented by Parry (1949) and other observers of delphinid locomotion. *Delphinus* (Fig. 1B) and many other derived odontocetes show a marked secondary reduction in centrum length anterior to the undulatory peak. The discoidal shapes of these vertebrae reduce intervertebral movement and provide a stable base against which the peduncle acts.

3. Vertebral Patterns and Locomotion in Archaeocetes

Relationships between vertebral structure and swimming pattern documented for living marine mammals are used below as tools in prediction of locomotor style in archaeocete cetaceans. These animals are traditionally included in the families Ambulocetidae (*Ambulocetus*), Remingtonocetidae (*Remingtonocetus*), Protocetidae (*Rodhocetus, Protocetus, Georgiacetus*), Basilosauridae, subfamily Dorudontinae (*Zygorhiza*), and Basilosauridae, subfamily Basilosaurinae (*Basilosaurus*).

3.1. *Ambulocetus natans*

The skeleton of *Ambulocetus natans* (H-GSP 18507) from the middle Eocene of Pakistan described by Thewissen *et al.* (1994, 1996a) includes only a few fragmentary vertebrae. Vertebral count is unknown, but almost certainly this animal possessed more thoracic than lumbar vertebrae, a torso limited to the lumbar vertebrae, and sacrals that articulated with a robust pelvis. Cervical centrum length is relatively long, although no CL/CH ratio can be computed with accuracy. The three thoracic vertebrae increase in relative width and length posteriorly. The single measurable lumbar centrum is broader (CW/CH = 1.4) and markedly longer (CL/CH = 1.6) than high. Two referred caudal centra are nearly circular in cross section (CW ~ CH) as are the caudals of many terrestrial (Figs. 1A, 2A) and semiaquatic (Fig. 5B,F–H) mammals. They display neither the lateral compression typical of a peduncle nor the vertical compression typical of a fluke. Proximal limb elements are relatively short, articular surfaces of the hind limb suggest only motion of knee and ankle, and the feet, which have hoofed digits, are "enormous" (Thewissen *et al.*, 1994, 1996a).

Thewissen *et al.* (1994, 1996a) have proposed that this animal swam by flexing and extending its lumbos to propel its large feet dorsoventrally through the water. As they note, this mode is reminiscent of aspects of both lutrine and phocid locomotion. The long lumbar vertebra and large feet clearly support their interpretation of a mobile lumbos and foot propulsors. The relatively long cervical vertebrae and hoofed digits signal a mobile neck, significant terrestrial competence, and at most a marginally aquatic lifestyle. *Ambulocetus* had less aquatic proficiency and more terrestrial ability than living phocids. It almost certainly lacked all of the typically cetacean traits of an elongate lumbos, flippers, peduncle, and fluke.

3.2. *Remingtonocetus* cf. *R. harudiensis*

Remingtonocetus is a middle Eocene archaeocete known from several localities in India and Pakistan. Its cranial morphology was described by Kumar and Sahni (1986). Vertebrae from several individuals originally placed in *Indocetus ramani* and now referred to *R.* cf. *harudiensis* were described by Gingerich *et al.* (1993, 1995), allowing a composite picture of the column. All known specimens are incomplete, but Gingerich *et al.* (1993) predict a vertebral count of 7 cervicals, 14 thoracics, 5 lumbars, 4 sacrals, and an unknown number of caudals. Cervical vertebrae are as long or longer than high (minimum CL/CH = 1.0). Thoracic vertebrae increase gradually in length posteriorly, and are noticeably broader than high (CW/CH = 1.4–1.7). The lumbar series has a smaller vertebral count than the thoracic series, and is composed of similarly broad centra that marginally exceed the thoracics, and equal the sacrals, in length. The four sacrals are fused not only to each other but also to the pelvis. The single reported sacral width is greater than that of any other vertebra (Gingerich *et al.*, 1993). The few known caudals are somewhat broader than high (CW/CH = 1.2–1.5) and shorter than the lumbars or sacrals. The hind limb is robust.

   Sacrum fusion and the robust hind limb argue that *Remingtonocetus* was able to locomote on land. Large cranial accessory air sinuses (Kumar and Sahni, 1986) indicate that the animal was capable of regulating pressure with changes in water depth, but the long cervical vertebrae suggest that the neck was mobile and that *Remingtonocetus* should be described as semiaquatic. Sacral fusion, robustness, and dimensional discontinuity with lumbar vertebrae argue against incorporation of the sacrals into an undulatory torso. Lumbar flexion with foot propulsion is the most likely mode of swimming. Nevertheless, caudal centra are relatively broader than in *Ambulocetus*, and may indicate a possible secondary role for the tail in aquatic locomotion. Peduncle and fluke were almost certainly absent.

3.3. *Rodhocetus kasrani*

   In 1994, Gingerich *et al.* described *Rodhocetus kasrani* from the early middle Eocene of Pakistan. The vertebral column of GSP-UM 3012 is nearly complete through the anterior caudals (Fig. 6). Cervical vertebrae are shorter than high (minimum CL/CH = 0.7) and unfused. Thoracic vertebrae (13) outnumber lumbars (6?), are approximately 1.5 times as broad as high, and show almost no increase in centrum length throughout the series. The most robust vertebrae are the lumbars, which show a progressive increase in width and height throughout the series. There are four unfused "sacralized" vertebrae, the first of which was articulated with the pelvis. Centrum height and width are both discontinuous across the lumbar/sacral transition. Centrum length gradually increases from the posterior thorax to a sacral peak. Centrum width declines and centrum height increases in the anterior tail (CW/CH of vertebra 31 = 1.19, of vertebra 34 = 1.08).

   Rodhocetus presents vertebral characteristics of both terrestrial and aquatic animals, as noted by Gingerich *et al.* (1994). Presence of significant centrum width and height discontinuities from lumbar to sacral vertebrae is reminiscent of quadrupedal animals and suggests that *Rodhocetus* had terrestrial ability. This is corroborated by the presence of a pelvic articulation with the first sacral. However, the progressive increase in lumbar centrum length that continues into and peaks in the sacrals suggests the presence of an undulatory torso that included lumbar, sacral, and anterior caudal vertebrae. Significantly, the posterior thoracics are not "lumbarized" as they are in many more aquatic cetaceans. Moderately foreshortened cervicals also reinforce an interpretation of a predominantly aquatic lifestyle, as does the reduced femur. The slight trend toward relatively taller, narrower centra in the

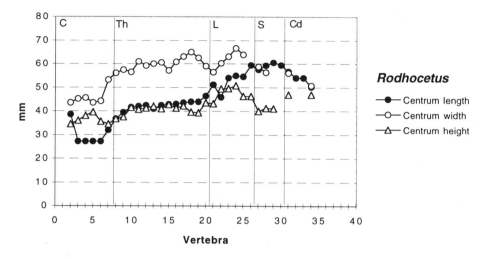

FIGURE 6. Centrum length, width, and height of *Rodhocetus kasrani*, GSP-UM 3012. Data from Gingerich (personal communication). Abbreviations as in Fig. 1.

anterior caudals suggests that the tail was sleeker than the torso. Although the propulsive surface was most likely the broad body and tail of the animal itself, lack of posterior caudals makes it impossible to rule out the presence of a poorly developed fluke similar to that of the manatee.

3.4. *Protocetus atavus*

The lower Lutetian archaeocete *Protocetus atavus* was described by Fraas (1904) and Kellogg (1936). It was also used by Slijper (1936) as the representative early cetacean. The 7 cervicals of *Protocetus* are moderately foreshortened (minimum CL/CH = 0.7). There are at least 12 thoracics, which display a gradual increase in length that continues through the lumbars and the single preserved sacral (Fig. 7). Although Kellogg predicted as many as 13 lumbars, only 7 are known and this may well be the full complement. The single vertebra identified as a sacral shows no sign of fusion to missing posterior vertebrae. However, its centrum is noticeably lower than that of the posterior lumbars and the single preserved transverse process is robust, almost as long as the centrum anteroposteriorly, and considerably thickened distally. Fraas (1904) interpreted this anatomy as indicative of its attachment to a pelvis in life, but Gingerich *et al.* (1994) have suggested only an indirect connection.

Vertebrae of *Protocetus* show marked similarity of centrum dimensions to those of *Rodhocetus*. Relative length of postaxis cervicals (CL/CH = 0.6–0.7) is nearly identical and implies an aquatic lifestyle. Also as in *Rodhocetus*, the trend toward increasing length and breadth of the lumbar and sacral vertebrae suggests that these comprised the undulatory unit, although dimensional discontinuity across the lumbosacral boundary indicates some difference between lumbar and sacral function. Nevertheless, there are indications that *Protocetus* was a much more fully aquatic animal than *Rodhocetus*. There is a single sacral in-

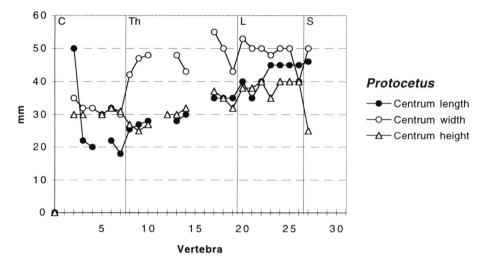

FIGURE 7. Centrum length, width, and height of *Protocetus atavus*. Data from Fraas (1904). Abbreviations as in Fig. 1.

stead of four, and in contrast to *Rodhocetus*, the undulatory torso was lengthened by the addition both of an additional lumbar and of posterior thoracics, which show the elongation typical of lumbar vertebrae. Unfortunately, the posterior end of the column is missing. Peduncle and fluke, if present, were likely only poorly defined.

3.5. *Georgiacetus vogtlensis*

The protocetid *Georgiacetus vogtlensis* was collected from the latest Lutetian or earliest Bartonian of the eastern coast of North America (Hulbert, 1991, 1993). Dimensions of 23 vertebrae from a single individual (GSM 350) and 3 from a second individual (GSM 351) are graphed here (Fig. 8). Hulbert (personal communication) suggests a total count of 7 cervical, 13 thoracic, 8 lumbar, 4 sacral, and an indeterminate number of caudal vertebrae. Only 3 of the 7 cervical vertebrae are preserved, and these show a moderate degree of foreshortening (minimum CL/CH = 0.7). Thoracics (13) barely outnumber lumbars (8) plus sacrals (4), and at least 2 thoracics are functional lumbars. The 4 sacrals are not fused to each other, are dimensionally continuous with the lumbars and caudals, and are interpreted here as "sacral lumbars." The innominate suggests the possibility of a soft tissue, but not a bony, connection to the sacrals. Centrum width modestly exceeds centrum height in the anterior thoracics (CW/CH = 1.2–1.3), markedly exceeds it in posterior thoracics and anterior lumbars (CW/CH = 1.5–1.6), decreases again in the posterior (sacral) lumbars (CW/CH = 1.1–1.3), and is exceeded by it in the single caudal (CW/CH = 0.9).

The lack of dimensional discontinuities from posterior thoracic to anterior caudal vertebrae identifies this region as the undulatory torso, with a caudal peak in centrum length. *Georgiacetus* was an axial undulator with a longer torso than *Rodhocetus* or *Protocetus*, and

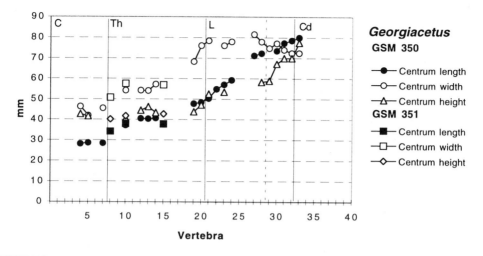

FIGURE 8. Centrum length, width, and height of *Georgiacetus vogtlensis*, GSM 350 (circles) and GSM 351 (triangles). Dotted line indicates boundary between true lumbars and "sacral lumbars." Data from Hulbert (personal communication). Abbreviations as in Fig. 1.

with no terrestrial competence. Centrum width/height relationships suggest that the animal was sleeker than either *Rodhocetus* or *Protocetus*. Caudal location of the centrum with maximal length and the reduction of centrum width relative to height in the posterior lumbar and caudal vertebrae signal the presence of an anteriorly located caudal peduncle that was at least somewhat laterally compressed and moved with reduced resistance through the water. Although posterior caudals are lacking, this peduncle must have borne a propulsive fluke. *Georgiacetus* was significantly sleeker in body form and a more proficient and rapid swimmer than either *Protocetus* or living sirenians.

3.6. *Zygorhiza kochi*

The upper Eocene dorudontine basilosaurid *Zygorhiza kochi* is known from multiple partial skeletons, and Kellogg (1936) relied on material from several individuals in his description of the vertebral column. Data from four of these individuals were used to create the profile presented in Fig. 9. Such a composite profile must assume that the animals differed to at least some extent in body size. As reconstructed, the column consists of 7 cervical, 15 thoracic, 15 lumbar (including sacral lumbar), and 21 caudal vertebrae. The cervical vertebrae are markedly foreshortened (minimum CL/CH = 0.4). Centrum length increases gradually throughout the thorax and lumbos, with a peak in the midtail. The functional torso therefore incorporated all vertebrae from the posterior thoracics to the anterior caudals. The last two lumbars were identified as sacrals by Kellogg, but they are not fused to each other and demonstrate no modification of the transverse processes for articulation with a pelvis. They are considered "sacral lumbars" here. Centrum width exceeds centrum height very modestly in the torso, and equals or is marginally less than it in some caudal

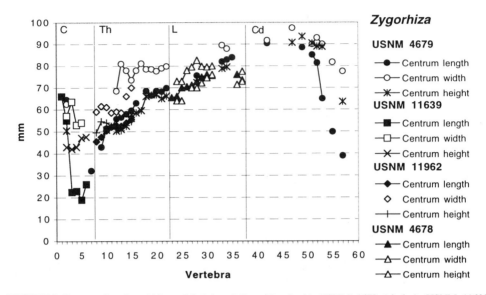

FIGURE 9. Centrum length, width, and height of *Zygorhiza kochi*, USNM 4679 (circles), USNM 11639 (squares), USNM 11962 (diamonds), and USNM 4678 (triangles). Data from Kellogg (1936). Abbreviations as in Fig. 1.

vertebrae (Fig. 10), suggesting the presence of a peduncle at least as well-defined as that of living sirenians. Although posterior caudals are very incomplete, the rapid drop in centrum length and the presence of centra that are wider than high signal the presence of a fluke.

    Zygorhiza was a fully aquatic animal, with short cervical vertebrae, a sleek body outline, and a caudal peduncle and fluke. The efficiency of its axial undulation over that of pro-

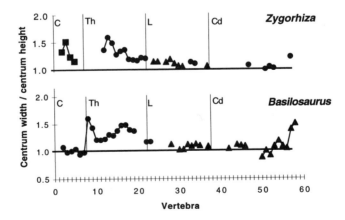

FIGURE 10. Relative centrum width in *Zygorhiza kochi* [composite of USNM 4679 (circles), USNM 11639 (squares), and USNM 4678 (triangles)] and *Basilosaurus cetoides* [composite of USNM 4675 (circles) and USNM 12261 (triangles)]. Data from Kellogg (1936). Abbreviations as in Fig. 1.

tocetids is indicated by its sleeker body profile and by the large relative length of the lumbos, to which a high lumbar count, posteriorly lengthening lumbar vertebrae, complete incorporation of sacrals into the lumbos, and lumbarization of posterior thoracic and anterior caudal vertebrae all contribute. Lumbars (true lumbars plus "sacral lumbars") equal thoracics in number. The absence of a sacral/pelvic articulation precludes the possibility of terrestrial locomotion. Of all of the specimens examined in this study, this animal has a vertebral profile, and by implication a locomotor mode, most like that of living nondelphinid odontocete cetaceans.

3.7. *Basilosaurus cetoides*

Partial skeletons of numerous individuals are assigned to the late Eocene basilosaurine basilosaurid *Basilosaurus cetoides*. Vertebrae belonging to USNM 4675 and USNM 12261 were described in detail by Kellogg (1936) and are graphed in Fig. 11. Cervical vertebrae are extensively foreshortened (minimum CL/CH = 0.4). The thorax is composed of 15 vertebrae, but of these the posterior 7 are elongate, bear ribs that articulate only with the corresponding vertebrae, and serve as functional lumbars. Although Kellogg reported the presence of 13 lumbar and 2 sacral vertebrae, the latter do not have articular facets for the ilia, were not fused to each other, and are dimensionally nearly identical with the posterior lumbars and anterior caudals. They are considered "sacral lumbars" here, bringing the lumbar count up to 15. At least the first 6 anterior caudals are also lumbar in dimension, and may be considered part of the functional torso (Fig. 11, inset). The posterior thoracics, lumbars, and anterior caudals are all extensively and nearly uniformly elongated (CL/CH ~ 2.0).

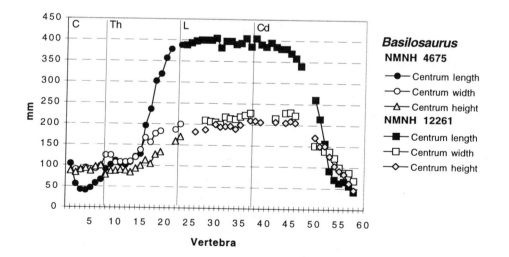

FIGURE 11. Centrum length, width, and height of *Basilosaurus cetoides* USNM 4675 (circles) and USNM 12261 (triangles). Inset shows average percent variation of centrum dimensions along column. Data from Kellogg (1936). Abbreviations as in Fig. 1.

Centrum width moderately exceeds centrum height in the thorax (CW/CH = 1.3–1.5), but only marginally exceeds it in the lumbos and anterior tail (CW/CH = 1.1, Fig. 10), indicating that the body was not only long but very sleek. A small number (up to six counting missing vertebrae) of midcaudal vertebrae show the reduction in centrum length and the reverse of width and height typical of the peduncle of living whales (Figs. 5, 11). The last six vertebrae are wider than high and show a reduction in centrum length.

This animal was fully marine. Its lack of terrestrial ability is signaled by the short cervicals, the lack of any dimensional discontinuity between the sacrals and adjacent vertebrae, and the lack of any sacral/pelvic articulation. The elongate torso is the result of extreme elongation of the posterior thoracics, sacrals, and anterior caudals, all of which were involved in undulation. As in *Zygorhiza*, lumbars plus "sacral lumbars" equal thoracics in number. The pattern of centrum length is distinctly different from that seen in other archaeocetes and living whales in that all torso vertebrae are subequal in length. The plateau of dimensionally uniform elongate vertebrae signals the presence of an undulatory wave of constant amplitude. This pattern is reminiscent of that of limbless lizards (*Ophisaurus*, Fig. 5E), and supports interpretation of an anguilliform swimming pattern with a dorsoventral orientation. As in living fish with similar body plans, these body proportions must have promoted maneuverability at the expense of speed (Webb, 1982).

Living tetrapods with anguilliform locomotion lack both a peduncle and a posterior fluke. Although not marked, the dimensional patterns of both in *Basilosaurus* are unequivocal. Presence of a fluke in *Basilosaurus* has also been proposed by Barnes and Mitchell (1978). Nevertheless, both peduncle and fluke are very short relative to length of the undulatory torso, and successive undulatory waves of the long torso must have generated most of the propulsive force. This animal has no living functional counterpart among mammals.

4. Discussion

Form and function of the postcranial skeleton changed radically in early cetacean history. Analysis of archaeocete skeletons reveals the following morphological trends in the regional variation of the vertebral column. Functional correlates of the structural trends are suggested.

1. Reduction in the number of sacral vertebrae. Sacral vertebrae are lost from the posterior end of the sacrum forward by lack of fusion to the preceding (sacral) vertebrae. Lack of fusion allows participation of sacral vertebrae in axial undulation. Vertebrae with a recent history of sacral function (e.g., *Protocetus*) may be recognized by dimensional discontinuities from preceding and/or following vertebrae.
2. Loss of the sacral/pelvic articulation. Terrestrial locomotion is hypothetically possible as long as an osteological sacral/pelvic articulation remains. Retention of at least one vertebra with a pelvic articulation in an animal (*Rodhocetus*) with no intrasacral fusion suggests that intrasacral fusion was lost historically before pelvic articulation and predicts that undulatory locomotion was possible in animals that still had some terrestrial ability.
3. Reduction in relative length of cervical vertebrae. Minimum cervical centrum (CL/CH) ratios decrease as the extent of aquatic adaptation increases, improving

hydrodynamic body form and increasing stability of the anterior column. Cervical centrum lengths less than centrum heights (CL/CH < 1.0) occur only in animals with predominantly aquatic lifestyles, and are most extremely reduced (CL/CH < 0.5) in those with obligate aquatic habits. Central cervicals are typically more extremely reduced than either anterior or posterior cervicals.

4. Elongation of the torso. Elongation of the torso occurs by increase in the length of lumbar vertebrae, by increase in the number of true lumbars, and by addition of functional lumbars to the series via lumbarization of sacral, anterior caudal, and posterior thoracic vertebrae. Torso elongation provides increased attachment areas for the axial musculature and increases the maximum possible extent of dorsoventral deflection of the column during undulation.

5. Posterior translocation of the portion of the column undergoing maximal dorsoventral displacement during undulation. The undulatory portion of the column can be recognized using the peak in centrum length as a guide. Maximum centrum length is lumbar in *Pachyaena*, sacral in *Rodhocetus* and *Protocetus*, and caudal in *Georgiacetus* and *Zygorhiza*.

6. Modification of the central tail into a laterally compressed peduncle and of the posterior tail into a well-defined fluke (*Georgiacetus, Zygorhiza, Basilosaurus*). A peduncle/fluke complex allows dorsoventral movement of the propulsor (fluke) through a wide arc because the peduncle cuts through the water with minimal resistance. Axially locomoting cetaceans without a well-defined peduncle/fluke complex (*Rodhocetus, Protocetus*) were markedly broad-bodied, and the propulsive surface must have been the broad body itself. Lumbar and sacral vertebrae lying anterior to the undulatory portion of the column are secondarily stabilized with disklike proportions in derived odontocete cetaceans.

Although archaeocete phylogenetic relationships are not yet well understood, they have been partially reconstructed by several authors (Hulbert, 1994; Thewissen, 1994; Thewissen *et al.*, 1996b), largely on the basis of cranial characters. The functional changes in the swimming styles of archaeocetes can be tentatively reconstructed by superimposing the structural changes onto this provisional phylogeny (Fig. 12). The historical course of the cetacean transition suggested by this exercise appears to be dependent primarily on the anteroposterior extent and location of dorsoventral undulation and the surface used as a propulsor. The portion of the column that constitutes the undulatory torso expands from the lumbar series (node 1, *Ambulocetus, Remingtonocetus*) to include first sacral and anterior caudal (node 2, *Rodhocetus*), and finally posterior thoracic vertebrae (node 4, *Protocetus, Georgiacetus, Zygorhiza, Basilosaurus*). Concurrently, the propulsive surface changes from the appendicular to the axial skeleton (node 2), and the inferred peak of the undulatory wave moves posteriorly from lumbar (node 1) to sacral (node 2) and finally to caudal (node 4) vertebrae.

A second suite of changes involves inferred body proportions. Primitive archaeocetes (*Ambulocetus, Remingtonocetus, Rodhocetus, Protocetus*) appear to have been broad-bodied, with the undulatory torso itself serving as the dominant propulsive surface. More derived genera (node 4, *Georgiacetus, Zygorhiza, Basilosaurus*) have sleeker torsos, narrow anterior tails (peduncles), and broad posterior flukes that serve as propulsive surfaces. *Zygorhiza* and *Basilosaurus* (node 5) both have an extremely reduced neck and a torso that

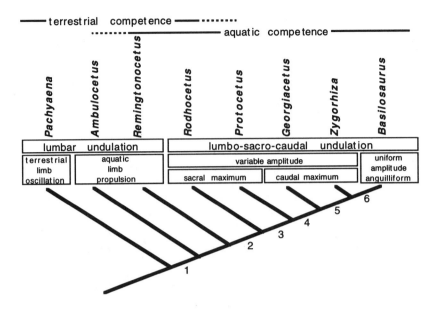

FIGURE 12. Changes in swimming pattern of archaeocete cetaceans inferred from regional variation in vertebral structure, superimposed on provisional phylogenetic relationships taken from Hulbert (1994), Thewissen (1994), and Thewissen *et al.* (1996b). Postcranial structural changes at numbered nodes include: 1, short robust limbs, elongate distal limb elements; 2, sacral/ilial articulation reduced to a single vertebra, sacral peak of centrum length, cervical centrum length less than centrum height, "lumbarization" of anterior caudals; 3, "lumbarization" of posterior thoracics; 4, sacral/ilial articulation absent, caudal peak of centrum length, sacrals dimensionally continuous with lumbars and caudals, centrum width and height pattern with signature of peduncle and fluke; 5, thoracic and lumbar centrum length exceeds centrum height, lumbars plus "sacral lumbars" equal thoracics in number, minimum cervical CL/CH $\leq$ 0.5; 6, thoracic and lumbar CL greatly exceeds CH, torso vertebrae subequal in length.

equals the chest in vertebral count and exceeds it in length. The stabilized base for fluke action provided by shortened torso vertebrae, typical of more derived odontocetes, is not seen in any of these taxa, but would be easily derivable from the pattern of *Zygorhiza*.

The elongate torso vertebrae and inferred anguilliform locomotion in *Basilosaurus* are unique, but are dependent on the undulation of the expanded torso and sleeker body also possessed by other derived archaeocetes. Its possession of a small terminal fluke appears to be contradictory, but may represent a historical retention rather than a fully functional unit.

5. Summary and Conclusions

Regional variation of vertebrae may be used as a tool in the interpretation of locomotor stages of extinct taxa. Comparison of archaeocete vertebral columns with those of living aquatic and semiaquatic tetrapods of known locomotor style suggests a sequence of transitional stages in the movement from land to water habitats. Vertebral morphology indicates that all cetaceans employed torso undulation during swimming, but that the propulsive surface changed from the feet to the broad torso, and finally to the terminal fluke. Concurrent changes included shortening of the neck and chest, elongation of the torso and tail, and reduction in body girth.

The following locomotor stages in the archaeocete transition from land to water are suggested by this analysis:

1. Quadrupedal terrestrial quadrupeds. Long neck, short stable lumbos, sacrum of multiple vertebrae with bony articulations to pelvis. Oscillatory limbs are the sole propulsors. Terrestrial with very little if any aquatic competence.
   Pachyaena and other mesonychids

2. Lumbar undulation with limb propulsion. Long neck, long thorax, short mobile lumbos, sacrum of multiple vertebrae with bony articulations to pelvis. Lumbar undulation helps to propel the hind limbs, especially the enlarged feet, which provide the dominant propulsive surface. Terrestrial and aquatic locomotion.
   Ambulocetus, Remingtonocetus

3. Lumbo-sacro-caudal undulation/variable amplitude (sacral maximum). A standing wave of lumbar, sacral, and caudal vertebrae peaks in the sacral or "sacral lumbar" vertebrae. Lumbos shorter than thorax, and sacrals dimensionally discontinuous from lumbars and caudals. Sacral articulation to pelvis limited to at most one vertebra. Tail without well-defined peduncle or fluke. CL/CH ratios of cervical vertebrae are reduced below 1.0. Efficient aquatic and very limited terrestrial locomotion.
   Rodhocetus, Protocetus

4. Lumbo-sacro-caudal undulation/variable amplitude (caudal maximum). A standing wave of lumbar, sacral, and caudal vertebrae peaks in the caudal vertebrae. The torso, composed of true lumbars as well as lumbarized thoracics and sacrals, exceeds the thorax in length and provides attachment for powerful muscles of axial undulation. Sacrals are dimensionally continuous with lumbars and caudals. Anterior tail is laterally compressed to form a peduncle; posterior tail is dorsoventrally compressed to form a fluke, which is the primary propulsor. No sacral/pelvic articulation. Fast, efficient swimmers with no terrestrial competence.
   Georgiacetus, Zygorhiza

5. Lumbo-sacro-caudal undulation (anguilliform). Sleek and extremely elongated torso composed of individually elongate posterior thoracic, lumbar, sacral, and anterior caudal vertebrae nearly identical in dimensions. Sequential undulatory waves of uniform amplitude pass down the torso during swimming, producing an eel-like, flexible body. Peduncle and fluke relatively small and only modestly set off from body. Slow swimmers with excellent maneuverability. No terrestrial competence.
   Basilosaurus

Acknowledgments

I thank Hans Thewissen, Phil Gingerich, John Heyning, Larry Barnes, Jim Mead, Richard Hulbert, and Charlie Potter for access to specimens and personal data compilations, and Nick Hotton for years of discussion about vertebrae. I also gratefully acknowledge the ideas and data analysis of the following students whose research projects have directly or indirectly contributed to this project: Vickie Chang, Denise Ching, Cynthia Efremoff, Kyleigh Geissler, Michelle Gillette, Jessika Turner, Kiran Ubriani.

References

Barnes, L., and Mitchell, E. 1978. Cetacea, in: V. J. Maglio and H. B. S. Cooke (eds.), *Evolution of African Mammals*, pp. 582–602. Harvard University Press, Cambridge, MA.

Beentjes, M. P. 1990. Comparative terrestrial locomotion of the Hooker's sea lion (*Phocarctos hookeri*) and the New Zealand fur seal (*Arctocephalus forsteri*): evolutionary and ecological implications. *Zool. J. Linn. Soc.* **98**:307–325.

Caldwell, D. K., and Caldwell, M. C. 1985. Manatees—*Trichechus manatus, Trichechus senegalensis*, and *Trichechus inunguis*, in: S. H. Ridgway and R. Harrison (eds.), *Handbook of Marine Mammals*, Volume 3, pp. 33–66. Academic Press, London.

Crovetto, A. 1991. Etude osteometrique et anatomo-functionelle de la colonne vertebrale chez les grands cetacés. *Invest. Cetacea* **23**:7–189.

Curren, K., Bose, N., and Lien, J. 1994. Swimming kinematics of a harbor porpoise (*Phocoena phocoena*) and an Atlantic white-sided dolphin (*Lagenorhynchus acutus*). *Mar. Mamm. Sci.* **10**(4):485–492.

Dawson, W. R., Bartholomew, G. A., and Bennett, A. F. 1977. A reappraisal of the aquatic specializations of the Galapagos marine iguana (*Amblyrhynchus cristatus*). *Evolution* **31**:89–97.

DeSmet, W. M. A. 1977. The regions of the cetacean vertebral column, in: R. J. Harrison (ed.), *Functional Anatomy of Marine Mammals*, Volume 3, pp. 58–80. Academic Press, New York.

English, A. W. 1976. Limb movements and locomotor function in the California sea lion (*Zalophus californianus*). *J. Zool.* **178**:341–364.

Feldkamp, S. D. 1987. Foreflipper propulsion in the California sea lion *Zalophus californianus*. *J. Zool.* **212**:4333–4357.

Fish, F. E., and Hui, C. A. 1991. Dolphin swimming—a review. *Mammal Rev.* **21**:181–195.

Fraas, E. 1904. Neue Zeuglodonten aus dem unteren Mitteleocän vom Mokattam bei Cairo. *Geol. Palaeontol. Abh. N. F.* **6**:199–220.

Gans, C. 1975. Tetrapod limblessness: evolution and functional corollaries. *Am. Zool.* **15**:455–467.

Gingerich, P. D., Wells, N. A., Russell, D. E., and Shah, S. M. I. 1983. Origin of whales in epicontinental remnant seas: new evidence from the early Eocene of Pakistan. *Science* **220**:403–406.

Gingerich, P. D., Raza, S. M., Arif, M., Anwar, M., and Zhou, X. 1993. Partial skeletons of *Indocetus ramani* (Mammalia, Cetacea) from the lower middle Eocene Domanda Shale in the Sulaiman Range of Punjab (Pakistan). *Contrib. Mus. Paleontol. Univ. Michigan* **28**(16):393–416.

Gingerich, P. D., Raza, S. M., Arif, M., Anwar, M., and Zhou, X. 1994. New whale from the Eocene of Pakistan and the origin of cetacean swimming. *Nature* **368**:844–847.

Gingerich, P. D., Arif, M., and Clyde, W. C. 1995. New archaeocetes (Mammalia, Cetacea) from the middle Eocene Domanda Formation of the Sulaiman Range, Punjab (Pakistan). *Contrib. Mus. Paleontol. Univ. Michigan* **29**(11):291–330.

Hartman, D. S. 1979. Ecology and behavior of the manatee (*Trichechus manatus*) in Florida. *Spec. Publ. Am. Soc. Mammal.* **5**:1–153.

Hildebrand, M. 1985. Walking and running, in: M. Hildebrand, D. M. Bramble, K. F. Liem, and D. B. Wake (eds.), *Functional Vertebrate Morphology*, pp. 110–128. Belknap Press, Cambridge, MA.

Hulbert, R. C. 1991. Innominate of a middle Eocene (Lutetian) protocetid whale from Georgia. *J. Vertebr. Paleontol.* **11**(Suppl. to No. 3):36A.

Hulbert, R. C. 1993. Craniodental anatomy and systematics of a middle Eocene protocetid whale from Georgia. *J. Vertebr. Paleontol.* **13**(Suppl. to No. 3):42A.

Hulbert, R. C. 1994. Phylogenetic analysis of Eocene whales ("Archaeoceti") with a diagnosis of a new North American protocetid genus. *J. Vertebr. Paleontol.* **14**(Suppl. to No. 3):30A.

Kaiser, H. E. 1974. *Morphology of the Sirenia*. Karger, Basel.

Kellogg, R. 1936. A Review of the Archaeoceti. *Carnegie Inst. Washington Publ.* **482**:1–366.

Kenyon, K. W. 1981. Sea otter, *Enhydra lutris* (Linnaeus, 1758), in: S. H. Ridgway and R. J. Harrison (eds.), *Handbook of Marine Mammals*, Volume 1, pp. 209–223. Academic Press, London.

Kumar, K., and Sahni, A. 1986. *Remingtonocetus harudiensis*, new combination, a middle Eocene archaeocete (Mammalia, Cetacea) from western Kutch, India. *J. Vertebr. Paleontol.* **6**:326–349.

Landes, R. 1976. Evolutionary mechanisms of limb loss in tetrapods. *Evolution* **32**:73–92.

Long, J. H., Pabst, D. A., Shepherd, W. R., and McLellan, W. A. 1997. Locomotor design of dolphin vertebral columns: bending mechanics and morphology of *Delphinus delphis*. *J. Exp. Biol.* **200**:65–81.

McShea, D. W. 1992. A metric for the study of evolutionary trends in the complexity of serial structures. *Biol. J. Linn. Soc.* **45**:39–55.

Nishiwaki, M., and Marsh, H. 1985. Dugong—*Dugong dugon*, in: S. H. Ridgway and R. Harrison (eds.), *Handbook of Marine Mammals*, Volume 3, pp. 1–32. Academic Press, London.

O'Leary, M. A., and Rose, K. D. 1995. Postcranial skeleton of the early Eocene mesonychid *Pachyaena* (Mammalia: Mesonychia). *J. Vertebr. Paleontol.* **15**(2):410–430.

Pabst, D. A. 1990. Axial muscles and connective tissues of the bottlenose dolphin, in: S. Leatherwood and R. R. Reeves (eds.), *The Bottlenose Dolphin*, pp. 51–67. Academic Press, New York.

Pabst, D. A. 1993. Intramuscular morphology and tendon geometry of the epaxial swimming muscles of dolphins. *J. Zool.* **230**:159–176.

Parry, D. A. 1949. The swimming of whales and a discussion of Gray's paradox. *J. Exp. Biol.* **26**:24–34.

Prothero, D. R., Manning, E. M., and Fischer, M. 1988. The phylogeny of the ungulates, in: M. J. Benton (ed.), *The Phylogeny and Classification of the Tetrapods*, Volume 2, pp. 201–234. Clarendon Press, Oxford.

Ray, G. C. 1963. Locomotion in pinnipeds. *Nat. Hist.* **72**:10–21.

Ritter, D. 1992. Lateral bending during lizard locomotion. *J. Exp. Biol.* **173**:1–10.

Slijper, E. J. 1936. Die Cetaceen. Vergleichend-Anatomisch und Systematisch. *Capita Zool.* **6–7**:1–590.

Slijper, E. J. 1946. Comparative biologic-anatomical investigations on the vertebral column and spinal musculature of mammals. *Verh. K. Ned. Akad. Wet. Afd. Natuurkd. Tweede Reeks* **42**:1–128.

Slijper, E. J. 1961. Locomotion and locomotory organs in whales and dolphins (Cetacea). *Zool. Soc. London Symp.* **5**:77–94.

Tarasoff, F. J., Bisaillon, A., Pierard, J., and Whit, A. P. 1972. Locomotory patterns and external morphology of the river otter, sea otter, and harp seal (Mammalia). *Can. J. Zool.* **50**:915–929.

Thewissen, J. G. M. 1994. Phylogenetic aspects of cetacean origins: a morphological perspective. *J. Mamm. Evol.* **2**(3):157–184.

Thewissen, J. G. M., Hussain, S. T., and Arif, M. 1994. Fossil evidence for the origin of aquatic locomotion in archaeocete whales. *Science* **263**:210–212.

Thewissen, J. G. M., Madar, S. I., and Hussain, S. T. 1996a. *Ambulocetus natans*, an Eocene cetacean (Mammalia) from Pakistan. *Cour. Forsch.-Inst. Senckenberg* **191**:1–86.

Thewissen, J. G. M., Roe, L. J., O'Neil, J. R., Hussain, S. T., Sahni, A., and Bajpai, S. 1996b. Evolution of cetacean osmoregulation. *Nature* **381**:379–380.

Webb, P. W. 1982. Locomotor patterns in the evolution of actinopterygian fishes. *Am. Zool.* **22**:329–342.

Webb, P. W., and Blake, R. W. 1985. Swimming, in: M. Hildebrand, D. M. Bramble, K. F. Liem, and D. B. Wake (eds.), *Functional Vertebrate Morphology*, pp. 110–128. Belknap Press, Cambridge, MA.

Whitaker, R., and Basu, D. 1982. The gharial (*Gavialis gangeticus*): a review. *J. Bombay Nat. Hist. Soc.* **79**:531–548.

Worthington, R. D., and Wake, D. B. 1972. Patterns of regional variation in the vertebral column of terrestrial salamanders. *J. Morphol.* **137**:257–277.

Zhou, X., Sanders, W. J., and Gingerich, P. D. 1992. Functional and behavioral implications of vertebral structure in *Pachyaena ossifraga* (Mammalia, Mesonychia). *Contrib. Mus. Paleontol. Univ. Michigan* **28**:289–319.

CHAPTER 12

Structural Adaptations of Early Archaeocete Long Bones

SANDRA I. MADAR

1. Background and Theory

1.1. Introduction

Fossil remains recovered during the past decade have provided the first glimpse of the appendicular skeleton of early cetaceans (Gingerich *et al.*, 1990, 1993, 1994, 1995; Hulbert and Petkewich, 1991; Aleshire, 1993; Madar and Thewissen, 1994; Hulbert, 1994, this volume; Thewissen *et al.*, 1994, 1996). When coupled with archaeocete craniofacial, dental, and axial remains, a much clearer picture is emerging of the morphological transitions that occurred during cetacean evolution. The ancestry of modern cetaceans is linked at present to the terrestrial mesonychian condylarths (Van Valen, 1966; Prothero *et al.*, 1988; Thewissen, 1994). Members of this group possess postcranial features linked to cursoriality, though emphasizing endurance rather than speed (Szalay and Gould, 1966; Zhou *et al.*, 1992; O'Leary and Rose, 1995). Given this ancestry, early archaeocete postcranial skeletons should document the series of structural modifications that occurred in a move from complete terrestrial competence, through an amphibious or semiaquatic stage, to the type of highly specialized aquatic locomotion that characterizes modern cetaceans (Thewissen *et al.*, 1996; Thewissen and Fish, 1997; Buchholtz, this volume).

In this chapter, I use the long bones of archaeocete cetaceans, mesonychians, and a series of extant aquatic, semiaquatic, and terrestrial mammals to explore the levels to which different archaeocete taxa have adapted their limbs to an aquatic niche. An organism's physical environment determines the magnitude and direction of forces that are necessary for movement, and thus the mechanical and anatomical requirements for swimming are dramatically different from those of terrestrial locomotion (Daniel and Webb, 1987; Webb and Buffrénil, 1990). Indeed, many studies have focused on the anatomical specializations that have appeared convergently among secondarily aquatic vertebrates (Howell, 1930; Slijper, 1946; Wall, 1983; Massare, 1994; Taylor, 1994; Fish, 1996).

SANDRA I. MADAR • Department of Biology, Hiram College, Hiram, Ohio 44234.
The Emergence of Whales, edited by Thewissen. Plenum Press, New York, 1998.

The move into a novel environment should also result in modifications of the structural properties of skeletal elements, as the forces transmitted through bones change (e.g., Wall, 1983; Currey and Alexander, 1985; Carter *et al.*, 1991; Domning and Buffrénil, 1991; Fish and Stein, 1991). According to Daniel and Webb (1987), the "density ratio" of substrate versus animal is the most important factor influencing the mechanisms used to power propulsion, and thus the shape of the propulsor. Differences in diaphyseal structure stem from a reduction in the required diaphyseal strength of obligate swimmers because of the buoyant properties of water and the absence of ground reaction forces incurred during terrestrial locomotion. In transitional taxa, the need to balance buoyancy concerns and load bearing should be reflected in skeletal structural properties.

This chapter builds on the groundwork of Currey and Alexander (1985), and focuses on the internal architecture of early archaeocete and extant aquatic mammalian limb elements. Currey and Alexander concluded that the design of tubular bones depends on their role in structural support and locomotion. Tubular bones must not yield under bending, not fail in fatigue, and be strong and stiff enough to withstand impact loading (Currey and Alexander, 1985). Each of these requirements can be associated with an optimum combination of shaft diameter and cortical thickness, characters that can then be used to assess the design features related to specific uses in locomotion. Weight bearing should be reflected in the cortex (external compact bone) and trabeculae (spongy or cancellous bone) of appendicular elements. Here I use radiography to examine the distribution of compact bone along the lengths of shafts, trabecular orientation patterns, and the presence or absence of open medullary cavities in order to examine the structural properties of limb elements. Long bone epiphyses are designed to withstand the loads that they incur during locomotion (Carter *et al.*, 1991). The architecture of trabecular bone struts, or densely packed columns of cancellous bone in epiphyses reflects the compressive and tensile loads endured by long bones (Enlow, 1964; Currey, 1984; Carter *et al.*, 1989, 1991). Therefore trabecular patterns provide an additional means of assessing limb use in cetaceans that could have been semi-aquatic.

Radiographic analysis can also be used to isolate features reflecting different buoyancy control strategies utilized by aquatic taxa (Felts and Spurrell, 1965, 1966; Domning and Buffrénil, 1991). Pachyostosis (overall bone thickening) and pachyosteosclerosis (defined as bone thickening combined with osteosclerosis, the replacement of trabecular bone with compact bone; Domning and Buffrénil, 1991) are both responses to aquatic living in vertebrates. Whereas pachyosteosclerosis has been documented in the thorax and forelimb of sirenians, to date it has only been identified in the ribs of highly derived archaeocetes of the middle and the late Eocene (Kellogg, 1936; Domning and Buffrénil, 1991; Thewissen *et al.*, 1996; Hulbert, this volume; but see Gingerich and Russell, 1994).

1.2. Locomotor Morphology and Functional Constraints of Aquatic Locomotion: Modern Analogues

Wall (1983; see also Stein, 1989; Fish and Stein, 1991) noted that shallow-diving mammals and highly aquatic birds have dense diaphyses relative to their terrestrial counterparts, allowing them to achieve neutral buoyancy at depth. Neutral depth is the depth that an organism can maintain without exerting physical effort. The most energetically efficient means of modifying neutral depth and thereby controlling buoyancy is to regulate lung volume, which does not require continuous muscular effort. This method is functionally equivalent to the use of gas bladders in many osteichthyans, and is employed by several secondarily aquatic reptiles as well as mammals (Felts and Spurrel, 1966; Taylor, 1994).

Relatively recent invaders of a marine habitat, such as penguins and otariid pinnipeds (and shallow marine mammals such as *Enhydra*), make use of several types of hydrostatic controls. In addition to modifying lung volume, some taxa use gastroliths (stomach stones) rather than pachyostosis (and/or osteosclerosis; Wall, 1983; Taylor, 1994). The pachyosteosclerosis of sirenians is consistent with their shallow swimming and herbivorous lifestyle, as increased skeletal mass balances the buoyant properties of air-filled lungs and soft tissues, allowing them to remain submerged to feed. Bone densities in related terrestrial and aquatic mammals diverge, although substantial overlap exists among semiaquatic forms (Wall, 1983; Currey and Alexander, 1985; Domning and Buffrénil, 1991; Fish and Stein, 1991). Rapid or sustained swimming below the surface is not part of the feeding strategy of any extant aquatic mammal exhibiting increased bone density. This is noteworthy, as increased density increases the drag on an organism, in effect increasing the energetic cost of locomotion (Taylor, 1994).

Unlike the systemic density increases of some aquatic vertebrates, such as crocodilians and sea otters (Wall, 1983), sirenians selectively increase the amount of bone in the thoracic cavity and forelimb in order to reposition the center of gravity anteriorly, close to the center of buoyancy (Buffrénil *et al.*, 1990; Domning and Buffrénil, 1991). Combined with the loss of their hind limbs, this gives sirenians a more or less horizontal orientation in the water, which is a hydrodynamic advantage (Buffrénil *et al.*, 1990; Domning and Buffrénil, 1991). Localized increased bone density appears to be a more derived adaptation for buoyancy control than systemic pachyostosis or the use of gastroliths, as Wall (1983) noted that skeletal density changes were typically systemic.

Horizontal posture may not be an issue for less committed swimmers where modifying neutral depth may be the primary concern (Wall, 1983). Though some late Eocene archaeocetes went through a period of increased bone density (Kellogg, 1936; Felts and Spurrell, 1965, 1966; Buffrénil *et al.*, 1990; Thewissen *et al.*, 1996; Hulbert, this volume), modern cetaceans have secondarily reduced bone density (Felts and Spurrell, 1966).

Hydrodynamic strategies used to control buoyancy in secondarily aquatic vertebrates are also described by Taylor (1994). Deep-diving animals such as cetaceans and pinnipeds actively maintain a desired depth by swimming. They also exhale before a dive, thus increasing the depth at which they are neutrally buoyant (Ridgway and Howard, 1979). The combination of continuous swimming and exhalation is an energetically efficient mechanism for predatory cetaceans and pinnipeds needing to feed at varied depths, as heavy skeletons increase drag and the subsequent costs of locomotion, and modifying lung volume alone would be insufficient to maintain the various depths required.

1.3. Predicted Archaeocete Bone Morphology Based on Behavioral Reconstructions

This chapter examines the structural properties of the appendicular remains from Eocene archaeocetes *Ambulocetus, Rodhocetus, Remingtonocetus, Dalanistes, Basilosaurus,*

Dorudon, and *Ancalecetus*. Using diaphyseal breaks and radiographs I will compare these taxa with a series of living terrestrial, semiaquatic, and aquatic mammals in order to determine whether the architecture of their long bones is consistent with published locomotor reconstructions.

Ambulocetus natans from the Upper Kuldana Formation of Pakistan is the only early Eocene archaeocete for which relatively complete appendicular remains are known (Thewissen *et al.*, 1994, 1996). *Ambulocetus* inhabited a brackish marine environment, and was semiaquatic. Given its dorsally positioned orbits facilitating vision above the water surface, and adaptations for hind-limb-propelled swimming somewhat analogous to modern lutrines, Thewissen *et al.* (1996) proposed that *Ambulocetus* fed like crocodiles, laying in wait beneath the surface to ambush prey at water's edge. This suggests that *Ambulocetus* should have the same increased density observed in semiaquatic mammals in order to prevent floating. Indeed, evidence of pachyostosis has already been noted in the ribs of *Ambulocetus* (Thewissen *et al.*, 1996). The lack of appendicular reduction that implied terrestrial competence suggests that the internal architecture of the long bones of *Ambulocetus* should display features associated with weight bearing.

Remingtonocetids are late early and middle Eocene marine archaeocetes known from the Domanda (Sulaiman Range) and Kohat (Kala Chitta Hills) Formations of Pakistan and the Harudi Formation of Kachchh, India (summarized by Williams, this volume). The unreduced femora and the fused sacral series referred to *Remingtonocetus harudiensis* and *Dalanistes ahmedi* suggest that these taxa, like *Ambulocetus*, were amphibious, using their hind limbs for both terrestrial and aquatic locomotion (Gingerich *et al.*, 1995). Hind limbs of these taxa may exhibit diaphyseal morphologies similar to *Ambulocetus*, combining features of weight bearing for terrestrial use with increased overall bone density as a mechanism of buoyancy control. A heavy skeleton would be more likely in these taxa than in the contemporary protocetids, as remingtonocetid axial skeletons lacked the mobility between functional series of vertebrae that characterized later archaeocetes and modern cetaceans (Buchholtz, this volume).

The protocetid *Rodhocetus kasrani* from the middle Eocene Domanda Formation of Pakistan inhabited a shallow marine environment (Gingerich *et al.*, 1994). *Rodhocetus* has a reduced hind limb, a shortened cervical vertebral series, and a sacrum composed of unfused vertebral elements, leading Gingerich *et al.* (1994, 1995) to conclude that it was a pursuit predator that swam via dorsoventral undulation of the spine. However, the sacroiliac synchondroses persisted, the first sacral vertebra remained distinct from lumbar elements, and the acetabulum was deep, suggesting that terrestrial competency persisted in *Rodhocetus* (Gingerich *et al.*, 1994). Its degree of terrestriality may have been similar to otariids and lutrines and their trabecular architecture should be similar as well. *Rodhocetus* may display one of two different strategies for buoyancy control, as most pinnipeds have increased bone density whereas some phocids (and modern cetaceans) have light skeletons, reflecting the use of hydrodynamic buoyancy control (Wall, 1983). The first hypothesis is consistent with the morphology of the ribs of North American protocetid *Georgiacetus* that combine pachyostosis and osteosclerosis (Hulbert, this volume).

I also examine the femora of the late Eocene archaeocete *Basilosaurus isis*, in which pachyostosis of the ribs similar to that of sirenians has been documented (Buffrénil *et al.*, 1990). Sirenian pachyostosis is restricted to anterior skeletal elements, leveling the horizontal trim of the animal by adding mass to the air-filled thoracic region. It is linked to hypothyroidism and a selective slowing of ontogenetic bone growth (Domning and Buffrénil, 1991). However, although pachyostosis associated with hypothyroidism would be consistent with a shallow marine adaptation, it is not consistent with predation, which requires a higher metabolic potential (Buffrénil *et al.*, 1990; Gingerich *et al.*, 1990). Although Buffrénil *et al.* (1990) note that *Basilosaurus* had a dentition designed for prey capture, its lengthened postthoracic vertebral column is indicative of an anguilliform swimming style, and therefore not with pursuit predation (Kellogg, 1936; Gingerich, this volume; Buchholtz, this volume). Thus, it is reasonable to expect that increased density in *Basilosaurus* was systemic, as its morphology and paleoecology suggest that speed and deep diving were not a significant component of its aquatic lifestyle.

Unlike basilosaurids, dorudontids *Zygorhiza kochii, Dorudon atrox*, and *Ancalecetus simonsi* have vertebral morphologies strikingly similar in form (and probably function) to modern cetaceans (Kellogg, 1936; Buffrénil *et al.*, 1990; Uhen, 1996; Buchholtz, this volume; Hulbert, this volume). This modern axial morphology is consistent with their proposed deep diving and predatory adaptation, distinguishing them from *Basilosaurus* (Gingerich and Uhen, 1996; Uhen, 1996, and this volume). Gingerich and Uhen (1996) noted that the ribs of *Ancalecetus simonsi* and *Dorudon* are not robust or pachyostotic, and that the thorax was not used as a buoyancy control mechanism. Therefore, neither is expected to have dense appendages. The histological manifestation of pachyostosis in the ribs of *Zygorhiza* is distinctly different from that of *Basilosaurus*; both have distally expanded ribs, but those of *Zygorhiza* have much less cortical thickness, and are instead filled with spongy bone (Domning and Buffrénil, 1991). This implies a secondary reduction in bone density in this taxon; therefore, like the other dorudontids, *Zygorhiza* should exhibit overall density reduction.

2. Materials and Methods

Specimen numbers from extant and fossil taxa used in this study are listed in Table I. Institutional abbreviations are: AMNH, American Museum of Natural History; CMNH, Cleveland Museum of Natural History; CGM, Cairo Geological Museum; GSP-UM, Geological Survey of Pakistan–University of Michigan; H-GSP, Howard University–Geological Survey of Pakistan; MNHN, Muséum National d'Historie Naturelle; USNM, National Museum of Natural History, Smithsonian Institution; UM, University of Michigan Museum of Paleontology; USGS, United States Geological Survey now housed at USNM; YPM-PU, Princeton University now housed at Yale Peabody Museum.

Measurements and observations were taken from femora, radii, and central metapodials, in order to include as many fossil taxa as possible in the analysis. Long bones from fossil and extant specimens were radiographed in anteroposterior and mediolateral views on a Hewlett Packard Cabinet Faxitron (Model A3855A) at the Cleveland Museum of Natural History and a Picker International 805D X-Ray unit in the Division of Fishes, Smithsonian Institution, using Kodak X-Omat TL double emulsion redipack film. Radiographs of different elements were taken at variable kilovoltages and exposure periods depending on size and mineral content of the bone. Total anteroposterior (AP) and mediolateral (ML) external diameter and cortical thickness along anterior, posterior, medial, and lateral axes were measured at midshaft from radiographs using dial calipers. In each case, the distribution of

Table I. Fossil and Extant Specimens Included in Analysis

Mesonychians
  Dissacus praenuntius (UM 75501; MNHN BR 12553)
  Mesonyx obtusidens (USNM 476342)
  Pachyaena ossifraga (USGS 25292; YPM-PU 14708; UM 94783; UM 95074; AMNH 16154)
  Pachyaena gracilis (USGS 25280)
  Pachyaena gigantea (USNM 14915)
Archaeocetes
  Ambulocetus natans (H-GSP 18507)
  Rodhocetus kasrani (GSP-UM 3012)
  Dalanistes ahmedi (GSP-UM 3106, GSP-UM 3115)
  Remingtonocetus cf. *R. harudiensis* (GSP-UM 3054)
  Ancalecetus simonsi (CGM 42290)
  Dorudon atrox (UM 100222)
  Basilosaurus isis (UM 93231, UM 97527)
Extant taxa
  Otariid pinnipeds
    Zalophus californianus (California sea lion: CMNH 17804; USNM 14410, 21736, 21735, 504991, 200847)
    Eumetopias jubatus (Steller sea lion: USNM 500840, 21537, 21523, 7140)
    Arctocephalus forsteri (South American fur seal: USNM 550479)
    Arctocephalus pusillus (South American fur seal: USNM 484928)
  Phocid pinnipeds
    Halichoerus grypus (gray seal: USNM 446405, 504481, 446409, 446408, 446406)
    Phoca vitulina (harbor seal: USNM 283568, 219344, 15276, 504298, 55060)
    Mirounga angustirostris (elephant seal: USNM 260876, 15270)
    Leptonychotes weddelli (Weddell's seal: USNM 504875, 550118)
  Fissiped carnivores
    Lontra canadensis—formerly *Lutra canadensis* (river otter: USNM 574356, 49902, 256976, 21232,
      267303; CMNH 387)
    Enhydra lutris (sea otter: USNM 11794, 20966, 21336, 49492, 13304)
    Pteronura brasiliensis (Brazilian river otter: USNM 304663, 319226)
    Ursus americanus (American black bear: CMNH 522)
    Ursus arctos (Alaskan brown bear: USNM 265076)
    Ursus maritimus (polar bear: USNM 2751241)
    Panthera leo (leopard: USNM 162919)
    Gulo gulo (wolverine: USNM 21493)
  Perissodactyla
    Tapirus terrestris (Brazilian tapir: USNM 292150, 261025)
  Artiodactyla
    Muntiacus muntjak (barking deer: CMNH 1697)
    Sus scrofa (European wild hog: USNM 15184)
    Hippopotamus amphibius (hippopotamus: USNM 162977)
    Hexaprotodon liberiensis (pygmy hippopotamus: USNM 549277)
  Sirenia
    Trichechus manatus (manatee: USNM 527909)
  Cetacea
    Inia geoffrensis (Amazon dolphin: USNM 395415)
    Phocoena phocoena (harbor porpoise: USNM 571762)
    Pontoporia blainvillei (La Plata dolphin: USNM 504920)
    Tursiops truncatus (bottle-nosed dolphin: USNM 504618)
    Globicephala macrorhynchus (pilot whale: USNM 550797)
    Steno bredanensis (rough toothed dolphin: USNM 550837)
    Orcinus orca (killer whale: USNM 571360)
    Delphinapterus leucas (beluga whale: USNM 571021)
    Mesoplodon stejnegeri (beaked whale: USNM 504731)
    Balaenoptera edeni (Bryde's whale: USNM 239307)

cancellous and cortical bone was noted, as well as the presence/absence of an open medullary cavity. Wherever possible, measurements of cortical thickness from fossils were taken from both natural breaks and radiographs in order to assess the accuracy of the radiographic method. All other measurements were obtained directly from individual elements and are identified in Table II. Craniocaudal length of the second (or third) thoracic vertebral centrum is used as a body-size estimator in this study, as it was available for most of the fossil taxa examined, and the thorax does not appear to have been significantly modified in length during early archaeocete evolution (Gingerich *et al.*, 1994; Gingerich, this volume).

I followed the method of Currey and Alexander (1985) for determining the total percentage of a given bone cross section that is devoted to cortical bone versus spongy bone or an open medullary cavity. However, radiographs are used instead of sections to determine cortical thickness. The radius R of each diaphysis was calculated by halving the midshaft diameter in the AP plane. The ratio R/t is obtained by dividing the radius by the average cortical thickness (t); t is the simple average of right and left cortical thicknesses. This ratio describes the relative wall thickness of a shaft. In addition, the variable K was calculated to reflect the percentage of the cross section encompassing the medullary cavity in the AP plane; K was calculated by forming a ratio of internal to external diameter. Although K

Table II.   Measurements Taken from Appendicular Elements

Femur
  Length—total articular length of bone from head to femoral condyles
  Head diameter—maximum diameter of femoral head
  Midshaft ML diameter—Mediolateral diameter of femoral midshaft
  Midshaft AP diameter—Anteroposterior diameter of femoral midshaft
  Distal ML—maximum breadth across femoral condyles
  Distal AP—maximum anteroposterior breadth, patellar groove to condyle
Radius
  Length—total articular length of bone
  Head ML diameter—maximum mediolateral diameter of radial head
  Head AP diameter—maximum anteroposterior diameter of radial head
  Midshaft ML diameter—mediolateral diameter of radial midshaft
  Midshaft AP diameter—anteroposterior diameter of radial midshaft
  Distal ML—maximum breadth across scapholunar articulation
  Distal AP—maximum anteroposterior articular breadth
Metapodials
  Length—maximum length
  Midshaft ML diameter—mediolateral diameter of midshaft
  Midshaft AP diameter—anteroposterior diameter of midshaft
All diaphyseal elements: from radiograph or from midshaft break
  Medial cortical thickness (t)—width of compact bone from medial aspect midshaft
  Lateral cortical thickness (t)—width of compact bone from lateral aspect midshaft
  Anterior cortical thickness (t)—width of compact bone from anterior aspect midshaft
  Posterior cortical thickness (t)—width of compact bone from posterior aspect midshaft
Calculated variables and metrics
  Radius (R)—one-half of the mediolateral thickness of a diaphyseal midshaft
  (R/t)—ratio of total cross-sectional size relative to cortical thickness (t)
  K—size estimator of noncortical component of a diaphysis. May be either marrow- or trabecular-filled region
    of shaft

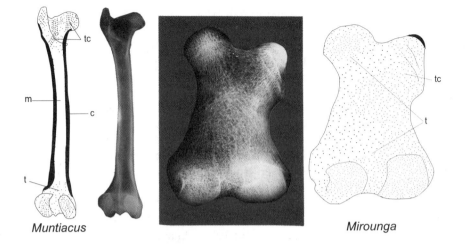

Muntiacus   *Mirounga*

FIGURE 1. Two extremes of femoral structural morphology. From left to right: Line drawing depicting the structural features of the femur of *Muntiacus muntjak*, the barking deer, with a radiograph of the femur from which the line drawing is derived. Radiograph and line drawing of the femur of phocid pinniped *Mirounga angustirostris* with same features identified. Abbreviations are as follows for structures defined in the text: (c) cortical bone, (m) open medullary cavity, (tc) trabecular column, (t) trabecular, or spongy bone. Although not all trabeculae are outlined in each figure, dominant orientation of trabecular bone is indicated where present. Density of trabecular infill (t) is determined by the concentration of points and dashed lines; cortical bone is black unless its presence is ambiguous because of poor preservation. In such cases, the cortical margin is indicated by a broken line.

is readily derived from R and t, it is included in the analysis as a simple descriptor of the medullary cavity size, though not its internal architecture.

Unlike Currey and Alexander's (1985) study that was designed to describe mechanical properties of tubular bones with air or marrow infill, some specimens included in this analysis have medullary cavities filled with trabecular bone. Although the trabecular infill undoubtedly affects the mechanical design of diaphyses, the measures obtained from their technique can still be used as a means of comparison of skeletal design and overall density. This makes it possible to examine and describe the diaphyses of sirenians and cetaceans, as well as many other semiaquatic mammals. Femoral, radial, and metapodial radiographs were analyzed to examine trabecular density and note the presence of trabecular columns, or struts of bone designed to withstand concentrated tensile and compressive forces. Cortical and trabecular distributions were traced directly from radiographs and scanned into a personal computer to produce figures representing radiographic images (Fig. 1).

3. Results

3.1. Modern Mammals

Modern terrestrial mammals appear to have a relatively tight range of values for K (amount of diaphyseal cross section comprising the medullary cavity) and R/t (ratio of midshaft radius to cortical thickness) for their long bones (Table III). Among the terrestrial mammals, *Sus, Muntiacus*, and *Panthera* have open medullary cavities in all of the diaphyses examined, whereas *Gulo* has open cavities in the pedal second, third, and fourth metapodials, and trabecular infill of the remaining pedal and manual elements. Femoral heads and necks of all fully terrestrial species have proximomedial-distolaterally oriented trabecular columns (Fig. 2). Vertically oriented columns occur in the radial head, as well as in femoral condyles of all terrestrial mammals examined. Among modern terrestrial taxa, trabecular columns are best developed in the large artiodactyl and carnivoran cursors such as *Panthera* and *Sus*.

The diaphyseal values of K and R/t for the diaphyses of *Hippopotamus* and the pygmy hippo *Hexaprotodon* fall in the range of those of terrestrial ungulates (Table III). Yet the overall bone densities of the hippos are undoubtedly greater than those of terrestrial forms, as the medullary cavities of their femoral, radial, and metapodial diaphyses are filled with dense, randomly oriented trabeculae (USNM 549277; figured in Wall, 1983). The trabecular columns in the femora and radii of hippos are more poorly defined in articular regions relative to similar-sized terrestrial ungulates. This may be related to their less graviportal adaptations. Metapodials in both taxa were filled with dense trabeculae.

Tapirs are also habitual swimmers (Nowak, 1991), and like hippos, they have trabecular-filled diaphyses. Tapir cortical thickness is similar to terrestrial mammals, but the added diaphyseal bone increases overall skeletal density. Structurally, they are most similar to *Muntiacus* and *Sus*, with vertical trabecular columns in the femoral head and condyles, and abundant proximolaterally and distomedially oriented trabeculae within the greater trochanter in the region of gluteal muscular insertion (Fig. 2). However, unlike *Sus* and *Muntiacus*, tapir metapodials are filled with dense trabeculae, like the hippos.

The femoral midshaft cortical thickness of *Ursus maritimus* matches that of the more terrestrial bears, *U. americanus* and *U. arctos* (Table III). The cortex of the radius and metapodials is thinner in the polar bear than in the other bears (Table III). All three species have similar trabecular distributions, with open medullary cavities in all diaphyses, and obvious trabecular columns in the heads and necks of femora and radii.

Lutrines have relatively similar values for R/t and K (Table III). The ratios depicting diaphyseal cross-sectional form were obtained from midshaft level, but some lutrines are distinct from the other aquatic animals examined thus far in the distribution of cortical bone in their diaphyses (Fig. 3). The highly aquatic species *Pteronura* and *Enhydra* resemble some pinnipeds in having much thicker femoral cortices in the proximal third of the femoral shaft than *Lontra* (Fig. 3). The cortex of *Pteronura* and *Enhydra* is thick, but does not extend deeply into the femoral neck and condyles to the same degree as in *Lontra* and the terrestrial taxa examined (see *Pachyaena* and *Muntiacus* in Fig. 2). In addition, the diaphyses of *Enhydra* can be distinguished from those of *Lontra* (Fig. 3) in that the femoral, radial, and metatarsal shafts of *Enhydra* are all filled with diffuse trabeculae, similar to the pinnipeds and archaeocetes described below. The second and third metacarpals have open diaphyses, and the distal two-thirds of the shift of MC I, IV, and V are filled with trabecular bone. *Pteronura* is similar to *Lontra* in having at least some portion of diaphyses free of trabeculae (open medullary cavity). Thicker cortices and greater distribution of trabeculae in *Enhydra* imply that its overall skeletal density is greater than in *Lontra* and *Pteronura*. This interpretation is corroborated by direct density measures taken of mustelid long bones by Fish and Stein (1991). All three lutrines possess trabecular columns extending from femoral head to neck, as well as in their proximal and distal radii, which is expected for load-bear-

Table III. Ratios of Diaphyseal Breadth versus Cortical Thickness (R/t)[a], and Percentage of Noncortical Contribution to Midshaft Sections (K)[a]

Specimen	Number	Femur R/t	Femur K	Radius R/t	Radius K	MC III R/t	MC III K	MT III R/t	MT III K	MT II R/t	MT II K
Archaeocetes											
Basilosaurus isis	UM 93231	1.69	0.41								
	UM 97537	1.08	0.25								
Rodhocetus	GSP-UM 3012	4.60									
Remingtonocetus	GSP-UM 3054	9.26	0.84								
Ambulocetus	H-GSP 18507	2.93	0.71	3.55	0.66	3.14	0.68	3.59	0.72		
Mesonychians											
Mesonyx	USNM 476342	3.15	0.79								
Dissacus	BR 12553			1.87	0.45						
	UM 75501			1.68	0.40						
Pachyaena gigantea	USNM 14915			2.20	0.55	2.39	0.57				
P. ossifraga	AMNH 4262			2.05	0.51						
	YPM-PU 14708	2.41	0.53								
	UM 97483	4.14	0.75								
P. gracilis	USGS 25292					2.50	0.61	1.61	0.38	2.02	0.51
	USGS 25280	2.28	0.56	1.54	0.35	1.95	0.49				
Modern aquatic taxa											
Zalophus	N = 2	2.52	0.62			3.27	0.69	2.74	0.63	3.71	0.73
Arctocephalus	N = 2	1.50	0.71	1.57	0.34	2.35	0.57	3.23	0.69		
Eumetopias	N = 2	1.62	0.77	8.01	1	—[c]	1	3.58	0.72		
Leptonychotes	N = 2	6.89	0.96	8.50	1	3.28	0.69	2.50	0.56		
Mirounga	N = 2	—[c]	0.98	—[c]	1	10.75	0.91				
Phoca	N = 2	1.85	0.46	6.55	0.84	3.52	0.70	3	0.67	1.54	0.35

Taxon	N								
Semiaquatic taxa									
Lontra	N = 4	2.17	0.60	1.50	0.31	1.45	0.30	2.32	0.48
Enhydra	N = 2	3.15	0.69	1.65	0.39	1.48	0.31		
Pteronura	N = 1	3.05	0.73	1.50	0.33				
Ursus maritimus	N = 1	2.77	0.61	3.89	0.74	2.29	0.56	2.83	0.65
Hippopotamus	N = 1	2.20	0.55						
Hexaprotodon	N = 1	3.01	0.65	1.88	0.47	2.63	0.62	2.05	0.51
Semiaquatic mammal average		2.73	0.64	2.08	0.45	1.96	0.45	2.40	0.55
Terrestrial taxa									
Tapirus	N = 2	3.24	0.67	1.87	0.46			4.83	0.79
Ursus arctos	N = 1	2.19	0.38	1.85	0.46	1.19	0.16	1.17	0.14
U. americanus	N = 1	3.11	0.59	2.29	0.56	1.20	0.16		
Muntiacus	N = 1	2.94	0.69	1.71	0.41				
Panthera	N = 1	2.29	0.45	1.41	0.29	1.71	0.41	1.89	0.47
Sus	N = 1	3.45	0.71	2.47	0.60	2.11	0.53	1.94	0.48
Gulo	N = 1	2.46	0.59	1.70	0.41			1.97	0.49
Terrestrial mammal average[b]		2.81	0.58	1.90	0.46	1.55	0.32	2.36	0.47
Alligator[b]			0.22						

[a] Average of measures taken from mediolateral and anteroposterior radiographs and sections, where possible.
[b] Data of Currey and Alexander (1985).
[c] The midshaft of this element was composed entirely of trabecular bone.

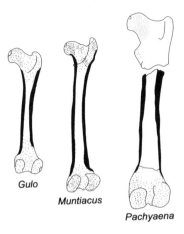

FIGURE 2. Femoral structure of terrestrial mammals: the wolverine *Gulo gulo*, barking deer *Muntiacus muntjak*, and mesonychid *Pachyaena ossifraga* (USGS 25280). Radiographs were reflected when necessary to provide the same orientation in each bone. See Wall (1983) for photographs of sectioned femoral diaphyses of large terrestrial (*Bison*) and semiaquatic artiodactyls (*Hexaprotodon*).

ing elements (Fig. 3). Nonetheless, each of the lutrines has smaller femoral head diameters and femoral lengths than the more terrestrial carnivores examined here (Table IV).

Radiographic analysis of otariid and phocid long bones confirms the results previously obtained in density studies of Wall (1983), and provides a structural framework for the quantitative differences that distinguish them from terrestrial mammals. All of the pinnipeds examined have diaphyses filled with trabeculae (Fig. 3). However, otariids differ from most phocids on the basis of their systemically thicker cortical bone (Table III), and their more compact trabeculae. Otariid femoral cortical bone is better developed along the lateral femoral margin, and is substantially thicker near midshaft than near epiphyses, unlike the relatively uniform cortical thickness of terrestrial mammals (Figs. 2 and 3). Otariids have

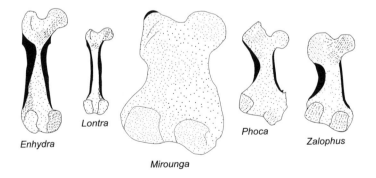

FIGURE 3. Femoral structure of modern aquatic and semiaquatic taxa: mustelid carnivores *Enhydra lutris* and *Lontra canadensis*, phocid pinnipeds *Mirounga angustirostris* and *Phoca vitulina*, and otariid pinniped *Zalophus californianus*.

relatively thick cortex along the leading edge of their hydrofoil, the medial margin of the radius. The ulnar margin of the radius lacks cortex distally. Proximodistally oriented trabecular columns are prominent in the otariid femoral and radial heads, extending into the proximal third of the shafts (Fig. 3).

Among the phocids included in this analysis, only *Phoca* (Fig. 3) is similar to otariids in having substantive cortex in its long bones. *Leptonychotes* exhibits thick cortex only in the proximal third of its femoral and radial shafts, and the central third of its metapodials. In the remainder of its diaphyses, the cortex is a thin shell. *Mirounga* lacks appreciable cortex in all bones examined (Figs. 1 and 3). Structurally, the non-weight-bearing phocid radii and femora lack regions of highly concentrated, patterned trabeculae and trabecular columns that are clearly defined in otariid taxa.

3.2. Mesonychians

Although many of the mesonychian diaphyses examined were too densely mineralized to permit clear radiography, several specimens revealed a typical terrestrial structural morphology. Mesonychian femora (Fig. 2) have clearly defined medullary cavities completely lacking trabeculae, moderately thick cortices extending proximally and distally into long bone epiphyses (Table III), and well-defined trabecular columns extending obliquely from the femoral head into the neck. The radii of mesonychians *Dissacus, Mesonyx,* and *Pachyaena* exhibit open medullary canals; all have vertically oriented trabecular columns in the radial head. *Pachyaena ossifraga* (USGS 25292) and *P. gracilis* (USGS 25280) have open medullary cavities in lateral metatarsals, and diffuse spongiosa filling the core of central metapodials. *P. gigantea* (USNM 14915) has trabecular infill of all of its manual metapodials. Both radial and femoral lengths relative to body size for mesonychians are closest to *Tapirus* and terrestrial carnivores (Table IV).

3.3. Archaeocetes

3.3.1. *Ambulocetus*

Radiographs of *Ambulocetus* (H-GSP 18507) femora are relatively radio-opaque; but they do indicate the presence of well-developed cortex and a dense distribution of cancellous bone near the epiphyses, consistent with weight bearing. The dense trabeculae of the femoral head are nearly vertical, with a column extending distolaterally into the neck (Fig. 4).

Diaphyseal breaks and radiographs show that *Ambulocetus* femora are amedullary. Instead, the cortex ends in transition to first a tightly packed, then a highly porous trabecula toward the central diaphyseal axis. The femoral cortical indices are at the high end of the range of the extant terrestrial mammals and the mesonychians, indicating that *Ambulocetus* had thinner cortical bone at midshaft (Table III). Although the trabecular-filled medullary cavities of *Ambulocetus* preclude direct comparison of diaphyseal mechanical properties with terrestrial taxa using Currey and Alexander's (1985) cortical indices, the infill of the *Ambulocetus* femora does indicate that they were denser than those of terrestrial mammals of similar size. The *Ambulocetus* radii have high R/t ratios (Table III), suggesting that ra-

Table IV. Femoral and Radial Length and Size Indices

Specimen	Number	Centrum length/ femoral head diameter	Femoral length/ centrum length	Anterior thoracic centrum length (T2 or T3)	Femur length	Femoral head diameter		Radial length	Radial length/ centrum length
						Pro.-Dis.	Ant.-Pos.		
Archaeocetes									
Basilosaurus	GSP-UM 93231				243.9				
Rodhocetus	GSP-UM 3012	1.31	4.38	40.6	178	31.1	30.8		
Dalanistes	GSP-UM 3106					41			
Remingtonocetus	GSP-UM 3015				250				
	GSP-UM 3054	1.00	6.72	35.7	240	35.8	32.3		
Ambulocetus	H-GSP 18507	1.13	6.59	44	290	39		180	4.09
Mesonychians									
Mesonyx	USNM 476342	1.18	9.47	24	227.3	20.4	23	156	6.5
P. gigantea	USNM 14915[b]			37.2				248	6.67
P. ossifraga	AMNH 4262[c]				319			201	
	YPM-PU 14708[c]								
	UM 97483[b,c]		11.08	26.5	293.6	28.5	30.2	196.2	7.40
	USGS 25292				293.6	33.6	33.6		
	AMNH 15729[b]				241			152	
P. gracilis	USGS 25280					22.1	25.8		
Pinnipeds									
Zalophus	USNM 14410	1.28	4.02	33	132.8	25.7		236.1	7.15
Phoca	USNM 219985, 283568	1.11	4.23	22.5	93.0	19.9		133	5.23

Eumetopias	USNM 21537	1.52	3.21	52	167.2	34.1		225.2	4.33
Halichoerus	USNM 446406	1.31	3.62	34	123	26		135	3.97
Mirounga	USNM 260876, 15270	1.31	3.0	51	152.5		39.6	280.1	5.39
Leptonychotes	USNM 504875	1.43	3.22	39.5	127		27.6	181	4.56
Arctocephalus	USNM 550479	1.55	3.27	30	98		19.3	161.9	5.40
Lutrines									
Lontra	USNM 81798, 49902, 256976	1.12	5.75	15	85.4	12.9		62.5	4.09
Enhydra	USNM 20966, 21336, 11794	0.99	5.64	17.2	118	20.4		89.6	4.31
Pteronura	USNM 304663	1.22	4.75	20.8	98.9	17		77.7	3.74
Ungulates and carnivores									
Hexaprotodon	USNM 549277	1.02	7.18	40.8	293		40	163	4.00
Tapirus	USNM 292150	0.78	9.98	33.2	331.5		42.6	220.5	6.64
	USNM 261025	0.81	9.55	33.6	321		41.5	208.4	6.20
Sus	USNM 15184	1.19	6.18	35.7	220.9		29.9	153.2	4.29
Ursus maritimus	USNM 2751241	0.90	9.15	35.1	321		39.2	300	8.55
U. arctos	USNM 265076	0.89	9.43	33.1	312		37.1	311	9.40
U. americanus	CMNH B687				315		36.1	265	
Gulo	USNM 21493	0.85	5.59	15.3	85.5		17.9	121.9	7.97
Panthera	USNM 162919	0.72	13.07	29	379		40.25	318	10.97

[a]From Thewissen and Hussain (1990). Femoral length is average of first three specimen numbers listed.
[b]From O'Leary and Rose (1995).
[c]From Zhou et al. (1992).

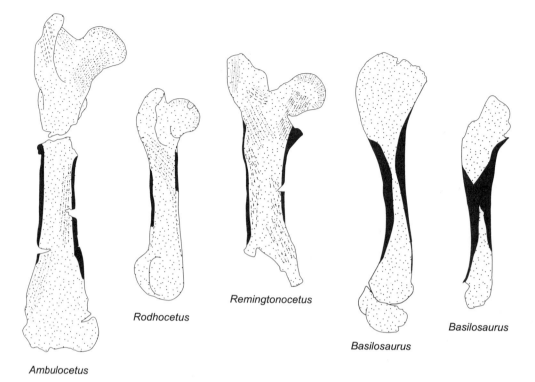

Ambulocetus

FIGURE 4. Femoral cortical and trabecular architecture of *Ambulocetus natans* (H-GSP 18507), *Rodhocetus kasrani (GSP-UM 3012), Remingtonocetus* (cf. *R. harudiensis*, GSP-UM 3054), and adult (UM 93231) and juvenile (UM 97527) *Basilosaurus* femora.

dial cortical thickness is also reduced relative to modern terrestrial mammals and mesonychians. The radiograph of the complete radius indicates that it, like the femur, had a diffuse trabecular core.

The central metapodials of *Ambulocetus* have relatively thin cortices along the distal two-thirds of the diaphyses, with tightly packed, longitudinal trabeculae filling most shafts. The fifth metatarsal was broken at midshaft, and preserves a hollow marrow cavity bordered first by loose trabeculae encapsulated by relatively thick cortex. The manual and pedal metapodials of *Ambulocetus* show reduced cortical thickness relative to terrestrial mammals, (Table III), but overall density is high as compact trabeculae are arranged longitudinally along some part of each of the shafts examined.

3.3.2. Remingtonocetids

The K and R/t values for the femur of *Remingtonocetus* (GSP-UM 3054) indicate that the cortex of this specimen is extremely thin near midshaft (Table III). The cortex thickens proximally along the medial margin of the shaft, but remains somewhat thin laterally (Fig. 4). The trabeculae in the femoral head and neck of the *Remingtonocetus* femur (GSP-UM

3054) are dense and oriented obliquely along the medial margin of the neck. This is indicative of vertical loading of the femoral head (Fig. 4). The distal femora attributed to *Dalanistes* (GSP-UM 3106 and GSP-UM 3115) were too densely mineralized to isolate trabecular columns in the condyles (which are poorly preserved), but they are clearly filled with uniform, tightly packed spongy bone.

3.3.3. *Rodhocetus*

Although there has been some postmortem crushing, the femoral diaphysis of the protocetid *Rodhocetus* (GSP-UM 3012) was filled with compact but thin spongiosa that is clearly distinguishable from the endocortical margin (Fig. 4). Its femoral R/t value indicates that the cortical bone was relatively thin at midshaft, but unlike *Remingtonocetus*, the cortex appears to remain thin as it extends proximally and distally along the length of the shaft. Although radiographs of the highly mineralized *Rodhocetus* femur are cloudy, the trabeculae of the femoral head and medial margin of the neck appear to be longitudinally oriented but do not form solid columns as in otariids, *Ambulocetus*, and terrestrial mammals. It is unclear whether the femoral condyles had clearly demarcated trabecular columns.

3.3.4. *Basilosaurus*

The femora of *Basilosaurus* (UM 93231 and UM 97527) are similar to those of all of the examined archaeocetes in lacking open medullary cavities, but are unique in having extremely thick diaphyseal cortices (Table III, Fig. 4). The preserved portions of the epiphyses of the juvenile specimen (UM 97527) are filled with extremely diffuse trabeculae, whereas the proximal third and distal quarter of an adult (UM 93213) bear randomly oriented trabeculae that are less compact but thicker than those of *Rodhocetus* and *Remingtonocetus*. Femoral heads are missing on both *Basilosaurus* femora examined here, but the cancellous bone extending into the region of the neck is randomly distributed, lacking the patterned orientation found in trabecular columns.

3.3.5. Dorudontids

Recently published radiographs of the radii of *Dorudon* and *Ancalecetus* (Gingerich and Uhen, 1996) show that these dorundontids have relatively thin layers of compact cortical bone lining the proximal margins of their diaphyses. The remainder of each radial diaphysis is composed of extremely diffuse trabeculae encapsulated by thin subperiosteal cortex. There are no well-defined trabecular columns in the proximal third of either radius.

4. Discussion

Whereas modern terrestrial mammals have a relatively tight range of values for K (medullary cavity size) and R/t (cortical thickness) for long bones, archaeocetes and extant semiaquatic taxa are highly variable in diaphyseal morphology. Terrestrial species have a diaphyseal structure that optimizes fatigue strength for withstanding repeated loading (optimal $K = 0.55$; $R/t = 3.0$), rather than ultimate strength against impact loading (optimal K

= 0.75; R/t = 2.3, Currey and Alexander, 1985). Although we cannot evaluate the mechanics of diaphyses with filled medullary cavities in the same fashion, values of K for the 40- to 50-m.y.-old archaeocetes clearly parallel the range seen in semiaquatic lineages of more recent origin, such as lutrines (Carroll, 1988) and pinnipeds (Berta et al., 1989).

Aquatic organisms need to balance the demands of maintaining neutral buoyancy at a depth necessary for finding food (tending to require increased skeletal mass), and the need for locomotor efficiency (tending to require decreased skeletal mass as body size increases to limit drag). The limb elements for early cetaceans all appear to have osteological modifications commensurate with highly aquatic adaptations, and in most cases support the predictions that were made for each taxon based on gross morphology and ecological reconstructions.

4.1. Mesonychians

The mesonychian diaphyseal structure falls almost entirely within the range of variation exhibited by the terrestrial mammal sample examined here. They have uniformly thick cortical bone, well-developed trabecular columns in femora and radii, and open medullary cavities in all but the central metapodials. Cross-sectional morphology, combined with diaphyseal proportional indices (Table IV), indicates that the mesonychians are most similar to ungulates. Although the long bone indices of mesonychians presented here are most similar to *Tapirus* (previously noted by O'Leary and Rose, 1995), their open diaphyses (with the exception of *P. gigantea* metacarpals) suggest that none of the taxa examined here spent a substantial amount of time in the water.

4.2. *Ambulocetus*

The presence of a sizable trabecular column in the femoral neck and a uniformly thick cortex in all long bones of *Ambulocetus* reinforces the idea that terrestriality was a significant part of its locomotor repertoire (Thewissen et al., 1994, 1996). Trabecular columns are similar to those of all terrestrial species examined (Fig. 2) and to the otariid pinnipeds (Fig. 3) which use their hind limbs for terrestrial locomotion. The columns indicate weight transmission between the femur and acetabulum, and are noticeably absent in phocid pinnipeds, which use spinal undulation rather than quadrupedal gaits during terrestrial locomotion (Backhouse, 1961).

The femora and radii of *Ambulocetus* parallel several semiaquatic taxa in lacking a medullary cavity. Sea otters, hippos, the tapir, all of the pinnipeds, and modern cetaceans exhibit trabecular infill of their diaphyses (Felts and Spurrell, 1965, 1966). The diaphyseal structure of *Ambulocetus* is similar to lutrines, *Enhydra* in particular, suggesting that its increased bone density is a hydrostatic response allowing it to overcome neutral buoyancy at the water surface. In addition, the osteological response that produced the density increase was systemic, as both ribs and appendicular bones exhibit pachyostosis and/or trabecular infill. Like *Enhydra*, the systemic higher bone density in *Ambulocetus* provided ballast allowing the animal to remain submerged (Fish and Stein, 1991). Unlike *Enhydra*, with its ballast aiding in diving and enlarged lungs keeping it afloat at the surface (Kooyman, 1973,

1989; Taylor, 1994), *Ambulocetus* may have used its mass to remain just beneath the surface while observing prey above the surface (Thewissen *et al.*, 1996).

The cortical thickness values for *Ambulocetus* metacarpals are similar to those for *Zalophus, Phoca*, and *Hexaprotodon*, and the values for its metatarsals are similar to *Eumetopias*, which uses its plantigrade foot in terrestrial locomotion and paddling, as well as to *Ursus maritimus* and *Tapirus* (Table III). The *Ambulocetus* metapodials are similar to those of *Pachyaena*, which also have trabecular-filled central elements, and open lateral elements. The diaphyseal structure of the *Ambulocetus* metapodials also differs from those of pinnipeds with their more tightly packed, uniformly oriented trabeculae; thus, they are most similar to the lutrines, bears, and tapirs, all of which have more diffuse infilling.

The metatarsals and phalanges of *Ambulocetus* are substantially longer than those of mesonychians, consistent with an emphasis on use of these elements in aquatic rather than terrestrial locomotion (Thewissen *et al.*, 1994, 1996). The increased length of the manual digits of *Ambulocetus*, and their ungual distal phalanges precluded their use in prey manipulation as in lutrines, but not in weight bearing as is the case in otariids (English, 1977).

The relative length of the femur of *Ambulocetus* is comparable to that of *Enhydra* and *Lontra*, but not pinnipeds and later archaeocetes (Table IV; Howell, 1930; Gingerich *et al.*, 1994; Thewissen *et al.*, 1996). This is clearly distinct from other more terrestrial carnivores and cursorial artiodactyls with long proximal hind-limb segments. The relative size of the femoral head of *Ambulocetus* is more similar to semiaquatic carnivores (*Lontra, Pteronura*) than to most of the pinnipeds with reduced diameters, or the terrestrial mammals (*Tapirus, Ursus, Panthera, Gulo*) with large diameters for load bearing (Table IV). Radial length of *Ambulocetus* is reduced relative to the mesonychians. Overall, *Ambulocetus* is most similar in diaphyseal proportions and structural morphology to lutrines (Table IV), and it clearly exhibits shifts in appendicular proportions linked to an aquatic mode of locomotion (Taylor, 1914; Thewissen *et al.*, 1996).

4.3. Remingtonocetids

Femoral diaphyseal architecture of *Remingtonocetus* is not like phocids such as *Mirounga* and *Leptonychotes*, which have the most similar femoral midshaft R/t values, but only a thin shell of subperiosteal cortex along their diaphyses. The *Remingtonocetus* femur is in fact more structurally similar to otariids (and some phocids such as *Phoca*), which have increased overall density by regionally expanding cortical thickness and augmenting cancellous density. As in *Ambulocetus*, hydrostatic buoyancy control is most consistent with the observed diaphyseal structure, as predicted earlier in the study.

A femoral length of 240 mm for *Remingtonocetus* can be inferred from comparison of GSP-UM 3054 and *Dalanistes* distal femur GSP-UM 3115. This estimate is slightly greater than that of Gingerich *et al.* (1995) and is based on the persistent robusticity of remingtonocetid distal femora. This estimated length suggests that remingtonocetid femora are reduced in length to the same modest degree as seen in *Ambulocetus* and modern lutrines (Table IV), as the *Remingtonocetus* femoral length to vertebral centrum length index is intermediate between lutrines and terrestrial carnivores and ungulates. Its femoral head diameter is also intermediate between lutrines and terrestrial mammals (Table IV), and is relatively larger than *Ambulocetus*. The preserved portions of the remingtonocetid distal femora indicate that the condyles and patellar region were robust, similar in proportion to *Ambulocetus*.

Remingtonocetids combined unevenly distributed but thick femoral cortices with well-developed trabecular columns in the femoral head and neck, suggesting that they were capable of hind-limb-supported terrestriality to the same degree as *Ambulocetus* and the most terrestrial otariids (*Zalophus, Arctocephalus*). The signs of weight bearing in the femur of *Remingtonocetus* remain distinct. The length of its femur, its large epiphyses, and solidly fused sacral centra suggest that the hind limb also persisted as a primary hydrofoil in remingtonocetids (Table IV; Gingerich *et al.*, 1995).

4.4. *Rodhocetus*

The high femoral *R/t* of *Rodhocetus* is most similar to pinnipeds *Zalophus* and *Phoca*, but in these pinnipeds the cortex thickens substantially above the midshaft and tapers sharply distally. The compact trabeculae filling the femoral diaphysis indicate that the skeletal density of *Rodhocetus* was probably high, which suggests use of this ballast as a means of hydrostatic buoyancy control. Unevenly distributed cortical bone, weak trabecular columns in the proximal femur, and a dense trabecular infill all point toward a degree of terrestrial hind-limb use intermediate between phocids and otariids, but certainly more limited than in the contemporary *Remingtonocetus*. These results are consistent with the predictions made by Gingerich *et al.* (1994) regarding the locomotor adaptations of *Rodhocetus*. The reduction of the femoral head in *Rodhocetus* is similar to phocids (Table IV), whereas the relative femoral length is intermediate between pinnipeds and lutrines. Diaphyseal architecture, reduced femoral length, persistent sacroiliac synchondroses, and well-developed acetabulum suggest that some weight-bearing capacity was likely (Gingerich *et al.*, 1994). However, terrestriality probably made up a small portion of its locomotor repertoire. During terrestrial excursions, load bearing and propulsion by limbs were likely facilitated by the axial skeleton as in pinnipeds (English, 1977).

4.5. *Basilosaurus*

Given the well-developed but shallow acetabulum, reduction of femoral condyles and diaphyseal length, and the large size of *Basilosaurus* (Gingerich *et al.*, 1990), the extreme thickness of the cortex and trabecular density are clearly not associated with terrestrial function in this taxon. The thick cortex of the femoral midshafts of *Basilosaurus isis* appears analogous to the humeral and costal cortices of sirenians as well as the diaphyses of aquatic reptiles (Wall, 1983). Its increased cortical development surpasses that of pinnipeds, lutrines, and semiaquatic ungulates. Extreme diaphyseal cortical thickness and rib pachyostosis in sirenians are adaptations for maintaining neutral buoyancy at shallow depths, and generating horizontal trim for efficient swimming (Domning and Buffrénil, 1991). *Basilosaurus* may also have lived in a shallow marine environment and is characterized as having weaker axial musculature than modern cetaceans (Barnes and Mitchell, 1978; Gingerich *et al.*, 1990), leading Buffrénil *et al.* (1990) to conclude that the pestle-shaped ribs were the result of a localized hormonal response for buoyancy control rather than the result of a systemic physiological process.

Contrary to the conclusions drawn from the analysis of rib pachyostosis in *Basilosaurus*, increased density of hind-limb elements noted here implies that unlike sirenians, its skeletal weight was systemically increased, as in the earlier archaeocetes examined. *Basilosaurus* is thus like pinnipeds, lutrines, and other semiaquatic mammals, although the degree of structural adaptation is advanced in this archaeocete. Because unevenly distributed body mass affects the horizontal trim of an object in water, it is otherwise unclear why an animal such as *Basilosaurus*, with its enormously lengthened axial skeleton, would maintain high bone density in its vestigial hind limb. The results presented here provide further evidence that sustained swimming and deep diving were not part of the locomotor repertoire of *Basilosaurus*.

4.6. Dorudontids

The dorudontid forelimbs exhibited in Gingerich and Uhen (1996) indicate that *Ancalecetus* and *Dorudon* possess a diaphyseal structure similar to modern cetaceans (Felts and Spurrell, 1965, 1966). Modern cetacean diaphyseal morphology is characterized by thin cortex throughout the length of diaphyses, and densely distributed but structurally finer trabeculae than those of the larger basilosaurids and the earlier archaeocetes. This is consistent with the hypothesis that dorudontids inhabited deeper waters than the basilosaurids (Kellogg, 1936; Buffrénil *et al.*, 1990). The more "limited development of pachyostosis" in the ribs of the North American dorudontid *Zygorhiza* relative to *Basilosaurus* also supports the contention that dorudontids used hydrodynamic buoyancy control (continuous swimming) to maintain depth rather than ballast (Buffrénil *et al.*, 1990).

5. Conclusions

The functional implications of structural characters identified in the appendicular skeletons of archaeocetes are summarized in Fig. 5. In general, the diaphyseal architecture of mesonychians supports previous locomotor reconstructions that include cursoriality, as elements are structurally indistinguishable from those of similar sized-terrestrial ungulates and carnivores (O'Leary and Rose, 1995). The diaphyseal structure of *Ambulocetus* incorporates struts for weight bearing in both fore- and hind limbs and loss of open medullary cavities indicative of increased skeletal density. Both characters are consistent with competency in swimming and walking, as was implied by skeletal reconstruction. Given the absence of significant reduction of limb lengths and epiphyses, *Ambulocetus* likely had a quadrupedal gait similar to lutrines, and the structural similarity between *Ambulocetus* and modern taxa such as *Enhydra* suggests that the earliest archaeocetes built and remodeled bone in a similar way.

Remingtonocetid diaphyseal morphology is intermediate between that of *Ambulocetus* and *Rodhocetus*. Although remingtonocetid femora do not exhibit significant reduction in length, their architecture is strikingly similar to otariids in having well-developed trabecular columns and an uneven distribution of cortical bone. Remingtonocetid diaphyseal

Taxon	Buoyancy Control Mechanism	Diaphyseal Cross-section	Weight Bearing Appendages	Reconstruction	Modern Analogue
Dorudontids	hydrodynamic exhalation?	thin / absent cortex dense trabecular infill	NO		phocids & modern cetaceans
Basilosaurus	systemic skeletal ballast	solid cortical midshaft dense infill of epiphyses	NO		sirenians & alligator
Rodhocetus	systemic skeletal ballast	relatively thin cortex dense trabecular infill	LIMITED		otariid pinnipeds & Phoca
Remingtonocetids	systemic skeletal ballast	cortex variably thick along shaft, dense trabecular infill	YES		otariid pinnipeds
Ambulocetus	systemic skeletal ballast	thick cortex diffuse trabecular infill	YES		lutrines, especially Enhydra
Mesonychians	none	thick cortex marrow filled shaft	YES		Tapirus & Sus

FIGURE 5. Summary of archaeocete diaphyseal architecture outlined in this chapter and the subsequent loco-motor implications drawn from internal structure. Reconstructions of individual taxa are not drawn to scale, but are meant to depict locomotor morphology reviewed in the text. Diagrams of diaphyseal cross sections for each taxon indicate the distribution of cortical and trabecular bone of a generic diaphysis at midshaft, thus the "vari-ably thick" cortex of some tax reflects the unevenly distributed cortical distribution along the length of the shaft. Modern analogues are similar to archaeocete taxa with respect to diaphyseal morphology and habitat use; they do not necessarily represent behavioral analogues.

structure is congruent with previous conclusions that imply an amphibious lifestyle (Gin-gerich *et al.*, 1995). When structure is considered in addition to features related to increased sacroiliac flexibility in remingtonocetids, it is likely that terrestrial quadrupedalism con-tributed less to remingtonocetid locomotor repertoires than in ambulocetids. Given the re-duction in cortical thickness, weight bearing on land was probably supported in part by ax-ial structures.

The diaphyseal morphology of *Rodhocetus* suggests that it may have been more aquat-ic in lifestyle than the contemporary remingtonocetids. The thin cortex, weak trabecular columns, and dense trabecular infill are reminiscent of some pinnipeds, particularly pho-cids. This is consistent with the significant femoral reduction seen in *Rodhocetus*. The ex-tent to which the hind limb was used as a weight-bearing strut in terrestrial locomotion re-mains unclear.

The variable architecture of late Eocene basilosaurid and dorudontid diaphyses match-es their morphological differences. The extremely dense femora of *Basilosaurus* were unexpected given the vestigial nature of the hind limb, and the added effect that a heavy hind limb would have on body trim given an already disproportionately large distal skeletal mass. The presence of rib pachyostosis and osteosclerosis indicates that the ontogenetic processes influencing bone density in *Basilosaurus* were systemic, rather than restricted to the anterior part of the body, as originally proposed (Domning and Buffrénil, 1990).

Dorudontids, on the other hand, have structural characters that are wholly modern in their appearance, which is again consistent with their highly derived locomotor adaptations. Their diaphyseal morphology is analogous to phocid pinnipeds and modern whales, implying that hydrodynamic buoyancy control was utilized in dorudontid taxa, unlike the basilosaurids, which utilized skeletal ballast in a form most similar to sirenians and crocodiles.

Felts and Spurrell (1966) postulate that modern cetacean bone structure reflects a "new mechanical organization befitting limb function" (p. 131), in which the lack of a marrow canal is an ancestral trait related to increased skeletal density that at one time occurred in a form similar to that of sirenians. The results of this analysis support their hypothesis (1965) that early whales should exhibit heavy bones before significant modifications of articular morphology. However, the evidence here does not support the idea that the ancestor to modern cetaceans possessed pachyostosis in a sirenian form. Ambulocetids, remingtonocetids, protocetids, and dorudontids lack medullary cavities as Felts and Spurrell predicted, although extreme cortical thickness is limited to *Basilosaurus* thus far.

As in the fetuses of modern cetaceans (Felts and Spurrell, 1965), archaeocetes did not secondarily resorb trabeculae that normally fill diaphyses during limb ontogeny. The latter occurs in terrestrial mammals (Carter *et al.*, 1991). Diaphyseal structure in cetaceans is thus the result of pedomorphosis. Although the hydrodynamic consequences appear relatively clear, the mechanical and physiological processes leading to this change in adult morphology are uncertain. The adaptive argument for the formation of tubular skeletal structures in terrestrial organisms is based on weight reduction, and the consequent energy savings for powering locomotion (Currey, 1984; Currey and Alexander, 1985). Thus, the diaphyseal architecture of archaeocetes and aquatic mammals suggests that variables such as diving depth, swimming speed, and metabolic rate may be influencing diaphyseal architecture to a greater degree than transportation costs (Buffrénil *et al.*, 1990). More study is needed to examine the parallels between pinniped and cetacean diaphyseal morphology that became apparent during this analysis.

Several approaches have been used to examine the functional constraints that influenced the shift from a terrestrial to an aquatic lifestyle in mammals, reptiles, and birds (Howell, 1930; Meister, 1962; Felts and Spurrell, 1965, 1966; Stein, 1989; Webb and Buffrénil, 1990; Fish and Stein, 1991; Massare, 1994; Fish, 1996). Although one cannot directly measure the actual bone density of fossils, archaeocetes examined radiographically do provide evidence that structural modifications affecting density did occur early in their history. The individual characteristics of the osseous adaptations of archaeocetes are as variable as those exhibited by modern aquatic and semiaquatic forms, and they outline a progressively increasing dependence on aquatic habitats in these taxa. The structural properties suggest that limb elements are particularly sensitive to changes in mechanical loading, which are not wholly discernible by examination of articular morphology and proportional indices. Therefore, it appears possible to utilize such analyses to detect architectural differences in fossil material for which locomotor behavior is otherwise ambiguous.

Acknowledgments

I would like to thank Hans Thewissen for inviting me to prepare this chapter, as well as for discussions resulting in significant improvements in content and appearance. I also thank K. D. Rose for helpful comments. I am grateful to P. D. Gingerich, K. D. Rose, and J. G. M. Thewissen for loan of or access to specimens included in this analysis, and I am especially indebted to L. Gordon, C. Potter, B. Latimer, and T. Mateson for providing access to extant mammals and radiographic equipment in their care. Study of these fossils has been supported by a Gerstacker-Gund Research Fellowship from Hiram College.

References

Aleshire, D. P. 1993. Functional morphology and locomotion in an early Eocene protocetid from Georgia. *J. Vertebr. Paleontol.* **13**:24A.

Backhouse, K. M. 1961. Locomotion of seals with particular reference to the forelimb. *Symp. Zool. Soc. London* **5**:59–75.

Barnes, L. G., and Mitchell, E. 1978. Cetacea, in: V. J. Maglio and H. B. S. Cooke (eds.), *Evolution of African Mammals*, pp. 582–602. Harvard University Press, Cambridge, MA.

Berta, A., Ray, C. E., and Wyss, A. R. 1989. Skeleton of the oldest known pinniped, Enaliarctos mealsi. *Science* **244**:60–62.

Buffrénil, V., de, Ricqlès, A. de, Ray, C. E., and Domning, D. P. 1990. Bone histology of the ribs of the archaeocetes (Mammalia, Cetacea). *J. Vertebr. Paleontol.* **10**(4):455–466.

Carroll, R. L. 1988. *Vertebrate Paleontology and Evolution*. Freeman, San Francisco.

Carter, D. R., Orr, T. E., and Fyhrie, D. P. 1989. Relationships between loading history and femoral cancellous bone architecture. *J. Biomech.* **22**:231–244.

Carter, D. R., Wong, M., and Orr, T. E. 1991. Musculoskeletal ontogeny, phylogeny, and functional adaptation. *J. Biomech.* **24**:3–16.

Currey, J. D. 1984. *The Mechanical Adaptations of Bones*. Princeton University Press, Princeton, NJ.

Currey, J. D., and Alexander, R. M. 1985. The thickness of the walls of tubular bones. *J. Zool. London (A)* **206**:453–468.

Daniel, T. L., and Webb, P. W. 1987. Physical determinants of locomotion, in: P. DeJours, L. Bolis, C. R. Taylor, and E. R. Weibel (eds.), *Comparative Physiology: Life in Water and on Land*, pp. 343–369. Liviana Press, New York.

Domning, D. P., and Buffrénil, V. de 1991. Hydrostasis in the Sirenia: quantitative data and functional interpretations. *Mar. Mamm. Sci.* **7**(4):331–368.

English, A. W. 1977. Structural correlates of forelimb function in fur seals and sea lions. *J. Morphol.* **151**:325–352.

Enlow, D. H. 1964. *Principles of Bone Remodeling*. Thomas, Springfield, IL.

Felts, W. J., and Spurrell, F. A. 1965. Structural orientation and density in cetacean humeri. *Am. J. Anat.* **116**:171–204.

Felts, W. J. L., and Spurrell, F. A. 1966. Some structural and developmental characteristics of cetacean (Odontocete) radii. A study of adaptive osteogenesis. *Am. J. Anat.* **118**:103–134.

Fish, F. E. 1996. Transitions from drag-based to lift-based propulsion in mammalian swimming. *Am. Zool.* **36**:628–641.

Fish, F. E., and Stein, B. R. 1991. Functional correlates of differences in bone density among terrestrial and aquatic genera in the family Mustelidae (Mammalia). *Zoomorph.* **110**:339–345.

Gingerich, P. D., and Uhen, M. D. 1996. *Ancalecetus simonsi*, a new dorudontine archaeocete (Mammalia, Cetacea) from the early late Eocene of Wadi Hitan, Egypt. *Contrib. Mus. Paleontol. Univ. Michigan* **29**(13):359–401.

Gingerich, P. D., Smith, B. H., and Simons, E. L. 1990. Hind limbs of Eocene *Basilosaurus isis*: evidence of feet in whales. *Science* **249**:154–157.

Gingerich, P. D., Raza, S. M., Arif, M., Anwar, M., and Zhou, X. 1993. Partial skeletons of *Indocetus ramani*

(Mammalia, Cetacea) from the lower middle Eocene Domanda Shale in the Sulaiman Range of Punjab (Pakistan). *Contrib. Mus. Paleontol. Univ. Michigan* **28**(16):393–416.

Gingerich, P. D., Raza, S. M., Arif, M., Anwar, M., and Zhou, X. 1994. New whale from the Eocene of Pakistan and the origin of cetacean swimming. *Nature* **368**:844–847.

Gingerich, P. D., Arif, M., and Clyde, W. C. 1995. New archaeocetes (Mammalia, Cetacea) from the middle Eocene Domanda Formation of the Sulaiman Range, Punjab (Pakistan). *Contrib. Mus. Paleontol. Univ. Michigan* **29**(11):291–330.

Howell, A. B. 1930. *Aquatic Mammals: Their Adaptation to Life in the Water*. Thomas, Springfield, IL.

Hulbert, R. C., Jr. 1994. Phylogenetic analysis of Eocene whales ("Archaeoceti") with a diagnosis of a new North American protocetid genus. *J. Vertebr. Paleontol.* **14**:30A.

Hulbert, R. C., Jr., and Petkewich, R. M. 1991. Innominate of a middle Eocene (Lutetian) protocetid whale from Georgia. *J. Vertebr. Paleontol.* **11**:36A.

Kellogg, R. 1936. A review of the Archaeoceti. *Carnegie Inst. Washington Publ.* **482**:1–366.

Kooyman, G. L. 1973. Respiratory adaptations in marine mammals. *Am. Zool.* **13**:457–468.

Kooyman, G. L. 1989. *Diverse Divers*. Springer, Berlin.

Madar, S. I., and Thewissen, J. G. M. 1994. Vertebral morphology of *Ambulocetus*, an Eocene cetacean from the Kuldana Formation (Pakistan). *J. Vertebr. Paleontol.* **14**:35A.

Massare, J. 1994. Swimming capabilities of Mesozoic marine reptiles: a review, in: L. Maddock, Q. Bone, and J. V. M. Rayner (eds.), *Mechanics and Physiology of Animal Swimming*, pp. 133–150. Cambridge University Press, London.

Meister, W. 1962. Histological structure of the long bones of penguins. *Anat. Rec.* **143**:377–388.

Nowak, R. M. 1991. *Walker's Mammals of the World*, 5th ed. Johns Hopkins University Press, Baltimore.

O'Leary, M. A., and Rose, K. D. 1995. Postcranial skeleton of the early Eocene mesonychid *Pachyaena* (Mammalia, Mesonychia). *J. Vertebr. Paleontol.* **15**:401–430.

Prothero, D. R., Manning, E. M., and Fischer, M. 1988. The phylogeny of the ungulates, in: M. J. Benton (ed.), *The Phylogeny and Classification of the Tetrapods*, Volume 2, pp. 201–234. Clarendon Press, Oxford.

Ridgway, S. H., and Howard, R. 1979. Dolphin lung collapse and intra-muscular circulation during free diving: Evidence from nitrogen washout. *Science* **206**:1182–1183.

Slijper, E. J. 1946. Comparative biologic-anatomical investigations on the vertebral column and spinal musculature of mammals. *Verh. K. Ned. Akad. Wet. Afd. Natuurkd.* **62**:1–128.

Stein, B. R. 1989. Bone density and adaptation in semi-aquatic mammals. *J. Mammal.* **70**:467–476.

Szalay, F. S., and Gould, S. J. 1966. Asiatic Mesonychidae (Mammalia, Condylarthra). *Bull. Am. Mus. Nat. Hist.* **132**:129–173.

Taylor, M. A. 1994. Stone, bone or blubber? Buoyancy control strategies in aquatic tetrapods, in: L. Maddock, Q. Bone, and J. M. V. Rayner (eds.), *Mechanics and Physiology of Animal Swimming*, pp. 151–209. Cambridge University Press, London.

Taylor, W. P. 1914. The problem of aquatic adaptation in the Carnivora, as illustrated in the osteology and evolution of the sea otter. *Bull. Dep. Geol. Univ. Calif.* **7**(25):465–495.

Thewissen, J. G. M. 1994. Phylogenetic aspects of cetacean origins: a morphological perspective. *J. Mamm. Evol.* **2**(3):157–184.

Thewissen, J. G. M., and Fish, F. E. 1997. Locomotor evolution in the earliest cetaceans: functional model, modern analogues, and paleontological evidence. *Paleobiology* **123**:482–490.

Thewissen, J. G. M., and Hussain, S. T. 1990. Postcranial osteology of the most primitive artiodactyl *Diacodexis pakistanensis* (Dichobunidae). *Anat. Histol. Embryol.* **19**:37–48.

Thewissen, J. G. M., Hussain, S. T., and Arif, M. 1994. Fossil evidence for the origin of aquatic locomotion in archaeocete whales. *Science* **263**:210–212.

Thewissen, J. G. M., Madar, S. I., and Hussain, S. T. 1996. *Ambulocetus natans*, an Eocene cetacean (Mammalia) from Pakistan. *Cour. Forsch.-Inst. Senckenberg* **191**:1–86.

Uhen, M. D. 1996. *Dorudon atrox* (Mammalia, Cetacea): form, function, and phylogenetic relationships of an archaeocete from the late middle Eocene of Egypt. Ph.D. dissertation, University of Michigan, Ann Arbor, 608 pp.

Van Valen, L. 1966. Deltatheridia, a new order of mammals. *Bull. Am. Mus. Nat. Hist.* **132**:1–126.

Wall, W. P. 1983. The correlation between high limb-bone density and aquatic habits in recent mammals. *J. Paleontol.* **57**(2):197–207.

Webb, P. W. 1988. Simple physical principles and vertebrate aquatic locomotion. *Am. Zool.* **28**:709–725.

Webb, P. W., and Buffrénil, V. de 1990. Locomotion in the biology of large aquatic vertebrates. *Trans. Am. Fish. Soc.* **119**:629–641.

Zhou, X., Sanders, W. J., and Gingerich, P. D. 1992. Functional and behavioral implication of vertebral structure in *Pachyaena ossifraga* (Mammalia, Mesonychia). *Contrib. Mus. Paleontol. Univ. Michigan* **28**: 289–313.

CHAPTER 13

Evolution of Thermoregulatory Function in Cetacean Reproductive Systems

D. ANN PABST, SENTIEL A. ROMMEL, and WILLIAM A. McLELLAN

1. Introduction

Modern cetaceans possess a suite of morphological adaptations that permit their existence in the marine environment (e.g., Howell, 1930; Slijper, 1936, 1979). Their streamlined body shape, hypertrophied axial musculoskeletal system, thick blubber layer, and *de novo* dorsal fin and flukes are morphological features that reduce the energetic costs of both swimming (e.g., Fish and Hui, 1991; Williams *et al.*, 1992; Fish, 1993a,b; Pabst, 1996) and whole body thermoregulation (e.g., Worthy and Edwards, 1990; Koopman *et al.*, 1996).

Interestingly, some of these morphological adaptations would appear to threaten the temperature-sensitive reproductive tissues of cetaceans. For example, males possess intra-abdominal testes—a condition identified as an adaptation for body streamlining (e.g., Howell, 1930; Slijper, 1936, 1979). As a consequence of streamlining and axial swimming style, the intra-abdominal testes of male cetaceans are literally wedged between thermogenic axial and abdominal locomotor muscles (Boice *et al.*, 1964; Arkowitz and Rommel, 1985) and could potentially be exposed to core or above core body temperatures (Fig. 1).

In many mammals, viable sperm production and epididymal storage requires temperatures below those found in the body core (e.g., Moore, 1926; Cowles, 1958, 1965; Bedford, 1977; Carrick and Setchell, 1977). Temperatures between 35 and 38°C can effectively block spermatogenesis (Cowles, 1958; Van Demark and Free, 1970) and abdominal temperatures can detrimentally affect long-term storage of spermatozoa in the epididymis (Bedford, 1977). Physical separation from the body core (e.g., scrota and cremaster sacs) and countercurrent heat exchange within the spermatic cord, between venous pampiniform plexus and spermatic artery (Harrison, 1948), are mechanisms that maintain below core temperatures at the mammalian testis. Interestingly, cetaceans possess high internal body temperatures (e.g., Wislocki, 1933; Harrison, 1948; Mackay, 1964; Ridgway, 1965, 1968, 1972;

D. ANN PABST and WILLIAM A. McLELLAN • Biological Sciences and Center for Marine Science Research, University of North Carolina at Wilmington, Wilmington, North Carolina 28403.   SENTIEL A. ROMMEL • Marine Mammal Pathobiology Laboratory, Florida Marine Research Institute, Florida Department of Environmental Protection, St. Petersburg, Florida 33711.
The Emergence of Whales, edited by Thewissen. Plenum Press, New York, 1998.

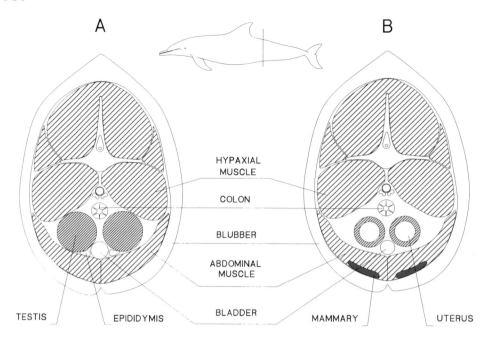

FIGURE 1. Schematic representation of cross sections through a sexually mature common dolphin, *Delphinus delphis*, at the level of the reproductive tract. The vertical line through the lateral view of the dolphin indicates the level of the cross sections in A (male) and B (female). Note that the testes and uterus are wedged between the hypaxial and abdominal locomotor muscles.

Whittow *et al.*, 1974; Brodie and Paasche, 1985; Rommel *et al.*, 1994; Pabst *et al.*, 1995), but lack a physical separation between testes and body core, and lack a pampiniform plexus (Harrison, 1948).

Body streamlining and axial locomotor style also impact the thermoregulation of the female reproductive system. Thermogenic muscle and insulating blubber surround the female reproductive system (Fig. 1); this arrangement suggests elevated temperatures at the uterus that could detrimentally affect fetal development. Because the mammalian fetal metabolic rate may be as much as twice that of maternal tissues (Power *et al.*, 1984), heat must be continuously transferred from the fetus to the mother so as to maintain a stable fetal temperature. Any physiological or anatomical condition that limits the ability of the fetus to transfer heat to the maternal environment will cause a potentially harmful increase in fetal temperature. Increases in fetal temperature are known to cause detrimental effects including low birth weights (Shelton, 1964), retarded fetal growth (Alexander *et al.*, 1987; Bell, 1987), skeletal and neural developmental anomalies (reviewed in Lotgering *et al.*, 1985), and ultimately acute fetal distress and death (Morishima *et al.*, 1975; Cephalo and Hellegers, 1978).

Under steady-state conditions in experimental mammals, approximately 85% of the heat developed by the fetus is transported to the placenta (Power *et al.*, 1984; Gilbert *et al.*, 1985; Gilbert and Power, 1986). This heat is then transferred to the maternal environment and subsequently lost to the external environment. The remaining 15% of fetal heat is transported away from the fetal skin surface, via the amniotic and allantoic fluids, to the uterine wall and to the maternal environment (Gilbert *et al.*, 1985; Bell, 1987). This heat is subsequently lost to the external environment through the "thermal window" of the maternal abdominal wall (Hart and Faber, 1985; Gilbert and Power, 1986). Heat can only be conducted away from the fetal surface, across the uterine wall, and into the maternal environment if, at each step, surrounding maternal tissues are cooler. In terrestrial mammals the relatively thin muscles and skin of the ventral abdominal wall, which are cooler than the maternal core temperature and cooler than other organs with which the uterus is in contact, function as a maternal "thermal window" (Hart and Faber, 1985; Gilbert and Power, 1986). The location of heat-producing, axial and abdominal locomotor muscles and insulating blubber suggests that cetaceans lack such a maternal "thermal window."

Thus, the streamlined body shapes of cetaceans appear to pose thermoregulatory threats to the reproductive systems of both males and females. How do cetaceans control the temperature of their reproductive tissues? We have recently described novel vascular arrangements in both males and females that function as a countercurrent heat exchanger (CCHE) deep within the caudal abdominal cavity (Rommel *et al.*, 1992, 1993). This CCHE brings cool venous blood returning from the superficial surfaces of the dorsal fin and flukes to a position juxtaposed to the arteries supplying the reproductive tissues (Rommel *et al.*, 1994; Pabst *et al.*, 1995).

In this chapter, we will describe the CCHE and offer physiological evidence that it functions to regulate the temperature of reproductive structures. As we describe the morphology of the male and female reproductive systems, we will discover that many morphological features of cetacean reproductive systems are shared by members of the clade Artiodactyla. That is, that the majority of reproductive structures reflect cetacean phylogenetic relationships and not novel adaptations to a marine environment. We will hypothesize that the unique features of the cetacean reproductive system—including testis position, position of reproductive structures relative to locomotor muscle, and the novel vascular plexuses of the CCHE—are paedomorphic. We suggest that the apparently complex suite of morphological changes that occurred during the evolution of cetaceans from a terrestrial to an aquatic form may be parsimoniously explained as representing a suite of arrested embryonic characters maintained in the adult.

Our specific goals are to describe the (1) gross morphology of the cetacean reproductive system, (2) vasculature associated with the reproductive system, and (3) physiological evidence that dolphins use a CCHE to regulate the temperature of their reproductive system. We will end with an evolutionary hypothesis, namely, that the CCHE is a mosaic of both embryonic vascular characters and vessels associated with the *de novo* dermal structure of the dorsal fin and flukes.

2. Reproductive Morphology of Cetaceans

2.1. Males

The morphology of the reproductive system of cetaceans has been described in some detail by Meek (1918), Ommanney (1932), Slijper (1936, 1966, 1979), and Harrison (1969). We briefly review the morphology, focusing mainly on odontocete cetaceans.

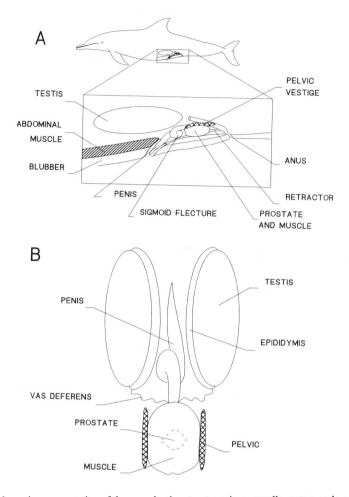

FIGURE 2. Schematic representation of the reproductive structures in a sexually mature male common dolphin, *Delphinus delphis*: lateral (A) and dorsal (B) views. The nonerect position of the fibroelastic penis is curved into a sigmoid flexure within the body wall. A retractor penis muscle (labeled retractor) originates on the superficial surface of the rectum and attaches to the ventral surface of the penis, just distal to the sigmoid flexure. The crura of the penis anchor onto the pelvic bones; the prostate gland and ischiocavernosus muscles lie between the pelvic bones.

The testes (Fig. 2) lie within the caudal abdominal cavity in adult cetaceans, a position we define as *intra-abdominal* (also called *cryptic, endorchid*, and *testicond*) (e.g., Slijper, 1936, 1966; De Smet, 1977; Rommel *et al.*, 1992). The testes of some *Mesoplodon* species lie within separate, tubelike, caudal extensions of the abdominal cavity (De Smet, 1977; personal observation). Although the testes of *Mesoplodon* maintain the same relationship to abdominal muscles and pelvic vestiges as found in other cetaceans, Anthony (1920, as cited in De Smet, 1977) described their position as "inguinal." We concur with De Smet (1977) and consider the position of *Mesoplodon* testes as intra-abdominal, but suggest that testicular morphology and position in ziphiid whales warrant further anatomic study.

Each testis is attached to the abdominal wall by a mesorchium that connects to the testis along its border with the epididymis (Fig. 2). The epididymis lies along the entire length of the testis and terminates in the ductus deferens, which joins the urethra distally via the ejaculatory duct (Harrison, 1969). The only accessory gland that has been described in cetaceans is the prostate (Slijper, 1936, 1966, 1979; Harrison, 1969; Simpson and Gardner, 1972), which lies at the base of the penis between the pelvic elements. The penis is anchored to the pelvic elements by two crura; these crura fuse in the body of the penis to form a single corpus cavernosum. The urethra travels through a poorly developed corpus spongiosum (Simpson and Gardner, 1972; termed *corpus cavernosum urethra* by Slijper, 1966).

The cetacean penis (Fig. 2), which can be retracted into the body wall (into a prepuce), is fibroelastic like that of most artiodactyls (Slijper, 1966, 1979). The nonerect position of the cetacean penis is curved into a sigmoid flexure within the body wall, as is seen in ruminants (Schummer *et al.*, 1979). Upon erection, the cetacean penis straightens and becomes turgid, but does not dramatically change its absolute length or diameter. A retractor penis muscle [apparently a ligament in some species (personal observation)] originates on the superficial surface of the rectum and attaches to the ventral surface of the penis, just distal to the sigmoid flexure (Slijper, 1966, 1979; Harrison, 1969). This anatomical arrangement is also seen in ruminants (Schummer *et al.*, 1979). The retractor penis ostensibly functions to maintain the position of the nonerect penis within the prepuce, although Slijper (1966) suggests that the retractor penis functions "as a brake in regulating the stretching of the penis during erection."

2.2. Females

The ovaries are paired, oval organs (Slijper, 1966; Harrison, 1969). Each ovary is surrounded by the fimbriae of a distal uterine tube (Fig. 3). The uterine tube has a small diameter, and joins a horn of the cetacean bicornuate uterus (described by Meek, 1918; Pycraft, 1932; Wislocki, 1933; Wislocki and Enders, 1941; Slijper, 1936, 1966, 1979; Harrison, 1969). Each ovary, uterine tube, and uterine horn are held in place by an extensive mesentery termed the *broad ligament*. The broad ligament has three regions, the mesovarium, mesosalpinx, and mesometrium. The mesovarium attaches the ovary to the dorsolateral abdominal wall. The mesosalpinx folds around the ovary forming an ovarian bursa, and attaches the uterine tube to the abdominal wall. The mesometrium attaches the uterus to the abdominal wall.

The placenta is diffuse and epitheliochorial, as in many artiodactyls (Slijper, 1966, 1979; Benirschke and Cornell, 1987). The bicornuate uterus terminates at a muscular, true cervix (Fig. 3). Just distal to the true cervix, the wall of the vaginal canal is thrown into annular folds, termed *pseudocervices* (Harrison, 1969), which have been identified as unique to cetaceans (Slijper, 1979; Schroeder, 1990). Schroeder (1990) calls the presence of pseudocervices a "remarkable anatomical adaptation for breeding in the marine environment." Slijper (1979) states that "these peculiar folds . . . are not found in any other mammal." However, the proximal vaginal canals of both cows and sows are also thrown into annular folds (Schummer *et al.*, 1979). Thus, this character is shared with artiodactyls, and not unique to cetaceans.

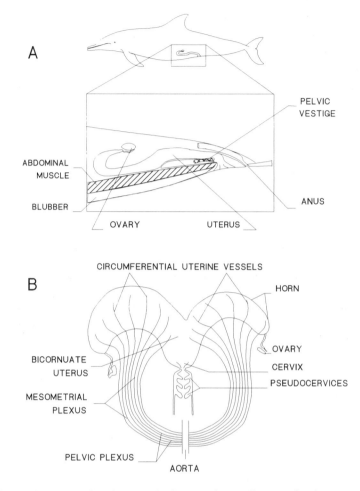

FIGURE 3. Schematic representation of the reproductive tract of a sexually mature female common dolphin, *Delphinus delphis*: lateral (A) and dorsal (B) views. In the highly schematized dorsal view, the left uterine horn is slightly enlarged, as it would be in early pregnancy. The aorta has been retracted caudally, and the pelvic and mesometrial plexuses have been laid flat to better illustrate their morphology. Note the pseudocervices, just distal to the cervix, in the vaginal tract.

The vaginal canal terminates at the vulva, which lies within a slit-shaped aperture in the body wall. The external genitalia include labia majora, labia minora, and a well-developed clitoris.

3. Vascular Structures Associated with the Cetacean Reproductive System

The vascular structures that supply and drain the reproductive tissues of cetaceans are homologous to those of other tetrapod mammals. The lumbocaudal venous plexus (described below) that is independent of, but juxtaposed to, the arterial supply to the reproductive tissues, is unique to cetaceans. Together, these vessels form a CCHE that delivers cooled venous blood deep within the abdomen, in a position to cool the arterial supply to the testes in males and the uterus in females. The CCHE flanks the dorsal aorta, between the kidney and the origin of the caudal artery.

We will describe the reproductive vasculature separately for males and females. Because the derived lumbocaudal venous plexus and superficial veins are similar in both sexes, we will describe these only in the male. Our descriptions are of the left side, but the vascular structures associated with the reproductive system in both males and females are bilaterally symmetrical. All arteries and veins are schematized—they are more complex and contorted than are presented here.

The morphological descriptions below are based on detailed dissections of over 30 fresh carcasses of animals that had either stranded or been killed incidentally in commercial fishing operations (see Rommel *et al.*, 1992, for detailed materials and methods, and United States National Museum accession numbers). The species included in this study are common (*Delphinus delphis*), bottlenose (*Tursiops truncatus*), and Atlantic white-sided (*Lagenorhynchus acutus*) dolphins, long-finned pilot whales (*Globicephala melas*), and harbor porpoises (*Phocoena phocoena*), although we have observed the CCHE in every odontocete cetacean that we have investigated. We have also observed the CCHE in fin whales (*Balaenoptera physalus*) (*n* = 2) during field dissections. We illustrate the CCHE in the common dolphin.

Some of the vascular patterns described below have been previously identified (e.g., spermatic arterial plexus of Slijper, 1936). Either because of the regional scope of the study (e.g., Ommanney, 1932; Scholander and Schevill, 1955; Harrison and Tomlinson, 1956) or errors in identification of specific vascular structures (e.g., Slijper, 1936; Walmsley, 1938), however, a countercurrent vascular design had not been previously identified (see Rommel *et al.*, 1992).

3.1. Males

3.1.1. Spermatic and Testicular Arteries and Veins

The blood supply to the testis is provided via the spermatic arterial plexus (Slijper, 1936), which is fed from the dorsal aorta (Fig. 4). Rather than originating as a single artery, as in most other mammals, approximately 40 individual spermatic arteries leave the aorta. Each artery is convoluted as it leaves the aorta, but straightens as it courses laterally toward the testis. The arteries, which are embedded in a thin connective tissue membrane, are organized into a single layer and are oriented roughly parallel to each other. They form a flat plexus of closely spaced, parallel arteries that we call the *spermatic arterial plexus*.

Near the ventrolateral margin of the plexus, the arteries begin to coalesce, and form a cone-shaped mass of vessels. These vessels anastomose into fewer, larger-diameter arteries as the cone tapers caudally. At the caudal terminus of the cone, a single testicular artery enters the tunic of the testis. A few branches of the spermatic arterial plexus bypass the cone and feed the epididymis directly. The spermatic arterial plexus is less well developed in sexually immature males than in mature males.

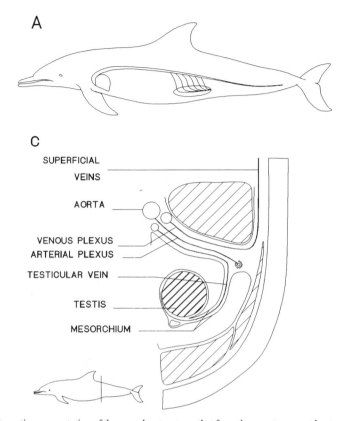

FIGURE 4. Schematic representation of the vascular structures that form the countercurrent heat exchanger (CCHE) at the testis of a common dolphin, *Delphinus delphis*. (A) Topography of the spermatic arterial plexus arising from the dorsal aorta. (B) Topography of the superficial veins that supply the lumbocaudal venous plexus. Blood in superficial veins of dorsal fin and flukes is cooled by exposure to ambient water. These extremities are drained by relatively thick-walled, large veins that remain superficial (i.e., just deep to the blubber layer and superficial to the axial muscles) as they coalesce and course toward the abdominal cavity. These veins enter the caudal abdominal cavity, near the pelvic vestiges, and feed directly into the lateral and caudal margins of the lumbocaudal venous plexus. Thus, relatively cool blood can be introduced into the deep caudal abdominal cavity near the testis. (C) Enlarged cross section through the dolphin at the level of the testis illustrating topography of the deep and superficial vessels. (D) Oblique left lateral view of CCHE. The spermatic arterial plexus is a unique arrangement of arteries that extend ventrolaterally from the dorsal aorta. The vessels are organized into a single layer and are oriented roughly parallel to each other. At the distal margin of the plexus, the arteries coalesce to form a cone-shaped structure, from which a single testicular artery continues caudally to enter the testis. Note the juxtaposition of the lumbocaudal venous plexus to the spermatic arterial plexus. Arrowheads indicate direction of flow. Cut ends at the lateral border of the lumbocaudal venous plexus are the sites where the superficial veins from the dorsal fin enter the abdominal cavity.

The testis is drained by large testicular veins. These veins emerge from the body of the testis, and course dorsomedially toward the paired venae cavae. In cetaceans, the lumbar venae cavae are paired (e.g., Slijper, 1936; Walmsley, 1938; Harrison and Tomlinson, 1956).

3.1.2. Lumbocaudal Venous Plexus

The spermatic and testicular arteries and testicular veins feed and drain the testis, and are homologous to the reproductive vasculature found in other tetrapods—the novel venous

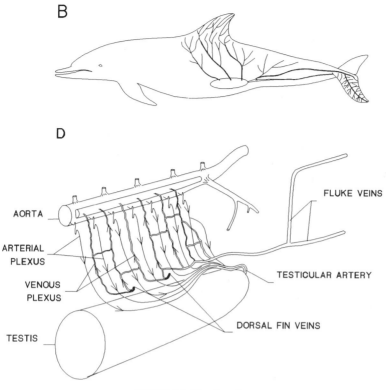

FIGURE 4. *(Continued).*

structure associated with the cetacean testis is the lumbocaudal venous plexus (Fig. 4). The lumbocaudal venous plexus is formed by a single layer of irregularly anastomosed, thin-walled vessels embedded in a connective tissue membrane. This plexus is affixed to the ventral surface of the hypaxial muscle, and lies dorsolateral and is juxtaposed to the spermatic arterial plexus. This juxtaposition puts arteries and veins in close proximity and is well suited for heat exchange.

The lumbocaudal venous plexus is supplied with blood from the superficial veins that drain the dorsal fin and the flukes (Fig. 4). An extensive system of large-diameter, superficial veins drain the dorsal fin; these lateral abdominal subcutaneous veins (named by Slijper, 1936) remain just deep to the blubber layer as they course ventrally. Along their course, they coalesce into three or four larger-diameter veins that remain superficial until they reach the dorsal border of the caudal abdominal wall. Here, these veins become deep, follow the lateral contour of the hypaxial muscle, and feed into the cranial, lateral edge of the lumbocaudal venous plexus. Similarly, the superficial veins that drain the surfaces of the flukes coalesce into the dorsal and ventral lateral, caudal subcutaneous veins (Slijper, 1936). These two veins coalesce and dive deep at the pelvic vestige. This vein joins the caudal, lateral margin of the lumbocaudal venous plexus. Thus, cooled blood from superficial veins is distributed along a portion of the dorsolateral wall of the abdominal cavity by the lumbocaudal venous plexus. The plexus, in turn, is drained via the ipsilateral vena cava.

3.2. Females

3.2.1. Uterovarian Vascular Plexuses

In contrast to the two or three vessels found in most other mammals (e.g., Schummer *et al.*, 1981), the uterovarian arteries and veins in cetaceans are flattened plexuses (Figs. 3 and 5) formed by many vessels (e.g., Walmsley, 1938). These plexuses are distinguished by two relatively distinct regions: a region proximal to the dorsal midline, and a more distal region within the mesometrium. The proximal region is juxtaposed to the lumbocaudal venous plexus described above. In the proximal region the arteries and veins are ordered in parallel channels with few branches or anastomoses. In the mesometrial region there are more branches and the arrangements of vessels are less regular. The mesometrium wraps around the lateral margin of the uterus and attaches along the ventrolateral margin of the uterine horn. Thus, the uterovarian plexuses are positioned between the wall of the uterus and the abdominal muscles, much like the mesorchium wraps the testis in the male.

The uterovarian arterial plexus is interposed between the dorsal aorta and the uterus. The uterovarian arterial plexus is composed of approximately 20–40 discrete arterial channels that exit the aorta or form branches immediately after exiting the aorta. A few branches from the common iliac artery (Walmsley, 1938) make up the caudalmost portion of the plexus (Figs. 3 and 5). These branches pass through or between the anastomoses connecting the dorsal and ventral parts of the vena cava adjacent to the lateral aspects of the lumbar aorta. The proximal region of the uterovarian arterial plexus lies juxtaposed to the lum-

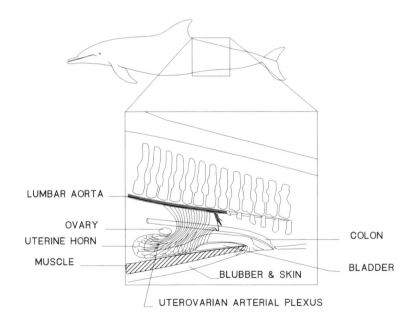

FIGURE 5. Schematic representation of uterovarian arterial plexus of common dolphin, *Delphinus delphis*. Note the highly ordered parallel vessels that form the plexus, an arrangement very different from that found in most mammals.

bocaudal venous plexus on the ventral surface of the hypaxial muscles. This arterial plexus is "loosely" attached (i.e., it can be moved by firm pressure) to the dorsolateral wall of the abdominal cavity. The uterovarian arterial plexus is sandwiched between the lumbocaudal venous plexus described above and the uterovarian venous plexus described below. The more distal part of the uterovarian arterial plexus is contained within the mesometrium. Immature specimens have both arterial and venous plexuses but in a less well-developed form than those found in pregnant or postpartum specimens.

On the ventromedial aspect of the uterovarian arterial plexus lies a uterovarian venous plexus formed by 20–40 parallel veins. This discrete array of vessels is in contrast to the irregularly anastomosing channels of the lumbocaudal venous plexus. This plexus provides venous return from the uterus to the vena cava. The veins of this plexus enter the ventral channel of the vena cava ventromedial to the uterovarian arterial plexus. As mentioned above, proximal to the dorsal midline this venous plexus is an orderly array of parallel vessels. Within the mesometrium there are more branches and anastomoses with a less regular pattern.

In females, the CCHE is formed by the proximal region of the uterovarian arterial plexus and the juxtaposed lumbocaudal venous plexus. In males, the CCHE is formed by the spermatic arterial plexus and the juxtaposed lumbocaudal venous plexus. Thus, the arterial supply to the reproductive tissues is juxtaposed to vessels carrying cooled venous blood returning from the superficial surfaces of the dorsal fin and flukes.

4. Function of the CCHE in *Tursiops truncatus*

The CCHE flanks a region of the bowel (Fig. 6) and influences colonic temperatures, thus permitting indirect assessment of the introduction of relatively cool venous blood into the abdominal cavity via the lumbocaudal venous plexus. Using a linear array of multiple, copper-constantan thermocouples housed in a rectal probe, we measured colonic temperatures in bottlenose dolphins (*T. truncatus*), at positions cranial to, within, and caudal to the region of the bowel flanked by the CCHE. We investigated colonic temperatures of both peripubescent and adult male dolphins while resting and of peripubescent males just before and after vigorous swimming (Rommel *et al.*, 1994; Pabst *et al.*, 1995).

In bottlenose dolphins under resting conditions, colonic temperatures measured in the region of the CCHE are cooler than temperatures measured cranial or caudal to this region (Fig. 6). The influence of the CCHE on colonic temperatures is dependent on a number of variables. For example, temperatures at the CCHE were 0.2–0.7°C cooler than positions cranial and/or caudal to the CCHE in peripubescent males, and were 0.9–1.3°C cooler in a sexually mature male (Rommel *et al.*, 1994). Temporary heating of the dorsal fin and flukes increased temperatures at the CCHE, but had little or no effect on temperatures caudal to its position. These results demonstrate that, under resting conditions, cooled blood is introduced into the deep abdominal cavity in a position to regulate the temperature of arterial blood flow to the dolphin testis.

Colonic temperatures after exercise were also position dependent (Fig. 7). Temperatures within the region of the colon flanked by the CCHE decreased after the dolphin swam vigorously (Pabst *et al.*, 1995). Temperatures at the cranial CCHE decreased by 0.5–0.6°C after the dolphin's most strenuous exercise period. When the dolphin was allowed to rest

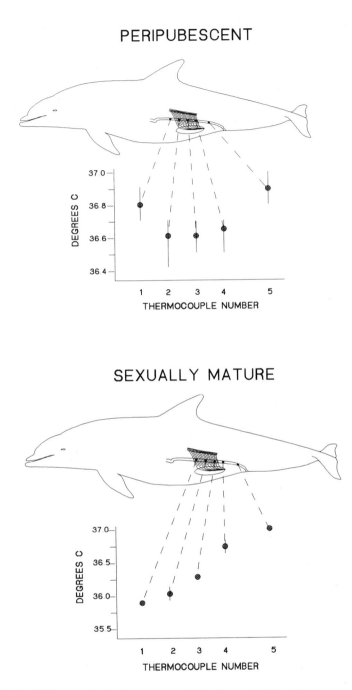

FIGURE 6. Regional differences in colonic temperatures in bottlenose dolphins, *Tursiops truncatus*, at rest (see methods in Rommel *et al.*, 1994). Temperature profiles of a (A) peripubescent (2.33 m) and (B) sexually mature (2.62 m) male bottlenose dolphin under resting conditions. Mean and maximum range of temperatures are reported for each position sampled continuously over a period of 5 to 15 min. In the peripubescent male, thermocouple #1 is cranial to, thermocouples #2–4 are within, and thermocouple #5 is caudal to the CCHE. In the sexually mature male, the thermocouple positions are shifted caudally: Thermocouples #1–4 are within the CCHE, and #5 is caudal to the region of the CCHE.

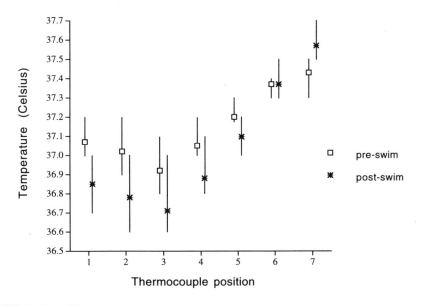

FIGURE 7. Regional differences in colonic temperatures in a 2.41-m peripubescent male bottlenose dolphin, *Tursiops truncatus*, after vigorous swimming reported as means and ranges. Results are pooled from four exercise sessions: Swim intervals ranged from 8 to 10 min at speeds of 9–12 km/hr. Colonic probe housed a linear array of seven thermocouples in these experiments. Temperatures in the region of the CCHE (thermocouples #1–5 in this male) decrease with exercise. Temperatures at the caudal most thermocouple increase with exercise.

for longer than 4–6 min after exercise, colonic temperatures at the region of the CCHE slowly increased.

Temperatures in the region of the colon flanked by the CCHE decrease with exercise. These data suggest that the CCHE has an increased ability to cool the arterial blood supply to the testes, and may thermally isolate the testes from adjacent locomotor muscles, when the dolphin is swimming vigorously. We hypothesize that the maximal cooling observed is the result of increased flow of cooled venous blood through the CCHE during exercise.

The venous blood supplying the CCHE returns from the surfaces of the dorsal fin and flukes (see Fig. 4). These appendages have two venous returns. Periarterial venous channels (Elsner *et al.*, 1974) are found deep within the fin and have been hypothesized as the heat-conserving, countercurrent heat exchanger (Scholander and Schevill, 1955). The superficial venous system has been hypothesized as a shunt for the deep countercurrent heat exchanger; blood routed through this venous system would be cooled by exposure to ambient water (Scholander and Schevill, 1995; Kanwisher and Sundes, 1966). Scholander and Schevill (1955) hypothesized that the mechanism for routing blood though these venous systems was "semiautomatic." If the dolphin needed to conserve heat, the rate of blood flow through the fin would be slow, and at lower pressure, and venous blood would be preferentially returned via the deep periarterial venous channels. If, on the other hand, the animal needed maximal cooling, blood flow and blood pressure through the fin would increase. The increased flow would swell the nutrient arteries, occlude the deep periarterial venous return, and force blood through the superficial venous system.

Heart rates of exercising dolphins are increased relative to resting dolphins (Williams *et al.*, 1992), suggesting that blood flow through the radiating surfaces of the dorsal fin and flukes would be increased during exercise. Heat loss from blood in the superficial veins would increase by increased convective heat exchange (Schmidt-Nielsen, 1990) at higher speeds of swimming. Thus, changes in blood flow patterns through the dorsal fin and flukes, coupled with increased convective heat loss from venous blood returning through the CCHE, may allow maximal cooling of the intra-abdominal testes during exercise.

5. Evolutionary Hypothesis: CCHE Is a Paedomorphic Vascular Design

To accurately identify those reproductive characters that are unique to cetaceans, we must first identify those characters shared with sister taxa (Wiley, 1981). Although there is broad acceptance that cetaceans are—or are sister taxa to—ungulates, the exact phylogenetic relationships of these clades are still controversial (e.g., Slijper, 1936; Van Valen, 1966; Prothero, 1993; Graur and Higgins, 1994; Thewissen, 1994; Arnason and Gullberg, 1996; Gatesy *et al.*, 1996; Smith *et al.*, 1996). It is outside the scope of this chapter to discuss the numerous and varied proposed phylogenetic relationships; suffice it to say that the two most commonly identified sister taxa to Cetacea are Artiodactyla and Perissodactyla (e.g., Slijper, 1936; Thewissen, 1994). We will use a cladogram that maps Cetacea and Artiodactyla as sister taxa, because it is the most parsimonious relationship based upon the reproductive characters considered here (Fig. 8).

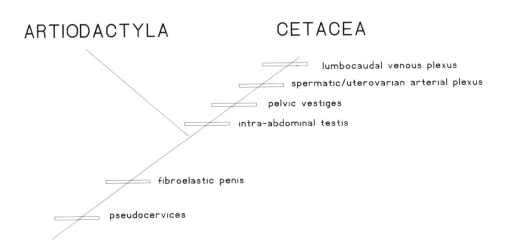

FIGURE 8. Hypothesized phylogenetic relationships between Cetacea and ungulates (based on one tree presented by Prothero, 1993). Note that the fibroelastic penis in males and pseudocervices in females are synapomorphies that unite Artiodactyla and Cetacea. Derived characters for Cetacea are intra-abdominal testis position in males; pelvic vestiges; elaborated, nutrient arterial plexuses (spermatic and uterovarian); and lumbocaudal venous plexus intervening between cutaneous veins and duplicated venae cavae in both males and females. We hypothesize that these novel reproductive features represent arrested embryonic characters, retained and specialized in the adult.

Many features of the cetacean reproductive system, such as the general morphology of the ovary and supporting ligaments, uterus, placenta, and vulva in females, and the morphology of the testis and duct system in males, are shared broadly across mammals. In males, the fibroelastic penis and associated penis retractor muscle appear to be synapomorphies for artiodactyls and cetaceans; the "pseudocervices" in females also appear to be uniquely shared by members of these two clades.

The reproductive features of cetaceans that are derived, relative to artiodactyls are: (1) intra-abdominal testis position in males, (2) pelvic vestiges, (3) elaborated, nutrient arterial plexuses (spermatic and uterovarian), and (4) lumbocaudal venous plexus intervening between cutaneous veins and duplicated venae cavae in both males and females. We hypothesize that these novel reproductive features are paedomorphic, i.e., represent arrested embryonic characters, retained and specialized in the adult. We briefly describe below supporting developmental evidence of this assertion.

1. Testis position. During embryonic or early postnatal development in most mammals, the testes descend from their original cranial position in the dorsal abdominal cavity and migrate through the inguinal canal to reside within the scrotum (e.g., Bailey and Miller, 1909; Hyman, 1942; Cartmill *et al.*, 1987; Evans and Christensen, 1993). The gubernaculum, a fold of mesentery that surrounds an undifferentiated mass of mesenchyme, is attached to the caudal pole of the testis and plays a critical role in testicular descent (Evans and Christensen, 1993). Although a gubernaculum appears early in fetal development in a variety of odontocetes (reviewed by van der Schoot, 1995) and Ommanney (1932) describes a "vestigial inguinal canal" in fin whales, the undescended and intra-abdominal position of the testes can be considered an arrested embryonic character.

2. Pelvic vestiges. Each half of the pelvis of other tetrapods develops from a cartilaginous template and ossifies as three initially distinct bones, the pubis, ischium, and ilium (e.g., Nickel *et al.*, 1986). To date, the exact identity and development of the elements of the pelvic vestige of extant cetaceans have not been established. Such identification is critical to fully understanding the events underlying the evolution of the cetacean pelvis. Because the crura of the penis are anchored to the pelvic vestiges, we hypothesize that the vestiges represent arrested developmental states of at least one bone, the ischium (see also van der Schoot, 1995). Further data are required to test the hypothesis that the pelvic vestige is an arrested embryonic character.

[Note: The pelvic bones of cetaceans are usually described as vestiges (e.g., Rommel, 1990) to denote that they are the product of a reduction from the condition found in their ancestors. The hypothesis presented here—that these bones represent an arrested embryonic condition—supports the usage of the term pelvic *rudiment* (*sensu* van der Schoot, 1990) rather than *vestige*.]

3. Spermatic and uterovarian arterial plexuses. Early in development, the blood supply to the gonads is by way of multiple, usually segmental vessels that arise off the dorsal aorta; later in development the majority of these vessels disappear (Bailey and Miller, 1909). The arterial supply to the testis and the ovary and uterus in cetaceans is by way of a planar plexus of many (approximately 40 in delphinids) vessels that arise directly off the dorsal aorta. We hypothesize that this arterial arrangement is best explained as a hypertrophied embryonic condition in the adult.

4. Lumbocaudal venous plexus intervening between cutaneous veins and duplicated venae cavae. An early step in the development of blood vessels is the formation of elabo-

rate networks or plexuses of capillaries (the "primordial vascular net" of Woollard, 1922; also reviewed by Hudlicka and Tyler, 1986). Eventually, larger vessels emerge from these nets to form the major, singular branches seen in adult vertebrates. At early stages of development, there are also extensive connections between deep and superficial vessels (Woollard, 1922) and the venae cavae are paired (e.g., Bailey and Miller, 1909). Thus, we hypothesize that the lumbocaudal venous plexus, the vascular connections between deep and superficial veins, and the paired venae cavae of cetaceans are embryonic characters retained and specialized in the adult. Slijper (1936) and Barnett *et al.* (1958) also identified the paired venae cavae as an embryonic character.

Thus, we posit that the evolution of the CCHE is most parsimoniously explained as a retention and modification of a suite of embryonic vascular structures. Interestingly, however, the function of the CCHE is completely reliant on the superficial veins returning cool blood from the dorsal fin and flukes. Although a connection between superficial and deep vessels is an embryonic character, these fins are *de novo* dermal structures, i.e., without evolutionary precursors in mammals. The dorsal fin and flukes are often identified as hydrodynamic control surfaces (e.g., Slijper, 1936)—without them, though, the ability to cool the reproductive system would appear to be severely limited. It is interesting to pose the question of how these vascular structure/function complexes influenced the evolution of these *de novo* fins.

Further studies on the development of reproductive organs, the pelvis and associated vasculature are required to test the hypothesis that novel reproductive features of cetaceans are paedomorphic. For example, van der Schoot's (1995) detailed analysis of the development of the gubernaculum demonstrated that although this organ appears early in development, it does not develop further into the "elaborate structures required for testis descent." This study provided strong evidence that cetaceans are "secondarily testicond," i.e., that the intra-abdominal position of the testes evolved from that of ungulate ancestors with descended testes. Comparative studies of the reproductive and vascular anatomy of finless cetaceans would yield valuable insights into thermoregulatory function in species lacking this important hydrodynamic and thermoregulatory control surface. It will only be through continued concerted research efforts of developmental and functional morphologists and paleontologists, though, that we will gain a better understanding of the evolution of functional designs critical to mammalian life in a marine environment.

6. Summary

The streamlined body shape, hypertrophied axial musculoskeletal system, and thick blubber layer of cetaceans are specializations that could pose thermoregulatory threats to temperature-sensitive reproductive tissues. Cetaceans possess a CCHE, deep within the caudal abdominal cavity, that functions to deliver cool venous blood returning from the superficial surfaces of the dorsal fin and flukes to a position to cool the arterial supply to the reproductive tissues. We posit that both the intra-abdominal position of the testes, and the vascular structures that form the CCHE in cetaceans, are derived relative to their putative sister taxa Artiodactyla. We hypothesize that these unique features are paedomorphic, and represent a suite of arrested embryonic character states maintained in the adult. Most morphological features of cetacean reproductive systems, however, are shared by members of the Artiodactyla; thus, the majority of the reproductive system reflects cetacean phylogenetic relationships and not novel adaptations to a marine environment.

Acknowledgements

This chapter is CMSR Contribution #194.

References

Alexander, G., Hales, J. R. S., Stevens, D., and Donnelly, J. B. 1987. Effects of acute and prolonged exposure to heat on regional blood flows in pregnant sheep. *J. Dev. Physiol.* **9**:1–15.

Anthony, R. 1920. L'exorchidie du *Mesoplodon* et la remontee des testicules as cours de la phylogenese des cetaces. *C. R. Acad. Sci.* **170**:529–531.

Arkowitz, R. A., and Rommel, S. A. 1985. Force and bending moment of the caudal muscles in the short finned pilot whale. *Mar. Mamm. Sci.* **1**(3):203–209.

Arnason, U., and Gullberg, A. 1996. Cytochrome b nucleotide sequences and the identification of five primary lineages of extant cetaceans. *Mol. Biol. Evol.* **13**(2):407–417.

Bailey, F. R., and Miller, A. M. 1909. *Textbook of Embryology*. William Wood & Co., York, PA.

Barnett, C. H., Harrison, R. J., and Tomlinson, J. D. W. 1958. Variations in the venous systems of mammals. *Biol. Rev.* **33**:442–487.

Bedford, J. M. 1977. Evolution of the scrotum: the epididymis as the prime mover, in: J. H. Calaby and C. H. Tyndale-Biscoe (eds.), *Reproduction and Evolution*, pp. 171–182. Australian Academy of Science, Canberra City.

Bell, A. W. 1987. Consequences of severe heat stress for fetal development, in: J. R. S. Hales and D. A. B. Richards (eds.), *Heat Stress: Physical Exertion and Environment*, pp. 313–333. Elsevier, Amsterdam.

Benirschke, K., and Cornell, L. H. 1987. The placenta of the killer whale, *Orcinus orca*. *Mar. Mamma. Sci.* **3**(1):82–86.

Boice, R. C., Swift, M. L., and Roberts, J. C., Jr. 1964. Cross-sectional anatomy of the dolphin. *Nor. Hvalfangst Tid.* **7**:178–193.

Brodie, P. F., and Paasche, A. 1985. Thermoregulation and energetics of fin and sei whales based on postmortem, stratified temperature measurements. *Can. J. Zool.* **63**:2267–2269.

Carrick, F. N., and Setchell, B. P. 1977. The evolution of the scrotum, in: J. H. Calaby and C. H. Tyndale-Biscoe (eds.), *Reproduction and Evolution*, pp. 165–170. Australian Academy of Science, Canberra City.

Cartmill, M., Hylander, W. L., and Shafland, J. 1987. *Human Structure*. Harvard University Press, Cambridge, MA.

Cephalo, R. C., and Hellegers, A. E. 1978. The effect of maternal hyperthermia on maternal and fetal cardiovascular and respiratory function. *Am. J. Obstet. Gynecol.* **131**:687–694.

Cowles, R. B. 1958. The evolutionary significance of the scrotum. *Evolution* **XII**:417–418.

Cowles, R. B. 1965. Hyperthermia, aspermia, mutation rates and evolution. *Q. Rev. Biol.* **40**:341–367.

De Smet, W. M. A. 1977. The position of the testes in cetaceans, in: R. J. Harrison (ed.), *Functional Anatomy of Marine Mammals*, Volume 3, pp. 361–386. Academic Press, London.

Elsner, R., Pirie, J., Kenney, D. D., and Schemmer, S. 1974. Functional circulatory anatomy of cetacean appendages, in: R. J. Harrison (ed.), *Functional Anatomy of Marine Mammals*, Volume 2, pp. 143–159. Academic Press, London.

Evans, H. E., and Christensen, G. C. 1993. The urogenital system, in: H. E. Evans (ed.), *Miller's Anatomy of the Dog*, 3rd ed., pp. 494–558. Saunders, Philadelphia.

Fish, F. E. 1993a. Power output and propulsive efficiency of swimming bottlenose dolphins (*Tursiops truncatus*). *J. Exp. Biol.* **185**:179–193.

Fish, F. E. 1993b. Influence of hydrodynamic design and propulsive mode on mammalian swimming energetics. *Aust. J. Zool.* **42**:79–101.

Fish, F. E., and Hui, C. A. 1991. Dolphin swimming—a review. *Mammal. Rev.* **21**:181–195.

Gatesy, J., Hayashi, C., Cronin, M. A., and Arctander, P. 1996. Evidence from milk casein genes that cetaceans are close relatives of hippopotamid artiodactyls. *Mol. Biol. Evol.* **13**(7):954–963.

Gilbert, R. D., and Power, G. G. 1986. Fetal and uteroplacental heat production in sheep. *J. Appl. Physiol.* **61**:2018–2022.

Gilbert, R. D., Schroder, H., Kawamura, T., Dale, P. S., and Power, G. G. 1985. Heat transfer pathways between fetal lamb and ewe. *J. Appl. Physiol.* **59**:634–638.

Graur, D., and Higgins, D. G. 1994. Molecular evidence for the inclusion of cetaceans within the order Artiodactyla. *Mol. Biol. Evol.* **11**(3):357–364.

Harrison, R. J. 1948. The comparative anatomy of the blood-supply of the mammalian testis. *Proc. Zool. Soc. London* **11**:325–334 (plates I–V).

Harrison, R. J. 1969. Reproduction and reproductive organs, in: H. T. Anderson (ed.), *The Biology of Marine Mammals*, pp. 253–348. Academic Press, New York.

Harrison, R. J., and Tomlinson, J. D. W. 1956. Observations on the venous system in certain pinnipedia and cetacea. *Proc. Zool. Soc. London* **126**:205–233.

Hart, F. M., and Farber, J. J. 1985. Fetal and maternal temperatures in rabbits. *J. Appl. Physiol.* **20**:737–741.

Howell, A. B. 1930. *Aquatic Mammals: Their Adaptations to Life in the Water*. Thomas, Springfield, IL.

Hudlicka, O., and Tyler, K. R. 1986. *Angiogenesis*. Academic Press, Orlando, FL.

Hyman, L. H. 1942. *Comparative Vertebrate Anatomy*. University of Chicago Press, Chicago.

Kanwisher, J., and Sundes, G. 1965. Physiology of a small cetacean. *Hvalradets Skr.* **48**:45–53.

Koopman, H. N., Iverson, S. J., and Gaskin, D. E. 1996. Stratification and age-related differences in blubber fatty acids of the male harbour porpoise (*Phocoena phocoena*). *J. Comp. Physiol. B* **165**:628–639.

Lotgering, F. K., Gilbert, R. D., and Longo, L. D. 1985. Maternal and fetal responses to exercise during pregnancy. *Physiol. Rev.* **65**(1):1–29.

Mackay, R. S. 1964. Deep body temperature of an untethered dolphin recorded by radio transmitter. *Science* **144**:864–866.

Meek, A. 1918. The reproductive organs of cetacea. *J. Anat.* **52**:186–210.

Moore, C. R. 1926. The biology of the mammalian testis and scrotum. *Q. Rev. Biol.* **I**:4–50.

Morishima, H. O., Glaser, B., Niemann, W. H., and James, L. S. 1975. Increased uterine activity and fetal deterioration during maternal hyperthermia. *Am. J. Obstet. Gynecol.* **121**(4):531–538.

Nickel, R., Schummer, A., Seiferle, E., Wilkens, H., Wille, K.-H., and Frewein, J. 1986. *The Locomotor System of the Domestic Mammals*, Volume 1. Springer-Verlag, Berlin.

Ommanney, F. D. 1932. The urogenital system of the fin whale (*Balaenoptera physalus*) with appendix: the dimensions and growth of the kidneys of blue and fin whales. *Discovery Rep.* **5**:363–465.

Pabst, D. A. 1996. Morphology of the subdermal connective tissue sheath of dolphins: a new fibre-wound, thin-walled, pressurized cylinder model for swimming vertebrates. *J. Zool.* (London) **238**:35–52.

Pabst, D. A., Rommel, S. A., McLellan, W. A., Williams, T. M., and Rowles, T. K. 1995. Thermoregulation of the intra-abdominal testes of the bottlenose dolphin (*Tursiops truncatus*) during exercise. *J. Exp. Biol.* **198**:221–226.

Power, G. G., Schroder, H., and Gilbert, R. D. 1984. Measurement of fetal heat production using differential calorimetry. *J. Appl. Physiol.* **57**(3):17–22.

Prothero, D. R. 1993. Ungulate phylogeny: molecular vs. morphological evidence, in: F. S. Szalay, M. J. Novacek, and M. C. McKenna (eds.), *Mammal Phylogeny*, pp. 173–181. Springer-Verlag, Berlin.

Pycraft, W. P. 1932. On the genital organs of a female common dolphin (*Delphinus delphis*). *Proc. Zool. Soc. London* **1932**:807–811 (3 plates).

Ridgway, S. H. 1965. Medical care of marine mammals. *J. Am. Vet. Med. Assoc.* **147**:1077–1085.

Ridgway, S. H. 1968. The bottlenosed dolphin in biomedical research. *Methods Anim. Exp.* **3**:416–417.

Ridgway, S. H. 1972. Homeostasis in the aquatic environment, in: S. H. Ridgway (ed.), *Mammals of the Sea: Biology and Medicine*, pp. 590–747. Thomas, Springfield, IL.

Rommel, S. A. 1990. Osteology of the bottlenose dolphin, in: S. Leatherwood and R. R. Reeves (eds.), *The Bottlenose Dolphin*, pp. 29–49. Academic Press, San Diego, CA.

Rommel, S. A., Pabst, D. A., McLellan, W. A., Mead, J. G., and Potter, C. W. 1992. Anatomical evidence for a countercurrent heat exchanger associated with dolphin testes. *Anat. Rec.* **232**(1):150–156.

Rommel, S. A., Pabst, D. A., and McLellan, W. A. 1993. Functional morphology of the vascular plexuses associated with the cetacean uterus. *Anat. Rec.* **237**(4):538–546.

Rommel, S. A., Pabst, D. A., McLellan, W. A., Williams, T. M., and Friedl, W. A. 1994. Temperature regulation of the testes of the bottlenose dolphin (*Tursiops truncatus*): evidence from colonic temperatures. *J. Comp. Physiol. B* **164**:130–134.

Schmidt-Nielsen, K. 1990. *Animal Physiology: Adaptation and Environment*, 4th ed. Cambridge University Press, London.

Scholander, P. F., and Schevill, W. E. 1955. Counter-current vascular heat exchange in the fins of whales. *J. Appl. Physiol.* **8**:279–282.

Schroeder, J. P. 1990. Breeding bottlenose dolphins in captivity, in: S. Leatherwood and R. R. Reeves (eds.), *The Bottlenose Dolphin*, pp. 425–446. Academic Press, San Diego, CA.

Schummer, A., Nickel, R., and Sack, W. O. 1979. *The Viscera of the Domestic Mammals*. Verlag Paul Parey, Berlin.

Schummer, A., Wilkens, H., Vollmerhaus, B., and Habermehl, K.-H. 1981. *The Circulatory System, the Skin, and the Cutaneous Organs of the Domestic Mammals*. Verlag Paul Parey, Berlin.

Shelton, M. 1964. Relation of environmental temperature during gestation on birth weight and mortality in lambs. *J. Anim. Sci.* **23**:360–364.

Simpson, J. G., and Gardner, M. B. 1972. Comparative microscopic anatomy of selected marine mammals, in: S. H. Ridgway (ed.), *Mammals of the Sea*, pp. 298–418. Thomas, Springfield, IL.

Slijper, E. J. 1936. *Die Cetaceen: Vergleichend-Anatomisch und Systematisch*. Asher & Co., Amsterdam, 1972 reprint.

Slijper, E. J. 1966. Functional morphology of the reproductive system in Cetacea, in: K. S. Norris (ed.), *Whales, Dolphins, and Porpoises*, pp. 277–319. University of California Press, Berkeley.

Slijper, E. J. 1979. *Whales*. Cornell University Press, Ithaca, NY.

Smith, M. R., Shivji, M. S., Waddell, V. G., and Stanhope, M. J. 1996. Phylogenetic evidence from the IRBP gene for the paraphyly of toothed whales, and mixed support for Cetacea as a suborder of Artiodactyla. *Mol. Biol. Evol.* **13**(7):918–922.

Thewissen, J. G. M. 1994. Phylogenetic aspects of cetacean origins: a morphological perspective. *J. Mamm. Evol.* **2**(3):157–184.

Van Demark, N. L., and Free, M. J. 1970. Temperature regulation and the testis, in: A. D. Johnson, W. R. Gomes, and N. L. Van Demark (eds.), *The Testis*, Volume III, pp. 233–312. Academic Press, New York.

Van der Schoot, P. 1995. Studies on the fetal development of the gubernaculum in Cetacea. *Anat. Rec.* **243**:449–460.

Van Valen, L. 1966. Monophyly or diphyly in the origin of whales. *Evolution* 22:37–41.

Walmsley, R. 1938. Some observations on the vascular system of a female fetal finback. *Contrib. Embryol.* **27**:109–178.

Whittow, G. C., Hampton, I. F. G., Matsura, D. T., Ohata, C. A., Smith, R. M., and Allen, J. F. 1974. Body temperature of three species of whales. *J. Mammal.* **55**:653–656.

Wiley, E. O. 1981. *Phylogenetics: The Theory and Practice of Phylogenetic Systematics*. Wiley, New York.

Williams, T. M., Friedl, W. A., Fong, M. L., Yamada, R. M., Sedivy, P., and Haun, J. E. 1992. Travel at low energetic cost by swimming and wave-riding bottlenose dolphins. *Nature* 355:821–823.

Wislocki, G. B. 1933. On the placentation of the harbor porpoise (*Phocoena phocoena* Linnaeus). *Biol. Bull.* **65**:80–98.

Wislocki, G. B., and Enders, R. K. 1941. The placentation of the bottle-nosed porpoise (*Tursiops truncatus*). *Am. J. Anat.* **68**(1):97–125.

Woollard, H. H. 1922. The development of the principal arterial stems in the forelimb of the pig. *Contributions to Embryology, Carnegie Institute of Washington* **70**:139–154 + plates 1 and 2.

Worthy, G. A. J., and Edwards, E. F. 1990. Morphometric and biochemical factors affecting heat loss in a small temperate cetacean (*Phocoena phocoena*) and a small tropical cetacean (*Stenella attenuata*). *Physiol. Zool.* **63**(2):432–442.

CHAPTER 14

Isotopic Approaches to Understanding the Terrestrial-to-Marine Transition of the Earliest Cetaceans

LOIS J. ROE, J. G. M. THEWISSEN, JAY QUADE, JAMES R. O'NEIL, SUNIL BAJPAI, ASHOK SAHNI, and S. TASEER HUSSAIN

1. Introduction

The fossil record is replete with examples of evolutionary transitions between marine and freshwater environments, in both directions. Perhaps the most striking and best documented example of such a transition is the evolution of cetaceans (whales, dolphins, and porpoises) from the extinct group of terrestrial mammals called *mesonychians*. This transition, first hypothesized by Van Valen (1966), occurred in the temporally and geographically restricted setting of the Paleogene remnant Tethyan epicontinental sea (Gingerich *et al.*, 1983) and adjacent terrestrial ecosystems. These environments lay in the zone of convergence between the Indian Plate and southern Eurasia during the early stages of the continent–continent collision that ultimately produced the Himalayas.

The evolution of cetaceans from terrestrial land-mammals included profound changes in many aspects of their anatomy and biology. Many of those changes, such as those that occurred in the locomotory and hearing systems, have been documented through studies of the morphology of fossils (e.g., Thewissen and Hussain, 1993; Gingerich *et al.*, 1994; Thewissen *et al.*, 1994). But the morphological transitions are only part of the story. In order to become fully marine and independent of freshwater, cetaceans had to evolve physiologically as well. They had to develop the ability to cope with the excess salt load associ-

LOIS J. ROE • Division of Ecosystem Sciences, University of California, Berkeley, Berkeley, California 94720.   J. G. M. THEWISSEN • Department of Anatomy, Northeastern Ohio Universities College of Medicine, Rootstown, Ohio 44272.   JAY QUADE • Desert Laboratory and Department of Geosciences, University of Arizona, Tucson, Arizona 85721.   JAMES R. O'NEIL • Department of Geological Sciences, University of Michigan, Ann Arbor, Michigan 48109.   SUNIL BAJPAI • Department of Earth Sciences, University of Roorkee, Roorkee 247667, Uttar Pradesh, India.   ASHOK SAHNI • Centre for Advanced Studies in Geology, Panjab University, Chandigarh 160-014, India.   S. TASEER HUSSAIN • Department of Anatomy, Howard University College of Medicine, Washington, D.C. 20059.
The Emergence of Whales, edited by Thewissen. Plenum Press, New York, 1998.

ated with seawater ingested either voluntarily (e.g., as drinking water) or incidentally (e.g., in the course of eating).

Most modern cetaceans are fully marine and never enter freshwater, but there are some species that are known to enter brackish waters of low salinities (e.g., *Phocoena phocoena*; Andersen and Nielsen, 1983), and the four species of highly endangered freshwater river dolphins known as *platanistoids* are exclusively freshwater. Although the platanistoids are secondarily evolved from marine species (Messenger, 1994; Muizon, 1994) and were not part of the original transition of cetaceans, they are important in the context of this study because they provide us with an appropriate modern freshwater group to compare with modern marine cetaceans.

When did the cetacean osmoregulatory system become able to handle the excess salt load associated with ingesting seawater and feeding in the oceans? The answer to this question is key to understanding the evolution of cetaceans, as this adaptation was probably prerequisite to the worldwide dispersal of cetaceans in the middle Eocene (e.g., Hulbert and Petkewich, 1991; Albright, 1996).

The stable isotope ratios of carbon ($^{13}C/^{12}C$) and oxygen ($^{18}O/^{16}O$) can provide such information, and have the potential to elucidate the timing and nature of the transition of early cetaceans from terrestrial to marine mammals. Our goal is to answer several fundamental questions: (1) When in geologic time did cetaceans become fully marine? (2) Was the transition to marine life a gradual or abrupt one? (3) Was the transition to seawater ingestion synchronous with the transition to a marine diet or did one precede the other? (4) To what extent did early whales differentiate ecologically?

Our approach involves five components: phosphate oxygen isotope analyses of the teeth and bones of (1) modern cetaceans and (2) Eocene cetaceans; carbon isotope analyses of structural carbonate from the teeth and bones of (3) modern and (4) Eocene cetaceans; and (5) an assessment of the degree to which diagenesis may have affected our isotopic results.

2. Background

2.1. Ion and Water Balance in Modern Cetaceans

2.1.1. Overview

To understand the terrestrial-to-marine transition of the earliest cetaceans, one must first appreciate how osmoregulation works in modern cetaceans and how it relates to the oxygen isotope composition of cetacean teeth and bones. Mammals, including those that are aquatic, lose water through respiration, skin evaporation, urination, defecation, lacrimation, and lactation (Bentley, 1971; Ridgway, 1972), and most drink freshwater to replace this loss. Some mammalian species have kidneys that can rid the body of salt and urea by producing urine more concentrated than their body fluids, but most mammals are not able to concentrate their urine very strongly. Many, including humans, die if they drink only seawater because their kidneys are unable to concentrate the ions sufficiently and too much water is lost (Schmidt-Nielsen, 1997). As a result, most mammals avoid drinking seawater and cannot survive without freshwater (Schmidt-Nielsen, 1964).

It is unclear how the osmoregulatory system of marine cetaceans handles ingested seawater. Several mechanisms involved in maintaining osmotic balance have been identified: (1) osmotically driven transfer of water but not ions across the integument (= percutaneous or transcutaneous flux; Hui, 1981; Andersen and Nielsen, 1983); (2) active drinking (Telfer *et al.*, 1970); (3) metabolism of food and/or body fat (Bentley, 1963) to produce freshwater; and (4) concentration of ions in urine to conserve water (Ridgway, 1972). In addition, water conservation may also be aided by concentration of lipids in milk during lactation (reviewed in Schmidt-Nielsen, 1997), lower respiratory rates, and undersaturation of water in expired air (Kirschner, 1991).

Kohn (1996) modeled the oxygen isotope systematics of cetaceans as being dominated by the transcutaneous flux of water. Using the results of Hui (1981), Andersen and Nielsen (1983), and Sokolov *et al.* (1994), he calculated that the transcutaneous flux constituted 98% of the total water flux. On this basis, he suggested that the phosphate oxygen isotope compositions of cetacean teeth and bones should record the isotopic composition of the water in which the animals live (Kohn, 1996). If the transcutaneous flux dominates the water fluxes in all cetaceans, measurements of the oxygen isotope compositions of cetacean teeth and bone phosphate would not reflect drinking water, diet, and species-specific physiological variables as they do in terrestrial animals (Bryant and Froelich, 1995; Kohn, 1996; Kohn *et al.*, 1996), but the environment in which the animals live (Kohn, 1996). But is this true of all cetaceans?

Until recently, it was thought that cetacean skin was impermeable to both solutes and water (Ridgway, 1972; see review in Kirschner, 1991), in part because sweating does not work underwater (Ridgway, 1972). Recent general physiology texts (e.g., Schmidt-Nielsen, 1997) refer to cetaceans and other higher marine vertebrates as more similar to terrestrial mammals than to fish, because, with respect to salt and water balance, they are physiologically isolated from the surrounding seawater. The results of some recent studies (Hui, 1981; Andersen and Nielsen, 1983; Sokolov *et al.*, 1994), however, suggest that this assumption of physiological isolation may be at least partially incorrect. Hui (1981) demonstrated experimentally that cetacean skin is impermeable to sodium, but permeable to water, which will flow into the animal when the osmotic pressure inside the animal is high enough to create the necessary gradient. Andersen and Nielsen (1983) demonstrated that the flow of water could occur in both directions, depending on the ionic concentrations of the external aqueous environment. The results of these experiments stand in contrast to the results of other experiments that seemed to indicate more important roles for drinking (Telfer *et al.*, 1970), urine concentration (Ridgway, 1972), and metabolism of food and body fat (Bentley, 1963; Telfer *et al.*, 1970). In most of these earlier experiments, however, skin impermeability was either implicitly (e.g., Bentley, 1963) or explicitly (Fetcher and Fetcher, 1942; Ridgway, 1972) assumed and was not included in calculations. Only Telfer *et al.* (1970) attempted a direct test of permeability, but their test involved the monitoring of a solute—radioactive ^{24}Na—rather than the water itself. Given Hui's (1981) demonstration that cetacean skin is permeable to water but not to sodium, the Telfer *et al.* test was not adequate to understand what is happening to the water, only to the solutes, and in this regard is quite consistent with Hui's (1981) results. The importance of drinking, metabolic water, and the concentration of ions in the urine may therefore need to be reevaluated.

2.1.2. Implications for Understanding Cetacean $\delta^{18}O_p$

Understanding which processes exert control over the osmotic balance in modern cetaceans is important for studies such as this one, because if the transcutaneous flux really dominates in modern cetaceans, it would be reasonable to infer that the first appearance of marine isotope signatures in cetacean teeth and bones would reflect evolution of a permeable integument. If, on the other hand, drinking dominates the maintenance of water and ion balances in modern cetaceans, then the same isotopic evidence would be interpretable as the beginning of cetacean seawater drinking. Similar arguments could be made for kidney evolution, and other possible mechanisms.

Despite the uncertainty regarding the mechanisms of osmoregulation in modern cetaceans, we can still determine when cetaceans became independent of freshwater and able to eat marine food. We accomplish this by determining empirically the nature and magnitude of oxygen and carbon isotope differences between modern freshwater and marine cetaceans. These measurements provide a framework for interpreting isotopic analyses of fossil species. The basis of these measurements in modern systems is discussed next.

2.2. Oxygen Isotope Variations in Fresh and Marine Waters

Oxygen isotope ratios ($^{18}O/^{16}O$) of fresh and marine waters differ because of the kinetic and equilibrium fractionations that result from the physical processes of evaporation and condensation operative in the hydrologic cycle (e.g., Epstein and Mayeda, 1953; Craig, 1961a). For example, $H_2^{16}O$ molecules preferentially evaporate (move from the liquid into the vapor phase) and $H_2^{18}O$ molecules preferentially condense out as precipitation. Evaporation is a kinetic process, and the water vapor produced by evaporation from the ocean has been found to be more depleted in ^{18}O than would be the case if the vapor formed in isotopic equilibrium with ocean water (Craig and Gordon, 1965). Despite the kinetic nature of this fractionation, the oxygen isotope compositions of water vapors formed over ocean waters have been found to be consistently 11 to 14‰ lower than those of the oceans (Craig and Gordon, 1965), with the exact fractionation dependent on the humidity and isotopic composition of water vapor already in the air at the time (Gat, 1981). As this vapor in the air mass moves inland, it progressively condenses out, thereby causing the clouds to become further depleted in ^{18}O (Epstein and Mayeda, 1953; Fig. 1). As the clouds become progressively more depleted in ^{18}O, the precipitation becomes more depleted in ^{18}O as well. In this way, the physical processes of the hydrologic cycle produce an isotopic difference between the oceans and freshwaters, such that freshwaters have lower ratios of $^{18}O/^{16}O$ than do the oceans. This difference is expressed in $\delta^{18}O$ values, where delta (δ) represents the deviation, in parts per mil (‰), of the $^{18}O/^{16}O$ of the sample from that of standard mean ocean water (SMOW). Samples with ratios of $^{18}O/^{16}O$ that are lower than those of seawater have lower (more negative) $\delta^{18}O$ values (see Methods section). The magnitude of this fractionation is partly a function of temperature, but the isotopic composition of any given body of freshwater also depends on the proportion of the clouds that have condensed out as precipitation by that point, how much evaporation has occurred, and what mixture of different waters and transport between phases has occurred (Whelan, 1987). In addition, $\delta^{18}O$ values decrease as one moves farther inland, higher in elevation, and also higher in latitude (Craig, 1961a; Dansgaard, 1964).

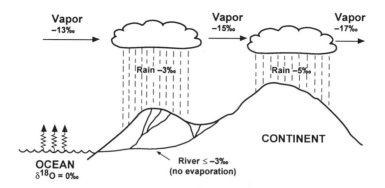

FIGURE 1. Schematic representation of the fractionation of oxygen isotopes in the hydrologic cycle as a result of evaporation and condensation (based on the work of Epstein and Mayeda, 1953; Craig, 1961a,b; and Craig and Gordon, 1965). Note resultant difference between the oxygen isotope compositions of the ocean and freshwaters. Delta (δ) values are in parts per mil (‰) deviation from standard mean ocean water (SMOW).

In this study, latitudinal variation is not a problem because the species that are most critical to distinguish as marine or freshwater lived in relatively close proximity to each other in, or around the margin of, the remnant Tethys. These early and middle Eocene deposits of the remnant Tethys all occur today at latitudes of 33°40' ±/10', and although their latitude may have been slightly different during the Eocene, their proximity to each other changed only slightly (on the order of a few kilometers) as a result of folding and faulting in the region. Our other samples—*Andrewsiphius, Gaviacetus, Indocetus,* and *Remingtonocetus* from Kachchh, India (23°30' ± 10' N latitude) and *Georgiacetus* from Burke County, Georgia (approximately 31° N latitude)—occurred in pelagic, open marine settings that are more readily distinguishable from terrestrial environments.

Also, although we do not know the Eocene elevations of the land surrounding the remnant Tethys, significant elevation would only have increased the isotopic difference between the local freshwaters and the Tethys, by adding water depleted in ^{18}O from precipitation at higher elevations. As a result, our lack of knowledge of paleoelevations is not a problem.

How do we determine what magnitude of difference to accept as indicative of a difference between marine and freshwaters? Although this difference depends in part on how far inland the meteoric water condensed out of the vapor phase, we can determine a minimum magnitude of the marine–freshwater difference by considering known fractionation factors. The mean $\delta^{18}O$ value of vapor forming over the ocean at 25°C is -13‰, or 13‰ more negative than SMOW (Fig. 1); the $\delta^{18}O$ value of the rain forming from that vapor would be 10‰ more positive or -3‰ (SMOW). The net result is that freshwaters consisting of pure, unevaporated rainwater would be about 3‰ more negative than the ocean (Fig. 1). Precipitation forming farther inland (or groundwater) would have lower $\delta^{18}O$ values, and could lower the values of freshwater near the coast by mixing. A river near the coast would thus have a $\delta^{18}O$ value of -3‰ if no evaporation takes place. Evaporation of water would increase the $\delta^{18}O$ value. At lower temperatures, the magnitudes of the differences would be slightly greater; at higher temperatures, they would be slightly less (for equilibrium fractionation factors between liquid water and water vapor at these temperatures, see Friedman and O'Neil, 1977).

An empirical test of this three per mil difference is a comparison with data on the isotopic composition of meteoric waters from small islands or stations very near the ocean, where only one cycle of evaporation and precipitation is possible. One such site is the station of Malan, South Africa, which is located at 33°97′ S latitude, and receives rain with a mean $\delta^{18}O$ value of approximately -3.5‰ (SMOW), at a mean and annual temperature of 15.9°C (International Atomic Energy Agency, 1986). The average elevation of Malan is 44 m above sea level, which is too low to produce any detectable fractionation in the precipitation. Similarly, meteoric water falling on Midway Island, which is located at 28°22′ N latitude and has an average elevation of 13 m above sea level, has an average $\delta^{18}O$ value of -1.81‰ (International Atomic Energy Agency, 1986).

Our strategy, therefore, is to exploit the persistence through geologic time of the hydrologic cycle (Kolodny and Luz, 1991), in order to document the habitat use (marine versus freshwater) of cetaceans in the fossil record. To do this, we need a proxy record of the water of the animals' habitats. This record exists in teeth and bones, whose oxygen isotope compositions are linearly related to the oxygen isotope composition of the water ingested, which is in turn related to the oxygen isotope composition of the environmental water in which an animal lives or that an animal primarily consumes (Longinelli and Nuti, 1973a,b; Kolodny *et al.*, 1983; Bryant and Froelich, 1995).

The precise nature of that linear relation varies with diet and physiology (Kohn, 1996; Kohn *et al.*, 1996), and possibly body size (Bryant and Froelich, 1995). The body water of most terrestrial mammals is derived primarily from the water ingested directly as drinking water (Luz *et al.*, 1984; Luz and Kolodny, 1985, 1989; Bryant and Froelich, 1995). Contributions from other sources of oxygen, such as food and atmospheric oxygen, are generally small compared with that from drinking water (Luz *et al.*, 1984; Bryant and Froelich, 1995). Some notable exceptions occur in arid environments, however, where animals must get much of their water from their food (Schmidt-Nielsen, 1964). Where an animal's food consists largely of leaves, from which substantial evaporation takes place, the oxygen isotope composition of an animal's teeth and bones reflects relative humidity (Ayliffe and Chivas, 1990; Luz *et al.*, 1990). There may also be species-specific physiological effects that influence the isotopic composition of the body water and therefore of teeth and bones (Kohn *et al.*, 1996).

How is the oxygen isotope composition of environmental water related to the oxygen isotope composition of the phosphate (PO_4^{3-}) in the teeth and bones of marine and freshwater cetaceans? Kohn (1996) inferred that the rapid, transcutaneous (= percutaneous) exchange of water documented in several cetacean species by Hui (1981), Andersen and Nielsen (1983), and Sokolov *et al.* (1994) dominates all other oxygen fluxes, and concluded that oxygen isotope compositions of body water should track the isotopic composition of ambient water perfectly. An implicit assumption of this model is that the gradient(s) needed to drive that flux, which is osmotic in nature, exist continuously throughout the life of the animals, which may not be the case.

Despite this and other possible complications, there is abundant empirical evidence of an isotopic difference between marine and freshwater cetaceans. Yoshida and Miyazaki (1991) and Barrick *et al.* (1992) presented evidence of a linear relation between the oxygen isotope composition of environmental water and the oxygen isotope composition of the teeth and bones of modern cetaceans. Yoshida and Miyazaki (1991) determined the oxygen isotope compositions of both marine and freshwater cetaceans and found a 3–7‰ difference between the $\delta^{18}O$ values of freshwater and marine species. The presence of a 3‰ minimum difference is in good agreement with our minimum estimate of the difference between marine and freshwaters based on hydrologic data and is further borne out by our recent analyses (Thewissen *et al.*, 1996b), which yielded a 2–3‰ minimum difference between both modern and fossil cetaceans.

2.3. Carbon Isotope Ratios ($^{13}C/^{12}C$) of Fresh and Marine Waters

Like the oxygen isotope compositions of natural waters, the carbon isotope compositions of both the organic and inorganic constituents of freshwaters are depleted in the heavier isotope—in this case ^{13}C—relative to the corresponding substances in the oceans (Clayton and Degens, 1959; Fig. 2). This difference is recorded in the carbon isotope compositions of both biogenic and abiogenic carbonates (Clayton and Degens, 1959; Keith *et al.*, 1964; Allen and Keith, 1965; Keith and Parker, 1965). Consequently, $\delta^{13}C$ values of carbonate phases are useful in distinguishing marine and freshwater environments and organisms (Keith *et al.*, 1964; Allen and Keith, 1965; see review in Fry and Sherr, 1984).

The teeth and bones of vertebrates contain structural carbonate and the carbon isotope composition of this carbonate has been demonstrated to reflect the carbon isotope composition of an animal's diet in terrestrial animals, with a 12‰ offset (e.g., Lee-Thorp and van der Merwe, 1987; Quade *et al.*, 1992). A relation between the carbon isotope composition of tooth and bone carbonate and diet has not been demonstrated in aquatic vertebrates, but that between diet and bone collagen of aquatic vertebrates has (DeNiro and Epstein, 1978; Schoeninger and DeNiro, 1984; Ames *et al.*, 1996; Hobson *et al.*, 1997). In modern ecosystems, carbon isotope differences between marine and freshwater animals are evident in

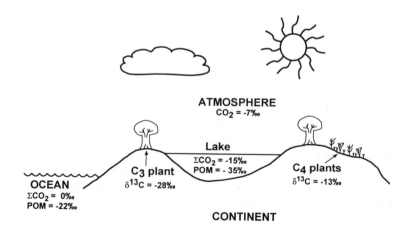

FIGURE 2. Schematic representation of the average carbon isotope composition of some major ecosystem components. Note the difference in isotopic composition of total inorganic carbon species (reported as CO_2) and organic matter between the ocean and freshwaters. POM, particulate organic matter. Delta (δ) values are in parts per mil (‰) deviation from the Pee Dee Belemnite standard (PDB). (After Peterson and Fry, 1987).

analyses of the soft tissues (Schoeninger and DeNiro, 1984; Keegan and DeNiro, 1988; Little and Schoeninger, 1995; Smith *et al.*, 1996), although overlap of the carbon isotope compositions of groups such as sea grasses and terrestrial C_4 plants must be taken into account (Keegan and DeNiro, 1988; Little and Schoeninger, 1995).

Here we employ analyses of the carbon isotope compositions of the structural carbonate in the teeth and bones of the earliest cetaceans to determine which of these animals ate marine and which ate freshwater or terrestrial food. We begin by making analyses of the structural carbonate of the teeth and bones of modern cetaceans in order to test empirically the existence of a difference between the carbon isotope composition of the tooth and bone carbonate of marine and freshwater animals. These analyses of modern cetacean tooth carbonate provide the framework for evaluating the results of our analyses of the carbon isotope compositions of fossil cetaceans. In addition, although the first fossil and isotopic evidence of C_4 plants does not appear until the Miocene, we analyzed pedogenic carbonates and the teeth and bones of herbivores in order to check for the presence of C_4 plants in the Eocene Tethyan systems.

3. Materials and Methods

3.1. Samples

3.1.1. Modern Cetaceans

Our aim was to include as broad a sampling as possible of modern freshwater and marine cetaceans, with a range of body sizes, locations, and diet. The modern cetaceans represented in our sample varied in typical body weight from 35 kg (the Tucuxi dolphin, *Sotalia*) to 40,000 kg (the sperm whale, *Physeter*), thus spanning the full range of cetacean body sizes and most of the range of body sizes considered by Bryant and Froelich (1995). Latitudinally, our specimens spanned temperate (*Phocoena*) to tropical (*Inia*) latitudes. Their diets varied from primarily invertebrates (*Physeter*) to predominantly fish (*Tursiops*), and a combination of fish and amniote prey (*Orcinus*).

Specimens of modern cetaceans were supplied by several institutions, including the American Museum of Natural History (AMNH), the U.S. National Museum (Smithsonian; USNM), and the National Oceanic and Atmospheric Administration's National Marine Fisheries Services (NMFS) laboratory in La Jolla, California. Because the methods we used to determine oxygen and carbon isotope ratios are destructive, we analyzed specimens of more common, less endangered species whenever possible. Specimens of some of the more common species we analyzed were not catalogued in regular museum collections and lacked provenance data. In sampling species that are found in both freshwater and seawater, however, we analyzed only specimens with good provenance data. Our specimen of *Sotalia* (USNM 571461), a genus that sometimes enters freshwaters, was collected in the sea off the coast of Trinidad. Similarly, we analyzed samples of freshwater dolphins of known provenance. These included one specimen of *Platanista gangetica* (NMFS uncatalogued specimen from the Ganges River); three of *Inia geoffrensis* (WEE 069 from the Orinoco River; USNM 396166 from the San Fernando River, Venezuela; and USNM 406801 from near San Juan, Venezuela); and one of *Lipotes vexilifer* (AMNH-M 57333) from Tung Ting

Lake in Hunan Province in the People's Republic of China. All of our modern specimens were teeth, except for our sample of the rare Yangtze River dolphin *Lipotes*, of which only a rib fragment was available.

3.1.2. Eocene Cetaceans

We analyzed fossil cetacean samples from the early Eocene Kuldana and middle Eocene Kohat Formations of Punjab, northern Pakistan, the middle Eocene Harudi Formation of Kachchh, in western India, and the late middle Eocene McBean Formation of Georgia, USA. Details of the stratigraphy and locality maps are provided by Williams (this volume). Cetaceans from the lower Kuldana Formation included *Ichthyolestes, Pakicetus*, and *Nalacetus*, all from a single locality in the Kala Chitta Hills of Punjab, Pakistan (Thewissen and Hussain, 1998). This locality (H-GSP Locality 62, West and Lukacs, 1979) has recently been interpreted by Aslan and Thewissen (1996) as a freshwater channel deposit that may be similar depositionally to the Barbora Banda locality described by Wells (1983). The associated fauna consists exclusively of terrestrial and freshwater vertebrates and planorbid gastropods.

Our cetacean specimens from the upper Kuldana Formation included the holotype of *Ambulocetus natans* (HGSP-18507; H-GSP Locality 9209), and several specimens referred to that species (H-GSP Locality 9207). H-GSP localities 9207 and 9209 are stratigraphically continuous and are approximately 90 m higher in the section than H-GSP Locality 62 and approximately 3 km away (Thewissen *et al.*, 1996a) from the latter. The depositional environment represented by localities 9207 and 9209 was nearshore marine (Wells, 1984; Thewissen *et al.*, 1996a). The noncetacean taxa (pycnodontid fish, oysters, and crabs) found at these localities are consistent with this environmental interpretation.

Our samples from the overlying Kohat Formation included a single specimen of *Attockicetus* (Thewissen and Hussain, 1998), the first whale recovered from that formation. This specimen includes the dorsal part of the skull with a few teeth.

Our specimens of middle Eocene (Lutetian) fossils from northern India come from the Rato Nala locality of the Harudi Formation of Kachchh District, Gujarat State, northern India. We have analyzed *Andrewsiphius, Gaviacetus, Indocetus*, and *Remingtonocetus* (Sahni and Mishra, 1975; Kumar and Sahni, 1986; Bajpai and Thewissen, this volume).

From the United States, we analyzed a tooth of the recently discovered protocetid whale *Georgiacetus* Hulbert 1995 from the middle Eocene McBean Formation of Burke County, Georgia (Hulbert and Petkewich, 1991; Hulbert, this volume).

3.2. Analytical Methods

Wherever possible, the enamel and dentine of fossil teeth were separated using a dental drill. Specimens that were too fragmentary to allow such separation are listed in Table II as enamel and dentine (e,d or d,e depending on the relative amounts of each). Most modern cetacean teeth have very thin ($\sim$1 mm) tooth enamel and/or are relatively small (e.g., the teeth of *Phocoena*). Because these teeth yield such small amounts of enamel, and diagenesis is not a concern when analyzing modern specimens, we analyzed most of the modern teeth whole.

Following separation, where applicable, samples were ground in a mortar and pestle, treated with approximately 3% sodium hypochlorite (NaOCl) to oxidize organic matter, and rinsed five times with deionized distilled water. Samples being prepared for phosphate analyses were then placed in a drying oven at approximately 80°C and, once dry, reground to a fine powder. Fossil samples that were used in carbonate analyses were subjected to a 1 M acetic acid [CH_3COOH (aq)] treatment to remove secondary, nonstructural carbonate, following the suggestion of Lee-Thorp and van der Merwe (1991) that this step solves the problem of diagenetic overprinting documented, for example, by Nelson *et al.* (1986). After the acetic acid leaching step, samples were again rinsed five times with deionized distilled water to ensure the removal of any acetate residue. Omitting this step results in the liberation of isotopically light CO_2 during the phosphoric reaction step (J. Quade, unpublished data).

Diagenesis is a concern when analyzing fossils, especially in settings where the depositional environments have repeatedly alternated between marine and freshwater through time, as happened during the Paleogene in the remnant Tethyan region of South Asia. It is generally agreed that tooth enamel is the best hard tissue to analyze because it is generally more densely crystalline and more resistant to alteration than are dentine, cementum, and bone (Lee-Thorp and van der Merwe, 1991; Ayliffe *et al.*, 1992, 1994; Koch *et al.*, 1994). For this reason, we analyzed tooth enamel wherever possible, but as some specimens of interest did not have tooth enamel, we chose not to limit ourselves to analyses of enamel in order to sample as many taxa as possible. To ascertain that our analyses of dentine and bone were meaningful, we made comparative analyses of bone, dentine, and enamel from several specimens to determine how our results would change if we used dentine and/or bone instead of enamel. Because we did not have adequate amounts of bone, dentine, and enamel from the cetaceans from the lower Kuldana Formation, we analyzed instead two anthracobunid specimens (H-GSP 83-31p and H-GSP 96214) and two brontothere specimens (H-GSP 18478 and H-GSP 96034). In addition, as a test of our pretreatment steps, we made analyses of treated and untreated aliquots of several samples and compared their isotopic compositions.

In order to determine whether the phosphate oxygen had been altered, we made analyses of the crystallinity of our specimens using Fourier transform infrared spectroscopy (FTIR) according to the method described by Shemesh (1990). We discovered, however, that this method is highly sensitive to the presence of absorbed water, which causes crystals to clump together and makes the sample appear less crystalline [= have a lower crystallinity index (C.I.)]. Progressive removal of absorbed water under vacuum increased the crystallinity of our samples, from below Shemesh's (1990) crystallinity index threshold of 4.0 to above it. Unfortunately, without a way of standardizing for the amount of absorbed water, we are unable to relate our results to Shemesh's (1990) crystallinity index.

Extractions of the carbon and oxygen in the carbonate (CO_3^{2-}) component of the teeth and bones of our samples were performed using the standard phosphoric acid [H_3PO_4 (aq)] reaction method of McCrea (1950) at 50°C, until the reactions ceased. Carbonate standards (IAEA-Cl, Carrara Marble and NBS-18, carbonatite) reacted completely within 30 min. Tooth enamel samples generally required a longer reaction time of 3 hours or more, but extensive regrinding of the tooth samples after pretreatments and prior to loading lessened their reaction times considerably. Oxygen isotope analyses of the phosphate (PO_4^{3-}) component were made using the trisilver phosphate (Ag_3PO_4) thermal decomposition method of O'Neil *et al.* (1994) at 1200°C. The end product of both the carbonate and phosphate extraction methods is CO_2, which we analyzed on Finnigan MAT Delta-S series 251 gas source mass spectrometers in the Stable Isotope Laboratories at the University of Arizona and the University of Michigan. Analytical precision of our samples is ± 0.06‰ or better for large (>50 μmol) samples; smaller samples had analytical uncertainties better than or equal to ±0.10‰.

We report our isotopic results in the standard delta (δ) notation as the deviation, in parts per mil (‰), of the isotopic ratios of the sample from those of the PDB and SMOW standards for carbon and oxygen, respectively, where $\delta = [R_{sample} - R_{standard}/R_{standard}]* 1000$ and $R = {}^{18}O/{}^{16}O$ or ${}^{13}C/{}^{12}C$. We have chosen not to treat the within-sample and within-taxon variation of the samples statistically, in recognition of the possibility that this variation may represent biologically real phenomena, such as the movement of the organisms between environments. In such a case, the ranges of delta values, rather than any measure of central tendency, will be the key to understanding the biology of these extinct species. We therefore present our data as ranges of values for each taxon, rather than as means with associated standard errors or standard deviations.

4. Results

4.1. Oxygen Isotope Compositions ($\delta^{18}O_p$) of Tooth and Bone Phosphate

4.1.1. Modern Cetaceans

Phosphate $\delta^{18}O$ values ($\delta^{18}O_p$) obtained for the modern cetaceans (Table I, Fig. 3A) range from +10.8 to +19.6‰ (SMOW), with a 2.4‰ gap between the ranges of the freshwater and marine species. The freshwater cetaceans have values in the range +10.8 to +15.7‰ and the marine cetaceans have values between +18.1 and +19.6‰. These results are very similar to those obtained by Yoshida and Miyazaki (1991), whose measurements of the $\delta^{18}O_p$ values of freshwater cetaceans ranged from +11.1 to +13.3‰, whereas those of marine cetaceans in their study ranged from +16.7 to +18.6‰, with a 3.4‰ gap between the freshwater and marine ranges.

4.1.2. Eocene Cetaceans

The Eocene fossil cetaceans we analyzed had $\delta^{18}O_p$ values spanning a range almost identical to that of the modern species (Table II, Fig. 3B). Two of the three cetaceans we analyzed from the lower Kuldana Formation, *Nalacetus* ($n = 3$) and *Pakicetus* ($n = 7$), had $\delta^{18}O_p$ values between +15.0 and +16.6‰; the third cetacean from the lower Kuldana, *Ichthyolestes* ($n = 3$), had a slightly larger range of values, from +13.8 to +16.3‰. Some of these values (4 out of 10) fall in the range of values of modern freshwater cetaceans, but slightly more (6 out of 10) were higher. In contrast, *Ambulocetus* and *Gandakasia* from the upper Kuldana Formation had a wide range of $\delta^{18}O_p$ values of the modern cetaceans, overlapping modern freshwater and marine values. *Ambulocetus* ($n = 13$) $\delta^{18}O_p$ values ranged from +13.0 to +20.1‰; *Gandakasia* ($n = 5$), which is also an ambulocetid (Thewissen *et al.*, 1996a), had a narrower range of $\delta^{18}O_p$ values, from +14.0 to +17.9, but that may reflect a smaller sample size.

Table I. Isotopic Compositions ($\delta^{18}O_p$, $\delta^{13}C_{sc}$) of the Teeth and Bones of Modern Cetaceans

Taxon	Specimen	Provenance	$\delta^{18}O_p$, ‰ (SMOW)	$\delta^{13}C_{sc}$, ‰ (PDB)
Delphinus	FKB-155B T1	C. Pacific	19.7	—
	FKB-155B T2	C. Pacific	19.2	−10.1
	FKB-155B T3	C. Pacific	—	−10.7
Inia	USNM 396166	San Fernando River, Venezuela	15.7	−11.6
	USNM 406801	Near San Juan, Venezuela	15.5	—
	WEE069	Orinoco River	13.6	−17.1
Lipotes	AMNH-M 57333	Tung Ting Lake, Hunan Prov., China	11.5	−12.8
Orcinus	USNM uncat.	Unknown	18.1	—
Phocoena	USNM uncat. 1c	Unknown	—	−8.7
	USNM uncat. 1d	Unknown	—	−8.9
Physeter	USNM uncat.	Unknown	19.0	−13.1
	FKB-508	Off San Diego	19.1	−8.0
Platanista	NMFS uncat.	Ganges River, India	14.9	—
Sotalia	USNM uncat. 1	Off Trinidad	18.9	—
	UDNM uncat. 2	Off Trinidad	19.3	−7.8
Stenella	USNM 571461	Off Trinidad	19.5	−8.5
Tursiops	USNM uncat. 1a	Off N. Carolina	18.3	−9.8
	USNM uncat. 2a	Off N. Carolina	18.7	−8.4

The Kohat Formation is the only pelagic marine unit in Pakistan from which we analyzed cetaceans. One cetacean, *Attockicetus* (a remingtonocetid), has been collected from this unit. This specimen had a $\delta^{18}O_p$ value of $+16.6‰$, at the high end of the range of values of *Pakicetus* from the lower Kuldana Formation.

Gaviacetus, Indocetus, and *Remingtonocetus* from the pelagic marine limestone of the Harudi Formation in western India had $\delta^{18}O_p$ values between $+18.2$ and $+21.8‰$. These values either overlap or are more positive than the $\delta^{18}O_p$ values of the modern marine cetaceans. They are also 1.5 to 5.2‰ higher than the $\delta^{18}O_p$ values of the cetaceans from the lower Kuldana Formation.

4.2. Carbon Isotope Compositions ($\delta^{13}C_{sc}$) of Tooth and Bone Carbonate

4.2.1. Modern Cetaceans

Our carbon isotope analyses of the structural carbonate ($\delta^{13}C_{sc}$ values) of modern cetacean teeth and bone (Table I, Fig. 4A) demonstrate that there is a difference in the carbon isotope composition of freshwater and marine cetaceans. The modern freshwater cetaceans have $\delta^{13}C_{sc}$ values between -17.1 and $-11.6‰$ (PDB), whereas with one exception, all of the modern marine cetaceans primarily had $\delta^{13}C_{sc}$ values between -10.9 and $-7.8‰$. The exception is the $\delta^{13}C_{sc}$ value of $-13‰$ (PDB) we obtained for a specimen of the sperm whale, *Physeter*. A second specimen of *Physeter* had a much lower $\delta^{13}C_{sc}$ value of $-8‰$, which is within the range of $\delta^{13}C_{sc}$ values of all of the other marine cetacean

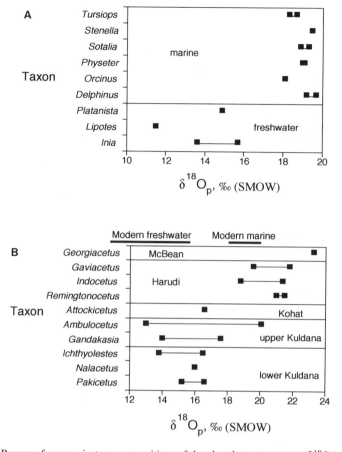

FIGURE 3. (A) Ranges of oxygen isotope compositions of the phosphate component ($\delta^{18}O_p$) of the teeth and bones of modern cetaceans. Note the 2–3‰ minimum difference between marine and freshwater taxa. (B) Ranges of oxygen isotope compositions of the phosphate component ($\delta^{18}O_p$) of the teeth and bones of Eocene cetaceans. Note the difference between the cetaceans from the lower Kuldana Formation and the Kohat Formation of Pakistan and the cetaceans from the Harudi Formation of India and the McBean Formation of Georgia. The range of $\delta^{18}O_p$ values of *Ambulocetus* and *Gandakasia* from the upper Kuldana Formation suggests that these animals may have been euryhaline.

species we analyzed. One possible explanation for this difference in values between the two individuals of *Physeter* is migration. It is possible that the specimen of *Physeter* with the more negative value migrated into and fed in waters having plankton at the bottom of the food chain with more negative $\delta^{13}C_{sc}$ values. It is known that carbon isotope composition of plankton in the oceans varies on the order of 5‰ (e.g., Rau *et al.*, 1982), and this difference is essentially identical to that observed between the two specimens. The plankton with the most negative $\delta^{13}C_{sc}$ values is found at the highest latitudes, although there is no single correlation between latitude and plankton $\delta^{13}C_{sc}$ that applies to all oceans (Rau *et al.*, 1982). We therefore conclude that, although there is some overlap in the ranges of $\delta^{13}C_{sc}$ values between freshwater and marine cetacean species, the general pattern, for cetaceans

Table II.   Isotopic Compositions ($\delta^{18}O_p$, $\delta^{13}C_{sc}$) of the Bone (b), Dentine (d), and Enamel (e)
of Eocene Cetaceans from Pakistan, India, and the United States

Taxon	Specimen	Formation	Type	$\delta^{18}O_p$ ‰ (SMOW)[a]	$\delta^{13}C_{sc}$ ‰ (PDB)
Ambulocetus	H-GSP 18472	u. Kuldana	b	13.8	—
	H-GSP 18473	u. Kuldana	b	14.0	−13.8
			d	14.4	−12.1
			e	—	−12.6
	H-GSP 18474	u. Kuldana	d	13.0	−12.3
			e	20.1	—
	H-GSP 18497	u. Kuldana	d	13.7, 19.2	−10.6
	H-GSP 18507	u. Kuldana	b	18.0	−13.6
			d,e	15.2	−10.8
			e	—	−14.2
	H-GSP 92148	u. Kuldana	b	13.7	—
	H-GSP 92151	u. Kuldana	b	14.6	−14.0
	H-GSP 96095	u. Kuldana	d	14.8	—
Andrewsiphius	VPL-1019	Harudi	d	—	−13.4
Attockicetus	H-GSP 96232	Kohat	d	16.6	−7.1
			b	—	−8.2
Gandakasia	H-GSP 96279	u. Kuldana	d	14.0	−12.4
			e	17.6	−14.0
	H-GSP 96331	u. Kuldana	d	14.9	—
	H-GSP 96333	u. Kuldana	e	—	−11.9
	H-GSP 96505	u. Kuldana	d	14.6	−12.2
Gaviacetus	VPL-1021	Harudi	b	20.4	−9.6
			d	19.6, 21.8	−9.8
			e	20.6	—
Georgiacetus	GSM-350	McBean	d	—	−7.0
			e	23.3	−8.4
Ichthyolestes	H-GSP 18395	l. Kuldana	d	14.0	−11.9
	H-GSP 18521	l. Kuldana	d	13.8, 16.5	—
	H-GSP 92163	l. Kaldana	b	16.3	−12.0
Indocetus	LUVP 11034	Harudi	d	18.8	−5.2
	VPL-1017	Harudi	b	21.4	−6.2
mesonychian	H-GSP 96134	l. Kuldana	d	—	−10.1
Nalacetus	H-GSP 18408	l. Kuldana	e,d	—	−13.2
	H-GSP 91036	l. Kuldana	e,d	16.0	−12.0
	H-GSP 91045	l. Kuldana	d	16.0	−12.6
Pakicetus	H-GSP 91014	l. Kuldana	d	—	−12.9
	H-GSP 91034	l. Kuldana	b	—	−12.4
			d	15.2	—
			e	15.2	—
	H-GSP 18410	l. Kuldana	d	15.8	−12.9
	H-GSP 18470-M3	l. Kuldana	e,d	—	−13.6
	H-GSP 18470-P4	l. Kuldana	e,d	16.6	−13.6
Remingtonocetus	RUSB 9621	Harudi	b	21.5	−10.6
	Rato Nala uncat.	Harudi	e	21.0	−7.8

[a]Where values obtained for a specimen span more than a 2‰ range, the minimum and maximum values of the range, rather than averages, are listed, separated by a comma.

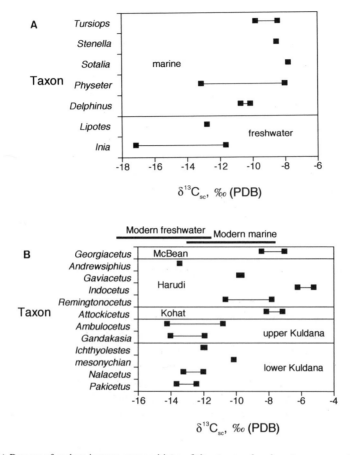

FIGURE 4. (A) Ranges of carbon isotope compositions of the structural carbonate component ($\delta^{13}C_{sc}$) of the teeth and bones of modern cetaceans. Note the difference in $\delta^{13}C_{sc}$ values between marine and freshwater taxa. The $\delta^{13}C_{sc}$ value of -13‰ obtained for one specimen of the sperm whale *Physeter* may reflect migration into higher-latitude waters and feeding on relatively ^{13}C-depleted food. (B) Ranges of carbon isotope compositions of the structural carbonate component ($\delta^{13}C_{sc}$) of the teeth and bones of Eocene taxa. Note the difference in $\delta^{13}C_{sc}$ values between the cetaceans from the Kuldana Formation and those from the Kohat, Harudi, and McBean Formations. The $\delta^{13}C_{sc}$ value of -13.4‰ of *Andrewsiphius* may be related either to a largely terrestrial/freshwater diet or to feeding on relatively ^{13}C-depleted plankton.

that do not migrate long distances, is that marine species have higher $\delta^{13}C_{sc}$ values than freshwater species.

4.2.2. Eocene Cetaceans

A difference between the earliest cetaceans from the Kuldana Formation of Pakistan and the slightly younger cetaceans from the Kohat, Harudi, and McBean Formations (Table II, Fig. 4B) is apparent in the carbon isotope composition of the structural carbonate of these animals. The earliest whales, *Ichthyolestes, Nalacetus,* and *Pakicetus,* from the lower Kuldana Formation, had $\delta^{13}C_{sc}$ values between -14 and $-12‰$ (PDB), in the range of modern freshwater cetaceans. The mesonychian had a higher $\delta^{13}C_{sc}$ value of $-10.1‰$, which is similar to the values obtained for Eocene herbivores from the lower Kuldana Formation, but that value also overlaps with marine values. The upper Kuldana cetaceans, *Ambulocetus* and *Gandakasia*, had very similar values. *Gandakasia* specimens ($n = 4$) had values between -14.0 and $-11.9‰$; *Ambulocetus* specimens ($n = 9$) had values between -14.2 and $-10.6‰$, with most (7 out of 9) of these falling in the range from -14.2 to $-12.0‰$, essentially identical to the cetaceans from the lower Kuldana.

In striking contrast to the Kuldana cetaceans, *Attockicetus* from the Kohat Formation, three of the four cetaceans from the Harudi Formation—*Gaviacetus*, *Indocetus*, and *Remingtonocetus*—and *Georgiacetus* from the McBean Formation all have $\delta^{13}C_{sc}$ values between -10.7 and $-5.2‰$, very similar to the range of values we obtained for modern marine cetaceans. The fourth cetacean from the Harudi Formation, *Andrewsiphius*, had a $\delta^{13}C_{sc}$ value of $-13.4‰$ ($n = 2$)—a value very similar to those of the Kuldana cetaceans.

4.3. Diagenetic Assessment of Bone, Dentine, and Enamel $\delta^{13}C_{sc}$

Our comparison of bone, dentine, and enamel (Fig. 5) reveals differences in the isotopic composition of the three phases for most specimens. Several important patterns are evident in this comparison. First, the differences among the phases are not all in the same direction, but where the values of dentine are more than $1‰$ different from bone or enamel, the dentine $\delta^{13}C_{sc}$ values are higher (*Gandakasia*, both *Ambulocetus* specimens, *Attockicetus*, and *Georgiacetus*) and bone $\delta^{13}C_{sc}$ values are closer to those obtained from enamel samples, a result similar to that of Koch *et al.* (1994). Second, the greatest variation among bone, dentine, and enamel $\delta^{13}C_{sc}$ values occurs in *Ambulocetus* and *Gandakasia*

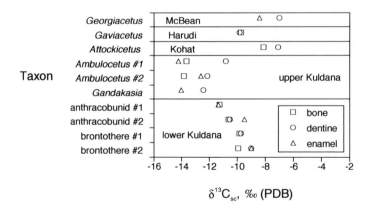

FIGURE 5. Comparison of the $\delta^{13}C_{sc}$ values of bone, dentine, and enamel of individual fossil specimens of Eocene taxa from the five geologic formations represented. The differences in $\delta^{13}C_{sc}$ values among the three phases may reflect variation in preservational integrity, original differences related to differential remodeling and time-averaging of mineralization or both. Note that the different phases all exhibit the same general pattern of isotopic values.

specimens from the upper Kuldana Formation. This variation may be related to diagenetic alteration in fluctuating environments or to real biological variation, and requires further investigation. Finally, regardless of whether one compares bone, dentine, or enamel, the relative differences among the lower Kuldana specimens, upper Kuldana, Kohat, Harudi, and McBean specimens are apparent.

5. Discussion

Potential differences in temperature and isotopic composition between the Eocene and modern oceans and rivers led us to expect that the actual $\delta^{18}O_p$ and $\delta^{13}C_{sc}$ values of the Eocene cetaceans might be quite different from the values of the modern species, and for this reason, we focus on relative differences within and between faunas rather than actual values. Nevertheless, not only does the 2–3‰ difference in $\delta^{18}O_p$ between putatively marine and freshwater Eocene cetaceans match the difference between modern marine and freshwater cetaceans, but most of the actual values of the fossils overlap the ranges of $\delta^{18}O_p$ values of the modern specimens (Figs. 3a and 3b). The notable exception to this pattern is *Georgiacetus*, which had a $\delta^{18}O_p$ value of 23.3‰. As *Georgiacetus* is the geologically youngest of the cetaceans we analyzed, it is tempting to explain this high value in terms of the ocean cooling recorded by foraminifera in the middle and late Eocene (Zachos *et al.*, 1994). This observation runs counter to the expectation that mammalian teeth should not record changes in temperature because they are thought to form in equilibrium with body water at a nearly constant temperature. Nevertheless, Barrick *et al.* (1992) also found a temporal trend in their cetacean fossils that paralleled the foraminiferal record. One possible explanation for this correlation in both studies is that because teeth are more exposed to the external medium than are bones and because cetaceans are nearly continuously submerged in that medium, teeth may record more environmental changes than has previously been hypothesized. One way to test this explanation would be to measure the $\delta^{18}O$ values of teeth and bones throughout the bodies of modern cetaceans over their full range of body sizes.

Similarly, the $\delta^{13}C_{sc}$ values of our Eocene cetaceans overlap those of our modern freshwater and marine cetaceans. In addition, the differences in $\delta^{13}C_{sc}$ values between freshwater and marine modern cetaceans correspond well to differences in $\delta^{13}C_{sc}$ values obtained in analyses of collagen (Schoeninger and DeNiro, 1984; Keegan and DeNiro, 1988; Little and Schoeninger, 1995). These values can also be used as a prediction of the carbonate $\delta^{13}C$ values ($\delta^{13}C_{sc}$) by adding 7‰ to the collagen $\delta^{13}C_{sc}$ values (Lee-Thorp and van der Merwe, 1987). Making this adjustment to the collagen values of marine cetaceans presented by Schoeninger and DeNiro (1984) yields expected $\delta^{13}C_{sc}$ values ranging from −9.4 to −5.6‰, in good agreement with our results. Unfortunately, the data of Schoeninger and DeNiro (1984) did not include any freshwater cetaceans or other freshwater mammals, but analyses of noncollagen soft tissues of freshwater and saltwater harbor seals (Smith *et al.*, 1996) yielded a 4 to 7‰ difference, consistent with the freshwater–marine $\delta^{13}C_{sc}$ values we obtained for both modern and Eocene cetaceans.

An examination of the isotopic compositions of the Eocene cetacean teeth and bones reveals a shift from more negative to more positive $\delta^{18}O_p$ and $\delta^{13}C_{sc}$ values (Figs. 3 and 4), suggesting that the transition of cetaceans from terrestrial/freshwater to marine life is preserved in our samples. Hypothetically, there are four broad ecological categories into

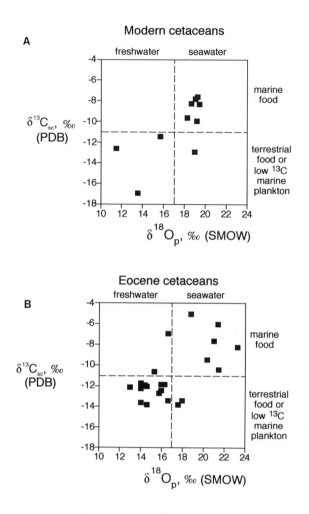

FIGURE 6. (A) Bivariate plot of the $\delta^{13}C_{sc}$ values and $\delta^{18}O_p$ values of modern cetaceans, showing the general pattern of lower values in freshwater cetaceans and higher values in marine taxa. (B) Bivariate plot of the $\delta^{13}C_{sc}$ values and $\delta^{18}O_p$ values of Eocene cetaceans. Note occurrence of taxa in all four ecological categories.

which a cetacean species may fall (Fig. 6A). These are general categories, delimited according to whether an animal: (1) eats terrestrial/freshwater food and ingests freshwater; (2) eats terrestrial/freshwater food and ingests seawater; (3) eats marine food and ingests freshwater; or (4) eats marine food and ingests seawater. In general, this categorization is consistent with the modern data because the ranges of $\delta^{18}O_p$ and $\delta^{13}C_{sc}$ values are largely nonoverlapping: only one specimen of the sperm whale *Physeter* has a $\delta^{13}C_{sc}$ value that overlaps those of the freshwater/terrestrial animals.

All modern cetaceans, except one, are either fully marine (eat marine food, ingest seawater) or fully freshwater (eat freshwater/terrestrial food and ingest freshwater) (Fig. 6A). Using the boundary values ($\delta^{18}O_p = +17\%o$ and $\delta^{13}C_{sc} = -11\%o$) of these categories in the modern data as a reference, we find that the $\delta^{18}O_p$ and $\delta^{13}C_{sc}$ values of the Eocene cetaceans place at least one taxon in each of the four categories (Fig. 6B). The presence of taxa in categories other than fully terrestrial and fully marine suggests that the transition from terrestrial to marine was not a simple progression, but rather mosaic in character and that some of the early cetacean species had ecological and physiological requirements that could not be inferred from morphology and depositional environments.

For example, it has been suggested that the remnant Tethys was rich in nutrients and an important source of food for *Pakicetus* (Gingerich *et al.*, 1983). However, our carbon isotope analyses of all three lower Kuldana cetaceans, *Ichthyolestes, Nalacetus,* and *Pakicetus*, suggest that these animals had a terrestrial or freshwater diet and were ingesting primarily, if not exclusively, freshwater. An alternative explanation is that these animals were consuming marine foods that were relatively low in ^{13}C in the remnant Tethys and that the increase in $\delta^{13}C_{sc}$ values evident in our data (Fig. 3B) represents a change in the isotopic composition of marine waters available through time. This seems unlikely because even cetaceans feeding on marine plankton with the lowest documented $\delta^{13}C_{sc}$ values of $-23.5‰$ (Rau *et al.*, 1982) would have tooth $\delta^{13}C_{sc}$ values of approximately -11.5, or 2–3‰ higher than the lowest $\delta^{13}C_{sc}$ values of the cetaceans from the Kuldana Formation. In addition, the low $\delta^{18}O_p$ values of these animals suggest that they were restricted to freshwater.

In contrast to the cetaceans from the lower Kuldana, the $\delta^{13}C_{sc}$ value of the mesonychian is on the border between marine and freshwater values. It falls within the range of values we have obtained from enamel of co-occurring herbivores, but a better understanding of the ecology of the Tethyan mesonychian must await the discovery and analyses of additional fossils.

The cetaceans from the upper Kuldana Formation, *Ambulocetus* and *Gandakasia*, appear to have had a primarily terrestrial diet, but their wide range of $\delta^{18}O_p$ values suggests that they ingested waters of a wide variety of isotopic compositions. These patterns are suggestive of a euryhaline physiology, and are consistent with the preservation of *Ambulocetus* and *Gandakasia* in nearshore marine depositional environments. *Ambulocetus* has yet to be found in any terrestrial deposits, but if *Ambulocetus* and *Gandakasia* obtained their food very near the mouths of rivers, their absence from terrestrial deposits is consistent with the available evidence.

Of all of the fossils analyzed here, those from the upper Kuldana Formation appear to have been most affected by postmortem alteration. Nevertheless, it is unlikely that diagenesis is responsible for the low $\delta^{18}O_p$ (apparent freshwater) values of the bone and dentine samples of *Ambulocetus*. Our analyses of diagenesis of the upper Kuldana samples indicate that the secondary (diagenetic) carbonate of our specimens had considerably higher $\delta^{13}C_{sc}$ and $\delta^{18}O_p$ values than did the structural carbonate of the specimens themselves. For example, an untreated aliquot of H-GSP 18473 bone had a $\delta^{13}C_{sc}$ value of $-5.6‰$ (PDB) and a $\delta^{18}O_p$ value of 33.4‰ (SMOW), whereas, after treatment, H-GSP 18473 bone had values of -13.7 and 28.8‰, respectively. These results are not conclusive, but strongly suggest that the lower values are primary. A further test of the reality of this intraspecimen isotopic variation would be to make fine-scale analyses along transects of the teeth of *Ambulocetus* and *Gandakasia* specimens. This could be used to determine whether the variation in the isotopic compositions of our samples represents real biological variation, such as an ontogenetic habitat shift.

One early cetacean species, *Attockicetus*, appears to have eaten marine food and ingested freshwater. *Attockicetus*, a remingtonocetid, was ecologically quite different from its close relative *Remingtonocetus* from the Harudi Formation.

By the middle Eocene, the first fully marine cetaceans had appeared. *Gaviacetus, Indocetus*, and *Remingtonocetus*, from the Harudi Formation, and *Georgiacetus* from the McBean Formation had $\delta^{18}O_p$ and $\delta^{13}C_{sc}$ values consistent with a fully marine existence. *Andrewsiphius*, which had a very negative $\delta^{13}C_{sc}$ value of -13.4‰, is the exception. Little is known of the morphology of *Andrewsiphius*, but one possibility is that it lived in the ocean and continued to feed on land, as *Ambulocetus* appears to have done.

Although diagenesis appears to have affected the carbon isotope composition of carbonate in our samples, acetic acid pretreatment reveals that the broad pattern of relative isotopic differences occurs regardless of whether the sample consisted of bone, dentine, or enamel. If these isotopic differences are a good guide, then bone is actually a better proxy of enamel composition than is dentine, a result in good agreement with those of Koch *et al.* (1994). On the other hand, intratooth variation on the order of 3‰ has been recorded in modern terrestrial mammals (Fricke and O'Neil, 1996) and could explain some of the variation observed among bone, dentine, and enamel of our Eocene fossils. Moreover, some of the isotopic variation in the fossils may also be related to differences in the timing and duration of mineralization. For example, the enamel of most mammal teeth forms over a short interval of 1–2 years (generally *in utero*) and is not remodeled during an animal's life as bone is, so depending on the age of the animal and the specifics of its physiology, bone may provide a longer-term average or a record from later in the animal's life than that preserved in its teeth.

6. Conclusions

The earliest cetaceans, *Ichthyolestes, Nalacetus*, and *Pakicetus*, from the lower Kuldana Formation of Pakistan, were closely tied to terrestrial sources of freshwater and food and apparently shared very similar ecologies in these two respects. *Ambulocetus* and *Gandakasia*, from the upper Kuldana Formation, also appear to have relied on terrestrial food sources, but may have been euryhaline and able to ingest seawater at least occasionally. The mesonychian from the lower Kuldana Formation probably fed on terrestrial animals, but could have been feeding in part on marine prey.

Attockicetus from the Kohat Formation of Pakistan appears to have eaten primarily marine food, but required freshwater, in contrast to its close relative, *Remingtonocetus*. *Gaviacetus, Indocetus*, and *Remingtonocetus* from the Harudi Formation, and *Georgiacetus* from the McBean Formation, appear to have been fully marine, both living and feeding at sea. *Andrewsiphius*, also from the Harudi Formation, appears to have had different ecological requirements from the other cetaceans from the Harudi Formation. It either fed on terrestrial/freshwater prey or marine plankton low in ^{13}C.

The first fully marine cetaceans appeared by the middle Eocene. The evolutionary transition of cetaceans from terrestrial to marine life was thus geologically rapid. The transition to life in seawater involved changes both in osmoregulatory physiology and in diet, but these changes were not strictly coupled. This apparent decoupling of food and water requirements may have facilitated niche differentiation, and as a result, the diversification of the earliest cetaceans.

Acknowledgments

We thank the Geological Survey of Pakistan, its Director General, Dr. M. Talib Hassan, Assistant Director Muhammad Arif, and Dr. S. Mahmood Raza of the Oil and Gas Development Corporation, for their support of our fieldwork in Pakistan. Financial support was provided by an NSF-funded Fellowship from the University of Arizona Research Training Group for the Analysis of Biological Diversification (to Roe), the National Geographic Society (to Thewissen), a Pioneer Award from the Northeastern Ohio Universities College of Medicine (to Thewissen), and by National Science Foundation Awards EAR-9005717 to O'Neil, EAR-9526686 to Thewissen, and EAR-9418207 to Quade. We thank R. D. E. MacPhee of the American Museum of Natural History, W. F. Perrin of the NOAA/National Marine Fisheries Service branch in La Jolla, and C. W. Potter of the United States National Museum (Smithsonian) for comparative recent material, H. Achyuthan for assistance with wet chemistry, D. Surge and T. Moore for assistance with illustrations, C. Beuchat, J. N. Stallone, and S. I. Madar for helpful discussions, and M. J. Schoeninger for a very helpful review.

References

Albright, L. B. 1996. A protocetid cetacean from the Eocene of South Carolina. *J. Paleontol.* **70**(3):519–523.

Allen, P., and Keith, M. L. 1965. Carbon isotope ratios and paleosalinities of Purbeck-Wealden carbonates. *Nature* **208**:1278–1280.

Ames, A., van Vleet, E. S., and Sackett, W. M. 1996. The use of stable carbon isotope analysis for determining the dietary habits of the Florida manatee, *Tricechus manatus latirostris*. *Mar. Mamm. Sci.* **12**(4):555–563.

Andersen, S. H., and Nielsen, E. 1983. Exchange of water between the harbor porpoise, *Phocoena phocoena*, and the environment. *Experientia* **39**:52–53.

Aslan, A., and Thewissen, J. G. M. 1996. Preliminary evaluation of Kuldana paleosols and implications for the interpretation of vertebrate fossil assemblages, Kuldana Formation, northern Pakistan, *Palaeovertebrata* **25**(2–4):261–277.

Ayliffe, L. K., and Chivas, A. R. 1990. Oxygen isotope composition of the bone phosphate of Australian kangaroos: potential as a palaeoenvironmental recorder. *Geochim. Cosmochim. Acta* **54**:2603–2609.

Ayliffe, L. K., Lister, A. M., and Chivas, A. R. 1992. The preservation of glacial–interglacial climatic signatures in the oxygen isotopes of elephant skeletal phosphate. *Palaeogeogr. Palaeoclimatol. Palaeoecol.* **99**:179–191.

Ayliffe, L. K., Chivas, A. R., and Leakey, M. G. 1994. The retention of primary oxygen isotope compositions of fossil elephant skeletal phosphate. *Geochim. Cosmochim. Acta* **58**(23):5291–5298.

Barrick, R. E., Fischer, A. G., Kolodny, Y., Luz, B., and Bohaska, D. 1992. Cetacean bone oxygen isotopes as proxies for Miocene ocean composition and glaciation. *Palaios* **7**:521–531.

Bentley, P. J. 1963. Composition of the urine of the fasting humpback whale (*Megaptera nodosa*). *Comp. Biochem. Physiol.* **10**:257–259.

Bentley, P. J. 1971. *Endocrines and Osmoregulation, a Comparative Account of the Regulation of Water and Salt in the Vertebrates*. Springer-Verlag, Berlin.

Bryant, J. D., and Froelich, P. N. 1995. A model of oxygen isotope fractionation in body water of large mammals. *Geochim. Cosmochim. Acta* **59**(21):4523–4537.

Clayton, R. N., and Degens, E. T. 1959. Use of carbon isotope analyses for differentiating fresh-water and marine sediments. *Am. Assoc. Petrol. Geol. Bull.* **43**:890–897.

Craig, H. 1961a. Isotopic variations in meteoric waters. *Science* **133**:1702–1703.

Craig, H. 1961b. Standards for reporting concentrations of deuterium and oxygen-18 in natural waters. *Science* **133**:1833–1834.

Craig, H., and Gordon, L. I. 1965. Deuterium and oxygen 18 variations in the ocean and the marine atmosphere, in: E. Tongiorigi (ed.), *Stable Isotopes in Oceanographic Studies and Paleotemperatures*, pp. 9–129. Consiglio Nazionale delle Ricerche, Pisa.

Dansgaard, W. 1964. Stable isotopes in precipitation. *Tellus* **16**(4):436–468.

DeNiro, M. J., and Epstein, S. 1978. Influence of diet on the distribution of carbon isotopes in animals. *Geochim. Cosmochim. Acta* **42**:495–506.

Epstein, S., and Mayeda, T. 1953. Variation of O^{18} content of waters from natural sources. *Geochim. Cosmochim. Acta* **4**:213–224.

Fetcher, E. S., Jr., and Fetcher, G. W. 1942. Experiments on the osmotic regulation of dolphins. *J. Cell. Comp. Physiol.* **19**:123–130.

Fricke, H. C., and O'Neil, J. R. 1996. Inter- and intra-tooth variation in the oxygen isotope composition of mammalian tooth enamel phosphate: implications for palaeoclimatological and palaeobiological research. *Palaeogeogr. Palaeoclimatol. Palaeoecol.* **126**:91–99.

Friedman, I., and O'Neil, J. R. 1977. Compilation of stable isotope fractionation factors of geochemical interest, in: M. Fleischer (ed.), *Data of Geochemistry, U.S. Geol. Surv. Prof. Pap.* **440-KK**.

Fry, B., and Sherr, E. B. 1984. $\delta^{13}C$ measurements as indicators of carbon flow in marine and freshwater ecosystems. *Contrib. Mar. Sci.* **27**:13–47.

Gat, J. R. 1981. Isotopic fractionation, in: J. R. Gat and R. Gonfiantini (eds.), *Stable Isotope Hydrology: Deuterium and Oxygen-18 in the Water Cycle*, pp. 21–33. IAEA Tech. Rep. Ser. No. 210.

Gingerich, P. D., Wells, N. A., Russell, D. E., and Shah, S. M. I. 1983. Origin of whales in epicontinental remnant seas: new evidence from the early Eocene of Pakistan. *Science* **220**:403–406.

Gingerich, P. D., Raza, S. M., Arif, M., Anwar, M., and Zhou, X. 1994. New whale from the Eocene of Pakistan and the origin of cetacean swimming. *Nature* **368**:844–847.

Hobson, K. A., Sease, J. L., Merrick, R. L., and Piatt, J. F. 1997. Investigating trophic relationships of pinnipeds in Alaska and Washington using stable isotopic ratios of nitrogen and carbon. *Mar. Mamm. Sci.* **13**(1):114–132.

Hui, C. A. 1981. Seawater consumption and water flux in the common dolphin *Delphinus delphis*. *Physiol. Zool.* **54**:430–440.

Hulbert, R. C., and Petkewich, R. M. 1991. Innominate of a middle Eocene (Lutetian) protocetid whale from Georgia. *J. Vertebr. Paleontol.* **11**(Suppl. to No. 3):36A.

International Atomic Energy Agency. 1986. *Environmental Isotope Data No. 8: World Survey of Isotope Concentration in Precipitation (1980–1983)*. Tech. Rep. Ser. No. 264.

Keegan, W. F., and DeNiro, M. J. 1988. Stable carbon- and nitrogen-isotope ratios of bone collagen used to study coral-reef and terrestrial components of prehistoric Bahamian diet. *Am. Antiq.* **53**(2):320–336.

Keith, M. L., and Parker, R. H. 1965. Local variation of the ^{13}C and ^{18}O content of mollusk shells and the relatively minor temperature effect in marginal marine environments. *Mar. Geol.* **3**:115–129.

Keith, M. L., Anderson, G. M., and Eichler, R. 1964. Carbon and oxygen isotopic composition of mollusk shells from marine and fresh-water environments. *Geochim. Cosmochim. Acta* **28**:1757–1786.

Kirschner, L. B. 1991. Water and ions, in: C. L. Prosser (ed.), *Environmental and Metabolic Animal Physiology: Comparative Physiology*, 4th ed., pp. 13–108. Wiley–Liss, New York.

Koch, P. L., Fogel, M. L., and Tuross, N. 1994. Tracing the diets of fossil animals using stable isotopes, in: K. Lajtha and R. H. Michener (eds.), *Stable Isotopes in Ecology and Environmental Science*, pp. 63–92. Blackwell, Oxford.

Kohn, M. J. 1996. Predicting animal $\delta^{18}O$: accounting for diet and physiological adaptation. *Geochim. Cosmochim. Acta* **60**(23):4811–4829.

Kohn, M. J., Schoeninger, M. J., and Valley, J. W. 1996. Herbivore tooth oxygen isotope compositions: effects of diet and physiology. *Geochim. Cosmochim. Acta* **60**(20):3889–3896.

Kolodny, Y., and Luz, B. 1991. Oxygen isotopes in phosphates of fossil fish; Devonian to Recent, in: H. P. Taylor, Jr., J. R. O'Neil, and I. R. Kaplan (eds.), *Stable Isotope Geochemistry; a Tribute to Samuel Epstein*, pp. 105–119. Geochemical Society Special Publication No. 3, University Park, PA.

Kolodny, Y., Luz, B., and Navon, O. 1983. Oxygen isotope variations in phosphate of biogenic apatites, I. Fish bone apatite—rechecking the rules of the game. *Earth Planet. Sci. Lett.* **64**:398–404.

Kumar, K., and Sahni, A. 1986. *Remingtonocetus harudiensis*, new combination, a middle Eocene archaeocete (Mammalia, Cetacea) from western Kutch, India. *J. Vertebr. Paleontol.* **6**:326–349.

Lee-Thorp, J. A., and van der Merwe, N. J. 1987. Carbon isotope analysis of fossil bone apatite. *S. Afr. J. Sci.* **83**:712–715.

Lee-Thorp, J. A., and van der Merwe, N. J. 1991. Aspects of the chemistry of modern and fossil biological apatites. *J. Archaeol. Sci.* **18**:343–354.

Little, E. A., and Schoeninger, M. J. 1995. The Late Woodland diet on Nantucket Island and the problem of maize in coastal New England. *Am. Antiq.* **60**(2):351–368.

Longinelli, A., and Nuti, A. 1973a. Revised phosphate–water isotopic temperature scale. *Earth Planet. Sci. Lett.* **19**:373–376.

Longinelli, A., and Nuti, A. 1973b. Oxygen isotope measurements of phosphate from fish teeth and bones. *Earth Planet. Sci. Lett.* **20**:337–340.

Luz, B., and Kolodny, Y. 1985. Oxygen isotope variations in phosphate of biogenic apatites, IV. Mammal teeth and bones. *Earth Planet. Sci. Lett.* **75**:29–36.

Luz, B., and Kolodny, Y. 1989. Oxygen isotope variation in bone phosphate. *Appl. Geochem.* **4**:317–323.

Luz, B., Kolodny, Y., and Horowitz, M. 1984. Fractionation of oxygen isotopes between mammalian bone-phosphate and environmental drinking water. *Geochim. Cosmochim. Acta* **48**:1689–1693.

Luz, B., Cormie, A. B., and Schwarcz, H. P. 1990. Oxygen isotope variations in phosphate of deer bones. *Geochim. Cosmochim. Acta* **54**:1723–1728.

McCrea, J. M. 1950. On the isotopic chemistry of carbonates and a paleotemperature scale. *J. Chem. Phys.* **18**:849–857.

Messenger, S. L. 1994. Phylogenetic relationships of palatanistoid river dolphins (Odontoceti, Cetacea): assessing the significance of fossil taxa. *Proc. San Diego Soc. Nat. Hist.* **29**:125–133.

Muizon, C. de. 1994. Are the squalodonts related to the platanistoids? *Proc. San Diego Soc. Nat. Hist.* **29**:135–146.

Nelson, B. K., DeNiro, M. J., Schoeninger, M. J., DePaolo, D. J., and Hare, P. E. 1986. Effects of diagenesis on strontium, carbon, nitrogen and oxygen concentration and isotopic composition of bone. *Geochim. Cosmochim. Acta* **50**:1941–1949.

O'Neil, J. R., Roe, L. J., Reinhard, E., and Blake, R. E. 1994. A rapid and precise method of oxygen isotope analysis of biogenic phosphate. *Isr. J. Earth Sci.* **43**:203–212.

Peterson, B. J., and Fry, B. 1987. Stable isotopes in ecosystem studies. *Annu. Rev. Ecol. Syst.* **18**:293–320.

Quade, J., Cerling, T. E., Barry, J. C., Morgan, M. E., Pilbeam, D. R., Chivas, A. R., Lee-Thorp, J. A., and van der Merwe, N. J. 1992. A 16-Ma record of paleodiet using carbon and oxygen isotopes in fossil teeth from Pakistan. *Chem. Geol. (Isot. Geosci.)* **94**:183–192.

Rau, G. H., Sweeney, R. E., and Kaplan, I. R., 1982. Plankton $^{13}C:^{12}C$ ratio changes with latitude: differences between the northern and southern oceans. *Deep Sea Res.* **29**(8A):1035–1039.

Ridgway, S. H. 1972. Homeostasis in the aquatic environment, in: S. H. Ridgway (ed.), *Mammals of the Sea, Biology and Medicine*, pp. 590–747. Thomas, Springfield, IL.

Sahni, A., and Mishra, V. P. 1975. Lower Tertiary vertebrates from western Kutch. *Monogr. Palaeontol. Soc. India* **3**:1–48.

Schmidt-Nielsen, K. 1964. *Desert Animals, Physiological Problems of Heat and Water*. Clarendon Press, Oxford.

Schmidt-Nielsen, K. 1997. *Animal Physiology: Adaptation and Environment*, 5th ed. Cambridge University Press, London.

Schoeninger, M. J., and DeNiro, M. J. 1984. Nitrogen and carbon isotopic composition of bone collagen from marine and terrestrial animals. *Geochim. Cosmochim. Acta* **48**:625–639.

Shemesh, A. 1990. Crystallinity and diagenesis of sedimentary apatites. *Geochim. Cosmochim. Acta* **54**:2433–2438.

Smith, R. J., Hobson, K. A,. Koopman, H. N., and Lavigne, D. M. 1996. Distinguishing between populations of fresh- and salt-water harbour seals (*Phoca vitulina*) using stable isotope ratios and fatty acid profiles. *Can. J. Fish. Aquat. Sci.* **53**:272–279.

Sokolov, V. E., Mashcherskii, I. G., Feoktistova, N. Y., and Klishin, V. O. 1994. Water balance of the Black Sea bottle-nosed dolphin. *Dokl. Akad. Nauk* **335**:396–398.

Telfer, N., Cornell, L. H., and Prescott, J. H. 1970. Do dolphins drink water? *J. Am. Vet. Med. Assoc.* **157**:555–558.

Thewissen, J. G. M., and Hussain, S. T. 1993. Origin of underwater hearing in whales. *Nature* **361**:444–445.

Thewissen, J. G. M., and Hussain, S. T. 1998. Systematic review of the Pakicetidae, early and middle Eocene Cetacea (Mammalia) from Pakistan and India. *Bull. Carnegie Mus. Nat. Hist.* **34**:220–238.

Thewissen, J. G. M., Hussain, S. T., and Arif, M. 1994. Fossil evidence for the origin of aquatic locomotion in archaeocete whales. *Science* **263**:210–212.

Thewissen, J. G. M., Madar, S. I., and Hussain, S. T. 1996a. *Ambulocetus natans*, an Eocene cetacean (Mammalia) from Pakistan. *Cour. Forsch.-Inst. Senckenberg* **191**:1–86.

Thewissen, J. G. M., Roe, L. J., O'Neil, J. R., Hussain, S. T., Sahni, A., and Bajpai, S. 1996b. Evolution of cetacean osmoregulation. *Nature* **381**:379–380.

Van Valen, L. 1966. Deltatheridia, a new order of mammals. *Bull. Am. Mus. Nat. Hist.* **132**:1–126.

Wells, N. A. 1983. Transient streams in sand-poor redbeds: early-middle Eocene Kuldana Formation of northern Pakistan. *Spec. Publ. Int. Assoc. Sedimentol.* **6**:393–403.

Wells, N. A. 1984. Marine and continental sedimentation in the early Cenozoic Kohat Basin and adjacent northwestern Indo-Pakistan. Ph.D. dissertation, University of Michigan, 465 pp.

West, R. M., and Lukacs, J. R. 1979. Geology and vertebrate-fossil localities, Tertiary continental rocks, Kala Chitta Hills, Attock District, Pakistan. *Milwaukee Public Mus. Contrib. Biol. Geol.* **26**:1–20.

Whelan, J. F. 1987. Stable isotope hydrology, in: T. K. Kyser, (ed.), *Stable Isotope Geochemistry of Low Temperature Fluids. Mineral. Assoc. Can. Short Course* **13**129–161.

Yoshida, N., and Miyazaki, N. 1991. Oxygen isotope correlation of cetacean bone phosphate with environmental water. *J. Geophys. Res.* **96**(C1):815–820.

Zachos, J. C., Stott, L. D., and Lohmann, K. C. 1994. Evolution of early Cenozoic marine temperatures. *Paleoceanography* **9**(2):358–387.

CHAPTER 15

Paleobiological Perspectives on Mesonychia, Archaeoceti, and the Origin of Whales

PHILIP D. GINGERICH

1. Introduction

Organisms living today are grouped together taxonomically because they are similar to each other and different from others. How similar organisms are within a group and how different the group is from other groups depends on the broader context of similarities and differences uniting and distinguishing groups. The rank to which a group is assigned depends in part on similarities and differences, but also on what we know about evolutionary history. Extant whales (order Cetacea) have long been known to be mammals because they share with other mammals such basic distinguishing characteristics as endothermy, lactation, large brains, and a high level of activity. Living cetaceans share, in addition, a suite of special characteristics related to life in water that distinguish them from land mammals: These include large body size, a reduced and simplified dentition, an audition-dominated sensory and communication system, a hydrodynamically streamlined body form with a muscular propulsive tail, and of course many ancillary anatomical, behavioral, and physiological differences.

Extant Mysticeti (baleen whales) and Odontoceti (toothed whales) are usually considered suborders of Cetacea, but they are sufficiently different from each other that some whale specialists in the past have regarded them as distinct orders. This illustrates the role context plays in determining how broadly taxonomic groups are drawn, and it also reflects the interdependence of morphology, classification, and evolutionary history: When mammals as different as mysticetes and odontocetes were classified in different orders, this was interpreted to reflect a long history of evolutionary independence (the history had to be long because of a general belief that evolution is so slow that differences take a long time to accumulate). We now know, thanks to the fossil record, that the modern suborders Mysticeti and Odontoceti have a fossil record extending back to the Oligocene epoch of the geological time scale, and they are thought to have diverged from each other sometime in the late

PHILIP D. GINGERICH • Museum of Paleontology, University of Michigan, Ann Arbor, Michigan 48109.

The Emergence of Whales, edited by Thewissen. Plenum Press, New York, 1998.

Eocene or early Oligocene, no more than about 40 m.y. ago (Fordyce and Barnes, 1994). Whales that are known from the earth's Eocene rivers and oceans all belong to a third sub-order, Archaeoceti, which is a group with much more generalized morphology. Archaeoceti includes the earliest aquatic whales.

No whales of any kind are known before the Eocene, and thus the evolutionary history of Cetacea is similar in length to that of other modern orders of mammals. And, like other modern orders (e.g., ungulate Artiodactyla), there are Paleocene land mammals, condylarthran Mesonychia in this case, that resemble Archaeoceti closely enough to suggest ancestor–descent relationship. Such a relationship is by no means proven as yet, but Mesonychia are clearly the best candidates for archaeocete ancestry by virtue of their morphological similarity and their overlapping temporal and geographic distributions. Mesonychia and Archaeoceti jointly are the subject of this topical perspective.

1.1. Study of Whale Origins

Most mammals live on land, and the aquatic specializations of Cetacea have long been viewed as derived characteristics acquired by whales when they made the transition from land to sea. This idea is reinforced by the long geological record of mammals on land (beginning in the late Triassic some 200 m.y. before present) and the relatively short geological record of cetaceans in the sea (beginning much later, in the early Eocene, some 50 m.y. before present). However, few would consider such inferential evidence of evolution from land mammals a satisfying solution to the problem of whale origins. I use *origins* advisedly here, not because whales had multiple or independent origins, but because their common origin had many equally important threshold stages—no single change made land mammals into whales.

What group of land mammals gave rise to whales? Where did it happen? When did it happen? How did it happen? What was the context? What were the consequences? These are all questions in the past tense, about a transformation we think happened in the past. All are paleobiological questions that group naturally into what might be considered the three broad objectives of study of fossil whales (or any group known from the fossil record):

1. Identification of the morphologically, geographically, and temporally intermediate stages of change (here the stages by which whales made the transition from land to sea). These intermediates, when known, are direct evidence (and the only direct evidence we have) telling us what happened in evolution.
2. Association of the times of acquisition of distinctive morphological specializations with other changes in morphology within the group of interest (here Cetacea) and with biotic- and physical-environmental changes outside the group of interest. These associations provide a context critical for understanding how any evolutionary transition took place.
3. Evaluation of consequences. What was the effect of any change on the group under study? This can be measured in terms of morphological disparity, taxonomic diversity, or taxon longevity.

It is difficult to appreciate that study of a group like Archaeoceti is still in its infancy. The first archaeocete to be studied and named, *Basilosaurus*, was collected in 1832, a year before Charles Lyell named the Eocene. *Basilosaurus* was recognized as a cetacean in 1841, the year that Richard Owen named Dinosauria. When the first archaeocete skeleton was mounted for exhibition at the U.S. National Museum in 1913 (again *Basilosaurus*), it was a composite and the number of vertebrae was unknown, the hands were reconstructed like flippers of a sea lion because they were not known (Lucas, 1900), the pelvis was mounted incorrectly, and the animal was assumed to have had no feet (Gidley, 1913). Remington Kellogg summarized all that was known at the time in his classic *Review of the Archaeoceti* (Kellogg, 1936), but there were still only three genera and species with reasonably complete skeletons (*Basilosaurus cetoides* and *Zygorhiza kochii* from the late Eocene, and *Protocetus atavus* from the middle Eocene; "Dorudon" *osiris* of Kellogg and others is a confusing composite including specimens of *Dorudon atrox*), and none of these had complete vertebral columns, hands, or feet.

Protocetus interested Kellogg largely because Fraas (1904) and Andrews (1906) regarded it as having the skull of an archaeocete and the dentition of a creodont (Fraas went so far as to remove archaeocetes from Cetacea, placing them in Creodonta). Kellogg retained Archaeoceti in Cetacea but concluded:

> In summation, it would appear that the evidence seems to point toward the concept that the archaeocetes are related to if not descended from some primitive insectivore-creodont stock, but that they branched off from that stock before the several orders of mammals that reached the flood tide of their evolutionary advance during the Cenozoic era were sufficiently differentiated to be recognized as such. Morphologically the archaeocetes seem to stand relatively near to the typical Mysticeti and Odontoceti, although all three suborders were separated from each other during a long interval of geologic time. It is not necessary to assume that any known archaeocete is ancestral to some particular kind of whale, for the archaeocete skull in its general structure appears to be divergent from rather than antecedent to the line of development that led to the telescoped condition of the braincase seen in skulls of typical cetaceans. On the contrary it is more probable that the archaeocetes are collateral derivatives of the same blood-related stock from which the Mysticeti and the Odontoceti sprang. (Kellogg, 1936, p. 343)

George Gaylord Simpson echoed these conclusions in his midcentury *Classification of Mammals:*

> Because of their perfected adaptation to a completely aquatic life, with all its attendant conditions of respiration, circulation, dentition, locomotion, etc., the cetaceans are on the whole the most peculiar and aberrant of mammals. Their place in the sequence of cohorts and orders [of mammalian classification] is open to question and is indeed quite impossible to determine in any purely objective way. (Simpson, 1945, p. 213)
>
> It is clear that the Cetacea are extremely ancient as such. . . . They probably arose very early and from a relatively undifferentiated eutherian ancestral stock. . . . Throughout the order Cetacea there is a noteworthy absence of annectent types, and nothing approaching a unified structural phylogeny can be suggested at present. . . . Thus the Archaeoceti . . . are definitely the most primitive of cetaceans, but they can hardly have given rise to the other suborders [Mysticeti and Odontoceti]. (Simpson, 1945, p. 214)

The first quotation from Simpson inspired Alan Boyden and Douglas Gemeroy to attack the problem of whale relationships serologically. Boyden and Gemeroy (1950) compared immunological cross-reactions of serum proteins of Cetacea with those of all other orders using precipitin tests. This was one of the first attempts to infer phylogenetic relationships from immunology. Boyden and Gemeroy found that interordinal reactions were generally weak, averaging about 2%, with the exception that the artiodactyl–cetacean comparisons were distinctly higher, averaging about 9–11%. This greater immunological reactivity Boyden and Gemeroy interpreted as indicating a close blood and genetic relationship of Cetacea to Artiodactyla.

Modern molecular gene sequencing has largely confirmed this Cetacea–Artiodactyla sister-group relationship. However, conflicting claims that (1) sperm whales are mysticetes (e.g., Milinkovitch *et al.*, 1993, 1995; Milinkovitch, 1995; but see Ohland *et al.*, 1995); (2) Cetacea originated *within* Artiodactyla as the sister group of extant camels, of extant hippopotami, or extant ruminants (Goodman *et al.*, 1985; Sarich, 1993; Irwin and Arnason, 1994; Graur and Higgins, 1994; Arnason and Gullberg, 1996; Gatesy, this volume); or (3) whales are the sister group of perissodactyls (McKenna, 1987), taken together, cast doubt on our ability to reconstruct past evolutionary history from living animals.

Van Valen (1966) approached the problem of cetacean relationships paleontologically:

> Only two known families need to be considered seriously as possibly ancestral to the archaeocetes and therefore to recent whales. These are the Mesonychidae and Hyaenodontidae (or just possibly some hyaenodontid-like palaeoryctid). No group that differentiated in the Eocene or later need be considered, since the earliest known archaeocete, *Protocetus atavus*, is from the early middle Eocene and is so specialized in the archaeocete direction that it is markedly dissimilar to any Eocene or earlier terrestrial mammal. It is also improbable that any strongly herbivorous taxon was ancestral to the highly predaceous archaeocetes. . . . Diverse and apparently equally valid objections exist for the various groups of Paleocene insectivores, one common to all being their small size. All marine mammals are large or rather large mammals. (Van Valen, 1966, p. 90)

Van Valen (1966, p. 92) drew attention to the late Eocene *Andrewsarchus* as a mesonychid having "a skull remarkably similar in shape to that of *Protocetus*, even to a largely longitudinal series of incisors" (the claim about remarkable similarity of skull shape is debatable). He reasoned (p. 93) that whales took to the sea in middle or late Paleocene times. And finally, he noted (p. 93) that Boyden and Gemeroy's serological argument for a special relationship between Cetacea and Artiodactyla is made more plausible by the evidence of an ancestral–descendant mesonychian-to-archaeocete relationship.

Although Boyden and Gemeroy's conclusions are consistent with those of Van Valen, it should be emphasized that a sister-group relationship between extant Artiodactyla and Cetacea like that hypothesized by Boyden and Gemeroy is different than a "mother-group" or ancestral–descendant relationship between Mesonychia and Archaeoceti like that hypothesized by Van Valen. The postulated divergence of proto-Artiodactyla from proto-Cetacea is not the same event as the transition from Mesonychia to Archaeoceti, nor is the time of divergence associated with the former likely to be equivalent to the time of transition of the latter. We shall return to this point later, and it is sufficient to note here that most authors now accept as a working hypothesis Van Valen's idea that Mesonychia gave rise to Archaeoceti.

1.2. Diversity and Morphology of Mesonychia

There are about 20–28 known genera of Mesonychia (depending on how these are counted), grouped in two, three, or four families: Hapalodectidae, Mesonychidae, and, questionably, Andrewsarchidae and Wyolestidae (Fig. 1). *Andrewsarchus* was included in Mesonychidae by Osborn (1924) and placed in a separate family-level group Andrewsarchinae by Szalay and Gould (1966). Van Valen (1978) considered andrewsarchines

MESONYCHIA	PALEOCENE			EOCENE			OLIG.
	Early	Middle	Late	Early	Middle	Late	Early
HAPALODECTIDAE							
Hapalodectes				A,N	A		
Hapalodectes?			A				
Hapalorestes					N		
MESONYCHIDAE							
Ankalagon		N	N				
Dissacus		A,N	A,E,N	E,N	E		
Dissacusium		A					
Harpagolestes					A,N	A,N	A
Hessolestes					N	N	
Honanodon					A		
Honanodon?				A			
Hukoutherium	A?	A					
Jiangxia			A				
Lohoodon					A	A	
Mesonyx			A?		A,N		
Metahapalodectes					A		
Mongolestes							A
Mongolonyx					A		
Olsenia						A	
Pachyaena			A	E,N			
Pachyaena?					A		
Plagiocristodon			A?	A?			
Sinonyx			A				
Synoplotherium					N		
?ANDREWSARCHIDAE							
Andrewsarchus					A		
Paratriisodon					A		
?WYOLESTIDAE							
Wyolestes				N			
?Mongoloryctes					A		
?Yantanglestes		A	A				
Key and total genera							
N = N. Am.:	0	2	2	4	5	2	0
E = Europe:	0	0	1	2	1	0	0
A = Asia:	?	4	6-8	2-3	11	3	2
World total:	?	5	7-9	5-6	15	4	2

FIGURE 1. Temporal and geographic distribution of Mesonychia based on published literature compiled by Zhou (1995) and by the author. Taxa preceded by a query are questionably included in the higher taxon in which they are listed. Generic names followed by a query probably represent additional diversity. Note that the first appearance of mesonychians is recorded as being in the early Paleocene of Asia, although the triisodontid arctocyonians that early mesonychians resemble closely are best known from the early and middle Paleocene of North America. Mesonychian generic richness is highest in Asia during all subepochs except the early Eocene (which is not yet as well sampled in Asia).

to be Arctocyonidae, and he may be right. Wang (1976) proposed that Didymoconidae are closely related to Mesonychidae. When describing *Wyolestes* (Gingerich, 1981), I was impressed by dental resemblances to *Yantanglestes* and *Mongoloryctes*, the former a mesonychian and the latter then classified as a didymoconid. Meng *et al.* (1994) have since shown that *Wyolestes* is unlikely to be a didymoconid and didymoconids are very different from mesonychians. However, dental resemblances of *Wyolestes* to *Yantanglestes* and *Mongoloryctes* still stand and I have grouped all here in Wyolestidae (with question marks re-

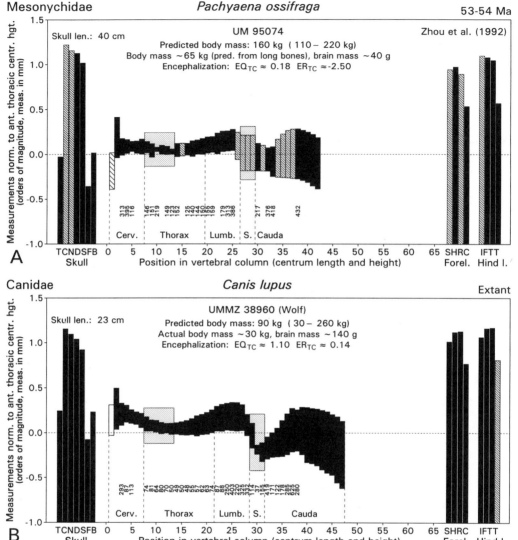

FIGURE 2. Diagrams of skeletal proportion comparing skull lengths, vertebral lengths and heights, forelimb long bone lengths, and hind limb long bone lengths of the early Eocene mesonychid *Pachyaena ossifraga* (A) to those of the skeletally similar extant wolf *Canis lupus* (B). All measurements are represented as a profile of bars, normalized to the mean height of the centrum of the six anterior thoracic vertebrae (dotted baseline). Two profiles are superimposed for vertebral measurements, representing centrum length and centrum height, and the bar shown is the *difference* between these (the position of the bar represents vertebral size and the length of each bar reflects measurement difference or shape): This is solidly filled when length exceeds height (as for most vertebrae here), and open when height exceeds length (as in the first cervical, vertebra 1, of *C. lupus*). Values for which reliable estimates can be interpolated or extrapolated are shown with hatching.

Note that *Pachyaena* has a slightly longer skull compared with the rest of its skeleton; *Canis* has a longer longest-cheek-tooth (T). Condylobasal skull length (C), external narial position (N), dentary length (D), and mandibular symphyseal position (S) decrease progressively in size; the greatest diameter of the mandibular foramen (F) is small in both; and the auditory bulla (B) is substantially longer than the mandibular foramen.

flecting uncertainty). Andrewsarchidae and Wyolestidae are regarded as families rather than subfamilies, paralleling Hapalodectidae, which was raised to family status by Ting and Li (1987).

Mesonychia range in age from early or middle Paleocene (ca. 63 Ma) through early Oligocene (ca. 33 Ma), and are found on all three of the northern continents. The number of mesonychian genera known from Asia exceeds that known from North America or Europe in every subepoch of the Paleocene through early Oligocene, save the early Eocene, which is not yet well sampled in Asia. Thus, Asia was possibly the center of origin of Mesonychia and Asia was certainly an important center of their evolutionary diversification.

Four genera of Mesonychidae are well known osteologically in being represented by complete or virtually complete postcranial skeletons: middle Paleocene *Hukoutherium* (Xue *et al.*, 1996; Xue, in preparation), late Paleocene *Sinonyx* (Zhou *et al.*, 1995; Gingerich *et al.*, in preparation), early Eocene *Pachyaena* (Matthew, 1915; Zhou *et al.*, 1992; O'Leary and Rose, 1995; Rose and O'Leary, 1995), and middle Eocene *Mesonyx* (Scott, 1886; Matthew, 1909). In contrast, very little is known about the postcranial osteology of hapalodectids, andrewsarchids, or wyolestids. It is perhaps possible that whales originated from one of these other families, but mesonychids are much better known and make a good model for cetacean ancestry.

Skeleton of Early Eocene *Pachyaena*

The skeleton of early Eocene *Pachyaena ossifraga* is represented in a *diagram of skeletal proportion* in Fig. 2, where it is compared with a skeleton of the extant wolf *Canis lupus*. Diagrams of skeletal proportion facilitate comparisons of functionally related cranial measurements, vertebral sizes and shapes, forelimb measurements, and hind limb measurements, all in terms of proportion. The common reference scale, average height of the six anteriormost thoracic vertebral centra (horizontal dashed line), is somewhat arbitrary. This reference scale was chosen to avoid any area of obvious functional specialization in the skeletons of mesonychians and cetaceans (skulls, necks, thoracolumbar vertebrae, tails, forelimbs, and hind limbs all have a range of different dimensions and proportions in the animals being compared). Because all diagrams of skeletal proportion are scaled in the same

◄───

Vertebral centrum length and height form three graphical vertebral arches, an anterior cervical arch rising from the anterior thorax where the forelimb originates (shaded box), a central thoracic and lumbar arch connecting this to the sacrum where the hind limb originates (second shaded box), and a posterior caudal arch. Size of each vertebral rectangle represents its proportions when viewed laterally (high solid rectangles represent vertebrae that are long and low, high open rectangles represent vertebrae that are short and high). Anterior thorax and sacrum (shaded boxes) are stable inflexible regions of the vertebral column characteristic of quadrupedal mammals.

Forelimbs (scapula S, humerus H, radius R, and longest metacarpal C) and hind limbs (innominate I, femur F, tibia T, and longest metatarsal T) of *Canis* are longer relative to the rest of the skeleton than those of *Pachyaena*, and the third segments (radius R and tibia T) are longer relative to other elements of the same limb. Compare these profiles with those of more aquatic mammals in Figs. 4–6.

Body masses predicted here are based on comparison with vertebrae of marine mammals (see text) to show that vertebral size of marine mammals overestimates the mass of terrestrial mammals by a factor of 2 to 3 (compare 160 kg with 65 kg, and 90 kg with 30 kg).

way, proportions can be compared between skeletons of different animals, even when these differ in absolute size as is true in the comparison of *Pachyaena* and *Canis*. All measurements are represented as a profile of bars, except for vertebral measurements where two profiles are plotted (centrum length and height) and the bar shown is the *difference* between these (the position of the bar represents vertebral size and the length of each bar reflects vertebral shape).

The skull of *Pachyaena* resembles that of *Canis* in relative size, with both being about an order of magnitude greater than the anterior-thoracic-height baseline. Anterior–posterior length of the longest cheek tooth (T in Fig. 2) is less in *Pachyaena* because it does not have the carnassial specialization of *Canis*. The relationships of cranial condylobasal length (C) to nasal position (N), dentary length (D), and mandibular symphysis position (S) are very similar in the two. The size of the mandibular foramen (F) is less than the baseline and less than auditory bulla length (B) in both.

The vertebral column of *Pachyaena* resembles that of *Canis* in relative size, and it has a pattern typical of cursorial land mammals. Postatlas cervical vertebrae (positions 2 through 7) show decreasing centrum length coupled with increasing and then decreasing centrum height in both *Pachyaena* and *Canis* (the atlas itself is difficult to measure in any functionally meaningful way). The important point is that cervicals are relatively long in *Pachyaena* and *Canis* (both have long necks compared with what we will see in archaeocetes), and the cervical series together with anterior thoracics forms an *anterior arch* supporting the skull anterior to and above the shoulder (shaded box) where the axial skeleton is connected to the forelimb. Note that when centrum height exceeds length, the normalized height and length measurements are connected by an open bar representing shape difference (the higher the open bar, the more height exceeds length). When centrum length exceeds height, the normalized length and height measurements are connected by a solid bar that again represents shape difference (but this time the higher the open bar, the more length exceeds height). A run of open bars represents a sequence of vertebrae with centra shorter than they are high, and a run of solid bars like the cervicals shown in Fig. 2 represents a sequence of vertebrae with centra longer than they are high.

Posterior thoracic, lumbar, and sacral centra in *Pachyaena* and *Canis* form a second arch or *central arch*, again similar in both, of increasing and then sharply decreasing length and slightly increasing and then decreasing height between the shoulder (first shaded box) and sacrum where the axial skeleton is connected to the hind limb (second shaded box). This is followed in both by a *posterior arch* of increasing and then slightly decreasing caudal centrum length, and slightly increasing and then sharply decreasing caudal centrum height in the tail. Bar segments of similar size represent vertebral centra of similar shape, whereas the position of the bar segment on the diagram is a measure of centrum size. Thus, the posterior caudal centra in both *Pachyaena* and *Canis* are similar in shape but decrease progressively in size. The anterior, central, and posterior arches shown here correspond to those in classic representation of the skeleton of a land mammal as a "bridge that walks" (e.g., Gregory, 1937).

The forelimbs of *Pachyaena* and *Canis* are similar in relative length of the scapula (S), humerus (H), radius (R), and longest metacarpal (C), and the hind limbs are similar in relative length of the pelvis or innominate (I), femur (F), tibia (T), and longest metatarsal (T). Fore- and hind limbs are similar in size relative to each other, and in size relative to the skull and vertebral column. However, fore- and hind limbs of *Pachyaena* differ in two important ways from those of *Canis*. The radius is shorter than the humerus in the forelimb and the tibia is shorter than the femur in the hind limb in *Pachyaena*, and the metacarpals and metatarsals of *Pachyaena* are shorter than those of *Canis*, indicating a slightly heavier build (*Pachyaena* is a larger and heavier animal) and somewhat less fully cursorial locomotor adaptation.

Important anatomical details of the teeth, vertebrae, and hands and feet of mesonychids cannot be represented on a diagram of skeletal proportion. Central cheek teeth of *Pachyaena* are not enlarged like those of many carnivorous mammals (including later archaeocetes) and they do not have the sharpness or the carnassial shearing specialization expected of predatory meat eaters. Lumbar vertebrae of *Pachyaena* and other mesonychids are unusual in having revolute zygapophyses like those of arctocyonid condylarths (Russell, 1964) and later artiodactyls (Slijper, 1947), making them stiff-backed runners (Zhou *et al.*, 1992). Terminal phalanges of *Pachyaena* are fissured ungules or hooves, which is consistent with nonpredatory behavior and with cursoriality. The overall skeletal similarity of early Eocene *Pachyaena* to extant *Canis* shown in Fig. 2 is interpreted as indicating similar behavior in life, recognizing that *Pachyaena*, with a metatarsal/femur ratio of just 0.31, cannot have been an active pursuit predator like a wolf (Janis and Wilhelm, 1993). Mesonychians are usually interpreted as solitary carrion feeders and scavengers that spent many of their waking hours trotting in search of dead animals and were best able to chew flesh after it was partially decomposed (Boule, 1903; Osborn, 1910; Scott, 1913; Zhou *et al.*, 1992). This is plausibly the kind of animal from which archaeocetes evolved.

1.3. Diversity and Morphology of Archaeoceti

There are about 25 known genera of Archaeoceti, grouped in six families: Ambulocetidae, Basilosauridae, Dorudontidae, Pakicetidae, Protocetidae, and Remingtonocetidae (Fig. 3). These range in age from latest early Eocene (ca. 49.5 Ma) through late Eocene (ca. 36 Ma), and are found on the margins of most of the world's oceans. The number of mesonychian genera known from Tethys exceeds that known elsewhere in every subepoch of the Eocene, save the late Eocene, which is not yet well sampled in Tethys. Thus, it appears that Tethys was possibly the center of origin of Cetacea and, more certainly, a center of their evolutionary diversification.

Six genera of Archaeoceti are well enough known osteologically to make meaningful comparisons using a diagram of skeletal proportion: Lutetian (early middle Eocene) *Rodhocetus* (Gingerich *et al.*, 1994; Gingerich, in preparation), *Dalanistes* (Gingerich *et al.*, 1995), and *Protocetus* (Fraas, 1904); late Bartonian latest middle Eocene *Dorudon* (Uhen, 1996); late Bartonian to Priabonian late middle to late Eocene *Basilosaurus* (Kellogg, 1936; Gingerich, in preparation); and Priabonian late Eocene *Saghacetus* (Gingerich, in preparation). These include four of the six archaeocete families: Protocetidae, Remingtonocetidae, Dorudontidae, and Basilosauridae. Ambulocetidae and Pakicetidae are known from important limb bones (*Ambulocetus*; Thewissen *et al.*, 1996) and cranial material (*Ambulocetus*, *Pakicetus*; Gingerich *et al.*, 1983; Thewissen *et al.*, 1996), but little is yet known of the vertebral skeleton, which is central to analyses of the kind presented here.

ARCHAEOCETI	PALEOCENE			EOCENE				OLIG.
	Early	Middle	Late	Ypresian	Lutetian	Barton.	Priabon.	Early
AMBULOCETIDAE								
Ambulocetus					T			
Gandakasia				T				
BASILOSAURIDAE								
Basilosaurus						M,T	W,A?	
Basiloterus						T	M?	
DORUDONTIDAE								
Ancalecetus						M		
Dorudon						W,M		
Pontogeneus						M	W	
Saghacetus							M	
Zygorhiza						E?	W	
PAKICETIDAE								
Ichthyolestes				T				
Nalacetus				T				
Pakicetus				T				
PROTOCETIDAE								
Babiacetus						T		
Cross whale						W		
Eocetus						M		
Gaviacetus					E?,T			
Georgiacetus						W		
Indocetus						T		
Pappocetus					E			
Protocetus					W?,E?,M			
Rodhocetus					T			
Takracetus					T			
REMINGTONOCETIDAE								
Andrewsiphius						T		
Dalanistes					T			
Remingtonocetus					T	T		
Key and total genera								
W = W. Atlantic:	0	0	0	0	?	3	3	0
E = E. Atlantic:	0	0	0	0	1-3	0-1	0	0
M = Med. Tethys:	0	0	0	0	1	5	1-2	0
T = E. Tethys:	0	0	0	4	6	6	0	0
A = Austral seas:	0	0	0	0	0	0	0-1	0
World total:	0	0	0	4	8	13	5	0

FIGURE 3. Temporal and geographic distribution of Archaeoceti based on a survey of the published literature. Note that the first appearance of archaeocetes, as Pakicetidae and Ambulocetidae, is in the latest Ypresian and earliest Lutetian, early middle Eocene, of eastern Tethys (Indo-Pakistan). Protocetidae and Remingtonocetidae predominate in the Lutetian early middle Eocene and are best known from eastern Tethys. Basilosauridae and Dorudontidae predominate in the Bartonian late middle Eocene and Priabonian late Eocene and are best known from the western Atlantic and mediterranean Tethys.

1.3.1. Skeletons of Early Middle Eocene *Rodhocetus* and *Dalanistes*

The skeleton of early middle Eocene *Rodhocetus* (Protocetidae) is represented in a diagram of skeletal proportion in Fig. 4, where it is compared to the skeleton of early middle Eocene *Dalanistes* (Remingtonocetidae). Skulls of both differ from those of *Pachyaena* (Fig. 2) in having the external nares (N) open at a position behind the front of the dentary, and both differ in having a much larger mandibular foramen (F)—this opening is now as high as the auditory bulla (B) is long. A mandibular foramen this large is typical of archaeocetes and later whales with a well-developed acoustic window and wave guide (Norris, 1968), indicating an auditory system specialized for hearing in water.

Vertebral centrum length tends to be more similar to centrum height in *Rodhocetus* and *Dalanistes* compared with what was seen in *Pachyaena*, and cervical vertebrae are relatively shorter compared with those of *Pachyaena*. Cervicals of *Rodhocetus* are particularly noteworthy in having centra that are even shorter than they are high, which is a characteristic of all later archaeocetes, mysticetes, and odontocetes. Shortening the neck is one component of hydrodynamic streamlining characteristic of living whales.

The thoracic, lumbar, and sacral centra of *Rodhocetus* increase progressively in both length and height, retaining just a hint of the central arch seen in land mammals that support their weight on land. There is a central arch in *Dalanistes*, but even here it is less conspicuously developed than in *Pachyaena* (Fig. 2A) because the sacral centra are higher. *Rodhocetus* and *Dalanistes* both have long neural spines on thoracic vertebrae, supporting the idea that they were able to use their forelimbs to lift much of their body weight on land. *Rodhocetus* retains a four-centrum sacrum, although the centra are not fused to each other (open box in Fig. 4A), whereas *Dalanistes* has a four-centrum sacrum with centra solidly fused (shaded box in Fig. 4B). The tail is poorly known in both, but proximal caudals in *Dalantistes* are longer than they are high, suggesting that it had a tail more like that of *Pachyaena*.

Forelimbs have not yet been found for either *Rodhocetus* or *Dalanistes*, but the presence of large pelves (I) and robust femora (F) is consistent with support and movement on land. *Rodhocetus* and *Dalanistes* differ from each other in relative length of the skull. The rostrum of *Rodhocetus* is normally proportioned for an archaeocete, whereas that of *Dalanistes* is unusually long and narrow. There is little doubt that these animals fed differently, and a shorter rostrum and more mobile vertebral column with unfused sacral centra means *Rodhocetus* was a better swimmer and probably an aquatic pursuit predator like later whales, whereas the long rostrum and fused sacrum of *Dalanistes* means it was a slower swimmer and probably an aquatic ambush predator like earlier *Ambulocetus* (Thewissen *et al.*, 1996). These behavioral differences are consistent with differences in cervical centrum lengths and differences in caudal centrum sizes and proportions.

1.3.2. Skeletons of Middle to Late Eocene *Dorudon* and *Basilosaurus*

The skeleton of late middle Eocene *Dorudon* (Dorudontidae) is represented in a diagram of skeletal proportion in Fig. 5, where it is compared with the skeleton of late Eocene *Basilosaurus* (Basilosauridae). Skulls of *Dorudon* and *Basilosaurus* differ from those of *Pachyaena* (Fig. 2) and from *Rodhocetus* and *Dalanistes* (Fig. 4) in having the external nares (N) opening farther back on the rostrum. In the vertebral column, cervical centra are much shorter, there is no sacrum, and the thoracolumbar or central and caudal or posterior vertebral arches together form a single long vertebral arch. These are different in detail but similar in overall functional conformation to vertebral arches of extant odontocetes and mysticetes (compare with profiles in Fig. 6). Forelimbs are now well known in *Dorudon* (Uhen, 1996) and *Basilosaurus* (Kellogg, 1936; Gingerich and Smith, 1990) and the humerus (H) is always much longer than forearm bones like the radius (R). The elbow joint

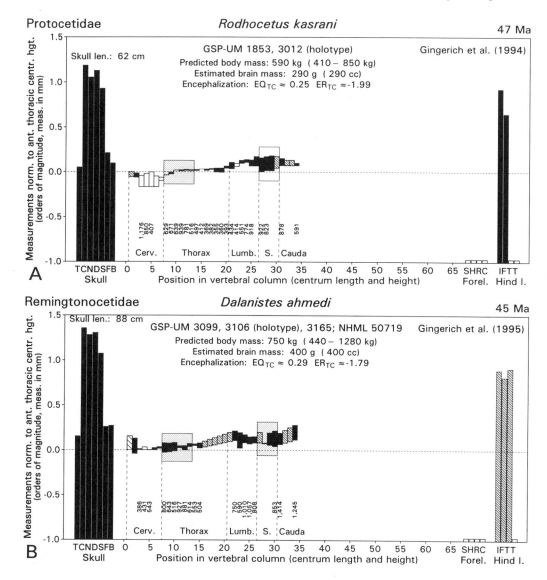

FIGURE 4. Diagrams of skeletal proportion comparing skull, vertebral, forelimb, and hind limb lengths and proportions of the early middle Eocene protocetid archaeocete *Rodhocetus kasrani* (A) with those of the contemporary and slightly later remingtonocetid archaeocete *Dalanistes ahmedi* (B). All measurements are those described
in Fig. 2. Note that *Dalanistes* has a longer skull compared with the rest of its skeleton, but proportions within the
skull are similar. Skulls of *Rodhocetus* and *Dalanistes* differ from those of *Pachyaena* and *Canis* (Fig. 2) in having the external nares and mandibular symphysis located more posteriorly. *Dalanistes* has shorter cervicals for its
size than *Pachyaena*, and those of *Rodhocetus* are shorter still. *Dalanistes* has a central vertebral arch connecting
stable anterior thoracics to a stable sacrum that is similar to those of *Pachyaena* and *Canis* but flatter, and the posterior arch appears to be *Pachyaena*-like as well. *Rodhocetus* has an even flatter thoracolumbar vertebral column
and no indication of the caudal arch. There is a four-centrum sacrum in *Rodhocetus* but these vertebrae are not
fused and hence they were not stable like those of land mammals (shown diagrammatically as an open box). Forelimbs of *Rodhocetus* and *Dalanistes* are not yet known. Both genera have substantial innominates and femora that
are similar in size relative to the rest of the skeleton, although the femur of *Rodhocetus* is shorter than that of

is still mobile in both. Hind limbs are present in both *Dorudon* and *Basilosaurus* though these are reduced in size both relative to the forelimbs and relative to hind limbs in earlier archaeocetes. Because there is no trace of a sacrum and the pelvis floated in muscle in the ventral body wall, it was impossible for either *Dorudon* or *Basilosaurus* to support its weight on land and both were clearly fully aquatic.

The principal differences between skeletons of *Dorudon* and *Basilosaurus* are the relative sizes and shapes of posterior thoracic, lumbar, and anterior caudal vertebrae. This shows clearly in the diagrams of skeletal proportion of Fig. 5. In *Basilosaurus* the ninth thoracic is notably longer than the eighth and it is notably longer than either is high. This increase in size affects both length and height separately and affects the two together as well, hence the marked change in shape. Elongation of thoracic through caudal centra gives *Basilosaurus* its anguilliform body shape, but the increase in centrum height (and width) over the proportion seen in *Dorudon* suggests more is happening than simple elongation, which must have affected swimming in important ways. Vertebrae of *Basilosaurus* are sometimes densely mineralized and heavy when found as fossils, but when these are not secondarily permineralized they are fragilely cancellous and light. In life they were marrow-filled, and surface-to-volume allometry means enlargement would have made them more buoyant. Large buoyant vertebrae suggest that *Basilosaurus* lived predominantly at the sea surface rather than being a three-dimensional diving swimmer like *Dorudon*, which is consistent with Slijper's (1946) interpretation, from vertebral metapophyses and neural spines, that *Basilosaurus* moved partly by lateral undulation.

1.4. Body Mass of Mesonychids and Archaeocetes

Mesonychians are typical land mammals for which body mass can be estimated in the usual ways, from long bone lengths and diameters (e.g., Gingerich, 1990) and tooth size (e.g., Legendre and Roth, 1988). Zhou (1995) found that body masses estimated from long bone lengths and diameters were closely collinear with log M_2 lengths and diameters for the 14 mesonychid specimens (representing six genera and ten species) having both, scaling like mass to carnassial size in felid Carnivora. From this he derived a regression equation for predicting body mass in kilograms from M_2 crown area in square millimeters: log $Y = 1.327 \log X - 1.457$. Zhou found that mesonychids ranged from about 7 to 194 kg in body mass. Gunnell and Gingerich (1996) extended this to Hapalodectidae and found body masses ranging from about 1 to 8 kg. Thus, Hapalodectidae all fall in the 0.5 to 10 kg range that is considered medium-sized in mammals, whereas most Mesonychidae fall in the 10 to 250 kg range of large mammals.

Archaeocetes may have been similar to mesonychids when they originated, but archaeocetes are not typical land mammals. Skulls, necks, tails, forelimbs, and hind limbs are

---

Dalanistes. Dalanistes and *Rodhocetus* are intermediate in skeletal proportions and vertebral profile between land mammals like early Eocene *Pachyaena* and later fully aquatic archaeocetes like *Dorudon* (Fig. 5). Body mass predictions here are based on comparison with vertebrae of marine mammals (see text), and brain masses are estimated from endocranial casts associated with skulls. Calculation of encephalization quotients (EQ) and residuals (ER) is explained in the text.

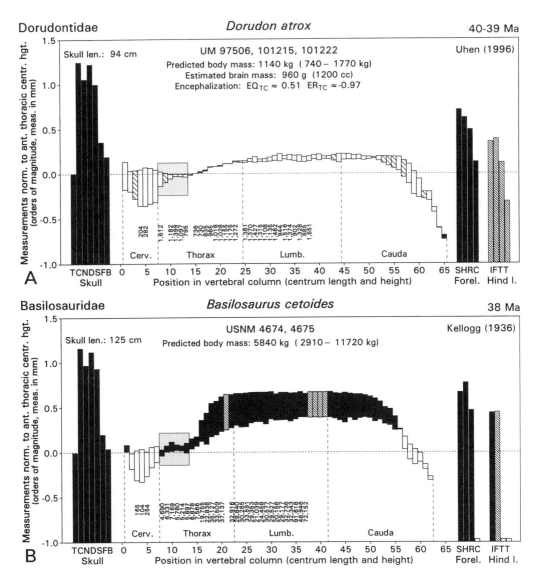

FIGURE 5. Diagrams of skeletal proportion comparing skull, vertebral, forelimb, and hind limb lengths and proportions of the late middle Eocene dorudontid archaeocete *Dorudon atrox* (A) with those of the slightly later basilosaurid archaeocete *Basilosaurus cetoides* (B). All measurements are those described in Fig. 2. Note that skulls of *Dorudon* and *Basilosaurus* are similar in relative size and proportions to that of *Rodhocetus*, though the external nares are positioned more posteriorly, closer to the position of the mandibular symphysis. Both have shorter cervicals for their size than *Rodhocetus*. The remaining vertebrae of *Dorudon* are almost all shorter than they are high (open rectangles), whereas most of the remaining vertebrae of *Basilosaurus* are much longer than they are high (solid rectangles). There is no real arch to the central vertebrae and although the anterior thoracics are a stable interval of the column anchoring the forelimbs (shaded box), there is no real sacrum and the rest of the vertebral column is flexible like that of later whales (Fig. 6). *Dorudon* and *Basilosaurus* both have vertebral centra that shorten rapidly after about vertebral position 55 (posterior caudals). Both probably had a caudal fluke of some kind. Forelimbs are large in *Dorudon* and *Basilosaurus,* and the humerus is long (the humerus is even longer than the scapula in *Basilosaurus*). Both genera had reduced innominates and femora. Lengths of hind limb elements

commonly enlarged or reduced relative to the rest of the skeleton in archaeocetes (as shown in the diagrams of skeletal proportion here), and for this reason a method for estimating body mass has been developed that is based on selected vertebrae. The reference sample includes eight cetaceans ranging in size from 150 to 23,500 kg and five pinnipeds ranging in size from 85 to 1210 kg. Body mass was regressed successively on vertebral centrum length, width, and height (simultaneously) at each vertebral position from vertebra 2 to vertebra 40, yielding a body mass estimate for each vertebra. These analyses showed that vertebrae 2–3, 6–7, and 38–40 have multiple regression coefficients of determination (r_2) less than 0.5 and estimates based on these were not used. Estimates based on the remaining vertebrae are written vertically at the appropriate position in Figs. 2, 4, 5, and 6. The predicted body mass (and its 95% confidence interval) for an archaeocete species, written near the top of each chart, is the median of all of the separate estimates. Use of the median rather than a mean ensures that outliers attributable to unusual vertebral sizes or proportions have no effect on the final estimate. Predicted body masses for Basilosauridae are based on the median of the first 12 estimates.

1.5. Brain Mass and Relative Brain Size of Mesonychids and Archaeocetes

Brain mass is now known for two mesonychids, middle Eocene *Mesonyx obtusidens* (Radinsky, 1976) and early Eocene *Pachyaena ossifraga* (Gingerich, in preparation) and these can be combined with estimated body weights to yield a measure of relative brain size.

As background, analysis of an extensive data base of $\log_{10}$ values of brain and body mass for 778 terrestrial mammalian species yielded a regression slope of 0.740 and a corresponding intercept of -1.205 (Gingerich, in preparation). Exponentiating, the brain mass E_p predicted for a given body mass P is $E_p = 0.062\ P^{0.740}$. Analysis order by order yields a weighted mean slope of 0.668 and intercept of 0.104 that are very close to the familiar values of 2/3 and 0.12 used by Jerison (1973) to calculate encephalization quotients (EQ). I interpret Jerison's scaling to be that expected for *terrestrial mammals analyzed at an ordinal scale*. Here (and in general) we are concerned with mammals as a class, comparing individual species or groups of species in relation to the whole. The corresponding EQ is EQ_{TC} (where T refers to *terrestrial* and C refers to a *class*-level taxonomic scale). Volant and marine mammals scale differently, but EQ_{TC} is appropriate as an initial baseline, even when volant or marine mammals are considered, because both evolved from terrestrial ancestors. EQ is calculated as the ratio of observed brain mass E (in grams) to brain size E_p predicted for a given body mass P (also in grams). Finally, EQs assume the asymmetrical and unequally scaled range of values typical of ratios, ranging from infinitesimally small to unity (when observation = prediction) to infinitely large. There are both practical and

---

for *Dorudon atrox* are scaled down from those of *Basilosaurus isis*, which they resemble closely in preserved parts (proximal half of femur, patella, and astragalus). Comparable hind limb elements of these genera are substantially shorter than those of *Rodhocetus* or *Dalanistes* (Fig. 4). *Dorudon* is intermediate in skeletal proportions and vertebral profile between earlier archaeocetes like *Dalanistes* and *Rodhocetus* (Fig. 4) and later odontocetes and mysticetes (Fig. 6). Body mass predictions here are based on comparison with vertebrae of marine mammals (see text), and brain mass of *Dorudon* is estimated from endocranial casts associated with skulls. Calculation of encephalization quotients (EQ) and residuals (ER) is explained in the text.

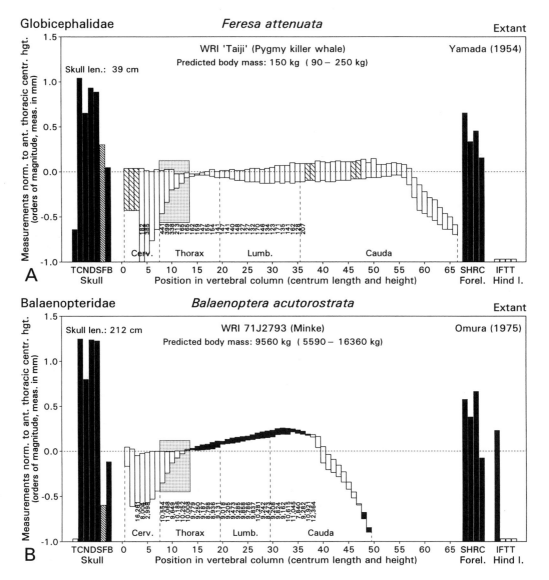

FIGURE 6. Diagrams of skeletal proportion comparing skull, vertebral, forelimb, and hind limb lengths and proportions of the extant globicephalid odontocete *Feresa attenuata* (A) with those of the extant balaenopterid mysticete *Balaenoptera acutorostrata* (B). All measurements are those described in Fig. 2. Note that *Balaenoptera* has a longer skull, compared with the rest of its skeleton, whereas *Feresa* has a larger mandibular canal and larger auditory bulla. *Feresa* has smaller teeth than those of archaeocetes, relative to other measures of size (and *Balaenoptera* of course has lost its teeth). Both have external nares positioned farther back on the skull than do archaeocetes. *Feresa* and *Balaenoptera* both have short cervical vertebrae. Thoracic, lumbar, and caudal vertebrae of *Feresa* are more numerous and more uniform in size than those of *Balaenoptera*. There is a single flattened vertebral arch behind the anterior thorax and there is no sacrum. Posterior caudal vertebrae decrease rapidly in size, as in *Dorudon* and *Basilosaurus*, and the tail is of course fluked. Forelimbs of *Feresa* and *Balaenoptera* are similar to those of advanced archaeocetes, but the humerus is shorter. Both have innominates as remnants of the hind limb (not plotted here), and *Balaenoptera* often retains a femur as well (again, not plotted here). *Feresa* and *Bal-*

theoretical advantages to comparing relative brain size in terms of ER, an encephalization residual on a doubling scale, where ER can be calculated from EQ as $ER = \log_2 (EQ)$. ERs have a symmetrical and equally scaled range from negative infinity to zero (when observation = prediction) to positive infinity. Familiar EQ and corresponding ER values are both reported here.

Radinsky (1976) estimated the endocranial volume of *Mesonyx obtusidens* to be 80 cm^2 representing a brain of about 80 g. He estimated the body mass of *M. obtusidens* from skull length and body length, and calculated body masses of about 40 and 55 kg. The corresponding EQ_{TC}s are 0.51 and 0.40, respectively (ER_{TC}s are -0.98 and -1.32). The endocranial volume of *Pachyaena ossifraga* is 40 cm^3, corresponding to a brain of about 40 g (one-half the size estimated for *M. obtusidens*). The body mass of *P. ossifraga* is abut 65 kg (Zhou *et al.*, 1992). These numbers yield an EQ_{TC} for *Pachyaena* of 0.18 ($ER_{TC} = -2.50$), which is less than one-half that of *Mesonyx* (the genera differ by more than one unit on a proportional halving-doubling or $\log_2$ scale).

Jerison (1973, 1978), following Dart (1923), reported that the endocranial volume (and corresponding brain mass) of "*Dorudon*" *osiris* was about 480 cm^3 (g); that of "*Dorudon intermedius*" was about 780 cm^3 (g); and that of "*Prozeuglodon atrox*" was about 800 cm^3 (g). Jerison estimated body masses of the three species as 350, 530, and 20,000 kg. EQ_{TC}s calculated from these numbers would be 0.61, 0.73, and 0.05, respectively ($ER_{TC} = -0.71$, -0.45, and -4.29), which seem seriously discrepant in archaeocetes that are sometimes classified in the same family and are the same or almost the same age geologically. Three problems are confounded here. The first is systematic, Dart's endocranial casts represent two species, not three. These are now identified (Gingerich, 1992) as *Saghacetus osiris*, for which Dart (1923) published endocranial volumes of 310 cm^3 (for a subadult) and 480 and 490 cm^3 for adults; and as *Dorudon atrox*, for which Dart (1923) published endocranial volumes of 785 and 800+ cm^3 for subadults. Jerison, following the literature (e.g., Barnes and Mitchell, 1978), associated the latter with postcranial remains now known to be *Basilosaurus isis* (Gingerich *et al.*, 1990), for which no brain casts or endocranial volumes were known at the time Jerison was writing. University of Michigan field parties working in Egypt have recovered new endocasts of all three species associated with identifiable skulls and postcranial skeletons: These average 485 cm^3 for *S. osiris*, 1200 cm^3 for *D. atrox*, and 2800 cm^3 for *B. isis*.

The second problem with Jerison's interpretation of brain size in archaeocetes is in conversion of endocranial volume to brain mass. Marples (1949) and Breathnach (1955) showed that much of what Dart (1923) interpreted as a massive cerebellum in Egyptian archaeocetes is really a large intracranial vascular rete mirabile. Uhen (1996) has recently estimated that this comprises some 20% of endocranial volume in *Dorudon atrox*. *Saghacetus*, *Dorudon*, and *Basilosaurus* are at a similar grade of brain expansion, with similar retia, and their endocranial volumes of 485, 1200, and 2800 cc are thus thought to correspond, respectively, to 388, 960, and 2520 g brain masses.

---

aenoptera illustrate the skeletal proportions of modern fully aquatic cetaceans, which are not very different from those of *Dorudon atrox* (Fig. 5A). Body mass predictions here are based on comparison with vertebrae of marine mammals (see text), and these agree closely with published body masses for each species (150 and 9000 kg, respectively; Watson and Ritchie, 1981).

The third problem with interpretation of brain size in Egyptian archaeocetes has to do with estimation of body mass for the three species. The method described here yields predicted body masses of 350 kg (identical to Jerison's estimate) for *Saghacetus osiris*, 1140 kg for *Dorudon atrox* (Fig. 5A; more than double Jerison's estimate), and 6480 kg for *Basilosaurus isis* (slightly heavier than the estimate of *B. cetoides* in Fig. 5B, but much less than Jerison's estimate of 20 metric tons). Combining these new estimates of brain mass with new estimates of body mass yield reasonably consistent EQ_{TC} estimates of 0.49, 0.51, and 0.37 ($ER_{TC} = -1.02, -0.97$, and -1.43). Middle-to-late Eocene archaeocetes have EQ_{TC} values averaging about 0.46, meaning that they have brains about 46% as large as those of an average terrestrial mammal of the same body mass living today. Stating the same result in terms of residuals, middle-to-late Eocene archaeocetes have ER_{TC} values averaging about -1.13, meaning that they have brains a little more than one unit smaller on a proportional $\log_2$ scale than expected for a terrestrial mammal of the same body mass living today.

For comparison, endocranial volumes of early middle Eocene *Rodhocetus kasrani* and *Dalanistes ahmedi* are given in Figure 4, along with body mass estimates. From the morphology of the endocasts it is clear that there was no substantial intracranial vascular rete in either of these genera, as there was none in *Indocetus* (Bajpai et al., 1996), and brain mass is thus essentially the same as endocranial volume. EQ_{TC}s calculated for *R. kasrani* and *D. ahmedi* are 0.25 and 0.29, respectively ($ER_{TC} = -1.99$ and -1.79). Early middle Eocene archaeocetes have EQ_{TC} values averaging about 0.27, meaning that they have brains about 27% as large as those of an average terrestrial mammal of the same body mass living today. Stating the same result in terms of residuals, middle-to-late Eocene archaeocetes have ER values averaging about -1.89, meaning that they have brains almost two units smaller on a proportional $\log_2$ scale than expected for a terrestrial mammal of the same body mass living today. This is 0.76 of a unit less than the encephalization of middle-to-late Eocene archaeocetes.

2. Trends in the Morphology and Adaptation in Archaeoceti

Eocene Archaeoceti are important for understanding the land-mammal ancestry of whales because they are intermediate in time, space, and form between slightly earlier Paleocene and early Eocene Mesonychidae and slightly later Oligocene Mysticeti and Odontoceti. Mesonychids are found in early Eocene sediments on the Central Asian and Indo-Pakistan continental masses bordering eastern Tethys to the north and south. Eastern Tethys is where we have the best record of early archaeocetes.

The morphological intermediacy of archaeocetes can be seen by comparing profiles of archaeocete skeletal proportions in Figs. 4A and 5A with those of *Pachaena* in Fig. 2A and extant cetaceans in Fig. 6A,B. Tooth size (T) was large throughout archaeocete evolution before teeth were reduced in odontocetes and lost in mysticetes. Nasals moved back on the skull through time, as reflected by the depth of the notch in the profiles over N compared with flanking skull length (C) and dentary length (D). Similarly, the height of the mandibular foramen (F) expanded to equal and exceed auditory bulla length (B, before receding again in mysticetes).

Three trends are clear in the vertebral column. The length of cervicals decreased at each stage relative to overall size while the cervical height remained almost constant, meaning that cervical shape changed greatly as well. Second, the central vertebral arch involving posterior thoracic, lumbar, and sacral vertebrae, and the posterior vertebral arch involving caudal vertebrae merged into a single relatively straight-sided arch with reduction and finally elimination of the sacrum as an intermediate point of support. Inflection of this single arch is effectively where it was in the posterior of the original arches. In addition, vertebrae composing the new arch assumed a more equidimensional shape as their functions became more uniform and more uniformly shared. Trends in the fore- and hind limbs are less clear in the diagrams of skeletal proportion, although the former were modified into flippers and the latter lost all external exposure.

I have quantified several other changes in the transition from mesonychids to cetaceans in Figs. 7 and 8. These involve body size, trophic specialization as indicated by size of the largest tooth, auditory specialization as indicated by bulla length (reflected too in size of the mandibular canal), hydrodynamic streamlining and limb reduction as indicated by femur length (reflected too in shortened cervical vertebrae).

2.1. Body Size

A graph of body masses for archaeocetes of different ages is shown in Fig. 7A, where these are compared with the body mass ranges of Mesonychidae and modern Cetacea. Archaeocetes for which body mass can be estimated reliably have masses similar to but larger than those estimated for most mesonychians. Thewissen *et al.* (1996) estimated that *Ambulocetus* weighed 141–235 kg from cross sections of long bones but I have used my own estimate of 720 kg based on vertebral size. Archaeocetes fall comfortably within the range of body masses of extant cetaceans (cf. Downhower and Blumer, 1988). There is a trend toward larger body mass through the course of the middle and late Eocene but archaeocetes of relatively small mass are known from the late Eocene (e.g., *Saghacetus*) and the trend involves an expansion of the range of body masses upward through time.

2.2. Trophic Specialization

Figure 7B shows a trend toward larger tooth size through time in archaeocetes, but this trend follows the trend toward larger body mass very closely and it can be explained largely as a consequence of the increasing range of body sizes (see Fig. 8B). The length of the largest cheek tooth in archaeocetes is no larger than that of mesonychids, but interestingly the largest cheek tooth in mesonychids is usually M_2 where as that in archaeocetes is usually P^3 or P_3. Furthermore, incisors of *Pakicetus* are delicate and pointed (Gingerich and Russell, 1990), unlike those of any known mesonychid. Both of these differences from mesonychids imply that the change to a characteristically archaeocete dentition and trophic specialization was achieved very early in archaeocete evolution. There was some experimentation with cranial and dental specialization in ambulocetids and remingtonocetids early in archaeocete evolution, presumably related to feeding. Later archaeocetes are remarkably uniform cranially and dentally, and trophic specializations, to the extent these were important, involved differences in swimming to catch prey.

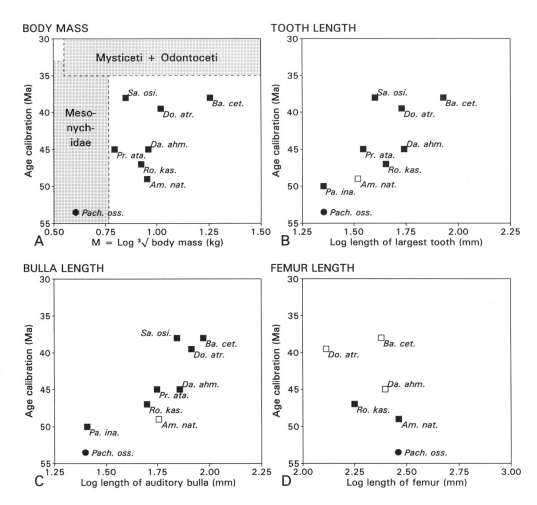

FIGURE 7. Trends of changing body mass (A), tooth length (B), auditory bulla length (C), and femur length (D) in the evolution of Archaeoceti. The same trends are normalized in Fig. 8 by subtracting body mass. Note that archaeocetes are generally larger than Mesonychidae, and hence much larger than Hapalodectidae and Wyolestidae, but lie within the broad range of body masses of extant cetaceans. Body masses of archaeocetes increased through time largely as a result of an expansion of their range of variation. Increase in tooth sizes largely reflects the increase in body masses. Bulla length increased in size through time, but this appears to have happened more rapidly early, e.g., between *Pakicetus* and *Ambulocetus*, and less rapidly later. Femur length decreased in size through time. Solid circle is mesonychid *Pachyaena ossifraga*. Solid and open squares are archaeocetes (*Am. nat., Ambulocetus natans; Ba. cet., Basilosaurus cetoides; Da. ahm., Dalanistes ahmedi; Do. atr., Dorudon atrox; Pa. ina., Pakicetus inachus; Pr. ata., Protocetus atavus; Ro. kas., Rodhocetus kasrani; Sa. osi., Saghacetus osiris;* symbol is open when values are estimated or, in the case of tooth length in *Ambulocetus*, when the tooth measured may not have been the longest).

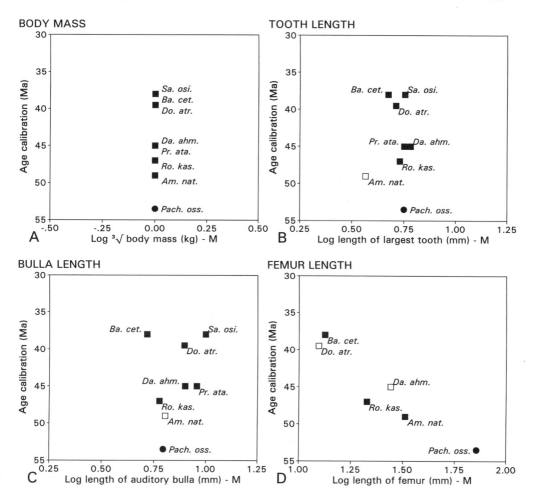

FIGURE 8. Size-normalized trends of body mass (A), tooth length (B), auditory bulla length (C), and femur length (D) in the evolution of Archaeoceti (compare with Fig. 7). A tightly clustered vertical distribution like that for tooth length means most observed variation can be explained by differences in body mass. Bulla size is only partially explained by increasing body mass, and the presence of relatively large bullae in small *Saghacetus osiris* and small bullae in large *Basilosaurus cetoides* suggests that there is some intermediate optimum size. Femur length decreased in size through time and this trend is stronger when body mass is taken into account. Symbols as in Fig. 7.

2.3. Auditory Specialization

Figure 7C shows the size of the auditory bulla in archaeocetes of different ages for which bulla size is known. This pattern too appears at first to be similar to that for tooth length, but there is more separation between *Pakicetus* and other archaeocetes. When *Pakicetus* is set aside, auditory bullae appear to increase slightly in size through time, but this change is not so great as would be expected from the expansion in range of known body masses. When the effect of body mass on bulla size (Fig. 8C) is subtracted as it was for tooth size, the remaining variance is larger than that for tooth size, the largest late archaeocete (*Basilosaurus*) has the smallest bullae, and the smallest late archaeocete (*Saghacetus*) has much larger bullae for its body size. This suggests that bulla size in archaeocetes is determined by a common factor like the density or some related acoustic characteristic of sea water in addition to archaeocete body size. The relatively small auditory bullae of mysticetes (Fig. 6B) support this idea.

2.4. Hydrodynamic Streamlining and Hind Limb Reduction

One of the most interesting changes in archaeocete evolution is reduction of the hind limb from a limb size typical of land mammals to the small limb of *Basilosaurus* and *Dorudon*. Reduced limbs in late archaeocetes probably retained a functional role in reproduction, but these are clearly too small to support the body on land (Gingerich *et al.*, 1990). Disarticulation of innominate bones of the pelvis from the sacrum is part of the same trend. Reduction of the hind limb can be represented by femur length, which shows a trend toward smaller absolute size through time while overall body mass is increasing (Fig. 7D). Subtracting the effect of changing body mass shows this to be an even stronger trend. Reduction of the hind limb coincides with shortening of cervical vertebrae, and both are part of the hydrodynamic streamlining necessary for efficient swimming. The greatest reduction relative to body size is seen in middle to late Eocene dorudontids and basilosaurids, which lack any articulation of innominate bones with the vertebral column and also have short cervical vertebrae. There is little doubt that *Dorudon* and *Basilosaurus* were fully aquatic, with all of the life history changes (e.g., precocial birth) that this implies. *Dorudon* appears to have been an efficient tail-powered swimmer like modern whales (Uhen, 1996), whereas *Basilosaurus*, though fully aquatic, was divergently specialized (possibly as more of a floating than diving mammal—see above).

2.5. Encephalization

The encephalization quotients (EQ_{TC}) and residuals (ER_{TC}) calculated above to represent relative brain size are most meaningful when interpreted in the context of similar values for mesonychids and for extant cetaceans. These comparisons are shown graphically in Fig. 9. Early archaeocetes appear to have had the encephalization of mesonychids, with ER_{TC} in the range of -1.5 to -2.5 indicating brains about one-quarter (two halvings) the size of those expected for a terrestrial mammal of the same size living today. Later archaeocetes had larger brains, but these were still smaller than expected for a mammal of the same size living today.

When we compare the relative brain size of extant whales, they span virtually the full range of sizes seen in terrestrial mammals living today. Mysticetes tend to have brains smaller than expected for their body mass, which can be explained in part by their unusually large body masses made possible by hydrostatic support in an aquatic medium. However, odontocetes tend to have brains larger than expected for their body mass, while their body mass tends to be large in comparison with terrestrial mammals. If a correction were to be made to brain-size-in-relation-to-body-size to account for increased body mass re-

RELATIVE BRAIN SIZE IN CETACEA

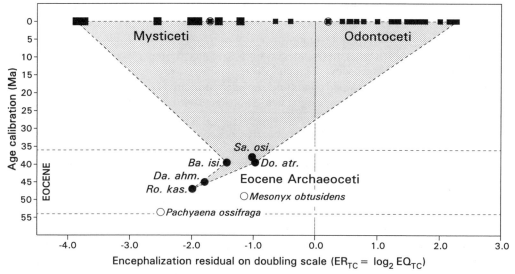

FIGURE 9. Pattern of change of relative brain size through geological time in Eocene Archaeoceti compared with relative brain sizes of Mysticeti and Odontoceti living today. Relative brain size is plotted as an encephalization residual (ER) on a $\log_2$ doubling scale, where the reference group is terrestrial mammals analyzed as a class. An encephalization residual has the advantage that zero is the null expectation and equivalent proportional differences are represented by equal arithmetic distances. Mysticetes are represented by larger solid squares, odontocetes by smaller solid squares, and physeterids (sperm whales) by small squares within circles. Note that all mysticetes have brains smaller than expected for an average terrestrial mammal living today and most odontocetes have brains larger than expected for an average terrestrial mammal living today (*Physeter*, with ER in the mysticete range, is a conspicuous exception). Archaeocete encephalization residuals (solid circles) lying between about −2.00 in *Rodhocetus* (Fig. 4A) and about −1.00 in *Dorudon* (Fig. 5A) span approximately the range of overlap of extant mysticetes and odontocetes. Taxonomic abbreviations are the same as in Fig. 7. Mesonychid encephalization residuals, where known (open circles), are shown for comparison.

sulting from an aquatic medium, odontocetes as a group would undoubtedly prove to have the largest brains of all mammals.

3. Summary and Conclusions

The most important thing that can be said about Eocene Archaeoceti is that they are beginning to fill the temporal, geographic, and morphological gap between Paleocene land mammals and Oligocene and later whales. The temporal and geographic distributions of Mesonychia and Archaeoceti in Figs. 1 and 3 support this. Size-adjusted comparisons of morphological characteristics of the skull, vertebral column, forelimb, and hind limb in Figs. 2, 4–6 show the general pattern of change from a wolflike mesonychid model ancestor through primitive remingtonocetid and protocetid archaeocetes to more advanced dorudontid and basilosaurid archaeocetes, to modern mysticetes and odontocetes. Characteristics of archaeocetes such as body mass, tooth size, auditory bulla size, femur length, and relative brain size are documented in Figs. 7–9, which are related to progressive trophic, auditory, locomotor, cognitive, and life history adaptation to life in the sea.

The fossil record does not support, indeed it positively refutes, Kellogg's (1936) idea that Archaeoceti, Mysticeti, and Odontoceti were separated during a "long interval" of geological time. We no longer view the archaeocete skull as divergent from the line of development that led to the telescoped condition seen later. Similarly, I do not think anyone working on archaic whales would now argue with Simpson (1945) that Cetacea arose very early from a relatively undifferentiated eutherian ancestral stock (i.e., Mesonychia—"very early" would still be during the Paleocene or early Eocene; Gingerich and Uhen, 1998). Boyden and Gemeroy's early serological association of Cetacea with Artiodactyla and Van Valen's linking of mesonychids to the ancestry of archaeocetes and later cetaceans are now widely accepted. Primitive archaeocetes are much better known than they were when Van Valen studied mesonychid ancestry of archaeocetes, and we can now cite much broader morphological trends in support of his general hypothesis.

Of the three broad objectives I list for study of fossil whales, we are beginning to have enough intermediates to say with confidence what happened in the transition of whales from land to sea. We can see that important changes took place at different times, with trophic dental changes happening at the beginning of the archaeocete radiation (Gingerich *et al.*, 1983), auditory changes happening next (Thewissen and Hussain, 1993; Luo and Gingerich, in press), and locomotor adaptation to a fully aquatic life happening later (Gingerich *et al.*, 1994), but there is still much to be learned about the context of the transition from land to sea. Finally, it is still too early to say very much about the consequences of change in archaeocetes for their morphological disparity, taxonomic diversity, or taxon longevity.

Acknowledgments

I thank the National Geographic Society for generous funding of field work in Egypt and Pakistan during the past decade, without which many of the archaeocetes studied here would not be nearly so well known and others would not yet have been discovered. Cooperation with successive directors of the Cairo Geological Museum and with Elwyn L. Simons and Prithijit Chatrath of Duke University from 1983 through 1993 made field research in Egypt possible, and William C. Clyde, Gregg Gunnell, Alex van Nievelt, William J. Sanders, and B. Holly Smith made fieldwork there a success. Research in Pakistan has been aided immeasurably by cooperation with S. Mahmood Raza, M. Talib Hasan, and Muhammed Arif of the Geological Survey of Pakistan, and Muhammid Arif, Mohammad Anwar, M. Akram Bhatti, W. C. Clyde, W. J. Sanders, and Xiaoyuan Zhou made the fieldwork a success. W. J. Sanders and B. H. Smith helped with measurement of comparative skeletons of extant mammals, and I thank Ross D. E. MacPhee of the American Museum of Natural History and Philip Myers of the University of Michigan Museum of Zoology for access to these. I thank X. Zhou for numerous discussions of mesonychids, and W. J. Sanders, J. G. M. Thewissen, and Mark D. Uhen for numerous discussions on archaeocetes. J. G. M. Thewissen, Mark D. Uhen, and two anonymous reviewers read and improved the text. W. J. Sanders prepared many of the archaeocete specimens described here, Jason Head prepared the braincase and endocast of *Pachyaena*, and Ms. Bonnie Miljour aided with preparation of illustrations.

References

Andrews, C. W. 1906. *A descriptive catalogue of the Tertiary Vertebrata of the Fayum, Egypt.* British Museum (Natural History), London.

Arnason, U., and Gullberg, A. 1996. Cytochrome *b* nucleotide sequences and the identification of five primary lineages of extant cetaceans. *Mol. Biol. Evol.* **13**:407–417.

Barnes, L. G., and Mitchell, E. D. 1978. Cetacea, in: V. J. Maglio and H. B. S. Cooke (eds.), *Evolution of African Mammals,* pp. 582–602. Harvard University Press, Cambridge, MA.

Boule, M. 1903. Le *Pachyaena* de Vaugirard. *Mem. Soc. Geol. Fr.* **28**:5–16.

Boyden, A., and Gemeroy, D. 1950. The relative position of the Cetacea among the orders of Mammalia as indicated by precipitin tests. *Zoolgica* **35**:145–151.

Breathnach, A. S. 1955. Observations on endocranial casts of recent and fossil cetaceans. *J. Anat.* **89**:533–546.

Dart, R. A. 19923. The brain of the Zeuglodontidae (Cetacea). *Proc. Zool. Soc. London* **1923**:615–654.

Downhower, J. F., and Blumer, L. S. 1988. Calculating just how small a whale can be. *Nature* **335**:675.

Fordyce, R. E., and Barnes, L. G. 1994. The evolutionary history of whales and dolphins. *Annu. Rev. Earth Planet. Sci.* **22**:419–455.

Fraas, E. 1904. Neue Zeuglodonten aus dem unteren Mitteleocän vom Mokattam bei Cairo. *Geol. Paläontol. Abh. N. F.* **6**:199–220.

Gidley, J. W. 1913. A recently mounted zeuglodon skeleton in the United States National Museum. *Proc. U.S. Nat. Mus.* **44**:649–654.

Gingerich, P. D. 1981. Radiation of early Cenozoic Didymoconidae (Condylarthra, Mesonychia) in Asia, with a new genus from the early Eocene of western North America. *J. Mammal.* **62**:526–538.

Gingerich, P. D. 1990. Prediction of body mass in mammalian species from long bone lengths and diameters. *Contrib. Mus. Paleontol. Univ. Michigan* **28**:79–92.

Gingerich, P. D. 1992. Marine mammals (Cetacea and Sirenia) from the Eocene of Gebel Mokattam and Fayum, Egypt: stratigraphy, age, and paleoenvironments. *Univ. Michigan Pap. Paleontol.* **30**:1–84.

Gingerich, P. D., and Russell, D. E. 1990. Dentition of early Eocene *Pakicetus* (Mammalia, Cetacea). *Contrib. Mus. Paleontol. Univ. Michigan* **28**(1):1–20.

Gingerich, P. D., and Smith, B. H. 1990. Forelimb and hand of *Basilosaurus isis* (Mammalia, Cetacea) from the middle Eocene of Egypt (abstract). *J. Vertebr. Paleontol.* **10A**:24.

Gingerich, P. D., and Uhen, M. D. 1998. Likelihood estimation of the time of origin of Cetacea and the time of divergence of Cetacea and Artiodactyla. *Palaentologica Electronica.*

Gingerich, P. D., Wells, N. A., Russell, D. E., and Shah, S. M. I. 1983. Origin of whales in epicontinental remnant seas: new evidence from the early Eocene of Pakistan. *Science* **220**:403–406.

Gingerich, P. D., Smith, B. H., and Simons, E. L. 1990. Hind limbs of Eocene *Basilosaurus isis:* evidence of feet in whales. *Science* **249**:154–157.

Gingerich, P. D., Raza, S. M., Arif, M., Answar, M., and Zhou, X. 1994. New whale from the Eocene of Pakistan and the origin of cetacean swimming. *Nature* **368**:844–847.

Gingerich, P. D., Arif, M., and Clyde, W. C. 1995. New archaeocetes (Mammalia, Cetacea) from the middle Eocene Domanda Formation of the Sulaiman Range, Punjab (Pakistan). *Contrib. Mus. Paleontol. Univ. Michigan* **29**(11):291–330.

Goodman, M., Czelusniak, J., and Beeber, J. E. 1985. Phylogeny of primates and other eutherian orders: a cladistic analysis using amino acid and nucleotide sequence data. *Cladistics* **1**:171–185.

Graur, D., and Higgins, D. G. 1994. Molecular evidence for the inclusion of cetaceans within the order Artiodactyla. *Mol. Biol. Evol.* **11**:357–364.

Gregory, W. K. 1937. The bridge that walks: the story of nature's most successful design. *Nat. Hist.* **39**:33–48.

Gunnell, G. F., and Gingerich, P. D. 1996. New hapalodectid *Hapalorestes lovei* (Mammalia, Mesonychia) from the early middle Eocene of northwestern Wyoming. *Contrib. Mus. Paleontol. Univ. Michigan* **29**:413–418.

Irwin, D. M., and Arnason, U. 1994. Cytochrome b gene of marine mammals: phylogeny and evolution. *J. Mammal. Evol.* **2**:37–55.

Janis, C. M., and Wilhelm, P. B. 1993. Were there mammalian pursuit predators in the Tertiary? Dances with wolf avatars. *J. Mamm. Evol.* **1**:103–125.

Jerison, H. J. 1973. *Evolution of the Brain and Intelligence.* Academic Press, New York.

Jerison, H. J. 1978. Brain and intelligence in whales. *Whales and Whaling: Report of the Independent Inquiry Conducted by Sir Sydney Frost,* Volume 2, pp. 162–197.

Kellogg, R. 1936. A review of the Archaeoceti. *Carnegie Inst. Washington Publ.* **482**:1–366.

Legendre, S., and Roth, C. 1988. Correlation of carnassial tooth size and body weight in recent carnivores (Mammalia). *Hist. Biol.* **1**:85–98.

Lucas, F. A. 1900. The pelvic girdle of Zeuglodon, *Basilosaurus cetoides* (Owen), with notes on the other portions of the skeleton. *Proc. U.S. Nat. Mus.* **23**:237–331.

Luo, Z., and Gingerich, P. D. In press. Transition from terrestrial ungulates to aquatic whales: transformation of the basicranium and evolution of hearing. *Univ. Mich. Papers Paleontol.*

Maples, B. J. 1949. Two endocranial casts of cetaceans from the Oligocene of New Zealand. *Am. J. Sci.* **247**:462–471.

Matthew, W. D. 1909. The Carnivora and Insectivora of the Bridger Basin, Middle Eocene. *Mem. Am. Mus. Nat. Hist.* **9**:289–567.

Matthew, W. D. 1915. A revision of the lower Eocene Wasatch and Wind River faunas. Part I—Order Ferae (Carnivora). Suborder Creodonta. *Bull. Am. Mus. Nat. Hist.* **34**:4–103.

McKenna, M. C. 1987. Molecular and morphological analysis of high-level mammalian interrelationships, in: C. Patterson (ed.), *Molecules and Morphology in Evolution: Conflict or Compromise?* pp. 55–93. Cambridge University Press, London.

Meng, J., Suyin, T., and Schiebout, J. A. 1994. The cranial morphology of an early Eocene didymoconid (Mammalia, Insectivora). *J. Vertebr. Palentol.* **14**:534–551.

Milinkovitch, M. C. 1995. Molecular phylogeny of cetaceans prompts revision of morphological transformations. *Trends Ecol. Evol.* **10**:328–334.

Milinkovitch, M. C., Ortí, G., and Meyer, A. 1993. Revised phylogeny of whales suggested by mitochondrial ribosomal DNA sequences. *Nature* **361**:346–348.

Milinkovitch, M. C., Ortí, G., and Meyer, A. 1995. Novel phylogeny of whales revisited but not revised. *Mol. Biol. Evol.* **12**:518–520.

Norris, K. S. 1968. The evolution of acoustic mechanisms in odontocete cetaceans, in: E. T. Drake (ed.), *Evolution and Environment,* pp. 297–324. Yale University Press, New Haven.

Ohland, D. P., Harley, E. H., and Best, P. B. 1995. Systematics of cetaceans using restriction site mapping of mitochondrial DNA. *Mol. Phylogenet. Evol.* **4**:10–19.

O'Leary, M. A., and Rose, K. D. 1995. Postcranial skeleton of the early Eocene mesonychid *Pachyaena* (Mammalia: Mesonychia). *J. Vertebr. Paleontol.* **15**:401–430.

Omura, H. 1975. Osteological study of the minke whale from the Antarctic. *Sci. Rep. Whales Res. Inst.* **27**:1–36.

Osborn, H. F. 1910. *The Age of Mammals in Europe, Asia, and North America.* Macmillan Co., New York.

Osborn, H. F. 1924. Andrewsarchus, giant mesonychid of Mongolia. *Am. Mus. Novit.* **146**:1–5.

Radinsky, L. B. 1976. The brain of *Mesonyx,* a middle Eocene mesonychid condylarth. *Fieldiana Field Mus. Nat. Hist. Geol. Ser.* **33**:323–337.

Rose, K. D., and O'Leary, M. A. 1995. The manus of *Pachyaena gigantea* (Mammalia: Mesonychia). *J. Vertebr. Paleontol.* **15**:855–859.

Russell, D. E. 1964. Les Mammifères Paléocènes d'Europe. *Mem. Mus. Nat. Hist. Nat. Paris Sér. C* **13**:1–324.

Sarich, V. M. 1993. Mammalian systematic: twenty-five years among their albumins and transferrins, in: M. J. Novacek, M. C. McKenna, and F. S. Szalay (eds.), *Mammal Phylogeny: Placentals,* pp. 103–114. Springer-Verlag, Berlin.

Scott, W. B. 1886. On some new and little known creodonts. *J. Acad. Nat. Sci. Philadelphia* **9**:155–185.

Scott, W. B. 1913. *A History of Land Mammals in the Western Hemisphere.* Macmillan Co., New York.

Simpson, G. G. 1945. The principles of classification and a classification of mammals. *Bull. Am. Mus. Nat. Hist.* **85**:1–350.

Slijper, E. J. 1946. Comparative biologic-anatomical investigations on the vertebral column and spinal musculature of mammals. *Verh. K. Ned. Akad. Wet. Afd. Natuurkd. Tweede Reeks* **42**:1–128.

Slijper, E. J. 1947. Observations on the vertebral column of the domestic animals. *Vet. J.* **103**:376–387.

Szalay, F. S., and Gould, S. J. 1966. Asiatic Mesonychidae (Mammalia, Condylarthra). *Bull. Am. Mus. Nat. Hist.* **132**:127–173.

Thewissen, J. G. M., and Hussain, S. T. 1993. Origin of underwater hearing in whales. *Nature* **361**:444–445.

Thewissen, J. G. M., Madar, S. I., and Hussain, S.T. 1996. *Ambulocetus natans*, an Eocene cetacean (Mammalia) from Pakistan. *Cour. Forsch.-Inst. Senckenberg* **191**:1–86.

Ting, S., and Li, C. 1987. The skull of *Hapalodectes (?Acreodi, Mammalia), with notes on some Chinese Paleocene mesonychids. Vertebr. PalAsiat.* **25**:161–186.

Uhen, M. D. 1996. *Dorudon atrox* (Mammalia, Cetacea): form, function, and phylogenetic relationships of an archaeocedte from the late middle Eocena of Egypt. Ph.D. dissertation, University of Michigan, Ann Arbor, 608 pp.

Van Valen, L. M. 1966. Deltatheridia, a new order of mammals. *Bull. Am. Mus. Nat. Hist.* **132**:1–126.

Van Valen, L. M. 1978. The beginning of the age of mammals. *Evol. Theory* **4**:45–80.

Wang, B. 1976. Late Paleocene mesonychids from Nanxiong Basin, Guangdong. *Vertebr. PalAsiat.* **14**:259–262.

Watson, L., and Ritchie, T. 1981. *Sea Guide to Whales of the World.* Dutton, New York.

Xue, X., Zhang, Y., Bi, Yue, L., and Chen, D. 1996. *The Development and Environmental Changes of the Intermontane Basins in the Eastern Part of Qinling Mountains.* Geological Publishing House, Beijing.

Yamada, M. 1954. An account of a rare porpoise *Feresa Gray* from Japan. *Sci. Rep. Whales Res. Inst.* **9**:59–88.

Zhou, X. 1995. Evolution of Paleocene–Eocene Mesonychidae (Mammalia, Mesonychia). Ph.D. dissertation, University of Michigan, Ann Arbor, 402 pp.

Zhou, X., Sanders, W. J., and Gingerich, P. D. 1992. Functional and behavioral implications of vertebral structure in *Pachyaena ossifraga* (Mammalia, Mesonychia). *Contrib. Mus. Paleontol. Univ. Michigan* **28**:289–319.

Zhou, X., Zhai, R., Gingerich, P. D., and Chen, L. 1995. Skull of a new mesonychid (Mammalia, Mesonychia) from the late Paleocene of China. *J. Vertebr. Palentol.* **15**:387–400.

CHAPTER 16

Cetacean Origins

Evolutionary Turmoil during the Invasion of the Oceans

J. G. M. THEWISSEN

1. Introduction

It has been more than half a century since Remington Kellogg published a detailed treatment of the earliest whales (1936). In it, and in an earlier more popular article (Kellogg, 1928), he established *Protocetus atavus* as the best available model for the cetacean archetype. This was an appropriate model for most decades to follow, and it was highly influential in shaping our understanding of the ancestry of cetaceans. *Protocetus* is known from many elements: a skull with teeth, vertebrae, and ribs (Fraas, 1904; Stromer, 1908). Compared with later cetaceans, *Protocetus* is remarkably primitive, it has heterodont teeth, a complete dental formula, and the caudal edge of its nasal opening is over P^1. Nonetheless, it was unambiguously a cetacean and separated by a large morphological gap from its four-footed terrestrial ancestors. Based on the derived nature of *Protocetus*, many authors in the first half of this century believed that cetaceans had a very ancient origin and that whales had no close relatives among the other modern orders. It was also commonly thought that mysticetes and odontocetes were derived independently from land mammals (e.g., Slijper, 1962).

In the 1960s, notions concerning cetacean origins that are still current originated: Mysticetes and odontocetes are derived from a common Cenozoic, aquatic ancestor (Van Valen, 1967), and the ancestors of cetaceans are found among a group of (presumed) carnivorous ungulates, the mesonychians (Van Valen, 1966). This opinion is widely held among morphologists, but it may not be consistent with the strong molecular evidence indicative of close cetacean–artiodactyl relations (Gatesy, this volume; Milinkovitch *et al.*, this volume).

Fossils only began to fill the gap between *Protocetus* and mesonychians in the 1970s and 1980s. In those decades, dental and cranial fossils from Pakistan and India were recognized as cetaceans that were more primitive than *Protocetus* (Sahni and Mishra, 1972, 1975; West, 1980; Gingerich and Russell, 1981; Gingerich et al., 1983). This shifted the at-

J. G. M. THEWISSEN • Department of Anatomy, Northeastern Ohio Universities College of Medicine, Rootstown, Ohio 44272.
The Emergence of Whales, edited by Thewissen. Plenum Press, New York, 1998.

tention from Africa as the birthplace of the order to Indo-Pakistan, an opinion that is now widely held.

The evolutionary history of cetaceans continues to unravel in the present decade. New discoveries include cranial and postcranial material for some of the earliest whales (Gingerich *et al.*, 1994; Thewissen *et al.*, 1994, 1996a; Bajpai and Thewissen, this volume) and *Protocetus*-like early whales in other continents, such as North America (Hulbert *et al.*, 1997; Hulbert, this volume; Williams, this volume). Our present understanding of cetacean origins has advanced significantly beyond *Protocetus*. The earliest cetaceans can be divided into four families: Pakicetidae, Ambulocetidae, Remingtonocetidae, and Protocetidae. Combined with the ancestors of cetaceans, most likely mesonychians, these families document the origin of aquatic life in cetaceans. The phylogeny of these taxa is the subject of much research (Geisler and Luo, this volume; O'Leary, this volume) and has generated controversy concerning cetacean affinities. The controversy is currently particularly urgent because of the publication of new molecular data (Gatesy *et al.*, 1996; Gatesy, 1997, this volume; Milinkovitch *et al.*, this volume; Shimamura *et al.*, 1997). Improved understanding of the phylogeny provides the framework for detailed studies on morphological adaptations (e.g., Buchholtz, this volume; Luo and Geisler, this volume; Madar, this volume) and for the use of new techniques that address questions that could previously not be answered (e.g., Roe et al., this volume). A better understanding of modern cetacean function and biology is also necessary for analyzing whale origins (Fish et al., this volume; Pabst et al., this volume), since it is of great importance in understanding the later parts of cetacean evolution (Uhen, this volume).

The aim of this chapter is to provide a heuristic outline of a field that is advancing quickly. I will start with an interpretative review of the players in cetacean origins, focusing on characterizations of the major groups and then briefly discussing some biological attributes of each. I will finish this chapter by looking ahead, attempting to identify those areas of research in cetacean origins that are most likely to advance significantly in the near future and pointing out the areas where serious gaps in our knowledge lie.

2. Early Cetaceans and Their Relatives

2.1. Paraxonia

Paraxonians have a foot with four toes and three phalanges each. The foot is paraxonic, digits III and IV being similar in length and longer than digits II and V (Fig. 1). The astragalar head is flat mediolaterally and convex anteroposteriorly. The crus breve of incus is longer than the crus longum.

Paraxonia includes Cetacea, Mesonychia, and Artiodactyla. Thewissen (1994) used the term "Cete" to refer to paraphyletic mesonychians and Cetacea. Geisler and Luo (this volume) use "Acreodi" for mesonychids (one of the families of mesonychians) and Cetacea, which is considered a monophyletic group by them.

Many authors in the first half of this century felt that the terrestrial relatives of cetaceans could not be identified because the order had too many derived characters (e.g., Simpson, 1945). It is now generally held that cetaceans are ungulates, and they have been formally included in this clade (McKenna, 1975; Prothero *et al.*, 1988). This has been con-

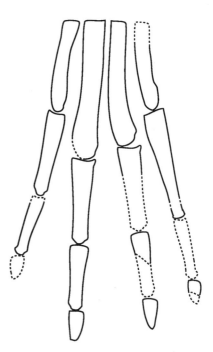

FIGURE 1. Outline of the foot of *Ambulocetus natans*, showing paraxony of early cetaceans (modified from Thewissen *et al.*, 1996a).

firmed by the hooflike morphology of the distal phalanx of the earliest whales (Thewissen *et al.*, 1996a). Early molecular evidence (Boyden and Gemeroy, 1950) supported a link between artiodactyls and cetaceans. Van Valen (1966) proposed that mesonychians are the terrestrial ancestors of cetaceans, and a (group of) mesonychian(s) is commonly found to be the sister group to cetaceans (Geisler and Luo, this volume; O'Leary, this volume).

Most modern authors agree that artiodactyls are the modern sister group to cetaceans, and new molecular data (Gatesy *et al.*, 1996; Gatesy, 1997, this volume; Milinkovitch *et al.*, this volume; Shimamura *et al.*, 1997) suggest that artiodactyls are paraphyletic if cetaceans are excluded.

Analyses by Gatesy (this volume) and Milinkovitch *et al.* (this volume) suggest that cetaceans are the sister group to hippopotamids. Milinkovitch and Thewissen (1997) pointed out that if hippopotamids are the modern sister group to cetaceans and if a mesonychian is the fossil sister or ancestor to cetaceans, then several defining characters of artiodactyls need to be reevaluated. Most significant among these is the unique trochlear shape of the astragalar head. If Cete (mesonychians and cetaceans) is included in artiodactyls, the trochlea was lost in this branch, as it is absent in mesonychians, but present in the earliest artiodactyls. A complete astragalus of an early whale will shed light on this matter. If the primitive cetacean astragalar head has the shape of a trochlea, then most mesonychians would probably be excluded from close cetacean (and artiodactyl) relations. The presence of a trochleated astragalus in early whales would probably lead most morphologists to accept the molecular phylogeny and consider cetaceans as artiodactyls. On the other hand, if the astragalar head of this primitive cetacean lacks a trochlea, several evolutionary scenar-

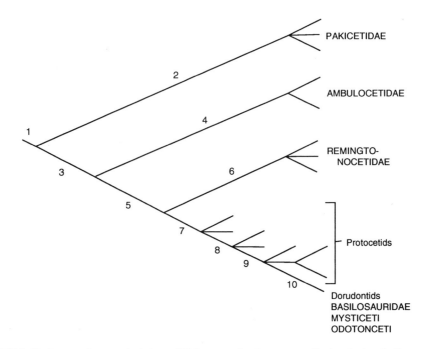

FIGURE 2. Cladogram of proposed relations of higher taxa of early cetaceans. Each endpoint of a line represents a genus, except for that of branch 10. Numbers refer to synapomorphies listed in the text. Branching pattern of higher nodes matches that of Fig. 3. Protocetid phylogeny based in part on unpublished work by R. Hulbert.

ios are possible. The molecular data could be correct, leading morphologists to question the degree of homoplasy present in the evolution of the trochleated astragalus. Molecular analyses presented so far (Gatesy, this volume; Milinkovitch *et al.*, this volume; Shimamura *et al.*, 1997) are strong but could be improved by including more taxa (e.g., gene sequences for only one hippopotamid are known in most cases). It is also interesting that much of the molecular data summarized by Gatesy (this volume) disagrees with other molecular work (Milinkovitch *et al.*, 1993) that suggests that odontocetes are paraphyletic. Among the morphological analyses, much remains to be done as there are no explicit morphological analyses that include a variety of species in all three relevant groups—early artiodactyls, mesonychians, and cetaceans—and because some critical fossil evidence (such as the astragalus) is lacking. The chapter by Geisler and Luo lays the groundwork for such an analysis.

2.2. Cetacea

Cetaceans are characterized by pachyosteosclerotic tympanics, petrosals, and auditory ossicles. Their tympanic has a sigmoid process and involucrum (see Luo, this volume, for a discussion of character transformations of these structures; a sigmoid process is here defined as a laterally projecting crest of the anterior crus of the tympanic that is formed by

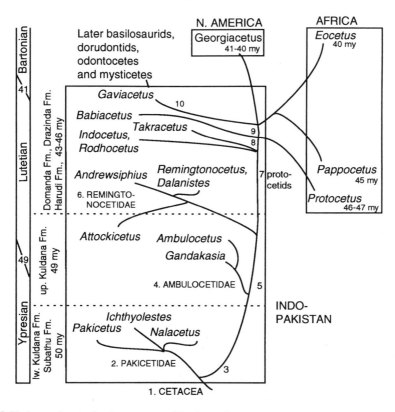

FIGURE 3. Phylogenetic tree of early cetaceans, with relevant formations in Indo-Pakistan. Synapomorphies that identify nodes are listed in the text. Branching pattern of higher nodes matches that of Fig. 2. For a discussion of individual genera, see Williams (this volume).

an undulation of the lateral wall). The cetacean middle ear is partly or fully rotated, the fossa for the malleus is distinct from the epitympanic recess, and there is a groove for tensor tympani between promontorium and tegmen tympani. Cheek teeth of Eocene cetaceans have reduced occlusal basins and pronounced shearing surfaces. P^4 lacks a protocone (branch 1 of Figs. 2 and 3).

Cetacea includes Odontoceti, Mysticeti, and a group of Eocene archaic cetaceans commonly referred to as archaeocetes. Archaeocetes are paraphyletic, and include the families Pakicetidae, Ambulocetidae, Remingtonocetidae, Protocetidae, Dorudontidae, and Basilosauridae. The monophyly of Basilosauridae and Remingtonocetidae is undisputed, but that of none of the remaining families is established beyond doubt.

The term *cetacean* was once considered to be synonymous with obligate marine, predatory mammals. Newly discovered taxa fill the gap between the late Eocene obligate marine cetaceans and their (presumed Paleocene) terrestrial forebears, and a more refined concept is necessary. At present, it is best to include pakicetids and its sister group in Cetacea. Most of the key adaptations of cetaceans are related to hearing and feeding

(Thewissen, 1994; Thewissen *et al.*, 1996a), but locomotor characters cannot be evaluated at present because of the absence of pakicetid postcranials. Geisler and Luo (this volume) provide a cladistic analysis of morphological characters that supports cetacean monophyly.

2.3. Pakicetids

Thewissen and Hussain (1998) listed an assemblage of primitive characters that characterizes pakicetids (branch 2 of Figs. 2 and 3) when combined with the cetacean synapomorphies: palatine fissures present, nasal opening over incisors, hypoglossal foramen present and well separated from jugular foramen, and mandibular foramen small. Paraconid and metaconid are present on lower molars, and the hypoconid is the only cusp on the talonid. Thewissen *et al.* (1996b) listed the great height of the paracone of P^4 as a possible synapomorphy, but this cusp in *Nalacetus* is not higher than in *Ambuloctus*. Pakicetids are the only cetaceans in which the middle ear ossicles are known to be partially rotated, but the condition is known in only one pakicetid and is not known in ambulocetids. Furthermore, the degree of ossicular rotation in *Pakicetus* forms an excellent intermediate for the full middle ear rotation of later cetaceans (Thewissen and Hussain, 1993) and could be considered a stage along a morphocline. The orbital configuration of two unpublished pakicetid specimens is highly derived, with the orbits positioned near the sagittal plane and facing dorsally. The supraorbital area is depressed into a deep furrow. This morphology is unlike any other cetacean or mesonychian, but it is not known if it occurs in all pakicetids. The sister group of pakicetids (branch 3 of Figs. 2 and 3) is characterized by the presence of an enlarged mandibular foramen, large parastyle on the upper molars, and loss of lower molar metaconid. Pakicetids and its sister group make up Cetacea.

Collections described by Guy Pilgrim include the oldest discovered pakicetids, but he (1940) did not recognize these as cetaceans. Philip Gingerich was the first to recognize that pakicetids were whales, but his publication on Chorlakki *Pakicetus* (Gingerich and Russell, 1981) postdates that of West on Kala Chitta specimens (1980). Pakicetids are commonly considered the most primitive family of cetaceans (e.g., Gingerich *et al.*, 1983; Thewissen, 1994; O'Leary, this volume; Geisler and Luo, this volume; Figs. 2 and 3). Published fossils of pakicetids include teeth (Gingerich and Russell, 1990; Thewissen and Hussain, 1998), several complete mandibles (West 1980; Thewissen and Hussain, 1998), and a braincase (Gingerich *et al.*, 1983). No postcranial remains can be confidently identified as pakicetid at present, and this is a major gap in our understanding of the morphology and ecology of the earliest cetaceans.

There are three genera of pakicetids, in decreasing order of size: *Pakicetus, Nalacetus*, and *Ichthyolestes*. The dentition and ear are the only parts that can be compared among all three pakicetids. *Pakicetus* was approximately as large as a coyote (*Canis latrans*) and has robust cusps on its molars that show heavy wear in the trigon basin, but little wear on the molar cusps (O'Leary, this volume). *Ichthyolestes* is smaller and has gracile incisors and narrow molars and probably ate less abrasive food, possibly fish. The dentition of *Nalacetus* is intermediate in size and morphology between *Pakicetus* and *Ichthyolestes*. There is a striking difference in the dental wear of pakicetids and mesonychians: The former develop shearing facets and leave the apices relatively intact, whereas wear in mesonychians is usually apical (O'Leary, this volume).

Despite early reconstructions showing pakicetids well adapted for locomotion in water (Gingerich *et al.*, 1983), there is no evidence to support this. The pakicetid ear has been interpreted as indicative of underwater hearing (Gingerich *et al.*, 1983; Thewissen and Hussain, 1993), but it is more likely that the unusual morphological features are indicative of inertial hearing on land or in water (Thewissen *et al.*, 1996a; see also Luo, this volume). If the inertial model for hearing in early cetaceans is upheld, it is possible that the earliest cetaceans evolved sophisticated hearing by using their mandibles as a bony interface between substrate (soil or possibly water) and middle ear. This system would be most useful for detecting low-frequency substrate-borne vibrations such as those produced by the footsteps of potential prey. Experimental studies of modern aquatic mammals are necessary to test this hypothesis.

Pakicetids are only known from deposits indicating a shallow freshwater environment (Aslan and Thewissen, 1997; Williams, this volume). The two localities that yield most pakicetids are H-GSP Locality 62 (West, 1980; Thewissen and Hussain, 1993, 1998b) and Chorlakki (Gingerich and Russell, 1981, 1990; Gingerich *et al.*, 1983); both represent channel lags. Large amounts of postcranial bones are present at H-GSP Locality 62, but these were concentrated by reworking and no anatomical associations are preserved (Aslan and Thewissen, 1997). The combination of sedimentological, isotopic, and faunal evidence on pakicetids (Aslan and Thewissen, 1997; Williams, this volume; Roe *et al.*, this volume) suggests that they were terrestrial or amphibious mammals, and did not inhabit marine environments. More detailed taphonomic and sedimentological studies will refine our understanding of the habitat of pakicetids.

Pakicetids are known from the same formations but also yield mesonychian remains (Ranga Rao, 1972; O'Leary, this volume). Only two teeth of Indo-Pakistani mesonychians are known, and it is possible that further recovery of mesonychian material from Indo-Pakistan may blur the morphological distinction between cetaceans and mesonychians.

2.4. Ambulocetids

Ambulocetids (branch 4 of Figs. 2 and 3) are characterized by two apomorphies: The ptyerygoid process is similar in height to the braincase, and the orbits face laterally and are positioned close to the sagittal plane. The supraorbital region is gently concave and there is no supraorbital shield. The mandibular foramen is large, approximately half as high as the height of the mandible. At present, these characters can only be verified in the type genus. Ambulocetids are the sister group of all nonpakicetid cetaceans, and form a good functional model for the ancestral marine cetacean. The sister group of ambulocetids (branch 5 in Figs. 2 and 3) is diagnosed by the large size of the mandibular foramen and the presence of a supraorbital shield.

Dehm and Oettingen-Spielberg (1958) were the first to publish an ambulocetid, but did not recognize it as a cetacean. A fragmentary tooth discovered by West (1980) represents *Ambulocetus*, but he thought that it was more derived than a protocetid. Ambulocetid morphology is mainly known from a relatively complete skeleton of *Ambulocetus natans* (Thewissen *et al.*, 1994, 1996a). This specimen made it possible to recognize that this family documents an amphibious stage in whale origins. Not all elements of the holotype have been described as the type locality could not be excavated for several years after 1992 for safety reasons. Excavations in 1996 resulted in the recovery of much of the vertebral column and pelvis of the holotype.

The body weight of *Ambulocetus* was estimated to be 141–235 kg (Thewissen *et al.*, 1996a). Gingerich (this volume) estimated it to be 720 kg based on the size of the vertebrae. This discrepancy in weight estimates could be the result of the unusual robust vertebrae in this taxon. A second genus of ambulocetid, *Gandakasia*, is known from dental material only and is smaller than *Ambulocetus*. Ambulocetids are only known from Indo-Pakistan and are always found in littoral sediments (Thewissen *et al.*, 1996a), but stable isotope data (Thewissen *et al.*, 1996b; Roe *et al.*, this volume) indicate that they were euryhaline and partly dependent on freshwater.

Dental wear in *Ambulocetus* is extensive and unusual, as in pakicetids (O'Leary, this volume). It is likely that *Ambulocetus* was a hard-object feeder. Robusticity of the skeleton, jaws, and teeth suggests that ambulocetids ate large and struggling prey, and Thewissen *et al.* (1996a) suggested that modern crocodiles could be the best ecological analogue. Crocodilelike forms occur as intermediates in the land–water transition of several tetrapods (Taylor, 1987), and Ahlberg and Milner (1994) proposed that a crocodilelike stage occurred in tetrapod origins. It is likely that *Ambulocetus* was an ambush predator in shallow water. The localities where *Ambulocetus* is found abound in marine plants and shallow marine invertebrates.

The locomotor morphology of *Ambulocetus* suggests that it was capable of land and aquatic locomotion and that it used its feet, not its tail, as a hydrofoil (Thewissen *et al.*, 1994, 1996a; Madar, this volume). This is consistent with the animal resembling a crocodile ecologically. Thewissen and Fish (1997) showed that *Lutra* and, to a lesser extent, *Pteronura* are modern analogues for swimming in *Ambulocetus*, and the range of swimming modes of early cetaceans may occur in modern mustelids.

Although well known skeletally, a number of important details of ambulocetid morphology remain obscure. Among these are much of the middle and inner ear morphology. This is especially disappointing because ambulocetid mandibles have a mandibular foramen much larger than pakicetids but much smaller than protocetids. This is probably the first morphological modification of the mandible to its role as a sound transmitter (Thewissen *et al.*, 1996a). Much of the tarsal morphology is also unknown, as in pakicetids. This is unfortunate as it bears heavily on the assessment of artiodactyl–cetacean relations, as indicated by Gatesy (this volume).

2.5. Remingtonocetids

Remingtonocetid (branch 6 of Figs. 2 and 3) synapomorphies include: the small size of the molar protocone, the extensive mandibular symphysis, the long rostrum, the small orbits, the strongly convex palate that extends ventrally beyond the crowns of the molars, the oblong tympanics, and the laterally placed ear region which causes the posterior aspect of the skull to be much wider than high.

Remingtonocetids from the District of Kachchh (State of Gujarat, western India) were the first whales to be described from South Asia (see discussion by Bajpai and Thewissen, this volume). Cranial remains of remingtonocetids are known from three areas in Indo-Pakistan (Bajpai and Thewissen, this volume; Gingerich *et al.*, 1995; Thewissen and Hussain, in press). Postcranial remains have been described for Sulaiman remingtonocetids (Gingerich *et al.*, 1995). Some of this material is not associated with clearly diagnostic cranial material and part of it was previously attributed to protocetids (Gingerich *et al.*, 1993). At present, the identifications of Gingerich *et al.* (1995) appear reasonable, but they should be confirmed by associated material. Despite excellent documentation of cranial material, very little dental material is known for remingtonocetids.

Of the four genera of remingtonocetids, *Remingtonocetus* and *Dalanistes* are similar in size and morphology. Gingerich (this volume) estimated the body weight of *Dalanistes* at 750 kg. *Andrewsiphius* is somewhat smaller than *Remingtonocetus*. *Attockicetus* is the smallest of its family, possibly similar in size to *Pakicetus*. Remingtonocetids are only known from Indo-Pakistan and probably left no descendants.

The sense organs of remingtonocetids are very different from other early cetaceans. The orbits are small, suggesting that the eyes were not the primary sense organ. The middle ear is large (Gingerich, this volume) and the basicranium wide. Left and right middle ears are well separated (Bajpai and Thewissen, this volume), possibly enhancing directional hearing by increasing the time difference at which sounds reach the two ears. The size of the middle ear cavity also correlates with improved hearing ability (Webster and Webster, 1976). It is likely that the reorganization of the hearing organ of remingtonocetids was a response to underwater sounds. Remingtonocetids are the most primitive cetaceans for which the inner ear is well known, and study of these specimens will elucidate the origin of underwater hearing in cetaceans.

If the proposed postcranial associations hold, remingtonocetids were peculiar animals. Their skeleton is not unlike that of ambulocetids in being robust and retaining large hind limbs. However, the skull of remingtonocetids is very different from ambulocetids: The eyes are small, the ear region is set far from the midline, and the snout is long and gracile. The jaw morphology is consistent with a diet of fast-swimming aquatic prey that is easily subdued, but the skeleton does not suggest that fast locomotion to pursue this prey was likely. It is possible that remingtonocetids were ambush predators, specializing on small prey, whereas *Ambulocetus* hunted large, struggling prey. This will remain speculative until there is a firm association of limb and cranial material for remingtonocetids, and until their diet is studied in detail. Remingtonocetids are mainly known from nearshore marine environments, although some lived in the littoral zone (Williams, this volume). Stable isotope data indicate that different remingtonocetids may have been more (*Remingtonocetus*) or less (*Attockicetus*) marine (Roe *et al.*, this volume).

2.6. Protocetids

Protocetids (branch 7 of Figs. 2 and 3) are a paraphyletic group that probably includes the sister group to late Eocene and younger cetaceans (Uhen, this volume). Protocetids can be characterized by a combination of primitive characters (distinguishing them from basilosaurids, dorudontids, odontocetes, and mysticetes) and derived characters (distinguishing them from pakicetids, ambulocetids, and remingtonocetids). Derived features are that the orbits face laterally and are covered by a supraorbital shield, that the sacrum is composed of less than four fused vertebrae, and that the occlusal outline of P^4 is much larger than that of M^1. Luo (this volume) lists several cranial features that may characterize protocetids and later cetaceans. The main primitive feature is the retention of a complete dental formula with M^3.

The phylogeny of protocetids and later cetaceans cannot be analyzed with explicit means, because a number of forms have not as yet been described in detail. Three grades of protocetids may be recognized, based on preliminary evidence compiled from the literature and Hulbert (personal communication). The most plesiomorphic protocetids are *Indocetus*, *Rodhocetus* (possibly included in *Indocetus*), and *Takracetus*. These genera display the protocetid synapomorphies listed above, but lack those of branch 8 (Fig. 2 and 3). Synapomorphies of branch 8 are the size reduction of the protocone of the upper molars (a parallel to remingtonocetids) and the shift of the posterior nasal opening from a position over the canine to a position over P^1.

Protocetus and *Babiacetus* display the synapomorphies of branch 8, but lack that of branch 9. Branch 9 is characterized by the presence of many accessory cuspules on the upper premolars. A third cluster of protocetids includes *Pappocetus*, *Georgiacetus*, and *Eocetus*. These share the main synapomorphy of branch 9, but lack that of branch 10: absence of M^3.

Late Eocene cetaceans, dorudontids and basilosaurids, are probably derived from protocetids. Here, I follow Miller (1923) and Simpson (1945) in considering basilosaurids and dorudontids distinct at the family level. Several recent workers (Barnes and Mitchell, 1978; Uhen, this volume) have considered these as subfamilies, although others have not (e.g., Gingerich, this volume). The great elongation of the posterior part of the body in *Basilosaurus* was probably related to a unique locomotor pattern (Buchholtz, this volume). This difference between basilosaurids and dorudontids is much larger than the differences between many of modern cetacean clades that are considered distinct at the familial level.

Protocetid diversity is highest in Indo-Pakistan, but this is the earliest cetacean family to be found in other continents. Protocetids apparently distributed across the Tethyan realm to Africa and North America. No complete skeletons are known for any protocetid, but dental and cranial material is abundant and many elements of the axial skeleton are known as well. The limb skeleton of protocetids is poorly known, although some elements are known for a number of taxa.

The locomotor organs of protocetids varied, but all had limbs more reduced than those of ambulocetids. The hind limbs were small and for no described form has a sacrum composed of multiple fused vertebra (Buchholtz, this volume; Hulbert, this volume). Modern cetaceans swim by means of dorsoventral oscillations of the tail fluke, a mode that probably evolved in protocetids. There is disagreement concerning the swimming modes of particular taxa and about the oldest taxon with a fluke (Gingerich *et al.*, 1994; Buchholtz, this volume).

Protocetids had a variety of lifestyles, but all seem firmly committed to marine environments (Williams, this volume; Roe, this volume). Some forms probably hunted larger, struggling prey (*Takracetus*, Gingerich *et al.*, 1995), but most have gracile skeletons and probably pursued small and agile aquatic prey. Protocetid dentitions indicate that the teeth were less suited for mastication than those of their early and middle Eocene relatives. This is evidenced by the reduction of the trigon basin and protocone. Endocasts suggest that the endocranial vascular retia evolved in protocetids (Bajpai *et al.*, 1996; Geisler and Luo, this volume).

3. Prospectus

Cetaceans originated when a Paleogene land mammal underwent a dramatic shift in biological attributes in order to accommodate an enormous shift in habitat. A variety of vertebrates have made the transition from terrestrial quadruped to obligate marine swimmer, but few have adapted to the sea as well as cetaceans have, and none are as successful as cetaceans in the recent.

The origin of cetaceans is already better documented than most other major evolutionary transitions, and no end is in sight for the cascade of discoveries. Morphology was clearly in crisis as the clade broke through the land–water boundary, as evidenced by the vast differences in morphological attributes between the transitional taxa, their terrestrial ancestors, and their Neogene descendants.

Evolutionary investigations, including those about cetacean origins, revolve around questions of physical setting (where and when did the transformation happen and how long did it last), questions of biological patterns (what changed during the transformation), and questions of evolutionary process (how and why did these changes occur). What unfolded in early whale evolution is known in broad outline, and many of the important players have been identified. There are several dozens of artiodactyls, mesonychians, and cetaceans that are possibly involved in cetacean origins, but the exact role that each played has yet to be determined (O'Leary, this volume; Gatesy, this volume; Geisler and Luo, this volume). A rigorous phylogenetic analysis of all early cetaceans has to await the detailed description of a number of new taxa (listed by Williams, this volume), but there is already some consensus, as evidenced by the chapters in this volume (summarized in Fig. 3). Given the numbers of early cetaceans discovered recently in areas that have only been cursorily investigated before, it is likely that more early cetaceans will be found in the near future and that the pattern of early cetacean evolution will be resolved.

Details of the anatomy, another suite of questions involving evolutionary patterns, remain unknown for many early cetaceans. Postcranials are not known for pakicetids and the limb skeleton is poorly known in all early cetaceans except *Ambulocetus*. Organ system evolution can rarely be studied in detail with available fossils, even for systems that change pervasively such as those related to hearing, locomotion, and mastication (Thewissen *et al.*, 1996a; Gingerich, this volume; Luo, this volume).

The physical setting of whale origins is known to some extent. Pakicetids, ambulocetids, and remingtonocetids are restricted to Indo-Pakistan and probably originated there. A major gap in our knowledge is that only two fragmentary mesonychian specimens are known from the subcontinent (O'Leary, this volume). Some aspects of the environment of the early cetacean families have been studied (e.g., Aslan and Thewissen, 1997), but no detailed environmental and taphonomic studies have been undertaken. These should greatly elucidate the ecological setting of whale origins and could provide a test for ecological hypotheses proposed on the basis of morphology.

Protocetids are known worldwide and were probably the family that conquered the oceans (Thewissen *et al.*, 1996b), although undescribed primitive cetaceans are also known from other continents (Gingerich, 1992). Gingerich (this volume) has studied the timing and rate of early cetacean evolution, but no detailed chronology of the localities of Indo-Pakistan has been undertaken. Such a study will not be easy because of the enormous structural complications related to the Himalayan Orogeny. If the known fossils present an accurate picture, evolution from a pakicetid to a protocetid stage took less than 5 million years.

The processes by which cetaceans modified their bodies are poorly known, but improved understanding of patterns will soon allow interpretation of processes. Fleischer (1978) and Lancaster (1990) showed that the cetacean middle ear is rotated, and Thewissen and Hussain (1993) found that the pakicetid middle ear is intermediate between land mammals and later whales. Kinkel *et al.* (1997) investigated middle ear morphology in embryos and suggested that the phylogenetic rotation is repeated in ontogeny and that partially rotated middle ears occur in the ancestral taxon. If confirmed, this is an example of hypermorphosis. Pabst *et al.* (this volume) found that pedomorphic development of the abdominal vascular system improved the thermoregulatory system of the gonads. That thermoregulatory system was probably disrupted as a result of improved locomotor efficiency (streamlining) or body insulation. Temperatures in different parts of the body cannot be assessed in fossil organisms at present, but it is conceivable that stable oxygen isotope methods can be refined for this purpose in the future (see Roe *et al.*, this volume). This could form a test for the hypothesis of Pabst *et al.* (this volume). Fluke design is a major contributor to efficient swimming in cetaceans (Fish, this volume) and its origin was probably a decisive factor in the morphology of the caudal portion of the protocetid skeleton. It is unlikely that most of the design features outlined by Fish can be studied in fossil whales, but Buchholtz (this volume) showed that some aspects of the fluke are preserved in fossils and that the origin of some soft structures can be traced by looking at bony landmarks.

The chapters by Fish and Pabst *et al.* show that understanding of form and function of recent cetaceans is of paramount importance in the interpretations of fossil sequences, and offer answers to questions of biological patterns. As a result, the high diversity and abundance of modern cetaceans is one of the reasons why cetacean origins can be understood in greater detail than that of other clades that made the land-to-water transition.

A question that cannot be answered satisfactorily is why cetaceans took to the water. It has been suggested that they took to the water to take advantage of a plethora of resources that had gone untapped since the extinction of Mesozoic marine reptiles (Fordyce and Barnes, 1994). This explanation is too simplistic. The earliest cetaceans lived in or near freshwater and it is unlikely that they profited from extinctions in the nearshore marine realm. Lack of competition for food is also an unlikely reason for the subsequent shift to the seas. The earliest marine cetaceans did not live like modern cetaceans, but resembled crocodiles ecologically (Thewissen *et al.*, 1996a). Crocodiles were not greatly affected by the K–T extinctions (e.g., Archibald, 1996) and they are abundant in the same deposits as those that yield the earliest cetaceans (Buffetaut, 1978). If early cetaceans were generalist feeders, they must have suffered considerable competition and predation from crocodiles.

Given the unusual dental morphology and dental wear of early cetaceans (O'Leary, this volume), it is more likely that early cetaceans were food specialists, tapping a resource that required a specialized masticatory morphology. This food source cannot be identified at present, but morphological analysis combined with stable isotope geochemistry (Roe, this volume) may assist in resolving this issue.

In sum, cetacean origins are already one of the best documented examples of major morphological change in the fossil record (Gould, 1994). There is great promise in the study of this morphological change in a rigorous phylogenetic context with understanding of the relation between form and function, and adequate data from soft anatomy and embryology.

References

Ahlberg, P. E., and Milner, A. R. 1994. The origin and early diversification of tetrapods. *Nature* **368**:507–514.

Archibald, J. D. 1996. Testing extinction theories at the Cretaceous–Tertiary boundary using the vertebrate fossil record, in: N. MacLeod and G. Keller (eds.), *Cretaceous–Tertiary Mass Extinctions: Biotic and Environmental Changes*, pp. 373–397. Norton, New York.

Aslan, A., and Thewissen, J. G. M. 1997. Preliminary evaluation of paleosols and implications for interpreting vertebrate fossil assemblages, Kuldana Formation, northern Pakistan. *Paleovertebrata* **25**:261–277.

Barnes, L. G., and Mitchell, E. 1978. Cetacea, in: V. J. Maglio and H. B. S. Cooke (eds.), *Evolution of East African Mammals*, pp. 582–602. Harvard University Press, Cambridge, MA.

Boyden, A., and Gemeroy, D. 1950. The relative position of Cetacea among the orders of Mammalia as indicated by precipitin tests. *Zoologica* **35**:145–151.

Buffetaut, E. 1978. Crocodilian remains from the Eocene of Pakistan. *N. Jb. Geol. Palaeontol.* **156**:262–283.

Dehm, R., and Oettingen-Spielberg, T. zu. 1958. Palaeontologische und geologische Untersuchungen im Tertiär von Pakistan. 2. Die mitteleocänen Saügetiere von Ganda Kas bei Basal in Nordwest Pakistan. *Abh. Bayer. Akad. Wiss. Math. Naturwiss. Kl. N. F.* **91**:1–54.

Fleischer, G. 1978. *Evolutionary Principles of Mammalian Middle Ear.* Springer-Verlag, Berlin.

Fordyce, R. E., and Barnes, L. G. 1994. The evolutionary history of whales and dolphins. *Annu. Rev. Earth Planet. Sci.* **22**:419–455.

Fraas, E. 1904. Neue Zeuglodonten aus dem unteren Mitteleocän vom Mokattam bei Cairo. *Geol. Palaeontol. Abh. N. F.* **6**:199–220.

Gatesy, J., Hayashi, C., Cronin, M., and Arctander, P. 1996. Evidence from milk casein genes that cetaceans are close relatives are of hippopotamid artiodactyls. *Mol. Biol. Evol.* **13**:954–963.

Gatesy, J. 1997. More DNA support for a Cetacea/Hippopotamidae clade: the blood-clotting protein gamma-fibrinogen. *Mol. Biol. Evol.* **14**:537–543.

Gingerich, P. D. 1992. Marine mammals (Cetacea and Sirenia) from the Middle Eocene of of Kpogamé-Hahotoé in Togo. *J. Vert. Paleont., Suppl.* **12**:29A–30A.

Gingerich, P. D., and Russell, D. E. 1981. *Pakicetus inachus*, a new archaeocete (Mammalia, Cetacea) from the early-middle Eocene Kuldana Formation of Kohat (Pakistan). *Contr. Mus. Paleont., Univ. Michigan* **25**:235–246.

Gingerich, P. D., and Russell, D. E. 1990. Dentition of the Early Eocene *Pakicetus* (Mammalia, Cetacea). *Contr. Mus. Paleont., Univ. Michigan* **28**:1–20.

Gingerich, P. D., Wells, N. A., Russell, D. E., and Shah, S. M. I. 1983. Origin of whales in epicontinental remnant seas: New evidence from the Early Eocene of Pakistan. *Science* **220**:403–406.

Gingerich, P. D., Raza, S. M., Arif, M., Anwar, M., and Zhou, X. 1993. Partial skeletons of *Indocetus ramani* (Mammalia, Cetacea) from the Lower Middle Eocene Domanda shale in the Sulaiman range of Punjab (Pakistan). *Contr. Mus. Paleont., Univ. Michigan* **28**:393–416.

Gingerich, P. D., Raza, S. M., Arif, M., Anwar, M., and Zhou, X. 1994. New whale from the Eocene of Pakistan and the origin of cetacean swimming. *Nature* **368**:844–847.

Gingerich, P. D., Arif, M., and Clyde, W. C. 1995. New archaeocetes (Mammalia, Cetacea) from the middle Eocene Domanda Formation of the Sulaiman Range, Punjab (Pakistan). *Contribu. Mus. Paleontol. Univ. Michigan* **29**(11):291–330.

Gould, S. J. 1994. Hooking Leviathan by its past. *Nat. Hist.* **103**:8–15.

Hulbert, R. C., Petkewich, R. M., Bishop, G. A., Bukry, D., and Aleshire, D. P. 1998. A new protocetid (Mammalia: Cetacea: Archaeoceti) and associated biota from the Middle Eocene of Georgia. *J. Paleontol.* **72**:905–925.

Kellogg, R. 1928. The history of whales—their adaptation to life in the water. *Q. Rev. Biol.* **3**:29–76.

Kellogg, R. 1936. A review of the Archaeoceti. *Carnegie Inst. Washington Publ.* **482**:1–366.

Kinkel, M. D., Thewissen, J. G. M., and Oelschläger, H. A. 1997. Repositioning of the incus and malleus during dolphin ontogeny. *J. Morphol.* **232**:275.

Lancaster, W. C. 1990. The middle ear of the Archaeoceti. *J. Vertebr. Paleontol.* **10**:117–127.

McKenna, M. C. 1975. Toward a phylogenetic classification of the Mammalia, in: W. P. Luckett and F. S. Szalay (eds.), *Phylogeny of the Primates*, pp. 21–46. Plenum Press, New York.

Milinkovitch, M. C., and Thewissen, J. G. M. 1997. Even-toed fingerprints on whale ancestry. *Nature* **388**:622–624.

Milinkovitch, M. C., Ortí, G., and Meyer, A. 1993. Revised phylogeny of whales suggested by mitochondrial ribosomal DNA sequences. *Nature* **361**:346–348.

Miller, G. S. 1923. The telescoping of the cetacean skull. *Smithson. Misc. Collect.* **76**:1–71.

Pilgrim, G. E. 1940. Middle Eocene mammals from north-west India. *Proc. Zool. Soc. B* **110**:127–152.

Prothero, D. R., Manning, E. M., and Fisher, M. 1988. The phylogeny of ungulates, in: M. J. Benton (ed.), *The Phylogeny and Classification of the Tetrapods*, Volume 2, pp. 201–234. Clarendon Press, Oxford.

Rango Rao, A. 1972. New mammalian genera and species from the Kalakot zone of Himalayan foot hills near Kalakot, Jammu and Kashmir State, India. Directorate of Geology Oil and Natural Gas Commission, Dehra Dun, India Special Paper No. 1:1–22.

Shimamura, M., Yasue, H., Ohshima, K., Abe, H., Munechika, I., and Okada, N. 1997. Molecular evidence from retroposons that whales form a clade within even-toed ungulates. *Nature* **388**:666–670.

Simpson, G. G. 1945. The principles of classification and a classification of mammals. *Bull. Am. Mus. Nat. Hist.* **85**:1–339.

Slijper, E. J. 1962. *Whales*. Basic Books, New York.

Stromer, E. 1908. Die Archaeoceti des Ägyptischen Eozäns. *Beitr. Paläontol. Geol. Österreich-Ungarns Orients* **21**:106–177.

Taylor, M. A. 1987. How tetrapods feed in water: a functional analysis by paradigm. *Zool. J. Linn. Soc.* **91**:171–195.

Thewissen, J. G. M. 1994. Phylogenetic aspects of cetacean origins: a morphological perspective. *J. Mamm. Evol.* **2**:157–184.

Thewissen, J. G. M., and Fish, F. E. 1997. Locomotor evolution in the earliest cetaceans: functional model, modern analogues, and paleontological evidence. *Paleobiology.* **23**:482–490.

Thewissen, J. G. M., and Hussain, S. T. 1993. Origin of underwater hearing in whales. *Nature* **361**:444–445.

Thewissen, J. G. M., and Hussain, S. T. In press. *Attockicetus praecursor*, a new remingtonocetid cetacean from marine Eocene sediments of Pakistan. *Nat. Hist. Mus. Los Angeles Cty. Sci. Ser.*

Thewissen, J. G. M., and Hussain, S. T. 1988. Systematic review of the Pakicetidae, early and middle Eocene Cetacea (Mammalia) from Pakistan and India. *Bull. Carnegie Mus. Natl. Hist.* **34**:220–238.

Thewissen, J. G. M., Hussain, S. T., and Arif, M. 1994. Fossil evidence for the origin of aquatic locomotion in archaeocete whales. *Science* **263**:210–212.

Thewissen, J. G. M., Madar, S. I., and Hussain, S. T. 1996a. *Ambulocetus natans*, an Eocene cetacean (Mammalia) from Pakistan. *Cour. Forsch.-Inst. Senckenberg* **191**:1–86.

Thewissen, J. G. M., Roe, L. J., O'Neil, J. R., Hussain, S. T., Sahni, A., and Bajpai, S. 1996b. Evolution of cetacean osmoregulation. *Nature* **381**:379–380.

Van Valen, L. 1966. Deltatheridia, a new order or mammals. *Bull. Am. Mus. Nat. Hist.* **132**:1–126.

Van Valen, L. 1967. New Paleocene insectivores and insectivore classification. *Bull. Am. Mus. Nat. Hist.* **135**:217–284.

Webster, D. B., and Webster, M. 1976. Auditory systems of Heteromyidae: functional morphology and evolution of the middle ear. *J. Morphol.* **146**:343–376.

West, R. M. 1980. Middle Eocene large mammal assemblage with Tethyan affinities, Ganda Kas region, Pakistan. *J. Paleontol.* **54**:508–533.

Index

The editor has left the use of classification levels (e.g., Dorudontidae, Dorudontinae) and the use of paraphyletic groups (archaeocetes) to the discretion of individual authors, and their usage is thus not consistent throughout this volume. In the index, presumed monophyletic clades are distinguished from paraphyletic groups by the use of Latin terms for the former (e.g., Remingtonocetidae), and of English terms for the latter (e.g.,., archaeocetes), regardless of their usage in individual chapters.